T0254315

CAMBRIDGE LIBRARY COLLECTION

Books of enduring scholarly value

Botany and Horticulture

Until the nineteenth century, the investigation of natural phenomena, plants and animals was considered either the preserve of elite scholars or a pastime for the leisured upper classes. As increasing academic rigour and systematisation was brought to the study of 'natural history', its subdisciplines were adopted into university curricula, and learned societies (such as the Royal Horticultural Society, founded in 1804) were established to support research in these areas. A related development was strong enthusiasm for exotic garden plants, which resulted in plant collecting expeditions to every corner of the globe, sometimes with tragic consequences. This series includes accounts of some of those expeditions, detailed reference works on the flora of different regions, and practical advice for amateur and professional gardeners.

A History of Garden Art

Marie Luise Gothein (1863–1931) published this scholarly two-volume history of garden design in German in 1913. Its second edition of 1925 was translated into English by Laura Archer-Hind, edited by gardening author Walter P. Wright (1864–1940), and published in 1928. The highly illustrated work is still regarded as among the most thorough and important surveys of its kind. It begins by examining evidence from both archaeology and literature, as well as climate and soil conditions, to discuss the gardens of ancient Egypt and Assyria, and continues to survey developments worldwide until the twentieth century. Individual gardens, technical innovations, and fashions in horticulture are all discussed in detail. Volume 1 surveys the ancient civilisations of the Near East, Greece and Rome, discusses Byzantine and Islamic gardens, and the importance of monastery gardens in western Europe, and ends with a review of gardening in Europe during the Renaissance.

Cambridge University Press has long been a pioneer in the reissuing of out-of-print titles from its own backlist, producing digital reprints of books that are still sought after by scholars and students but could not be reprinted economically using traditional technology. The Cambridge Library Collection extends this activity to a wider range of books which are still of importance to researchers and professionals, either for the source material they contain, or as landmarks in the history of their academic discipline.

Drawing from the world-renowned collections in the Cambridge University Library and other partner libraries, and guided by the advice of experts in each subject area, Cambridge University Press is using state-of-the-art scanning machines in its own Printing House to capture the content of each book selected for inclusion. The files are processed to give a consistently clear, crisp image, and the books finished to the high quality standard for which the Press is recognised around the world. The latest print-on-demand technology ensures that the books will remain available indefinitely, and that orders for single or multiple copies can quickly be supplied.

The Cambridge Library Collection brings back to life books of enduring scholarly value (including out-of-copyright works originally issued by other publishers) across a wide range of disciplines in the humanities and social sciences and in science and technology.

A History of Garden Art

From the Earliest Times to the Present Day

Volume 1

Marie Luise Gothein
Edited by Walter P. Wright
Translated by Laura Archer-Hind

CAMBRIDGE
UNIVERSITY PRESS

University Printing House, Cambridge, CB2 8BS, United Kingdom

Cambridge University Press is part of the University of Cambridge.
It furthers the University's mission by disseminating knowledge in the pursuit of
education, learning and research at the highest international levels of excellence.

www.cambridge.org
Information on this title: www.cambridge.org/9781108076142

This edition first published 1928
This digitally printed version 2014

ISBN 978-1-108-07614-2 Paperback

A HISTORY OF GARDEN ART

FROM THE EARLIEST
TIMES TO THE PRESENT DAY

IN TWO VOLUMES

PART OF THE ROCK GARDEN AT WISLEY, THE SURREY GARDEN OF THE ROYAL HORTICULTURAL SOCIETY—LOOKING UP THE STREAM

MARIE LUISE GOTHEIN

A HISTORY OF GARDEN ART

Edited by
WALTER P. WRIGHT

Translated from the German by
MRS. ARCHER-HIND, M.A.

With over Six Hundred Illustrations

VOLUME I

LONDON & TORONTO
J. M. DENT & SONS LIMITED
NEW YORK: E. P. DUTTON & CO. LTD.

First Published in English . 1928

PRINTED IN GREAT BRITAIN

PREFACE TO THE ENGLISH EDITION

BY THE EDITOR

A PRELIMINARY examination of Marie Luise Gothein's *Geschichte der Gartenkunst* revealed a work not only remarkably erudite but also of so distinct a type and so strongly individual in spirit and execution as to merit deep respect and undeviating loyalty. In considering an English edition, it was obvious that there was but one course to pursue: namely, to cling as closely as the exigencies of different languages would permit to the genius of the original. Translation of words—yes. "Translation" of spirit—no.

The *History of Garden Art* is a work of supreme ability. To those who have made a lifelong study of gardening—and the number is probably greater than is commonly supposed—it must make an irresistible appeal, for its pages present a panorama beginning in that dim past in which history itself took birth, and extending through ages crowded with vital events and heroic figures. It is at once significant and enthralling to realise how closely gardening has been interwoven with those decisive occurrences and those momentous personalities which have affected human destinies the most profoundly.

To others—to the uncounted thousands who approach gardening books in no spirit of virtuosity, but rather under the simple though praiseworthy impulse of a love of flowers—the work will impress from another angle, portraying as it does the means and methods by which in all ages and in all climes garden-lovers have sought beauty. Thus, while the *History of Garden Art* is in soul and structure a record of gardens and not a manual on gardening, it nevertheless combines instruction with inspiration, for in description and in picture it conveys lessons which no discriminating reader can overlook.

The *History of Garden Art* easily takes its place as a garden classic—one might almost say, so all-embracing is its scope, as a historical and social classic. Mighty imperial, political, clerical, literary and artistic figures pass through its pages. We hear the proud and autocratic monarch Louis XIV. smilingly speaking of kisses from his gardener. We walk with a contemplative Wolsey through the grounds of Hampton Court. We listen sympathetically to the impassioned outbursts of the youthful Goethe, intoxicated as he is with beauty. Powerful women like Elizabeth and Maria Theresa display their softer sides. Unhappy Mary Tudor presents a crown-piece to the humble gardener who brings her a nosegay. Marie Antoinette flits her butterfly existence through the avenues of Versailles. Pliny and Virgil and Cicero; Evelyn and Pepys; Boccaccio and Ronsard; Schiller and Winckelmann; the Medici; Rousseau and Voltaire; Augustus and Napoleon—all these great figures have their place.

Only the most fussy and pedantic of editors would misinterpret his duty in dealing with a work so spacious and at the same time so meticulous as Frau Gothein's. My part is obvious—to present her to readers in English, with competent aid in translation, as accurately and with as few changes as possible.

I have, however, thought it desirable to add two supplementary chapters, one on modern English, the other on American, garden art.

Under Frau Gothein's scheme English gardening is dealt with in several different chapters. While this is not as illogical as it might seem if superficially considered, it yet has the effect of partially obscuring the remarkable development of gardening in Britain during the past half-century under the leadership of William Robinson, Mr. Reginald Farrer, Miss Gertrude Jekyll, Miss Ellen Willmott, the Hon. Vicary Gibbs, Sir Harry J. Veitch and others. One hopes that the chapter on Modern English Gardening goes some way towards remedying that defect.

And there have been great recent movements in America and Canada, equally significant, although mainly on different lines. I have been fortunate in finding so able an exponent of these developments as the well-known American landscape architect, Professor Frank A. Waugh.

I desire to pay a warm tribute to Mrs. Archer-Hind, not only for her care in the task of translation, but also for her unwearying patience in examining the many difficult points which were presented to her. And I wish to acknowledge gratefully the assistance rendered by the Reverend Hilderic Friend in reading the proofs and in the transliteration of names. His own studies in the history of garden art, and also in Oriental languages, made his help particularly valuable.

WALTER P. WRIGHT.

PREFACE TO THE FIRST EDITION

IN May 1904, at a meeting of the New Philological Society at Cologne, in an address on the Origin of English Landscape Gardening, I offered the first fruits of those studies out of which this book has grown.

In Germany at that date it may have seemed somewhat strange—and not only to a philological audience—that Science should concern itself with the affairs of Gardening. Even the historians of art had only shown a perfunctory interest. Jacob von Falk's book, *The Garden: its Art and History*, which at the time he wrote it (1884) was really meritorious, had received little attention outside a small circle; for its appearance coincided with a phase of empty and meaningless art. In more recent years, however, partly through the influence of England, partly also by a decided inclination of fashion in the direction of the formal garden, the attention of leading artists and practical men, of laymen and scientists, has been drawn to this branch of art, so long neglected. All historians of art have been busy with the subject of modern gardening.

Certain monographs have been warmly greeted, notably Hugo Koch's *Saxon Garden Art*, 1910; and the first comprehensive work written in German, *The Garden*, by August Grisebach, which, however, though called in a sub-title *A History of its Formation*, almost entirely neglects to describe the historical development through the earlier centuries. Beginning with the "Vegetable and Pleasure-Gardens of the Middle Ages," the author makes a creditable attempt to separate the individual types.

Other countries, especially England, have year by year added to the increasing list of books on the subject, but they are only of first-rate importance as supplying a wonderful wealth of material to look at. I have dealt with these books, especially *Formal Gardens in England and Scotland*, by H. Inigo Triggs (1902), and also his *Art of Garden Design in Italy*, in my text. Triggs's later book, *Garden-Craft in Europe*, is not an account based upon his own personal research, any more than Fouquier's book, *De l'Art des Jardins du XVe aux XXe Siècle*. In this respect the book by the Hon. Alicia Amherst, *A History of Gardening in England* (1896), is far better; although it gives to garden art proper only a secondary position, it includes much personal observation and research. The second edition (1910) has moreover an enlarged and valuable bibliography of writers on English gardens up to the year 1830.

The indispensable condition for an adequate presentation of garden art is a critical examination of sources. In every epoch the material is new, and very different; and it has had to be extracted from literary records—often widely scattered—partly in writings, deeds and accounts, partly in paintings, sketches and engravings. In my story I have laid particular stress on the actual historical moment, and on the indications, when the types first appear, of the way they take and the changes they suffer. By following this method one finds a surprising continuity from antiquity to the present day.

The study of old gardens that still exist, which I have tried to pursue in the course of much travel extending over more than ten years, is peculiarly difficult; what you actually see with your eyes has to be "restored," like a corrupt text, into its original context, and then compared with traditions and ancient examples. This has been the only possible way of finding the right place for each individual subject in the large canvas of the whole.

On the other hand, the picture could not be really alive unless the most important gardens were treated separately as individuals, with as few gaps as possible. It is true of all art, but especially of ours, that its life is closely interwoven with the life of society, and the history of the one forms part of the history of the other. All important currents of thought have affected the fate of the garden, and some of the outstanding figures in history appear in a new light when they are fostering and furthering its ends. The garden has an important determining influence on our interpretation of other arts, and is itself affected thereby, more especially villa architecture, and at certain periods sculpture.

In relation to the first half of my undertaking—the history prior to the Middle Ages in Europe—I found no previous work worth mentioning, and I could not have attained my goal without the friendly counsel of certain distinguished men of letters. For Egypt I owe the warmest thanks to Hermann Ranke, for the Asiatic section I received important hints from Carl Bezold, and for the Greek invaluable advice from my friend Georg Karo. Alfred von Domaszewski has helped me with the Roman gardens, Carl F. Becker with the account of Islam, and Friedrich Sarre and Ernst Herzfeld have both put material at my disposal. I owe to Albrecht Haupt my best authorities for the Portuguese section. For the Middle Ages I have had much help from the untiring friendliness of Paul Clemens, and I here thank Erich Frank for his careful kindness in the heavy business of proof correcting.

I found much kind help in various libraries, especially in the Hof und Landes Bibliothek at Karlsruhe, the Bibliothek des Kunstgewerbemuseums in Berlin, and in the Bibliothèque d'Art et d'Archéologie in Paris. I should also like to thank the gentlemen in the library at Heidelberg for their constant kind help. But before all others is my gratitude due to the Königlich Preussischen Akademie des Bauwesens for accepting this book as one of its own publications, and for issuing it with the fine illustrations which it could not otherwise have had, though to a work of this kind they are nearly indispensable. I am also very glad to thank the publisher for his unselfish assistance in many ways.

The book should appeal to readers of many sorts: to the archæologist, to the historian of art, to the historian, indeed, of any kind of civilisation, to all of whom it should be useful in paving the way to a special subject of research which, one hopes, will no longer be neglected. But perhaps the best result that my lengthy work can hope for is to get a grip on the actual life and conduct of to-day. Among non-professional people a natural love for the garden may be stimulated by the story of past days; and if it should come about that practical artists revert to the formal style, some knowledge and understanding of the chances and changes of thousands of years should be helpful in their work. My wish is that they may find not so much a storehouse of the ideas of great masters of the past, as an abundant harvest for their own creations in the present day.

MARIE LUISE GOTHEIN.

HEIDELBERG, *September* 1913.

ABRIDGED PREFACE TO THE SECOND EDITION

THE first edition of the *History of Garden Art* has been out of print for several years. I hope the book has had the effect I spoke of in my former preface. A new edition had necessarily to wait till more ordinary economic conditions in Germany made it possible, and I wish to thank my publisher for taking the first opportunity of bringing out a second edition.

After mature consideration I have decided to leave the text of the former edition without fundamental alteration. My intention was to write a bit of history, not to produce a handbook. An examination of the literature of the last ten years, especially on the subject of Italian gardens (which has attracted most writers), has proved to me that a complete alteration of my text is unnecessary.

Of course the text has been examined, in case errors might have crept in. I was prevented from correcting the proofs myself after the first volume, because I went to India on a journey that had long been planned. But I know that the corrections of the second volume were in the best of hands; and I thank Professor Salin and Dr. Gertrud Bettmann of Heidelberg for their kindness and trouble. Also to my other good friends who have found me new material I here tender my most hearty thanks.

MARIE LUISE GOTHEIN.

BANJOEMAS, JAVA,

June 1925.

CONTENTS OF VOLUME I

PAGE

CHAPTER I. ANCIENT EGYPT 1

Nature of the soil—Reverence for trees—Cultivation of trees: sycomore, palm, acacia—Vineyards—Arbours—Hieroglyphs—Vegetable cultivation—Flower-beds—Meten's Villa—Tomb inscription—Plans of Egyptian houses and gardens—Garden of Amenhotep (Amenophis) III.—Dwelling-houses of El-Amarna—Visiting scene in a garden, taken from a Theban tomb—Second visiting scene in a garden—The garden of Apoui—Suburban villa of the Setna Romance—The high priest Mērirē—The king's palace at El-Amarna—Queen Hatshepsut—Temple at Deir-el-Bakhiar—Expedition to the Land of Punt—Importation of incense trees—Rameses III.—Acclimatisation of plants—Presents to sanctuaries—Consecration to religion—Cultivation of gardens—Wall-paintings in the tomb of Sennefer—Festival of the dead in a garden—Grave of the writer Inny—Necropolis gardens—Primitive gardens of the dead—Love of flowers among the Egyptians—Flower-beds—Garlands—Garden song.

CHAPTER II. WESTERN ASIA IN ANCIENT TIMES 27

The age of civilisation—Parks—The cedar wood in the Gilgamesh Epic—The earliest useful plantations—Vineyards under the Sumerian kings—Tiglath-Pileser's acclimatisation—Early pictures of trees—Parks "like the Amanus mountains"—Garden temples—Sennacherib's parks, and his temple to the God Assur—Irrigation—Hanging gardens of Semiramis—Ancient accounts: Diodorus, Strabo—Trees more important than flower culture—Hunting park of Assurbanipal—Reverence for trees among the Persians—The park of Tissaphernes, "Alcibiades"—Xenophon's word "Paradise": his account of Persian parks—Ruins of Persepolis—Grove of tombs, and the tomb of Cyrus—The Israelites: their burials in gardens—India: cultivation of trees, trees of Buddha, gardens of Mâyâ-Divi—Presents of parks to the monks of Buddha—Gardens of Buddha in Ceylon—The holy tree, and its imitations—Excavations in Ceylon—Pokuñas—Bôrô-Budur, Java.

CHAPTER III. ANCIENT GREECE 51

Cretan and Mycenæan civilisation—Gardens in Homeric poetry—Garden of Laertes—Garden of Alcinous—Care of flowers—Planted courts—Sacred groves—Nymph sanctuaries—Country life in the fifth century: custom of garlanding; flowers of this period—Little gardens of Adonis—Life in cities hostile to the garden—Park of Gelon with "Horn of Amalthea"—Temple gardens and sacred groves—Hero sanctuaries and games—The gymnasium—The Academy at Athens—Gardens in towns—Vitruvius's description of a palæstra—Meaning of *xystos* and *xysta*—Gymnasium at Delphi—Gymnasium at Pergamon—Importance of these sites in garden history—Private gymnasiums and philosophers' gardens—Age of Alexander the Great—Influence of Oriental gardening—Rise of the great Hellenistic towns—Alexandria—Town sites—Necropolis gardens—Tombs in private gardens—Antioch—Garden street—The suburb Heraclea—The park of Daphne—Private houses in Greek cities—The peristyle—Grand tent of Ptolemy Philadelphus—Grottoes—Grand ship of Hiero II.—Gardens in Greek romances—Longus—*Daphnis and Chloe*—The Geoponica.

CHAPTER IV. THE ROMAN EMPIRE 81

Importance of *res rustica* in Rome—The two homes of a townsman—The *hortus* (or villa)—The elder Scipio's villa—Seneca on the villas of the republic—Cicero's villas—Increase of luxury—*Villa urbana* and *villa rustica*—Local distribution of villa sites—Objections to luxury—Cicero's descriptions of villas in speeches and letters—Revival of gardens of Greek philosophers: their lay-out, statues, irrigation—Hunting-parks after Scipio's time—The park of Q. Hortensius—Roman gardens in imperial times—Estate of Augustus—Portico of Livia—Tomb of Augustus—Stroll through the gardens of Rome: on the south, gardens of Cæsar; of Antony; Naumachia of Augustus; Vatican territory: garden of the older Agrippina; circus of Caligula; gardens of Domitia with the tomb of Hadrian; Field of Mars: *villa pubblica*; property of Pompey; Scipio's villa; baths of Agrippa; Collis Hortorum: gardens of Valerius Asiaticus; gardens of the Domitians; gardens of the Acilians; between the Collis Hortorum and the Quirinal: gardens of Sallust; on the Quirinal: house of Atticus; Esquiline: house of Mæcenas; Auditorium of Mæcenas; houses of Virgil, Propertius, Martial, and the younger Pliny; between the Palatine and the Esquiline: the Golden House of Nero—*Villa urbana*—The form of villas—Hellenistic influence—Egyptian influence—Wall-paintings—Sacred trees—Pictures of gardens on walls of rooms: in the villa of Livia near Porta Prima; in the Frigidarium at the Stabian Baths—Villa of the younger Pliny—Garden hippodrome on the Palatine—Villa Laurentinum of the younger Pliny—Other types of villas—Want of steps in the

PAGE

open—The emperor's villa on the Lake of Albano—Hadrian's villa at Tivoli—Old Italian and Greek houses—Pompeii: the house of Sallust, of Epidius Rufus, of Pansa, of the Silver Wedding, of the Vettii, of Diomedes—Balcony-gardens at Rome—African villas—Mosaics—Baths of Pompeianus—Women's gardens—Villas in the northern provinces—Poem of Ausonius on the Moselle—Villas at Metz, Ruhling, Ulrich, Teting—Villas on the Rhine: Wittlich, Welschbillig—Migration of nations into Spain and Southern France—Sidonius's account of the Villa Avitiacum—Growth of the *burgus* from the villa—Italy: Ravenna.

CHAPTER V. BYZANTINE GARDENS AND THE COUNTRIES OF ISLAM 135

The imperial palace at Byzantium—Triconchos—Anadendradion of the Magnaura—Mesokepion of the Lausiacus—Balcony gardens—Mesokepion near the new church—Byzantine fountains at graves and in love-romances — Mechanical water-works — Tree fountains — Byzantine villas—Dastogard—Downfall of the Persian kingdom—Chosroes and his carpet—Importance of carpets in Persian garden art —Traditions of Asiatic art—Views of Ibn Chaldun—Place of the Arabs in garden art—Samaria: foundation and duration; Balkuwara Palace; caliphs' palace—Bagdad: Taj Palace on the east; Palace of Paradise; Hasani Villa—The Byzantine Embassy—House of the Tree—The New Kiosk—Garden at castle of the Tulunids at Cairo—Arabian description—Hellenistic-Roman influence—Court-gardens and garden luxury among the Arabs—Spain under the Omiads—Abd-ur-Rahman I.—Villa Bussafa at Cordova—Az-Zahra and Abd-ur-Rahman III.—Alhambra: former and present condition—Garden-courts: Myrtle Court; Court of the Lions; Patio de Daraxa—The Generalife: Navagero's description; condition to-day—Importance of water-stairs and water arts—Water-stairs in the Park of Sahum in Egypt—Other pleasure-houses at the Alhambra—Sicily: La Zisa; Cuba—Later fate of Persian territory: Paradise of the "Old Man of the Mountain"—Bokhara and Samarcand—Timur's gardens—The madhouse garden of Bayezid II. at Adrianople—Persia under the Abbassids: Ispahan: Maidan, Tshehar-Bagh, Tshihil-sutun-Pavilion—Palace surroundings—Pleasure-castle Ashraf: position of the garden, Bagh-i-Shah, chief garden—Shiraz—Bagh-i-Takht—Garden houses—Garden in Fin-nahē at Kashan.

CHAPTER VI. THE MIDDLE AGES IN THE WEST 169

Beginnings of monkish life—Foundation of Augustine at Hippo—Ancient forms of the cloister—Paradise of St. Maria Laach—Planted crossways—Plan of monastery at St. Gall—Love of flowers in early cloisters —Plan of the Canterbury cloisters—Gardens at Clairvaux and Thorney—*Hortus Deliciarum*—Walafried Strabo's *Hortulus*—Rhabanus Maurus' didactic poem—Rose-tree at Hildesheim—Carthusian monasteries —Early gardens of the laity—Site of town gardens—Earliest arrangement of lay gardens: walls, gates, fencing, arbours, labyrinths—Animal-and tree-gardens—Influence of the East—The Golden Tree and the artificial voices of animals in the West—Gardens in the *Romaunt of the Rose* and its imitations—Illustrations—Awakening of the desire for knowledge—Albertus Magnus and the gardens at Cologne—Town gardens—Public places in the towns—*Pratum commune*—Gardens of Brotherhoods—Ascendancy of Italy—Villas—Petrus Crescentius—Boccaccio—Petrarch.

CHAPTER VII. ITALY IN THE TIME OF THE RENAISSANCE AND THE BAROQUE STYLE 205

Leon Battista Alberti—Rucellai's Villa Quaracchi—Colonna's *Sogno di Polifilo*—Pontanus' *De Hortis Hesperidum*—Careggi—Poggio a Cajano—Poggio Reale—Bembo's *Gli Asolani*—Soderini—Scamozzi—Deer parks—Filarete—Hermitages and ruins—Gardens of statues—Terraces—Boccaccio—Beginnings at Rome—Bramante's court in the Vatican—Villa Madama—Palazzo del Te—Other buildings of the Gonzaga family—Villa Imperiale at Pesaro—Other villas of Urbino—San Vigilio—Botanic Garden at Padua—Villa di Castello—Palazzo Doria, Genoa—Villa d'Este, Tivoli—Villa Pia—The Vatican Hill—Farnese Gardens—Villa Papa Giulio—Villa Lante, Bagnaja—Caprarola, castle, casino—Villa Campi at Signa—Pratolino—Giardino Boboli—Importance of nepotism in Roman Gardens—Papal gardens on the Quirinal—Villa Medici—Sixtus V. and Villa Montalto—Villa Mattei—Villa Corsini—Frascati—Villa Falconieri—Villa Aldobrandini—Villa Ludovisi (Tolonia) Villa Lancelotte—Villa Muti—Villa Belpoggio—Villa Mondragone—Villa Borghese, Frascati—Villa Borghese, Rome—Villa Ludovisi Rome —Villa Pamfili—Isola Bella—Villa Collodi—Villa Albani.

CHAPTER VIII. SPAIN AND PORTUGAL IN THE TIME OF THE RENAISSANCE . . 351

Mudejar style in Spain—Alcazar in Seville—Charles the Fifth's palace at the Alhambra—Charles V. at San Yuste—The Escorial—The Castle of Madrid under Philip II.—Casa del Campo—San Jeronimo—Aranjuez under Philip II.—The house of Pilatus at Seville—Castles of Duke of Alva: Alva de Tormes, Lagunilla—Foreign influences in Spanish gardens—Cosimo Lotti at Aranjuez—Hunting-castles round Madrid: El Pardo, Torre de la Parada, Zarzuela, Buen Retiro—Portugal: Philip II.—Portuguese Crusades—Penha verde—Quinta di Bacalhão—Bibafria—Villa Bemfica—Quinta da Ramalhão—Queluz —Staircase of the "Stations" at the Bussaco Monastery, at Bom Jesus do Monte, Braga, and at Lamego.

PAGE

CHAPTER IX. FRANCE IN THE TIME OF THE RENAISSANCE 389

March of Charles VIII. into Italy; importance for the French Renaissance—Italian artists in France, and French in Italy—Italian villa and French castle: castle-garden at Amboise; at Blois; at Gaillon—The French canal-garden—Fontainebleau—Chantilly—Leonardo da Vinci's plan of a castle-garden—Bury—Dampierre—Valleri—Chambord—"Madrid"—Chenonceaux—Anet—Verneuil—Charleval—Montargis—Tuileries—Rabelais, Abbey of Thelemites—Bernard Palissy as writer on gardens—Chartreuse of Gaillon; the earliest hermitage—Garden theorists in Henry the Fourth's time—Olivier de Serres—Claude Mollet—Du Pérac, and the introduction of *parterres de broderie*—Influence of floral embroidery—Boyceau—Garden art under Henry IV.: Saint-Germain-en-Laye; Fontainebleau; Luxembourg Gardens; Ruel—Evelyn's description of French gardens—René Rapin's poem on gardens.

CHAPTER X. ENGLAND IN THE TIME OF THE RENAISSANCE 433

Nature of the transition—Later beginning of Renaissance art—James I. of Scotland, description of the garden at Windsor—Cardinal Wolsey and Hampton Court—Hampton Court under Henry VIII.—Nonsuch—Elizabeth's connection with garden art—Kenilworth: description of Laneham and Sir Walter Scott—Theobalds—Elizabethan garden literature—Introduction of foreign plants: Holinshed's *Chronicle*—Gardens of plants—Bacon's essay—Masque in the garden at Whitehall—Garden art under James I.—Scotch gardens—Exchange of Theobalds for Hatfield House—Ben Jonson—Hatfield House and the first Lord Salisbury—Montacute—Salomon and Isaac de Caus—*Hortus Pembrochianus*—Botanic Garden at Oxford—Moor Park and Sir William Temple—The Revolution and gardening—Fate of Hampton Court, Nonsuch, and Theobalds—Wimbledon.

LIST OF ILLUSTRATIONS

The Editor and Publishers desire to draw attention to the fact that due acknowledgments of the permission to use various copyright pictures are made in the lists of illustrations, where they are appended to the appropriate subjects.

PART OF THE ROCK GARDEN AT WISLEY, THE SURREY GARDEN OF THE ROYAL HORTICULTURAL SOCIETY —LOOKING UP THE STREAM. (Photograph by courtesy of The Royal Horticultural Society) . *Frontispiece*

CHAPTER I. ANCIENT EGYPT

FIG. PAGE

1. A WELL-SWEEP IN ANCIENT EGYPT. (Wilkinson, *The Manners and Customs of the Ancient Egyptians*, I.). 4
2. A WELL-SWEEP IN MODERN EGYPT. (Wilkinson, *The Manners and Customs of the Ancient Egyptians*, I.). 4
3. PEASANT DOING HONOUR TO A SYCOMORE TREE. (Maspero, *Histoire ancienne des Peuples de l'Orient*) . 4
4. MONKEYS HELPING TO GATHER FIGS, BENI-HASSAN. (Lepsius, *Monuments of Egypt*, II.) . . . 5
5. GODS PAINTING THE KING'S NAME ON THE YSHIT TREE, THEBES: TEMPLE OF RAMESES II. (Lepsius, *Monuments of Egypt*, III.) 6
6. WINE ARBOUR OF THE OLD KINGDOM. With hieroglyphics. Sakhara, Grave of Ptah-hotep. (Paget and Pirie, *The Tomb of Ptah-hotep*) 7
7. A ROUND VINE ARBOUR, BENI-HASSAN. (Lepsius, *Monuments of Egypt*, II.) 8
8. GARDENERS IN THE VEGETABLE GARDEN, BENI-HASSAN. (Lepsius, *Monuments of Egypt*, II.) . . . 8
9. A VINE ARBOUR AND VEGETABLE GARDEN AT EL-BERSHEH. (Newberry, *El-Bersheh*) 9
10. THE GARDEN OF THE HIGH OFFICIAL OF AMENHOTEP (AMENOPHIS) III., THEBES. (Rosellini, *Monumenti Civili*, II.) . . . . *Inset* 10–11
11. VISITING SCENE IN THE GARDEN OF A VILLA AT THEBES. (Erman, *Egypt*) 11
12. A VILLA WITH GARDEN, THEBES. (Wilkinson, *The Manners and Customs of the Ancient Egyptians*, I.) . 12
13. THE GARDEN OF APOUI, THEBES. (Maspero, *Mémoires publiées par les Membres de la Mission française au Caire*, V.) 12
14. THE PLACE OF MĒRIRĒ THE HIGH PRIEST, EL-AMARNA. (Lepsius, *Monuments of Egypt*, III.) . . 13
15. FRONT GARDEN AT THE PALACE OF AMENHOTEP (AMENOPHIS) IV., EL-AMARNA. (Maspero, *Mémoires publiées par les Membres de la Mission française au Caire*, V.) 15
16. A PAINTED FLOOR IN THE PALACE OF AMENHOTEP (AMENOPHIS) IV. (From Professor Sir Flinders Petrie, *Tel-el-Amarna*) 16
17. THE TEMPLE OF DEIR-EL-BAKHARI—A RECONSTRUCTION. (Naville, *Deir-el-Bakhari*, VI.) . . . 17
18. TRANSPORT OF INCENSE TREES FROM THE LAND OF PUNT. (Naville, *Deir-el-Bakhari*, III.) . . . 18
19. INCENSE TREES FROM THE LAND OF PUNT IN THE TEMPLE OF DEIR-EL-BAKHARI. (Naville, *Deir-el-Bakhari*, III.) 19
20. FESTIVAL OF THE DEAD IN THE GARDEN OF REKHMARA. (Maspero, *Mémoires publiées par les Membres de la Mission française au Caire*, V.) 20
21. A LITTLE GARDEN AT A GRAVE—A STELE AT CAIRO. (Erman, *Egypt*). 21
22. TOMB OF OSIRIS WITH SACRED TAMARISK. (Wilkinson, *The Manners and Customs of the Ancient Egyptians*) . 22
23. FOOD AND DRINK IN THE TOMB OF THE DEAD. (From the papyrus of Ani in *The Book of the Dead*) . 22
24. PRIMITIVE GARDEN OF THE DEAD—THE TOMB OF RENNI. (J. J. Tylor and F. L. Griffith, *The Tomb of Renni*. By permission of Mrs. Sydney Morse) 23
25. A PRIMITIVE GARDEN OF THE DEAD. (J. J. Tylor and F. L. Griffith, *The Tomb of Paheri at El Kab*. By permission of Mrs. Sydney Morse) 23
26. FLOWERS FROM THE FLOOR OF AMARNA. (From Professor Sir Flinders Petrie, *Tel-el-Amarna*) . . 24

CHAPTER II. WESTERN ASIA IN ANCIENT TIMES

27. PRIMITIVE DRAWINGS OF TREES. (Rawlinson, *Five Great Monarchies*, I.) 30
28. A PRIMITIVE DRAWING OF A SACRED TREE. (Rawlinson, *Five Great Monarchies*, I.) 31
29. A TEMPLE AND ARTIFICIAL HILL IN A PARK AT KHORSABAD. (Botta and Flandin, *Monuments de Ninive*) . 32
30. THE BANQUETING HALL OF SENNACHERIB AT ASSUR. (*Mitteilungen der deutschen Orientgesellschaft*, 1907, No. 33) 33
31. A GARDEN SURROUNDED BY WATER. (A. H. Layard, *Discoveries in the Ruins of Nineveh*) . . . 34
32. A HANGING GARDEN. (A. H. Layard, *Discoveries in the Ruins of Nineveh*) 35
33. BAGH-I-TAKHT—THE GARDEN OF THE THRONE AT SHIRAZ. (Photograph, Herzfeld) 36
34. A TEMPLE AND HANGING GARDEN AT KUYUNDJIK. (Rawlinson, *Five Great Monarchies*, I.) . . . 37

FIG. PAGE
35. WOMEN'S GARDEN, FROM A PICTURE ON A CYLINDER OF JASPER. (Jean B. F. Lajarde, *Recherches sur les Mystères de Mithra*) . . . 37
36. LILIES FROM THE NORTHERN PALACE AT KUYUNDJIK. (Rawlinson, *Five Great Monarchies*, I.) . . . 38
37. AN ASSYRIAN HUNTING-PARK WITH VINE-BRANCHES AND LILIES, KUYUNDJIK. (Place, *Ninive et l'Assyrie*) . 38
38. ASSURBANIPAL AT A MEAL IN THE PARK OF KUYUNDJIK. (Place, *Ninive et l'Assyrie*) . . . 39
39. DARIUS HUNTING IN A GROVE OF PALMS. (From an Engraving on a mountain crystal. Jean B. F. Lajarde, *Recherches sur les Mystères de Mithra*) . . . 41
40. PERSEPOLIS—THE SIDE WALL OF THE GREAT OPEN STAIRWAY. (Stolze, Andreas, and Nöldeke, *Persepolis*) . 42
41. THE TOMB OF CYRUS AT PASARGADÆ. (Marcel Dieulafoy, *L'Art antique de la Perse*) . . . 43
42. THE TOMB OF SAADI AT SHIRAZ . . . 44
43. ENTRANCE TO THE PLACE OF THE HOLY TREE, CEYLON. (Cave, *Baudenkmäler aus ältester Zeit in Ceylon*) . 47
44. POKUÑA—BATH POND IN CEYLON. (Cave, *Baudenkmäler aus ältester Zeit in Ceylon*) . . . 49
45. RELIEF FROM THE TEMPLE OF BÔRÔ-BUDUR, JAVA. (F. C. Wilson and Leemans, *Bôrô-Boudour*) . . 50

CHAPTER III. ANCIENT GREECE

46. CRETAN FLOWER-PAINTINGS FROM HAGIA TRIADA. (From *Monumenti Antichi*, XIII.) . . . 54
47. PALACE OF KNOSSOS—THE GREAT HALL AND TERRACE . . . 55
48. SILVER VESSEL FROM MYCENÆ, WITH FLOWER ORNAMENTATION. (In the Museum at Athens. After Gilliéron) . . . 56
49. ADONIS GARDEN. (From an Aryballos in the Museum at Karlsruhe) . . . 59
50. ADONIS GARDEN ON A VASE. (From the Museum at Athens) . . . 60
51. A NYMPH GROTTO—RELIEF. (From the Lateran Museum) . . . 62
52. A YOUTH BEFORE A TEMPLE WITH A SACRED TREE. (From Schreiber, *Hellenistische Reliefs*) . . . 64
53. GENERAL VIEW OF THE GYMNASIUM AT DELPHI. (Photograph) . . . 67
54. BATH BASIN AT THE GYMNASIUM, DELPHI. (From *Athenische Mitteilungen*, 1910) . . . 68
55. THE GYMNASIUM AT PERGAMON. (From *Athenische Mitteilungen*, 1910) . . . 69
56. THE FARNESE BULL—A PARK GROUP. (From the Museum at Naples) . . . 75

CHAPTER IV. THE ROMAN EMPIRE

57. A GROTTO—A FRESCO FROM BOSCOREALE. (From Sambon, *Fresques de Boscoreale*, 1903) . . . 87
58. PORTICUS LIVIÆ, TOWN PLAN OF ROME. (Fragments 9 and 10, from Lanciani, *Bolletino communale*, 1886) . 89
59. THE TOMB OF AUGUSTUS IN THE TIME OF THE RENAISSANCE. (Engraving by Scaichi in *I vestigi dell' Antichità di Roma*) . . . 91
60. A ROMAN VILLA—FROM A WALL-PAINTING AT POMPEII. (From Rostovtzeff, "Pompejanische Villen," *Röm. Mitt.*, 1904) . . . 95
61. SANCTUARY, CHAPEL, AND TREE—FROM A WALL-PAINTING AT POMPEII. (From Comparetti, *Le Pitture di Ercolano*) . . . 96
62. GARDEN-ROOM AT THE VILLA OF LIVIA—A WALL-PAINTING (AD GALLINAS), PORTA PRIMA, ROME. (From *Antike Denkmäler*, I.) . . . 97
63. FRIGIDARIUM OF THE STABIAN BATHS, POMPEII. (From Niccolini, *Le Case ed i Monumenti di Pompéi*) . 98
64. A WALL-PAINTING OF A GARDEN AT THE FRIGIDARIUM OF THE STABIAN BATHS, POMPEII. (From Niccolini, *Le Case ed i Monumenti di Pompéi*) . . . 99
65. PLAN OF THE YOUNGER PLINY'S VILLA TUSCI. (Plinius, ep. v. 6) . . . 100
66. A GARDEN AT POMPEII—FROM A WALL-PAINTING. (From Comparetti, *Le Pitture di Ercolano*) . . . 101
67. A GARDEN PICTURE IN THE PERISTYLE OF A POMPEIAN HOUSE. (Museo Borbonico, XII.) . . . 102
68. GROUND-PLAN OF THE HIPPODROME GARDENS ON THE PALATINE. (From Marx, "The so-called Stadium on the Palatine," *Jahrbuch des Inst.*, 1895) . . . 105
69. A VILLA ON THE SEA—FROM A WALL-PAINTING AT POMPEII. (Comparetti, *Le Pitture di Ercolano*) . . 106
70. VILLA ON THE SEA OR ON A LAKE—A WALL-PAINTING AT POMPEII. (Rostovtzeff, *Jahrbuch d. Röm. Inst.*, 1904) 106
71. VILLA WITH XYSTUS—A WALL-PAINTING AT POMPEII. (Rostovtzeff, *Röm. Mitt.*, 1904) . . . 107
72. GROUND-PLAN OF A VILLA WITH XYSTUS. (Rostovtzeff, *Röm. Mitt.*, 1904) . . . 107
73. DESIGN FOR A STAIRWAY APPROACH TO MONTE PINCIO. (Drawing by Ligorio, in *Monumenti Antichi della Reale Accademia dei Lincei*, 1889) . . . 108
74. HEMICYCLIUM IN THE GARDEN OF THE SACRÉ CŒUR, ON MONTE PINCIO. (Middleton, on MSS. notes and drawing by Ligorio, in *Archæologia*, 1888) . . . 109
75. NYMPHÆUM AT THE IMPERIAL VILLA, ALBANO. (Photograph from an engraving by Piranesi) . . 110
76. GENERAL PLAN OF HADRIAN'S VILLA AT TIVOLI. (From Winnefeld, *Arch. Jahrbuch Ergänzungsheft*, 1904) . 111
77. HADRIAN'S VILLA—THE WALL OF THE SO-CALLED POIKILE. (Gusman, *La Villa Imperiale de Tivoli*, 1904) . 112
78. „ „ —THE SO-CALLED NATATORIUM. (Gusman, *La Villa Imperiale de Tivoli*, 1904) . . 113
79. VARRO'S ORNITHON. (Reconstruction by Laurus, *Antiquæ Urbis Splendor*) . . . 114

FIG. PAGE
80. HADRIAN'S VILLA—A PLAN OF THE NYMPHÆUM. (From *Arch. Jahrbuch Ergänzungsheft*, 1904) . . 116
81. FLOWER-BOXES IN THE ATRIUM OF A HOUSE AT TIMGAD. (From Boeswillwald, Cagnat et Ballu, *Timgad, une Cité africaine*) 117
82. GARDEN IN THE HOUSE OF SALLUST AT POMPEII. (From Mazois, *Les Ruines de Pompéi*) . . . 118
83. A WALL-PAINTING IN THE HOUSE OF SALLUST AT POMPEII. (From Mazois, *Les Ruines de Pompéi*) . . 119
84. THE HOUSE OF EPIDIUS RUFUS, POMPEII. (Ground-plan by Mau, Pompeii) 119
85. THE HOUSE OF THE SILVER WEDDING, POMPEII. (Ground-plan by Mau, Pompeii) 119
86. A SMALL GARDEN TERRACE AT THE HOUSE OF MARIUS LUCRETIUS, POMPEII 120
87. FLOWER-GARDEN—A WALL-PAINTING AT POMPEII. (From Comparetti, *Le Pitture di Ercolano*) . . 120
88. PERISTYLE IN THE HOUSE OF THE VETTII, POMPEII 121
89. PLAN OF THE VILLA OF DIOMEDES. (From Niccolini, *Le Case ed i Monumenti di Pompéi*) . . . 122
90. A VILLA WITH GARDEN DECORATION—A FRESCO FROM BOSCOREALE. (From Sambon, *Fresques de Boscoreale*, 1903) 123
91. AN AFRICAN-ROMAN VILLA—A MOSAIC FROM THE BATHS OF POMPEIANUS. (From Blanchère et Gauckler, *Catalogue du Musée de Alaoui*) 125
92. AN AFRICAN-ROMAN VILLA WITH GARDENS—A MOSAIC. (From Blanchère et Gauckler, *Catalogue du Musée de Alaoui*) 125
93. A LADY IN A GARDEN—A MOSAIC FROM THE BATHS OF POMPEIANUS. (From Tissot, *Géographie d'Afrique*) . 126
94. PLAN OF A VILLA AT RUHLING. (From Grenier, *Habitations gauloises*) 128
95. GROUND-PLAN OF A VILLA AT TETING. (From Grenier, *Habitations gauloises*) 129
96. PLAN OF A VILLA AT WITTLICH. (From A. Eberle, *Westdeutsche Zeitschrift*, 1906) 130
97. A ROMAN BASIN AT WELSCHBILLIG, RECONSTRUCTED. (From F. Hettner, Director of the Provincial Museum at Trier) 131

CHAPTER V. BYZANTINE GARDENS AND THE COUNTRIES OF ISLAM

98. A TERRACE-GARDEN IN A BYZANTINE MINIATURE. (*Ménologe du Vatican*, Eleventh Century. Beylié, *L'Habitation byzantine*) 138
99. A GARDEN AND PINEAPPLE FOUNTAIN. (*Homelies de Jacques*, Twelfth Century. Beylié, *L'Habitation byzantine*) 138
100. FOUNTAIN AND ARBOUR—A MOSAIC FROM THE DAPHNE CLOISTER NEAR ANTIOCH 140
101. A SILVER TREE FOUNTAIN. (Beylié, *L'Habitation byzantine*) 141
102. PERSIAN CARPET OF A GARDEN DESIGN, ABOUT 1600 142
103. " " " " 143
104. A GARDEN IN A PERSIAN MINIATURE. (Musée des Arts décoratifs) 144
105. SEAT IN THE BRANCHES OF A TREE—A PERSIAN PAINTING. (In the possession of F. Sarre) . . . 145
106. GROUND-PLAN OF THE PALACE OF BALKUWARA, SAMARIA (Herzfeld, *Vorbericht von Samarrâ*, 1912) . 146
107. CONJECTURAL PLAN OF DAR-EL-KHALIF. (Viollet, *Description du Palais de Al-Moutasim à Samarrâ*) . 147
108. THE COURT OF THE MYRTLES, ALHAMBRA, GRANADA 153
109. THE COURT OF THE LIONS, ALHAMBRA, GRANADA 154
110. COURT OF LINDERAJA (BOUDOIR OF THE SULTANA), ALHAMBRA, GRANADA 155
111. THE GENERALIFE, GRANADA—GARDEN COURT 157
112. " " A VIEW OF THE GARDEN COURT 158
113. " " A VIEW OF THE GARDEN 159
114. ISPAHAN—PART OF THE TOWN PLAN. (Sarre, *Denkmäler persischer Baukunst*) 161
115. TSHEHAR-BAGH, ISPAHAN. (Chardin, *Voyage en Perse*) 162
116. TSHIHIL-SUTUN, ISPAHAN—GROUND-PLAN. (Sarre, *Denkmäler persischer Baukunst*) 163
117. ASHRAF, PERSIA—GROUND-PLAN OF THE GARDENS. (Sarre, *Denkmäler persischer Baukunst*) . . . 164
118. PAVILION IN THE SHAH'S GARDEN, ASHRAF. (Sarre, *Denkmäler persischer Baukunst*) 165
119. A POND IN THE SHAH'S GARDEN, ASHRAF. (Sarre, *Denkmäler persischer Baukunst*) 166
120. A GARDEN HOUSE AT SHIRAZ. (Photograph, Herzfeld) 167
121. GARDENS AT FIN-NAHĒ. (Photograph, Herzfeld) 168

CHAPTER VI. THE MIDDLE AGES IN THE WEST

122. A PARADISE AT THE MONASTERY OF ST. MARIA LAACH 172
123. THE CROSSWAY AT S. PAOLO FUORI LA MURA, ROME 173
124. PROPOSED PLAN FOR A MONASTERY AT ST. GALL. (Keller, *Plan of St. Gall*) . . . *Inset* 176-7
125. PLAN OF THE ABBEY AT CANTERBURY. (Lenoir, *Architecture monastique*) 177
126. HORTUS EREMITÆ. (From the *Hortus Deliciarum* of Herrad von Landsperg) 179
127. THE CLOISTER COURT OF THE THERMÆ MUSEUM, ROME. (Letarouilly, *Les Edifices de Rome*) . . . 181

FIG. PAGE
128. THE GARDEN DOOR—A MINIATURE. (From *Roman de la Rose*. British Museum, *Haarlem MSS*. 4425) . 182
129. CASTLE OF VINCENNES, PARIS. (Du Cerceau, *Les plus beaux Bâtiments de France*) 183
130. THE CASTLE GARDEN AT THE BELFRY IN THE PICTURE OF "CIVITAS DEI." (From the *Breviarium Grimani*. Cliché from *Bibliothéque d'Art et d'Archéologie*) 184
131. TURF SEAT BY A WALL. (From *Roman de la Rose*. British Museum, *Haarlem MSS*. 4425) . . . 185
132. WOMEN SITTING ON GRASS WEAVING WREATHS. (From *Heures d'Anne de Bretagne*. Cliché from *Bibl. d'Art et d'Arch.*) 186
133. A GARDEN PICTURE FROM *Roman de la Rose*. (British Museum, *Haarlem MSS*. 4425) 186
134. GARDEN WITH CLIPPED TREES AND FLOWER-PLOTS. (From the *Breviarium Grimani*. *Bibl. d'Art et d'Arch.*) 187
135. A MAY TREE WITH ARTIFICIAL FRUITS. (From *Heures d'Anne de Bretagne*. *Bibl. d'Art et d'Arch.*) . . 188
136. A ROSE ARBOUR. (From *Roman de la Rose*. *Bibl. d'Art et d'Arch.*) 189
137. AN ARBOUR USED AS A DINING-ROOM—A CARPET DESIGN. (South Kensington Museum) . . . 189
138. THE GARDEN HOUSE OF A LEARNED MAN. (Brussels Library) 190
139. SUSANNA AT THE BATH—A CARPET. (South Kensington Museum) 191
140. BATH IN A CASTLE GARDEN. (From *Horæ Beatæ Mariæ Virginis*. British Museum, *Rothschild MSS.*) . 192
141. WATER BASIN WITH CANAL. (From *Roman de la Rose*. British Museum, *Haarlem MSS*. 4425) . . 193
142. A ROSE HEDGE AND A WATER TOWER. (From *Roman de la Rose*. British Museum, *Haarlem MSS*. 4425) 193
143. A HOUSE GARDEN WITH A TURF SEAT. (From a Painting by Dierk Bouts) 194
144. THE GARDEN OF LOVE—FROM "TRIONFO DELLA MORTE." (A Wall-painting at Campo Santo, Pisa) . . 195
145. THE GREAT GARDEN OF LOVE. (Copperplate, "Meister des Liebesgarten." Lützow, *Geschichte des deutschen Kupferstiches und Holzschnittes*) 196
146. A LITTLE GARDEN OF PARADISE—FROM THE HISTORICAL MUSEUM AT FRANKFORT 197
147. HOUSE GARDEN WITH LATTICE FENCE. (From *Heures de Notre Dame*, attributed to Hennessy. *Bibl. d'Art et d'Arch.*) 198
148. SUBURBAN GARDEN AND SMALL FARMYARD. (From a Painting by P. de Hoogh. Photograph by Bruckmann) 199
149. GARDEN IN THE DAVID PICTURE. (From a French MS. of *Horæ Beatæ Mariæ Virginis*. British Museum, 14805) 200
150. LE JARDIN D'ALENÇON. (Marcel Fouquier, *De l'Art des Jardins du XV*[e] *au XX*[e] *Siècle*, Paris, 1911). . 201

CHAPTER VII. ITALY IN THE TIME OF THE RENAISSANCE AND THE BAROQUE STYLE

150*a*. A ROOF GARDEN AT VERONA. (Joh. Christoph Volkamer, *Nürnbergische Hesperides*, Nuremberg, 1708, with Continuation in 1714) 208
151. PAVILION WITH FOUNTAIN. (From *Hypnerotomachia, Il Sogno di Polifilo*) 210
152. CLIPPED TREES. (From *Hypneroromachia, Il Sogno di Polifilo*) 211
153. MEDICEAN VILLA AT CAREGGI 212
154. VILLA SALVIATI, FLORENCE 213
155. FOUNTAIN: BOY WITH DOLPHIN; BY VERROCCHIO, ORIGINALLY AT VILLA CAREGGI 214
156. POGGIO A CAJANO—VIEW OF THE VILLA IN THE DISTANCE 215
157. „ „ WITH STEPS LEADING TO THE GARDEN 216
158. PICTURE OF RUINS. (From *Hypnerotomachia, Il Sogno di Polifilo*) 221
159. THE GARDEN COURT OF THE BELVEDERE. (Engraving by H. van Cleef) 223
160. THE GARDEN OF STATUES AT THE VILLA FIRST CALLED DELLA VALLE AND LATER CAPRANICA, 1553. (Engraving by Cock, in *Arch. Jahrbuch*, 1891) 225
161. COURT OF THE BELVEDERE, ROME—SITE-PLAN. (Paul Letarouilly, *Les Edifices de Rome moderne ou Recueil des Palais, Maisons, Eglises, etc., de la Ville de Rome*, Paris, 1860) 227
162. THE BELVEDERE, ROME—CROSS-SECTION OF THE COURT. (Letarouilly, *Les Edifices de Rome moderne ou Recueil des Palais, Maisons, Eglises, etc., de la Ville de Rome*, Paris, 1860) 228
163. COURT OF THE BELVEDERE, ROME, AT THE TIME OF BRAMANTE'S DEATH. (Letarouilly, *Les Edifices de Rome moderne ou Recueil des Palais, Maisons, Eglises, etc., de la Ville de Rome*, Paris, 1860) . . 229
164. VILLA MADAMA, ROME—THE UPPER TERRACE WITH THE POOL GROTTO. (Th. Hofmann, *Raffael als Architekt*, Zittau, 1900) 230
165. „ „ THE ELEPHANT ALCOVE. (Hofmann, *Raffael als Architekt*, Zittau, 1900) . . 231
166. „ „ RECONSTRUCTION OF THE EAST GARDEN. (Hofmann, *Raffael als Architekt*, Zittau, 1900) 232
167. „ „ RAPHAEL'S DESIGN FOR THE SOUTH GARDEN. (Hofmann, *Raffael als Architekt*, Zittau, 1900) 233
168. THE PALAZZO DEL TE, MANTUA—GROUND-PLAN. (Bottani, *Descrizione Storica*, 1783, from Patzack, *Villa Imperiale in Pesaro*, 1908) 234
169. VILLA IMPERIALE, PESARO—GROUND-PLAN AND CROSS-SECTION. (Patzack, *Villa Imperiale in Pesaro*, 1908. After Bertulli) 237

FIG. PAGE
170. San Vigilio, on the Lake of Garda. (Thode, *Somnii Explanatio*, published by Grote, Berlin, 1909) . 240
171. „ „ „ „ Rondel in the Garden 241
172. The Garden of Citrons, San Vigilio 242
173. The Botanic Garden at Padua—Old Part. (P. A. Saccardo, *L'Orto Botanico di Padova*, 1895. After Prosperini) 243
174. The Villa di Castello, Florence—Ground-plan. (H. Inigo Triggs, A.R.I.B.A., *The Art of Garden Design in Italy*, London, 1906. Longmans, Green & Co. Ltd.). 244
175. „ „ „ the Lower Garden 245
176. „ „ „ the Antæus Fountain. 246
177. Fountain with Nymph, formerly at the Villa di Castello, Florence; now at the Villa dei Petraia 247
178. A Seat in an Oak-tree at Pratolino. (Sgrilli, *Pratolino*, 1742. Engraving by Della Bella) . . . 248
179. The Palazzo Doria, Genoa—Ground-plan. (Triggs, *The Art of Garden Design in Italy*, London, 1906. Longmans, Green & Co. Ltd.) 249
180. „ „ „ the Chief Parterre 250
181. Villa d'Este, Tivoli. (Engraving showing the Axial Line. Venturini in *Le Fontane del Giardino Estense*) . 251
182. „ „ „ the Dragon Fountain 253
183. „ „ „ in the Sixteenth Century—Ground-plan. (Engraving by Dupérac) . . 255
184. „ „ „ the Water Basins. (Engraving by Venturini in *Le Fontane del Giardino Estense*). 256
185. „ „ „ the Water Organ. (Venturini in *Le Fontane del Giardino Estense*) . . . 256
186. „ „ „ the Water Organ as it is To-day 257
187. „ „ „ the Water Theatre Ruins 258
188. „ „ „ the Main (Fountain) Walk. (Venturini in *Le Fontane del Giardino Estense*) . 259
189. „ „ „ the Lower Rondel of Cypresses. (Venturini in *Le Fontane del Giardino Estense*). 259
190. „ „ „ the Rondel of Cypresses as it is To-day 260
191. The Vatican Gardens, Rome—General Plan. (Falda, *Giardini di Roma*) 261
192. Villa Pia, Rome—Ground-plan. (Percier et Fontaine, *Les plus beaux Bâtiments de l'Italie*, 1809) . 262
193 and 193*a*. Villa Pia, Rome—the Approach to the Court from the Nymphæum; and the Interior of the Court. (Charles Latham, *The Gardens of Italy*. Copyright: *Country Life*) . . . 263
194. Villa Pia, Vatican. (Falda, *Giardini di Roma*) 264
195. The Farnese Gardens, Rome—General Plan. (Falda, *Giardini di Roma*) 264
196. „ „ „ the Approach from the Forum. (Letarouilly, *Les Edifices de Rome moderne ou Recueil des Palais, Maisons, Eglises, etc., de la Ville de Rome*, Paris, 1860) 266
197. „ „ „ Upper Terraces. (Falda, *Giardini di Roma*) 267
198. Villa Papa Giulio, Rome —Ground-plan of the Middle Gardens. (Letarouilly, *Les Edifices de Rome moderne ou Recueil des Palais, Maisons, Eglises, etc., de la Ville de Rome*, Paris, 1860) 268
199. „ „ „ the Nymphæum. (Falda, *Giardini di Roma*). 269
200. Villa Lante, Bagnaja—Ground-plan. (Percier et Fontaine, *Les plus beaux Bâtiments de l'Italie*, 1809) . 270
201. „ „ „ Principal View 271
202. „ „ „ Sixteenth Century. (Laurus, *Antiquæ Urbis Splendor*) 272
203. „ „ „ Parterre with Basins seen from Above. (Latham, *The Gardens of Italy*. Copyright: *Country Life*) 273
204. Caprarola—Ground-plan of the Castle and Garden. (Engraving by Giuseppe Vasi) . . . 274
205. „ Caryatides on the Garden Bridge 275
206. „ Ground-plan of the Casino. (Percier et Fontaine, *Les plus beaux Bâtiments de l'Italie*, 1809) 276
207. „ Principal View of the Casino 277
208. „ Stairway of the River-horses near the Casino 278
209. „ the Approach to the Casino from the Side 279
210. „ Casino and Garden Boundary at the Back 280
211. Villa Campi, Florence—Ground-plan. (Triggs, *The Art of Garden Design in Italy*, London, 1906. Longmans, Green & Co. Ltd.) 281
212. „ „ „ View of the Parterre. (Triggs, *The Art of Garden Design in Italy*, London, 1906. Longmans, Green & Co. Ltd.) 282
213. Pratolino, Florence—Ground-plan. (Sgrilli) 283
214. „ „ Apennino at the Fountain. (Sgrilli, after Della Bella) 283
215. „ „ Statue of Apennino 284
216. Boboli Gardens, Florence—Ascent to the Ammanati Court 285
217. „ „ „ Ground-plan. (Triggs, *The Art of Garden Design in Italy*, London, 1906. Longmans, Green & Co. Ltd.) 286
218. „ „ „ Fountains and Amphitheatre 287
219. „ „ „ the Amphitheatre 289

FIG. PAGE
220. Boboli Gardens, Florence—the Neptune Fountain 290
221. „ „ „ the Cypress Avenue 291
222. „ „ „ the Covered Walk to the Porta Romana 291
223. „ „ „ View from the Isolotto 292
224. „ „ „ the Crossways 293
225. „ „ „ View Inside the Grotto 294
226. The Gardens of the Quirinal on Monte Cavallo, Rome—General Plan. (Falda, *Giardini di Roma*) . 296
227. Villa Medici, Rome—General Plan. (Falda, *Giardini di Roma*) 297
228. „ „ „ Present Appearance 298
229. „ „ „ Statues in the Porch. (Venturini, *Le Fontane di Roma*) 299
230. „ „ „ Bird's-eye View of the Chief Gardens. (Falda, *Giardini di Roma*) . . . 300
231. „ „ „ Earliest Design. (Scaichi) 301
232. Villa Montalto, Rome, with Gardens. (Falda, *Giardini di Roma*) 302
233. „ „ „ the Approach to the Casino. (Venturini, *Le Fontane di Roma*) . . . 303
234. „ „ „ Principal Entrance and Cypress Avenue. (Percier et Fontaine, *Choix des plus célèbres Maisons de Plaisance de Rome*) 304
235. „ „ „ General Plan of the Gardens. (Falda, *Giardini di Roma*) 306
236. Villa Mattei, Rome, in the Seventeenth Century—General Plan (Falda, *Giardini di Roma*) . . 307
237. „ „ „ Palace with Prato. (Falda, *Giardini di Roma*) 308
238. „ „ „ in the Sixteenth Century—General Plan. (Laurus, *Antiquæ Urbis Splendor*) . 308
239. „ „ „ the Eagle Fountain. (Venturini, *Le Fontane di Roma*) 309
240. Villa Falconieri, Frascati—the Approach to the Lake 311
241. Villa Aldobrandini, Frascati—Ground-plan. (Percier et Fontaine, *Choix de plus célèbres Maisons de Plaisance de Rome*) 312
242. „ „ „ the Approach to the Front. (Engraving by Specchi) 314
243. „ „ „ the Fountain of the Little Ships. (Falda, *Le Fontane delle Ville di Frascati*) 315
244. „ „ „ Pool in the Flower-garden 316
245. „ „ „ the Theatre 317
246. „ „ „ the Water Stairway. (Falda, *Le Fontane delle Ville di Frascati*) . 318
247. „ „ „ Rustic Fountain. (Falda, *Le Fontane delle Ville di Frascati*) . . 318
248. „ „ „ Natural Fountain. (Falda, *Le Fontane delle Ville di Frascati*) . . 319
249. Villa Ludovisi (Torlonia)—the Artificial Water Arrangements. (Falda, *Le Fontane delle Ville di Frascati*. Engraving by Venturini 320
250. „ „ „ Pool above the Water Stairway. (Falda, *Le Fontane delle Ville di Frascati*. Engraving by Venturini) 320
251. „ „ „ the Upper Pool 321
252. „ „ „ the Fountain on the Terrace. (Falda, *Le Fontane delle Ville di Frascati*) 322
253. Villa Lancelotti, Frascati—the Parterre 322
254. Villa Muti, Frascati—the Middle Terrace 323
255. Villa Belpoggio, Frascati—Ground-plan. (Percier et Fontaine, *Choix des plus célèbres Maisons de Plaisance de Rome*) 324
256. Villa Mondragone, Frascati—Ground-plan of the Flower-gardens. (Percier et Fontaine, *Choix des plus célèbres Maisons de Plaisance de Rome*) . . . 325
257. „ „ „ Flower-garden with Theatre Behind. (Falda, *Giardini di Roma*) . 326
258. Villa Borghese, Rome—General Plan in the Seventeenth Century. (Falda, *Giardini di Roma*. Engraving by Simon Felice) 327
259. „ „ „ a Bird's-eye View. (Falda, *Giardini di Roma*) 328
260. „ „ „ the Fountain of the River-horses 329
261. „ „ „ the Front Façade. (Falda, *Giardini di Roma*) 330
262. „ „ „ Fountain of Narcissus. (Venturini, *Le Fontane nei Giardini di Roma*) . . 331
263. Villa Ludovisi—General Plan. (Falda, *Giardini di Roma*) 332
264. A Tree in the Old Garden at Villa Ludovisi, Rome 333
265. Villa Ludovisi, Rome—a View of the Garden. (Falda, *Giardini di Roma*) 334
266. Villa Pamfili, Rome—General Plan. (Falda, *Giardini di Roma*. Engraving by Simon Felice) . . 335
267. „ „ „ the Sunk Parterre. (Falda, *Giardini di Roma*. Engraving by Simon Felice) . 336
268. „ „ „ the Site of the Theatre. (Venturini, *Le Fontane nei Gardini di Roma*) . . 337
269. Isola Bella—Ground-plan. (Triggs, *The Art of Garden Design in Italy*. London, 1906. Longmans, Green & Co. Ltd.) 338
270. „ „ 339
271. „ „ the Ten Terraces of the Garden 340
272. „ „ Grotto at the End of the Terraces, Land Side 341
273. „ „ the Side Terraces 341

FIG. PAGE
274. Villa Collodi near Pescia, Lucca—General Plan. (Triggs, *The Art of Garden Design in Italy*, London, 1906. Longmans, Green & Co. Ltd.) . . 342
275. " " " " the Terrace Garden 343
276. Villa Albani, Rome—Ground-plan of the Garden between the Casino and the Coffee-house. (Percier et Fontaine, *Choix des plus célèbres Maisons de Plaisance de Rome*) . 344
277. " " " a Garden Walk in the Park 345
278. " " " View of the Coffee-house from the Casino Terrace 346
279. " " " the Principal Parterre seen from the Side 347
280. " " " Wall of Cypresses, with Statues 348
281. " " " a Covered Walk or Pergola. (Percier et Fontaine, *Choix des plus célèbres Maisons de Plaisance de Rome*) 349

CHAPTER VIII. SPAIN AND PORTUGAL IN THE TIME OF THE RENAISSANCE

282. The Alcazar, Seville—Ground-plan of the Gardens. (From the *Architektonische Rundschau*). . 353
283. " " Patio de las Doncellas. 354
284. " " View over the Gardens. 355
285. " " Another View over the Gardens 356
286. " " the Gallery of Pedro I. 357
287. " " the Pavilion of Carlos V. 358
288. The Alhambra, Granada—Fountain of Carlos V. 359
289. " " Jardin de los Ardives 360
290. The Escorial—General View. (Engraving by Meunier; *Diversas vistas de las casas y jardines de placer da Rei d'España*) 361
290*a*. " " Patio de los Evangelistas. (Schubert, *Geschichte des Barock in Spanien*, Paul Neff Verlag (Max Schreiber), Esslingen) 363
291. Casa del Campo, Madrid—View of the Garden. (Meunier, *Diversas vistas de las casas y jardines de placer da Rei d'España*) 365
292. Aranjuez, Madrid—from the Nuremberg Plan. (Engraving by J. B. Homann) 366
293. " " Jardin de la Isla, known as Salon de los Reyes Católicos 367
294. " " Entrance to the Jardin de la Isla, with Hercules Fountain 368
295. " " the Fountain of the Tritons. (From a Painting by Velasquez. C. Justi, *Velasquez*, I.) 369
296. A Noble Fountain at Aranjuez, Madrid 370
296*a*. Aranjuez—Fountain and Lattice-work Avenues. (Meunier, *Diversas vistas de las casas y jardines de placer da Rei d'España*) 371
297. Trellis from the Royal Gardens, Madrid. (After Drawing by Beaumont; Fouquier, *De l'Art des Jardins*) 372
298. Aranjuez, Madrid—a View of the Parterre and the Tagus 373
299. El Pardo—View of the Terraces and Theatre Wall 374
300. Buen Retiro, Madrid—Plan of the Park. (From the Town Plan of Nuremberg. Engraving by J. B. Homann) 376
301. " " " Jardin de la Reina. (Meunier, *Diversas vistas de las casas y jardines de placer da Rei d'España* 377
302. " " " General View. (From the Town Plan of Nuremberg. Engraving by J. B. Homann) 378
303. " " " Hermitage of St. Paul. (Meunier, *Diversas vistas de las casas y jardines de placer da Rei d'España*) 378
304. Quinta di Bacalhão—Plan of the Site. (J. Rasteiro, *Palacio e Quinta di Bacalhão*, 1898) . . 381
305. " " Pool in the Garden and Pillared Pavilion. (From *Arte e Naturezze em Portugal*) 382
306. Bemfica: Quinta of the Marquez de Fronteira—the Chief Parterre. (From a Photograph in the possession of A. Haupt) 383
307. " " " " " " (From a Photograph in the possession of A. Haupt) . 384
308. Queluz, Lisbon—the Principal Parterre. (From *Arte e Naturezze em Portugal*) 385
309. Stairways in the Park at the Monastery of Bussaco. (From *Arte e Naturezze em Portugal*) . . 386
310. Avenue of Cedars in the Park at the Monastery of Bussaco. (From *Arte e Naturezze em Portugal*). 387
311. Braga, Bom Jesus do Monte—Steps of the "Stations." (From *Arte e Naturezze em Portugal*) . . 388

CHAPTER IX. FRANCE IN THE TIME OF THE RENAISSANCE

312. Amboise—General Plan. (Du Cerceau, *Les plus beaux Bâtiments de France*) 392
313. Blois—General Plan of the Castle and Gardens. (Du Cerceau, *Les plus beaux Bâtiments de France*) . 393

FIG. PAGE
314. Blois—Chapel of Anne of Bretagne 394
315. Gaillon—General View. (Du Cerceau, *Les plus beaux Bâtiments de France*) 395
316. „ the Great Fountain. (Du Cerceau, *Les plus beaux Bâtiments de France*) 397
317. Fontainebleau at the Time of Francis I. (Du Cerceau, *Les plus beaux Bâtiments de France*) . . 399
318. „ Grotte des Pins. (Ward, *The Architecture of the Renaissance in France*. Drawing by P. D. Hepworth) 400
319. Chantilly in the Sixteenth Century. (Du Cerceau, *Les plus beaux Bâtiments de France*) . . . 401
320. A Plan for a Castle, by Leonardo da Vinci. (Geymüller, *Geschichte der Französischen Renaissance*) . 402
321. Bury—a General View of the Castle. (Du Cerceau, *Les plus beaux Bâtiments de France*) . . . 403
322. Dampierre—General View. (Du Cerceau, *Les plus beaux Bâtiments de France*) 404
323. Valleri—General View of the Castle and Gardens. (Du Cerceau, *Les plus beaux Bâtiments de France*) 404
324. „ the Principal Parterre. (Du Cerceau, *Les plus beaux Bâtiments de France*) 406
325. Anet. (Du Cerceau, *Les plus beaux Bâtiments de France*) 407
326. Verneuil—General Plan of the Castle and Gardens. (Du Cerceau, *Les plus beaux Bâtiments de France*) 408
327. Charleval—General Plan of the Gardens. (Du Cerceau, *Les plus beaux Bâtiments de France*) . . 410
328. Montargis—General Plan of the Castle and Gardens. (Du Cerceau, *Les plus beaux Bâtiments de France*) 411
329. „ Trellis in the Castle Garden. (Du Cerceau, *Les plus beaux Bâtiments de France*) . . 412
330. The Tuileries—Plan of the Gardens. (From the Town Plan of Paris, 1652, by J. Gomboust) . . 413
331. Meudon—the Grotto and the Parterre in the Time of Louis XIV. (After E. N. Langlois) . . 415
332. Gaillon—the Hermitage and the White House. (Du Cerceau, *Les plus beaux Bâtiments de France*) . 417
333. An Old Plan for a Botanic Garden. (O. de Serres, *Le Théâtre d'Agriculture*). 419
334. Luxembourg Gardens—Plan for a Parterre. (Boyceau, *Traité du Jardinage*) 421
335. The Tuileries Gardens—the Parterre. (Engraving by Perelle) 422
336. Saint-Germain-en-Laye—Plan of the Gardens. (Furttenbach, *Architectura civilis*) . . . 423
337. Fontainebleau—the Orangery. (Engraving by Israel Silvestre) 425
338. „ at the Time of Henry IV. (Furttenbach, *Architectura civilis*) 426
339. Luxembourg—the Gardens in 1652. (From the Town Plan of Paris, 1652. Furttenbach, *Architectura civilis*) 427
340. „ the Great Parterre. (Engraving by Perelle) 428
341. „ the Fountain of the Medici. (Alphand, *Les Promenades de Paris*) 429
342. Ruel—the Grotto. (Engraving by Perelle) 430

CHAPTER X. ENGLAND IN THE TIME OF THE RENAISSANCE

343. Hampton Court—Old Garden Scenes. (From the Picture of Henry VIII. and his Family, in the Museum at Hampton Court) 438
344. „ „ the Old Pond Garden 440
345. Nonsuch Castle. (Georg Braun and Franciscus Hogenbergius, *De præcipuis totius Universi urbibus, ap. Phil. Gallæum* 1575–1618 *in Polis*) 442
346. A Hedged Garden. (Thomas Hill, *The Gardener's Labyrinth*) 444
347. A House Garden. (Th. Hill, *The Gardener's Labyrinth*) 445
348. Hatfield House, Hertfordshire—Garden and Labyrinth. (Triggs, *Formal Gardens in England and Scotland*. By permission of B. T. Batsford, Ltd.) 450
349. „ „ „ the Old Garden. (Triggs, *Formal Gardens in England and Scotland*. By permission of B. T. Batsford, Ltd.) 451
350. Montacute, Somersetshire—Plan of the Site. (Triggs, *Formal Gardens in England and Scotland*. By permission of B. T. Batsford, Ltd.) 452
351. Haddon Hall—Cedars on the Upper Terrace. (After Latham, *Gardens Old and New*. Copyright: *Country Life*) 453
352. The Garden at Wilton House, near Salisbury ("Hortus Pembrochianus.") (Engraving by Isaac de Caus) 454
353. The Thicket at Wilton House ("Hortus Pembrochianus.") (Engraving by Isaac de Caus) . . 455
354. Oxford Botanic Garden. (*Les Délices de la Grande Bretagne*). 456
355. The Old Garden at Wimbledon—Ground-plan. (From the Parliamentary Survey, No. 72) . . 458

CHAPTER I

ANCIENT EGYPT

A HISTORY OF GARDEN ART

CHAPTER I

ANCIENT EGYPT

UNDER primitive conditions, the cultivation of gardens could only begin when a wandering tribe settled. The nomad drove his herds across the open unfenced pasture-land. When, however, with the aid of the pickaxe he broke the ground, providing something in the nature of cultivated land and a fixed dwelling-place, he was compelled to erect a fence, in order to protect the homestead from the inroads of enemies and wild beasts. This enclosure, rudimentary as it was, formed the first garden. It was earlier than the farm, which could only be developed after fields had been formed with further fencing. It is in this particular sense that Herder was right when he called the garden older than the farm; nevertheless it was not a garden proper, because it was not yet separated from the rest of the land. Such as it was, it was quite close to the house, and contained fruit-trees and vegetables. The different kinds of plants had to be set out in regular beds because of the sort of attention they needed, for only by observing a certain order in arrangement was it possible to secure a satisfactory return.

We cannot, of course, study the earliest stages of gardening at first hand, and the prehistoric discoveries about cultivated plants throw no light on the manner or order of their actual planting. One can only say with certainty that the cultivation was not equal everywhere, nor according to any one plan; and that in a country won from the primeval forest the gardens had quite a different aspect from the oasis gardens still girt by the sandy desert.

In the very cradle of all human civilisation is a land which from the peculiar character of its soil and climate was bound to show an early and important development of garden cultivation. This land is Egypt. Here the Nile by its own independent work has wrested from the desert a valley, narrow and long. Every year its waters, pouring forth blessing and fruitfulness, take on themselves that work which in other places has to be carried out by the care and pains of men themselves—the work of conveying to the earth the nourishment first obtained therefrom. But this narrow strip of land, so specially favourable to the production of succulent stalks, was ill adapted for the cultivation of the larger kinds of trees or for long-lived vegetation. Whether or not Egypt had extensive woodlands in prehistoric times, as would appear from palæontological discoveries, we cannot consider here, for it presupposes a different temperature and climate. We can only remark in passing that a rainless sky must have checked the development of forest, especially in Upper Egypt. If the kings wanted to enjoy the pleasures of the chase, they had to be content with the neighbouring desert. In pictures of the Old and the Middle Period we find hurdles, set up for hunting purposes, apparently only with the object of a battue. Among the desert creatures we find stags, and this almost implies forest.

The establishment of large trees could only be aimed at in high places, or at the edges of the valley where the waters of the stream did not reach during the inundations.

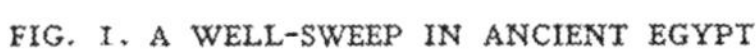

FIG. 1. A WELL-SWEEP IN ANCIENT EGYPT

FIG. 2. A WELL-SWEEP IN MODERN EGYPT

It could only succeed from the very start through man's most diligent care. Watering and the provision of nourishment for the land and its crops needed skill and artifice. The waters of the Nile were brought to the higher and more distant parts of the country by an elaborate network of canals, regulated by dams, terraces, and sluices, and were then drawn up by the help of a well-sweep (shadoof or shaduf). On one arm of the pump-handle hung a weight, on the other a bucket, and the water was poured out on plants, trees, and fields. Just as we see it in pictures, thousands of years old, so can it be seen now (Figs. 1 and 2). It is obvious that with so much labour entailed, only such trees and plants would be considered as were found useful enough to repay a man for his trouble. It was precisely from the profit-making care of plants that all horticulture arose. Edible fruits, timber, and shade—these the Egyptian demanded and obtained from his garden.

Though it is only the New Period that has given us pictures of a systematic arrangement and grouping of different plantations in one, that is, of enclosures which can be considered gardens in our sense, there is yet undoubted evidence in the Fourth and Fifth Dynasties, and still more in the Middle Period, of plots laid out for trees, vines and vegetables. These pictures show the individual work of the garden. They show how the gardener waters the land, how he plucks the fruits from the trees, how he gathers the ripe bunches of grapes, and how he pours libations to the gods for a blessing on his work.

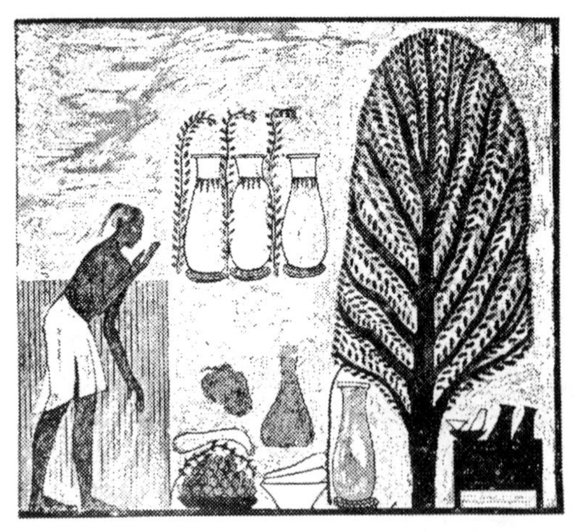

FIG. 3. PEASANT DOING HONOUR TO A SYCOMORE TREE

First among the trees which belonged to the earliest form of Egyptian garden is the sycamore [EDITOR'S NOTE: Or sycomore, for doubtless the sycomore of Luke xix.4 is meant. The "sycamore fig" is *Ficus sykomoros*. The common foliage sycamore, or "false plane," is *Acer Pseudo-platanus*]. It is very often mentioned, and in the old records the sign for Sycamore often stands for Tree in general. It seems that, according to a very ancient belief, a sycomore stood under the canopy of heaven beside both the rising and the setting of the sun; it was supposed to be of malachite, perhaps to indicate its imperishable green hue. The fruit and the wood of this tree are both of use; in its shade the living rejoice as well as the dead, and the peasant honours it as especially sacred, and sacrifices to it the fruits of the earth (Fig. 3). There is a feeling of close sympathy between the dweller in the Nile Valley and this tree, expressed in a little poem which depicts the tree as the friend of lovers. On the day when "the Garden holds its festival," when the tree is in all the glory of its flowering, it sends a message to the maiden:

The little sycomore,
Which she planted with her own hand,
She moves her lips to speak.
How fair are her lovely branches!
She is laden with fruits
That are redder than the jasper.
Her shade is cool,
She lays a little letter in a girl's hand,
The head gardener's daughter,
She bids her hasten to the well-beloved:
"Come and stay among thy maidens.
We are drunken if we would go to thee,
Ay, before we have tasted aught.
The servants who obey thee
Are coming with their vessels;
Beer of every kind they bring
And every kind of bread,
Many flowers of to-day and yesterday
And all refreshing fruits.
Come, and make it fine to-day,
To-morrow and next day, three days long. . . .
Sit in my shade."
Her friend sits on her right hand,
She makes him drunken
And yields to what he says. . . .
But I am dumb,
And say not what I see.
I will say no word."

The pictures of the Old Period show how they planted the trees at regular intervals, how they collected the fruit, then cut the tree down and disposed of its wood, or else when it was ready to cut down allowed the young goats to eat the foliage. And in early days they seem to have used animals to help in their harvesting. The monkeys had to help gather figs (Fig. 4), and were allowed to enjoy some of the fruit themselves so long as they left the men's share in the baskets. The fig-tree is no less important to the ancient Egyptians than the sycomore. Near to these in importance stand two kinds of palm,

FIG. 4. MONKEYS HELPING TO GATHER FIGS, BENI-HASSAN

the date and the doum (*Dom*), the date especially being very much used, and in many ways. There was also the acacia, with which perhaps the tree (*Mimusops Schimperi*) must be counted that the ancients called *Persea*. [EDITOR'S NOTE: The Sunt-tree (*Acacia nilotica*) must be remembered. Together with the sycomore and the tamarisk, it was abundant. The Lebbek (*Albizzio* or *Acacia Lebbek*) was imported. As regards the Persea, the *Persea*

FIG. 5. GODS PAINTING THE KING'S NAME ON THE YSHIT TREF, THEBES: TEMPLE OF RAMESES II

gratissima of modern botanists is the avocado or alligator pear, a West Indian tree. Neither it nor the modern Mimusops is of the *Leguminosæ*, like the true Acacia.] Together with the sycomore it belongs to the group which in the earliest times fixes the character of Egyptian tree-culture, although their native land was not Egypt. Besides these there was also a sacred tree held in high honour, which is called in the writings *Ished*—a fruit-tree whose identity is not really fixed. This ished (yshit) is the tree of history, on which the gods depict the name and the deeds of the king (Fig. 5).

In the New Period, the species are far more numerous, and the writer Ennene, who lived about 1500, enumerates on the inscription intended for his tomb the trees which he planted in his lifetime. He gives a list of twenty different sorts, of which some are not yet identified. The old kinds are also given in an extra list, and comprise no fewer than 73 sycomores, 170 date palms, and 120 doum palms. In the New Period the want of forest was in part compensated for by a great number of sacred groves. Every one of the forty-two districts, into which Upper and Lower Egypt were divided, had its own temple with its sacred grove attached. From very early times trees were held sacred by the Egyptians, but now each of these temples had a particular tree sacred to itself, which was chiefly, if not exclusively, cultivated in the temple garden. If we may judge by the inscriptions, the greater temples must have owned very extensive lands.

Next to trees the Old Period seems to have valued vineyards. We do not know the actual beginnings. Pictures of the Fourth and Fifth Dynasties show the whole story of the grape, how it begins to ripen, how guards chase away the birds with sticks, how it is gathered, how it is trodden, and how the must is poured into tall, pointed vessels without feet. In the earliest days they seem to have grown the vines on arbours; that is, they stuck up stakes and set a transom across. The oldest hieroglyph for Wine shows this (Fig. 6). But even in the Old Period they began to substitute posts for the rough wooden props, and from these there came in the New Period their beautiful, finely painted pergolas. The arbours were often round, which made them prettier (Fig. 7), and a second hieroglyph shows that this form was a deviation from the original straight lines. In the Middle Period, and still more in the New, these ideas were developed in a more and more pleasing way. Right on into the New Period vineyard arbours were the centre and chief ornament of all gardens.

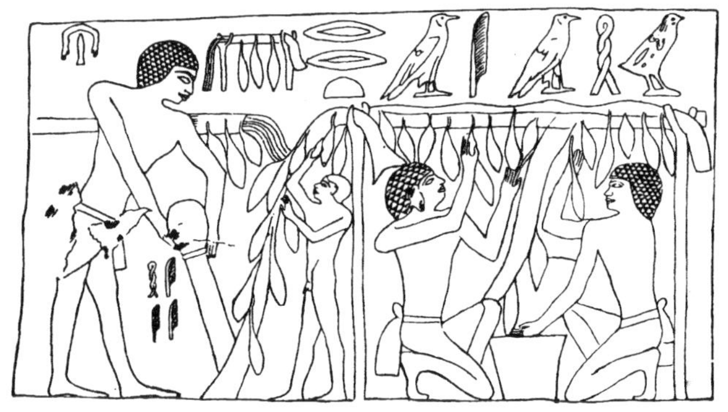

FIG. 6. WINE ARBOUR OF THE OLD KINGDOM

As to the abundance of vegetable products among the ancient Egyptians, we need only consult their sacrificial vessels, their baskets of fruits, and their festival processions. There are long rows of men carrying baskets laden with fruit on their shoulders, and ears of corn or papyrus in their hands; between them are other beautifully woven baskets, piled high with every sort of garden food. But the Middle Period has the fewest pictures of vegetable gardens: one picture of Beni-Hassan shows (in very poor perspective) square beds, planted with green vegetation (Fig. 8). A canal ending in a round pond has green tendrils all about it, as though to indicate that it lies within the garden; beside it two men are working, bringing water and emptying it over the beds. They would seem to be important persons in the service of their master, for both their names are perpetuated, not only in an inscription over their heads, but in the large picture of the servants bringing gifts to place on the sacrificial tables of the dead, where our two gardeners appear with their names again noted, Neter-Necht and Nefer-Hetep; they are bringing to their master the fruits of the earth obtained by their own labour, in baskets carried over their shoulders on a crossbar.

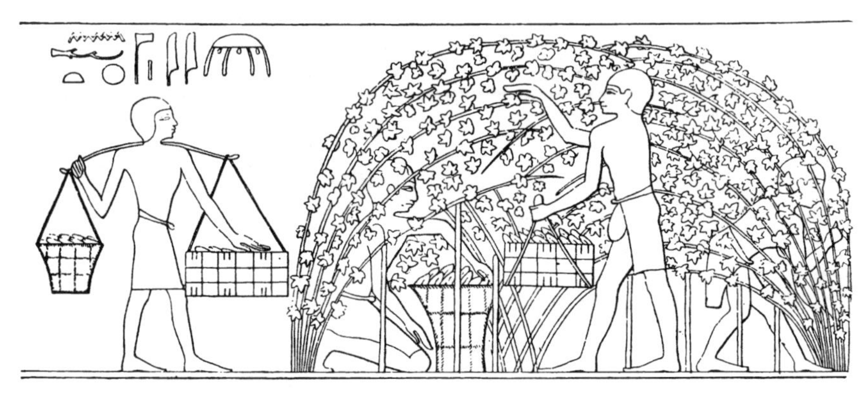

FIG. 7. A ROUND VINE ARBOUR, BENI-HASSAN

FIG. 8. GARDENERS IN THE VEGETABLE GARDEN, BENI-HASSAN

On another picture of the same period the beds are marked out in squares like a chess-board, with many kinds of plants in them; as gardeners are here watering the land, there has to be a canal or a pond to complete the scene (Fig. 9). Above stands a row of pot plants, an early example of ornament that became common later on. To the left of the garden is a vine arbour, where men are in the act of plucking ripe grapes. The same wine harvest is shown at the Beni-Hassan tomb, but here the arbour is most elegantly furnished, and covered over—which is rare in so old a work.

Much is told by pictures of the time before the New Period; but even at that time utility was not everything, and rich people, from the earliest days, made gardens round their houses in the country, where they meant to retire after their busy careers; we see this from an inscription on the tomb of Meten, an official and high priest who lived under the last king of the Third, and the first of the Fourth Dynasty. Meten built himself a wonderful villa with a wall round it, which enclosed a square 105 metres in length each way. His house was well built, and supplied with all the necessaries of life; it was enclosed by trees both for ornament and use—palms, figs, and acacias. Several ponds with green surroundings made a home for water-fowl. Before the house stood arbours, and two fields of vine plantations yielded him every year an abundant supply of their fine wines. There is no picture that shows the actual arrangement of these plantations, which afforded so much pleasure to their owner; we must pass over 1500 years, to find descriptions like

FIG. 9. A VINE ARBOUR AND VEGETABLE GARDEN AT EL-BERSHEH

Meten's in the New Period, which will, however, present precisely the same pictures of the order of plants and buildings, though in a highly advanced state of culture.

The plans given in Egyptian paintings of houses and grounds are not easy to understand without further information. The wish to put into one picture all the things they liked and thought good, and so to have them all at the same time, led this happy-go-lucky folk into absurdly poor perspective: they wanted to see the house inside and outside, front view and back view, everything that grew in the garden, at the back or at the side, all at once. So a drawing would be partly a ground-plan, partly an elevation, and again partly a bird's-eye view. But if one's eye has once learned to change over these drawings and interpret them in our own style, the instruction one gets from them is very many-sided.

It is then quite easy to understand a picture that comes from the tomb (at Thebes) of a high official who served under Amenhotep (Amenophis) III. (Fig. 10), although there are certain difficult points in its interpretation. It presents a complete plan of the villa, mostly as a bird's-eye view. As it was at Meten's country house, the square of the piece of land is circumscribed by a wall, which in this case has round tiles on the top. You enter from the front by a large entrance gate, or by one of two little side wicket-gates. A shady avenue follows the wall outside, and a canal outside that; this adds to the feeling of complete seclusion which the picture suggests. You step through the door straight into the house, which is shown much too large in comparison with its surroundings, the doors being the only break in the façade. No doubt the artist wished to suggest that the owner was a very rich person, by emphasising the beauty of his front gate. Here too was the porter's lodge, perhaps also a reception-room for such visitors as were not allowed in the main building, which was hidden away in the garden. Between the gate and the house, occupying the whole of the middle space, was the vineyard. It consisted of four arched arbours, their rafters supported by posts. A path is left open in the middle, forming the chief approach to the house from the gate; and from this path two side-walks lead directly to the covered ways.

It is hard to reconstruct the master's house, which is drawn too small. It perhaps had a front hall, like most Egyptian houses, with three rooms below and an upper story. There were flower-beds at the side, and shady avenues around. The whole upper garden is intersected in symmetrical lines by avenues of sycomores, and different palms —date and doum palms.

In other ways, too, a feeling for strict symmetry is shown, for the garden is carved out by walls into eight separate similar parts, only differing in their size. The chief section (with the house on it) includes the vineyard; and besides avenues it has two small and attractive open pavilions with flower-beds in front, overlooking two rectangular ponds, whose banks appear to be bordered with green grass; on the water there are lotuses swaying, and ducks swimming about. Two similar ponds, but pointing in a different direction, are nearer the front, on either side of the gate, and here again a space is filled by two plantations of trees, fenced in a peculiar way. There are two more avenues composed of all three kinds of trees, planted alongside the walls, standing alone and cut off by a low wall. Thus we have here (1) a square of land surrounded by lofty walls, (2) the dwelling-house, carefully hidden away, shaded by trees, enlivened by the pond and its water-fowl and green border, (3) the vineyard in the middle with all the trees of different kinds grouped about it in avenues. We find nearly all these designs in Meten's garden; and if we look at the beautiful regular plan, the fine alternation of trees, planted with prudent forethought, the elegant shape of the sunk ponds, the judiciously disposed buildings in the garden—we recognise with astonishment that we here have a formal garden in an advanced state of development on the very threshold of our history. Rhythm, symmetry, and a happy combination of elegance and utility—a blend often desired in later days of hope and struggle—these have been fully attained, and with them a delight in quiet communion with Nature, expressing as she does the sense of beauty in orderliness. Moreover there is a tendency to separate particular parts—a scheme often met with in later times. The next pleasing feature is the complete supremacy of the garden, to which all buildings—the dwelling-house included—are subordinate.

With the later excavations at El-Amarna the form of the town house becomes more clear. In suburban houses the pylon-like entrance gate hardly ever leads into the garden. Straight ahead you get the large pond, which at El-Amarna seems to have been always filled from underground water. Behind the pond, and often at the side as well, we find in the best houses a kiosk with pillared hall and a terrace leading up. The dwelling-house is generally at the side, nearer to the street.

By the help of a ground-plan like this, it is easy to complete and explain many other garden pictures. There is a Theban picture, from a tomb of the same date as the one just described, which (Fig. 11) gives a visiting scene at a great house; here too the house stands inside the garden enclosed by a wall. The entrance door leads straight into the garden; and thence the visitors proceed to the house, which has garden ground all round it (this the artist indicates by one large tree and two little plants behind the house); there is a second way in at the back. The house has an open porch in front, where the guests, who have walked through the garden, are politely received. The most conspicuous place seems to have been allotted to the vine arbour, where the ripe grapes hang on a pergola with pretty columns, and among the trees are figs, pomegranates, and sycomores, which in their own avenues provide fruit and grateful shade. The picture shows no pond; but water is a necessity—an Egyptian garden is unthinkable without it—and it is unusual not to have it shown in a picture.

There is another garden of an opulent kind where guests are being received (Fig. 12). From the canal which waters the villa one arm reaches out into the garden, and widens at the end into a tank. Around it are avenues of trees, which are also continued up to the house. Such T-shaped canals are often found in gardens. We can see from the Apoui garden in a tomb at Thebes how tall things grew with care and good watering (Fig. 13).

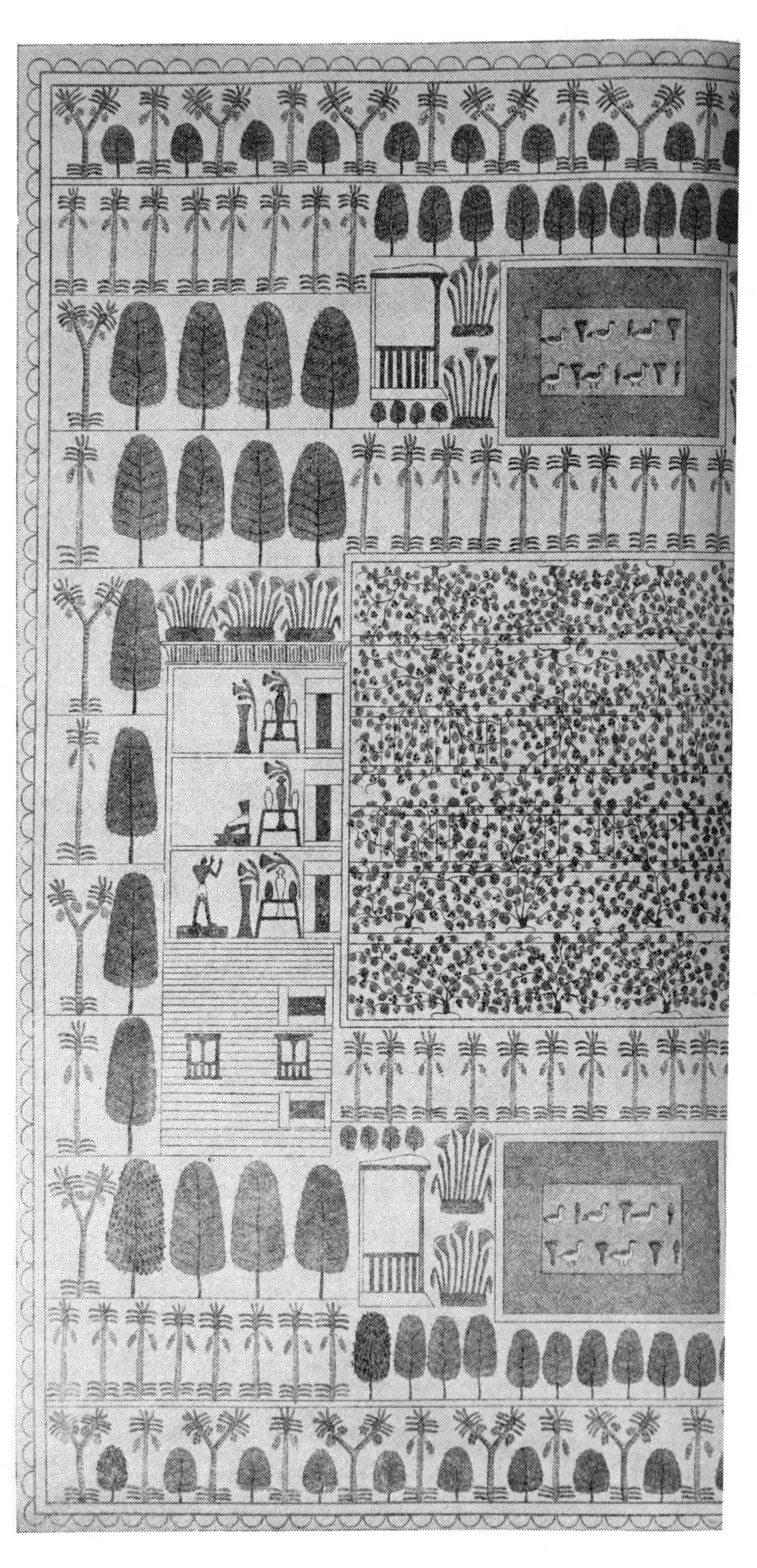

FIG. 10. THE GARDEN OF THE HIGH OFFICIAL

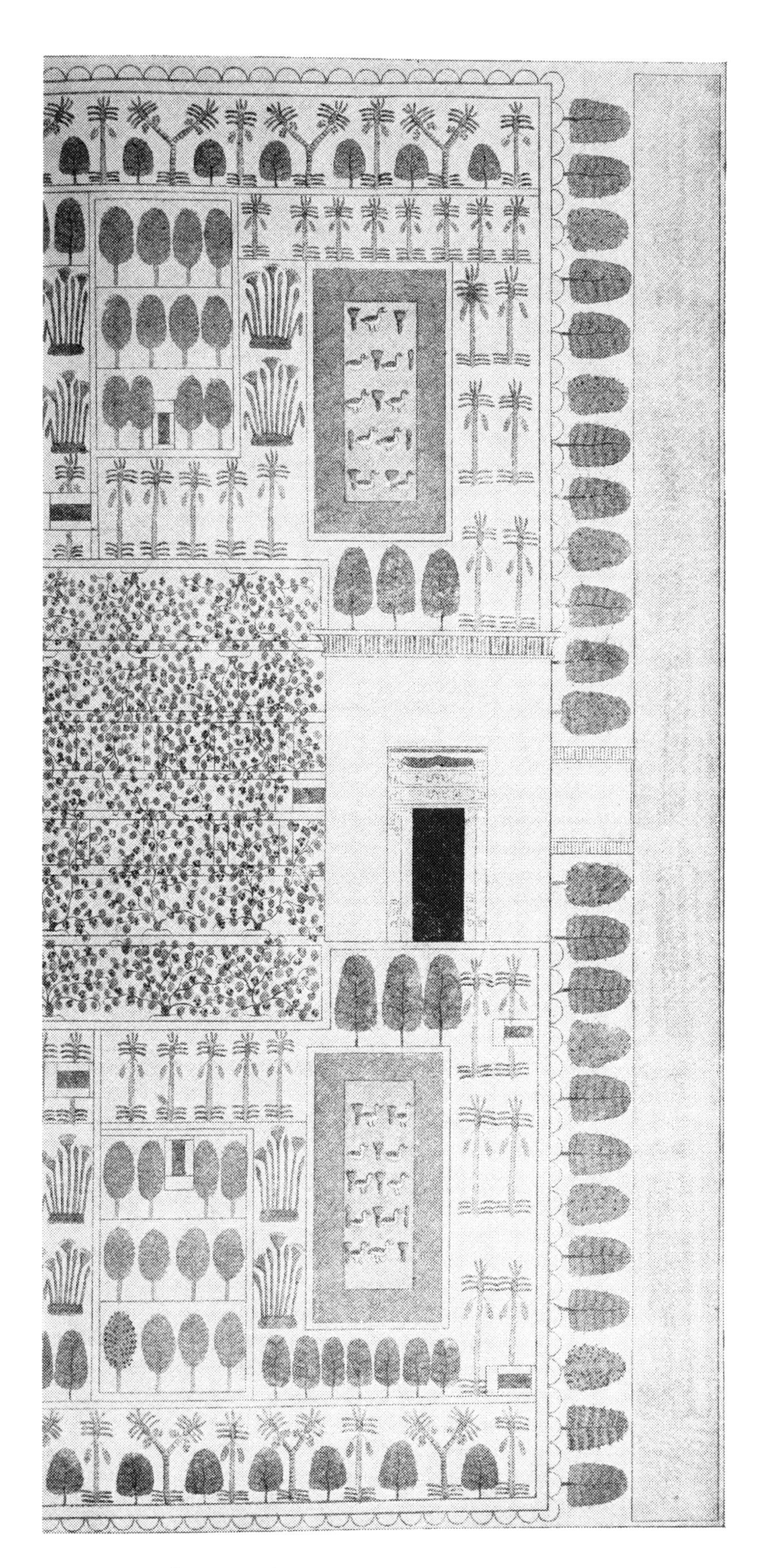

OF AMENHOTEP (AMENOPHIS) III., THEBES

FIG. 11. VISITING SCENE IN THE GARDEN OF A VILLA AT THEBES

FIG. 12. A VILLA WITH GARDEN, THEBES

A very attractive building stands in the middle of a large garden, to which one goes down from the gate by an open stair; on both sides is a canal, and all along it are beds of cornflower, poppy, papyrus, and the like. Great avenues of fruit-trees with splendid fruits are standing among the flowers. Gardeners are busy with shadoofs, fetching up water to pour over the plants.

Most of these places are small dwelling-houses built in suitable proportions, and may possibly have only been intended, like the villas of Italian grandees at a later date, as temporary homes at a favourable time of year: this is suggested by the fact that there is no sign of kitchen arrangements; certainly nothing of the sort appears on the ground-floor of our plan of the enclosed estate. Again, the Setna Romance only describes a suburban villa: Setna, we are told, walked to the west side of the town, where he came upon a very beautiful house. It was surrounded by a wall, and had a garden, and before the doors there was a terrace.

But the tombs show us larger places too. The high priest Mērirē from El-Amarna was one of the friends of King Amenhotep IV.—that great innovator who introduced into Egypt the cult of the sun's disk as the state religion; he changed his residence from Thebes to the plain of El-Amarna. The picture that adorns the tomb of Mērirē shows a number of buildings, all with their different purposes (Fig. 14). Some are living-rooms for the high priest, some for the inferior clergy, and others are store-rooms or treasure-houses for

FIG. 13. THE GARDEN OF APOUI, THEBES

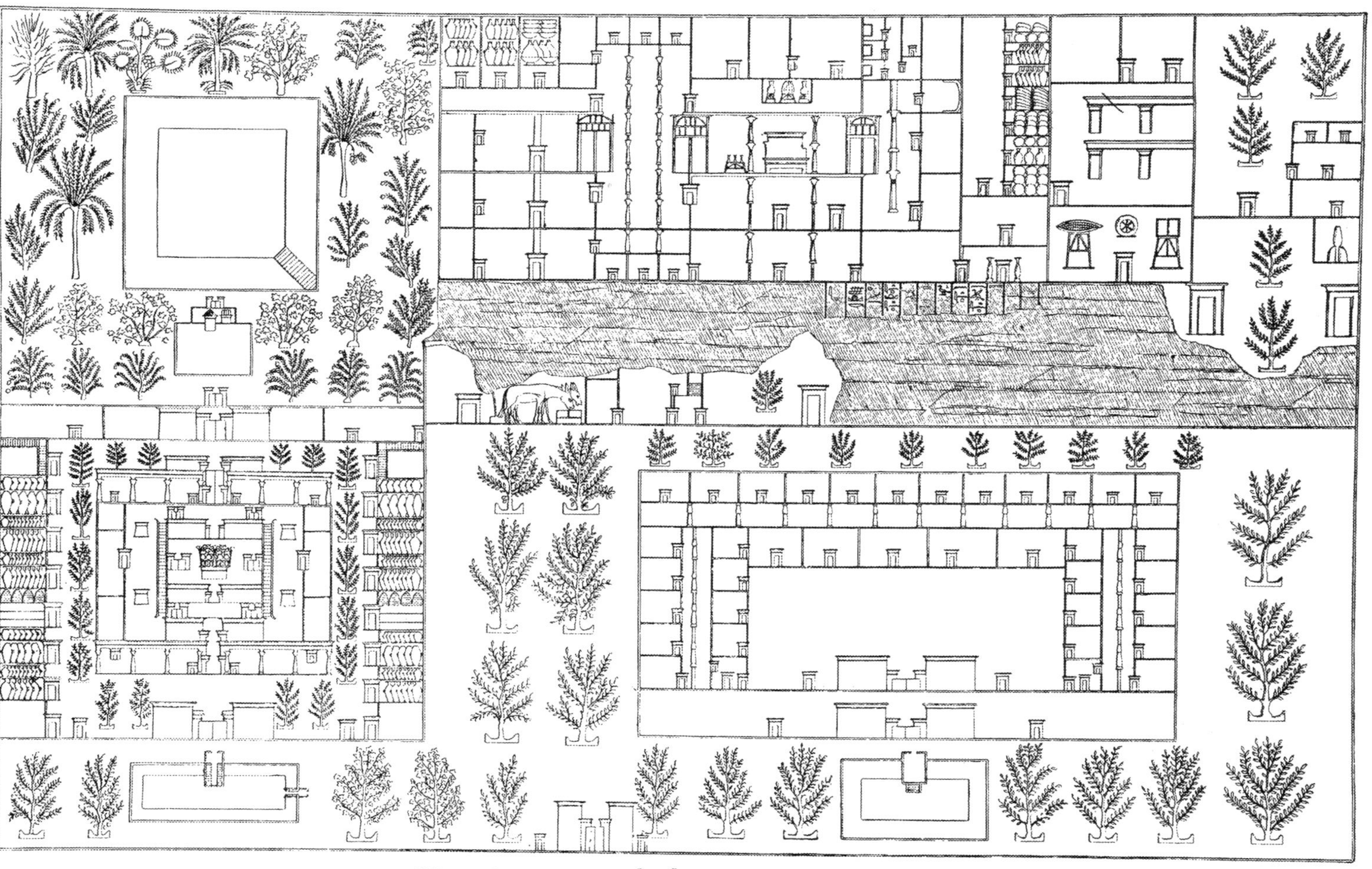

FIG. 14. THE PLACE OF MĒRIRĒ THE HIGH PRIEST, EL-AMARNA

the Temple properties: they are all inside garden-grounds. Even between the granary and the treasure-house there are plants, flowering in beautifully decorated pots. The entrance too is particularly charming with a porch like a baldachin, and in the front court there is a pretty pavilion.

The Egyptian still likes, as he did at an earlier period, to have special parts separated off by walls and doors; so the courts are further subdivided, and most of them planted with rows of trees. Single trees have a little wall of earth, and a ditch round them, to hold the water better. The chief garden-site includes the corners at the extreme end. Before the house, amid avenues of trees, is a sunk pond with steps leading down to it. There is a similar pond in front of the neighbouring house, which shows no ornamentation. The garden-house has a concentric form. There is a circle of trees round the basin, and there are pillared halls beyond. In the middle there appears to be a raised platform with a path sloping upward and an altar on the top. In a straight line behind this is the main garden, separated from the house by an intermediate court and double gates.

The central part of the garden includes a very large rectangular sunk pond, and in the middle—possibly with a view to times of drought—is a deep well, beside which stands a shadoof. Round the basin are trees of different kinds; the familiar sorts of palm, sycomore, and pomegranate are particularly neatly and prettily drawn. A second shadoof stands on the farther bank, and in front of it among the trees can be perceived yet another building which may be a summer-house, or perhaps a second altar.

Far too little do we know even now of this land—where everything culminates in the Pharaoh, and all eyes are fixed on him—as to the individual nature of those great palaces which the kings built to live in; for here we shall have to learn more from excavations than from tombs. If the houses of official persons had such fine gardens, one has a right to expect that the king's own gardens shall excel them in size and beauty, just as the ruler himself must excel his subjects. Even at El-Amarna, where the immediate relation of king and people is most clearly expressed in the pictures—his winning, gracious personality seems to have had great weight—the king's palace is often painted on the tombs of his greater subjects, but there is nothing positive about the garden, and we must wait for what excavations can, little by little, reveal. In the direction of the Nile there was no doubt a fine flower-garden in front of the palace, obviously on a terrace. The king and queen had each a kiosk for their own boats at the landing-place, and from thence different paths led by way of the strand and the front garden to a pillared court, which was the boundary on the palace side (Fig. 15).

The whole ornamentation of the palace proves that Amenhotep IV. was a great lover of flowers: his bedroom is painted all round with flowers, most of the little gardens in the court have a pond with lovely flowers growing round it. A painted floor, brought to light by excavations, shows most realistically the beauty of such a pond and its flower-beds, out of which spring the pillars of the chamber, like flower-stalks, supporting the roof of the hall. Probably Amenhotep IV. also planted large park-like gardens round his palace, to vie with those his father had laid out at Thebes. To get an idea of size, we may look at the great pond that Amenhotep III. dug out for his wife Teje. This was one and a half kilometres long, and more than three hundred metres wide. The king had it filled with water on the twelfth anniversary of his coronation. He and his queen then went on

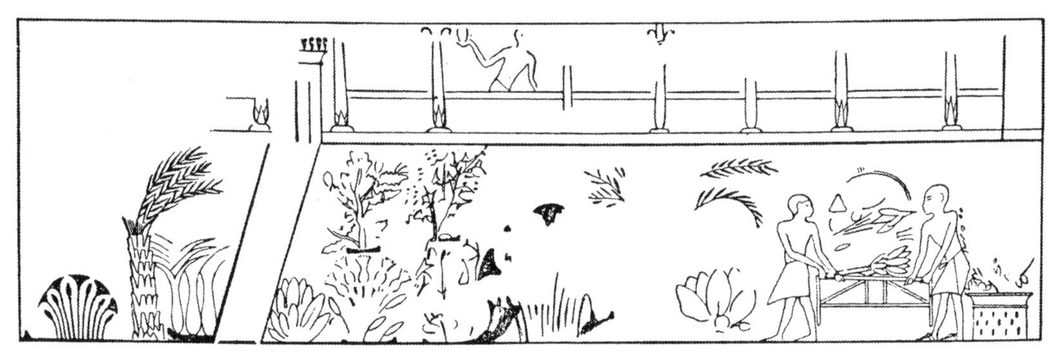

FIG. 15. FRONT GARDEN AT THE PALACE OF (AMENHOTEP) AMENOPHIS IV., EL-AMARNA

the water for the first time in the royal boat, and made this the leading feature of a grand festivity.

The evidences of shrines, built by the kings in honour of their gods, are more intelligible. Here too it is from the New Period that we first get pictures, though in the Middle Period the temples were certainly surrounded by gardens. It was at the time when Egypt was nearing with great strides the headship of the world, that in the long row of kings there steps forth from the darkness the first important figure of a woman, Queen Hatshepsut. She had insisted on her recognition as monarch, and during her sovereignty had greatly advanced her country in a material sense. She erected a splendid memorial of herself and her great deeds in the Temple of Deir-el-Bakhari, in honour of the god Amon. All buildings as well as gardens in Egypt had to be on high ground, because of the peculiar nature of the country. The floods, slowly rising, had from early days been encroaching upon the temples (which the kings had erected to the honour of the gods, and for their own souls), and also upon the private dwellings.

The edifices themselves, temples and palaces, were often built upon terraces, and as early as the Eleventh Dynasty a temple was set up in the beautiful valley, with rocks all round it, erected on the top of a structure that was really a pillared corridor. It was this temple that Queen Hatshepsut had in her mind when she built near it the temple for her own tomb; but here there stands out for the very first time in the history of art a most magnificent idea — that of building three terraces, one above the other, each of their bordering walls set against the mountain-side, and made beautiful with pillared corridors, the actual shrine in a cavity in the highest terrace which was blasted out of the rock (Fig. 17). What a wonderful view one would come upon, arriving from the river, passing through a long double row of sphinxes—which on either side seem to have had an avenue of tall acacias—till one stood before the majestic pylon-gate that gave entrance to the lowest of the three terraces! Excavations in front of this gate have exposed the square walled pits, which were filled with earth taken up from the Nile with a view to giving the best possible nourishment to the trees. They were watered by an arrangement of pipes from the side. In these holes there have also been discovered traces of Persea trunks.

With the same care the trees were planted in the gardens on the three terraces; and there too other remains of tree-trunks have been found, by excavation, in two round walled-in pits. A gently-inclined path ascends in the middle from terrace to terrace, cutting

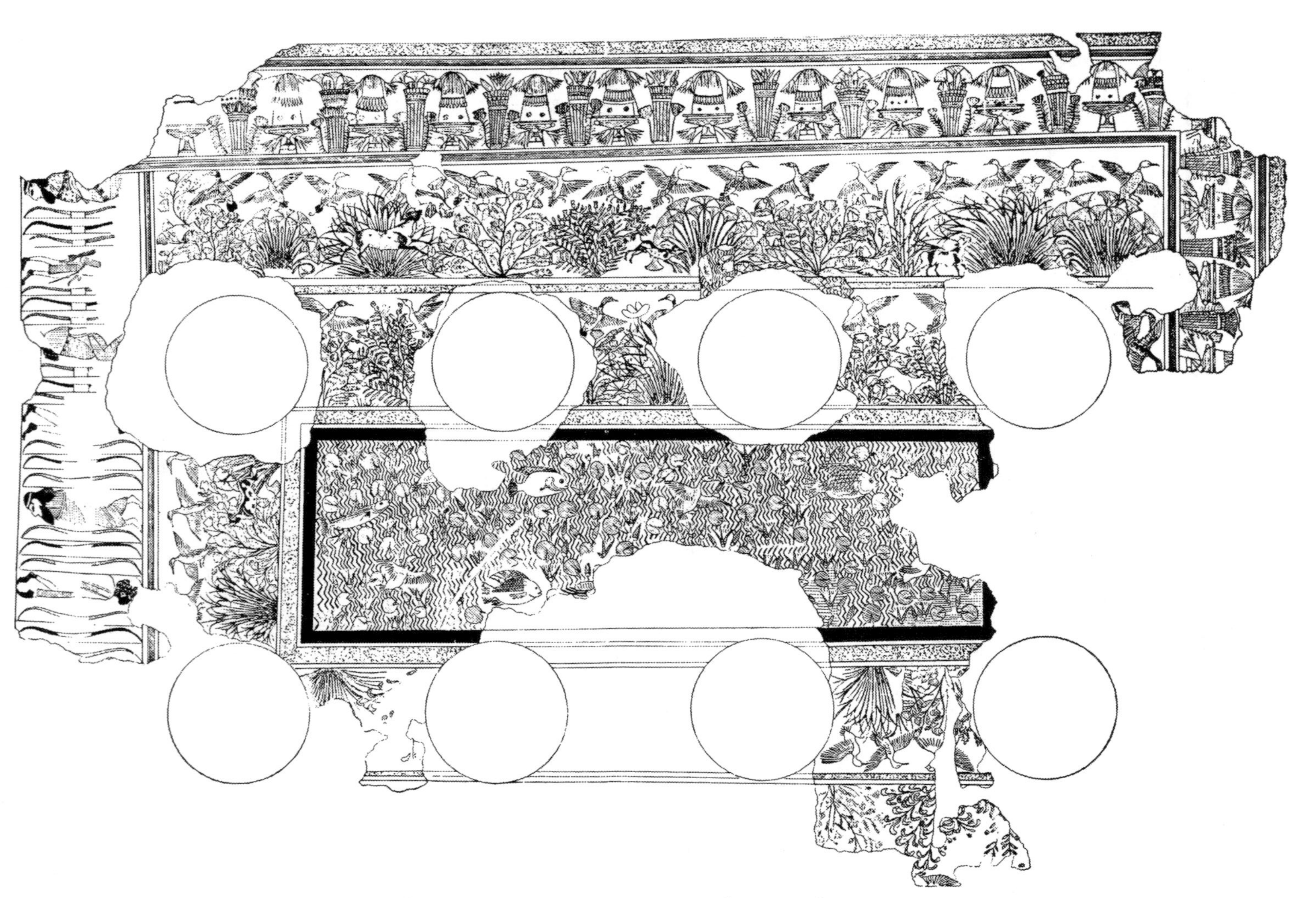

FIG. 16. A PAINTED FLOOR IN THE PALACE OF (AMENHOTEP) AMENOPHIS IV.

through the boundary walls. In the corridors that are at the back of the last and highest garden, the queen has given the story of her deeds, both in words and in pictures. First of all she tells how the god Amon bade her make at his house a garden so large that he could walk therein. And at his command she equipped an expedition to fetch incense-bearing trees from the land of Punt, the country of the Gods, in obedience to the word of Amon to make "a Punt in my house (temple)."

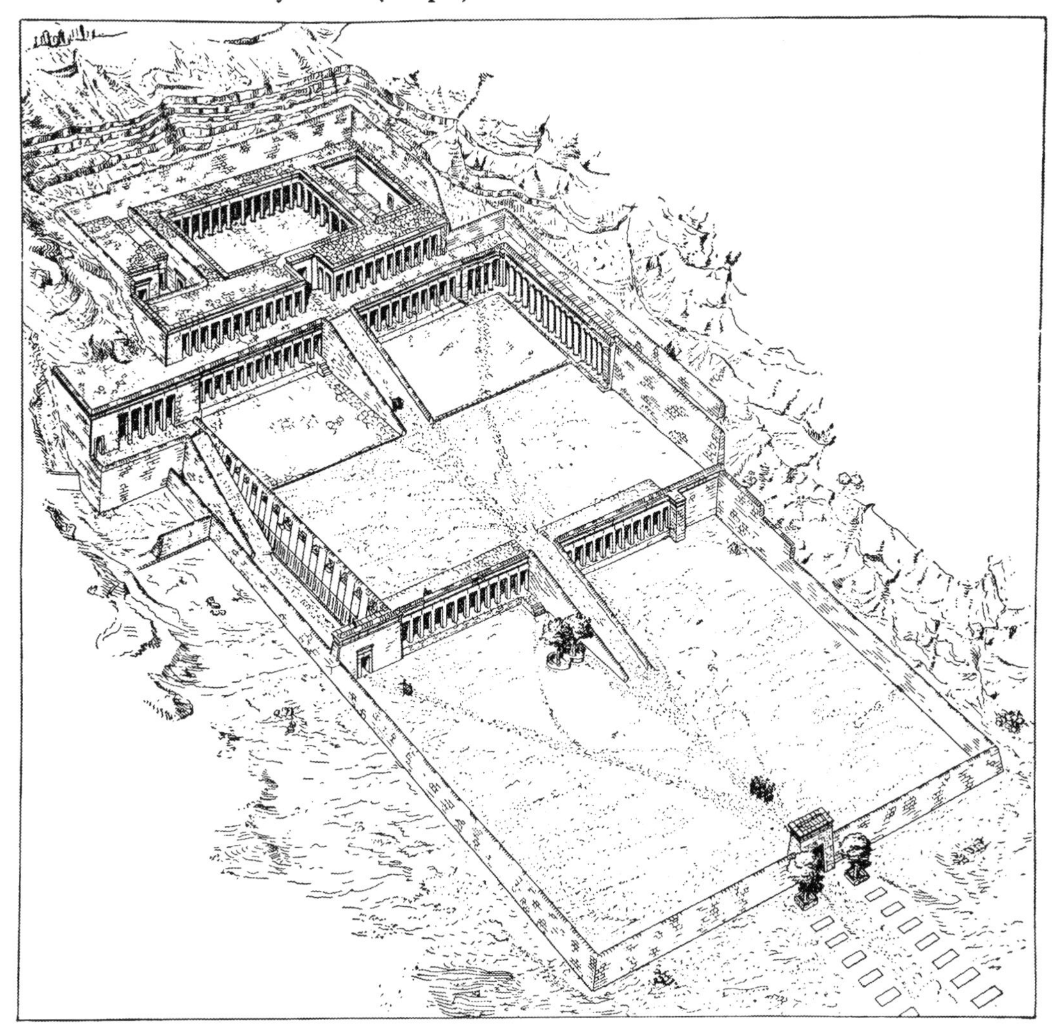

FIG. 17. THE TEMPLE OF DEIR-EL-BAKHARI—A RECONSTRUCTION

Nobody at that time knew the myrrh-terraces of the land of Punt, but the queen's ships arrived there, and saw the country and its wonderful inhabitants. The pictures on the wall, that illustrate this story, show us a land of many trees, to the south-east of Egypt. The queen's people collected all sorts of treasures for this place, and among the most valuable were the thirty-two incense-bearing trees for the god's garden. They were dug up, earth and all, apparently in spring, before they began to bud, and were planted in pots. They were carried to the ships by sailors with straps passed over a pole, four to six men according to the size of the tree (Fig. 18). Wherever we find ships laden, there are trees on them. After a joyful return home, the precious freight is unpacked, fresh and beautifully green, and at once planted by the queen in the pleasure-garden of Amon. This trouble was richly rewarded, for in a picture we see the trees finely grown; so well have they thriven that cattle can graze under them; and other votive offerings are piled up beneath their branches (Fig. 19). Beside the trees incense-gum is collected in immense heaps. [Editor's Note: The land of Punt would be Puoni on the coast of Somaliland, and the tree would probably be the Boswellia, commonly called the olibanum, a member of the Order Burseraceæ. The dried gum of this tree is frankincense.] For this huge structure of terrace-gardens a very elaborate water system would be needed throughout.

Afterwards the kings often organised peaceful expeditions of the same nature. If their goal was the land of cedars, it was the valuable wood that was to be brought back. The cedar was never acclimatised in Egypt, but Rameses III. relates how he (also in honour of the god Amon) imported and naturalised foreign plants. Whether or no he planted these in his garden at Medinet-Habû cannot be decided in these days, but there is no doubt that here too the great front court was covered with plots of flowers. "I dug a pond before it," says he, praising the site of the garden at this temple, "where the ocean of Heaven flows over it, and planted it with trees and green growths as they are in Lower Egypt. Gardens of vines, of trees, fruits and flowers, are around thy temple, before thy face." In another place he tells of pleasure-houses, in front of which he had dug out a tank for lotus flowers—and no doubt added pavilions, and everything that we know to have belonged to the garden of this period.

FIG. 18. TRANSPORT OF INCENSE TREES FROM THE LAND OF PUNT

Rameses was pre-eminently a friend to the garden, and a generous one. In a new construction on the delta, he made large gardens of the first importance, "wide places to walk in with all kinds of trees that bear sweet fruit, a sacred way glowing with the flowers of every country, with lotus and papyrus, numberless as the sands." He also gave very rich gifts to the other temples: we find in a list of benefactions among the gifts to Heliopolis: "I give to thee great gardens, with trees and vines in the temple of Atuma, I give to thee lands with olive-trees in the city of On. I have furnished them with gardeners, and many men to make ready oil of Egypt for kindling the lamps of thy noble temple. I give to thee trees and wood, date-palms, incense, and lotus, rushes, grasses, and flowers of every land, to set before thy fair face." In the fifth section, which contains a complete catalogue of the royal gifts to all the religious houses, we find that, together with 107,180 measures of arable land, there are 514 gardens and tree sites, and 19,130,032 nosegays.

Royal gifts like these encouraged the gardens at the temples; and the sacred groves, of which we hear in the inscriptions mentioned above, increased and spread. But in a certain sense gardens were always held sacred by Egyptians. Indeed, everything any man did

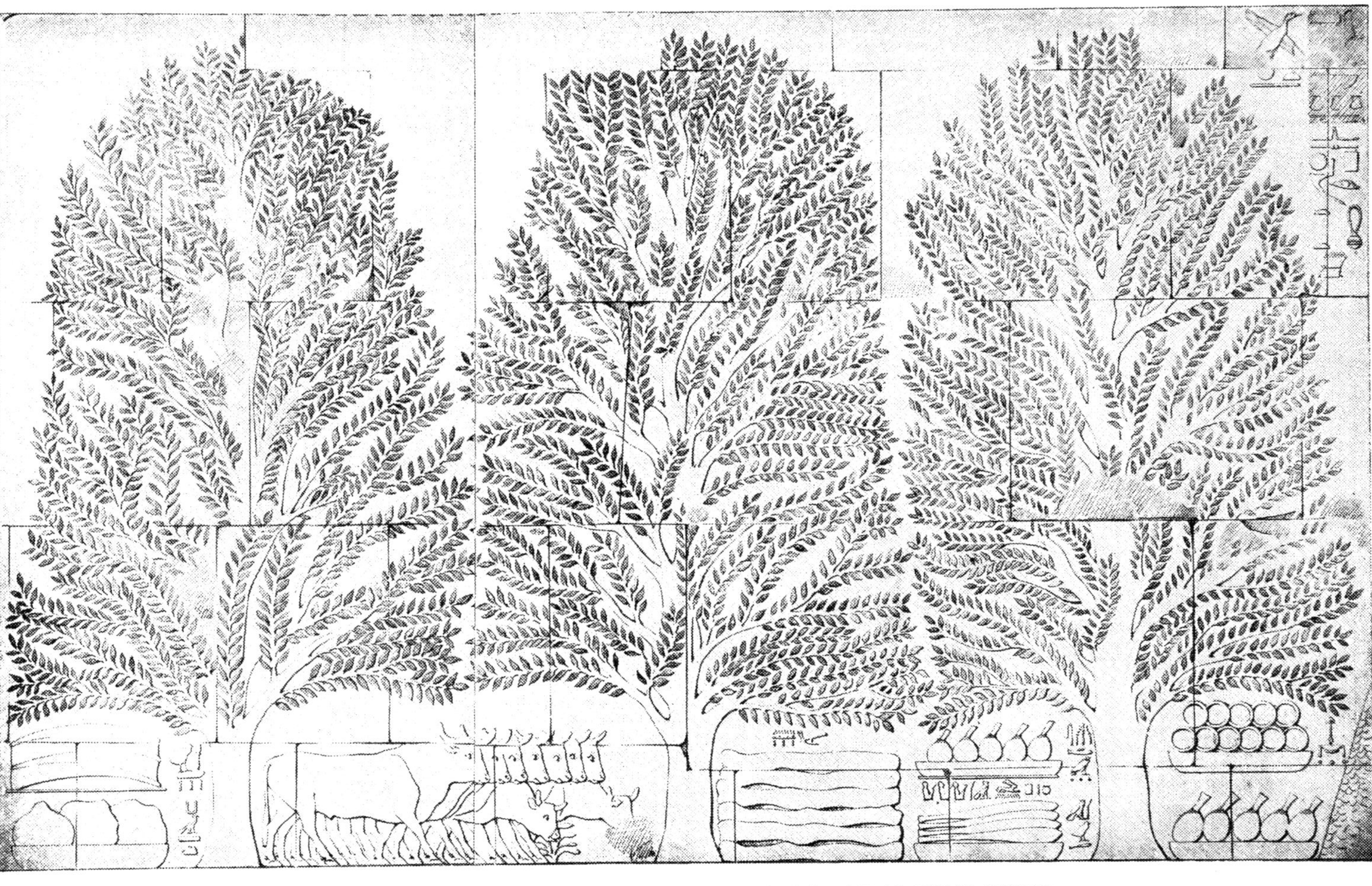

FIG. 19. INCENSE TREES FROM THE LAND OF PUNT IN THE TEMPLE OF DEIR-EL-BAKHARI

had a religious significance, for what he accomplished in this life was of immediate utility to his soul on the other side. Every private person sought, above all things, if it were in any way feasible, to put shady trees round his home; for what he planted here—to enjoy the shade and to bathe in the perfume of its flowers—was really an act of thoughtful kindness to his soul. In times of heat his soul will be able to step forth from the grave wherein it dwells, and enjoy the cooling shade. The inscriptions on the tombs lay stress on this idea: "That each day I may walk unceasingly on the banks of my water, that my soul may repose on the branches of the trees which I planted, that I may refresh myself under the shadow of my sycomore."

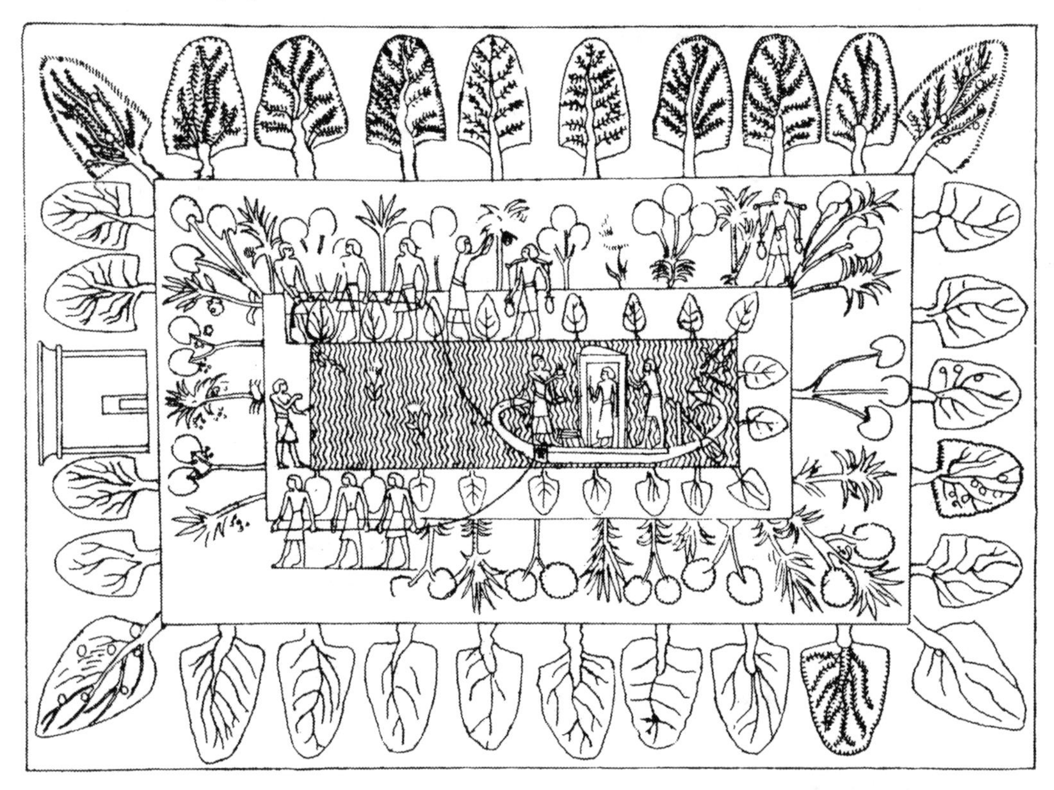

FIG. 20. FESTIVAL OF THE DEAD IN THE GARDEN OF REKHMARA

The Egyptians took this forecast quite positively and literally. With a view to it, and with a kind of symbolism as well, they painted their gardens (which they did not expect to have in the actual material sense) on the walls of the death-chamber; for we must bear in mind that the greater number of garden pictures known to us are taken from tombs. Less often an attempt is made in the paintings to place the graves themselves in the garden as though they were part of it. In the tomb of Sennefer, the overseer of Amon's cattle, herds and gardens, there is a beautiful decoration representing a vine, which covers the whole roof, and makes it like an arbour. If we transfer this idea from the grave-chamber to an ordinary living-room, we see in it an early attempt to bring, by the help of plant-paintings, a bit of our gardens into our rooms. We are constantly meeting with this endeavour, from the earliest times to the present day.

The importance of gardens in connection with the dead is shown by the very numerous drawings in which the mummy or the statue of a dead man is shown taking leave of his beloved possession, while he is once again rowed over the water, accompanied by his servants and met by them (Fig. 20). This probably is part of the special death-feast, which was held in the garden: the coffin was placed where the dead man had best loved to be when living, perhaps in the garden plot in front of the store-room, or on the island in the pond with the votive offerings, while the train of mourners is assembling on the bank. For the offerings special arbours were set up in the gardens. But it is always the actual garden that is represented, the very place that he himself had made, even though the dead man may appear in the picture as though he was master still.

In the tomb of the writer Ennene, who (as was mentioned before) gives a proud list of the trees he had planted, the garden is depicted as well as the house. There are several tiers, one above the other: first comes the house with a granary and a wall round it with two entrance doors; next, a pond among trees, then several other rows of trees, and lastly in the top tier an open kiosk, in which the dead man is sitting with his wife, and is saluted

FIG. 21. A LITTLE GARDEN AT A GRAVE—A STELE AT CAIRO

by a servant who is seen passing along. Thus Ennene has attained what he asks for in his inscription: "He treads once more his gardens in the west, he is cooled under his sycomore, he gazes on plots of fair tall trees, which he himself had planted when on earth." And he also looks upon his real home, where work is quietly proceeding, where gardeners who are still on earth are tending and watering his garden.

But besides these gardens that are pictured inside the tombs, to indicate that the soul on the other side can enjoy its possessions, there are also other little gardens of the dead laid out in front of the graves. In the story of Sinuhe the fugitive comes home from a foreign land, and thankfully rejoices that Pharaoh has made ready a grand tomb for him: "It was a Necropolis-garden, with fields lying before the town, as it might be for his best friend." This tale belongs to the Middle Period, and in the New we find on two sepulchral columns an account of gardens of the kind. At the entrance to the tomb there are two or three palms and a sycomore. On one of these columns there is also a table with votive offerings (Fig. 21).

These gardens were certainly not large, for in the desert land they could not raise great plantations owing to lack of water. Perhaps they were only intended as a sort of symbolical picture of a garden, to show that the soul, when it came forth from the grave, really finds a tree to sit beside. It seems certain that the description of Osiris is of this kind. In the form of the bird Bennu (the Greek Phœnix) Osiris sits on the sacred tamarisk

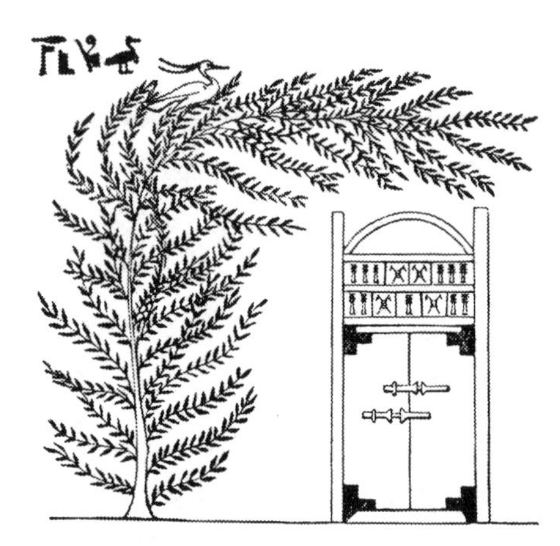

FIG. 22. TOMB OF OSIRIS WITH SACRED TAMARISK

tree in front of the tomb (Fig. 22). There is a scene, often shown, of the way hunger and thirst are stilled in the garden. The goddess is seen with half of her body stretching out from the sycomore, handing food and drink to the soul. It is often only an arm, which reaches out as though it were one of the branches. The Book of the Dead also gives one picture where a dead man and his wife are collecting water into their hands, and drinking (Fig. 23).

Perhaps a presentation of the more ordinary graves is given in a peculiarly formal kind of scene, that is again found with very slight differences in various pictures of the Eighteenth Dynasty. It shows palms, seven, five, or fewer, with a water-tank passing through the middle of all their trunks, as though to indicate that there is water on both sides of them. There are vegetable beds drawn like a chess board at the side, and generally two other trees, perhaps meant for sycomores (Fig. 24). These gardens are indirectly connected with the actual feasts for the dead (Fig. 25). One of them is at the tomb of the writer Inny, close to his own beautiful garden. Surely by this was intended some sort of ceremonial or symbolic picture of the garden of the dead, and maybe Sinuhe's Necropolis-garden should be regarded in a similar way, fields and all. It seems just as likely that it is a very ancient garden type, handed down deliberately in the books that record sepulchral monuments. In the Middle Period we certainly found gardens of the living not very different from these, at any rate in the case of vegetable gardens. But it is quite possible that our pictures really go back to a still older time, of which as yet no descriptions have been discovered.

The Egyptian shows his whole relation to Nature in his love for the garden—not an extravagant Nature, but one who deserves the care and pains of man, because of her

FIG. 23. FOOD AND DRINK IN THE TOMB OF THE DEAD

great beauty, her protecting shade, her wonderful flowers, and her costly fruits. To Nature the dweller in the Nile valley linked all that was dear to him: his happiest fêtes, poetry, and love—all were bound up with the garden and its products, especially flowers. Few Oriental nations can think of a festival without flowers, but nowhere are they so completely a part of human life, and so essential, as in Egypt. In architecture they give the forms for columns, for the growth of capitals, and all ornaments; in the house they are the chief decoration. The inside of a room often shows nothing but great bunches of flowers in prettily decorated vases, or votive tables decked with flowers. But if a feast is held in a house, or made ready for the gods, everything is clothed in flowers. Garlands are on head and neck, the guest is presented with a nosegay, or at least a single flower, as he enters. The servants are not only busy with preparing the dishes, but are always weaving new wreaths, and with each kind of refreshment new flowers are handed round. The guests carried them in their hands all the time, and also enjoyed their perfume.

FIG. 24. PRIMITIVE GARDEN OF THE DEAD—THE TOMB OF RENNI

Naturally the lotus comes first among the flowers—that native of the Nile valley which in old days, growing wild, had covered in its superabundant wealth every stream, canal, or lake, and had changed the whole country in the months of inundation into an immense field of flowers, whitish-blue and shining red. Every picture of the Old Period has lotus, but in the garden decorations of the Middle and New Periods, both pictures and inscriptions have a number of other flowers to show: after foreign plants had been brought in from all the countries that the Pharaoh's ships could reach, the lotus still held its own, and also the papyrus—that other wild marsh plant of Egypt: indeed, these two became a sort of symbol for flowers in general.

When at last the painter had learned to copy individual flowers with real and admirable truth, as in the palace floors at Tel-el-Amarna, he adhered to the time-honoured method of drawing lotus and papyrus conventionally (Fig. 26). In the same way the sycomore, the oldest tree represented in art, has maintained its former significant character among other realistically painted trees. Thus in the pavement of Amarna the papyrus shrub is severely conventionalised, while the other plants, anemone, poppy, thistle, and reed, are drawn so as to suggest natural form and movement.

FIG. 25. A PRIMITIVE GARDEN OF THE DEAD

The Middle Period shows most of

these plants in the tomb excavations. In addition there are frequently found chrysanthemums and cornflowers, which did not need to be raised in every garden.

It is not known how the beds were arranged in private gardens. Generally flowers are represented in separate clumps, but the vegetables in the Middle Period were planted in square beds. Unger describes a picture from the tomb of Rameses III. at Thebes, in which he saw a bed of crescent shape alongside a large canal. The flowers—always one kind to one bed—were apparently planted in rows. The need for flowers demanded skilled gardeners, especially in the later days. Rich people very likely had a great many men to attend to their gardens, as they had for all services, and the head gardener often has his name written in the inscription above the picture. In one papyrus the same word is used both for the gardener and the garden: the man's duty was to attend to the vegetables

FIG. 26. FLOWERS FROM THE FLOOR OF AMARNA

in the morning, and the vines in the evening. At a review of the royal garden of Seti II., twenty-one gardeners appeared, men, women, and children. One wonderful art, that the gardener was bound to learn, was the weaving of wreaths. This was an art that was much esteemed, for the Egyptian wreaths were complicated, and demanded real skill: they were used both for the living and the dead, and in great numbers.

The sincerity and depth of the love for gardens and flowers is shown once again in Egyptian poetry. The love poems, like those of the Jews, are full of playfulness, and (in praise of the beloved) full of comparisons with flowers and trees. The lovers meet in a garden, as in the song about the sycomore. There is a love song in the Turin papyrus which, in the fashion of an Italian ritornello, attaches each strophe to a flower, and indicates the meaning by a play on words. One of its strophes gives a particularly graceful garden picture:

I am thy first sister,
I am for thee like the garden,
Which I have planted with flowers
And all sweet-smelling herbs.
Fair is the canal therein,
Which thy hand dug,
When the north wind blows cool.
The fair place, where we walk,
When thy hand lies in mine
And my heart is full of joy,
Because we walk together.
It is an intoxication, to hear thy voice,
And I live because I hear it.
When ever I behold thee,
It is better than meat and drink.

CHAPTER II

WESTERN ASIA IN ANCIENT TIMES

CHAPTER II

WESTERN ASIA IN ANCIENT TIMES

HISTORY has nothing to show that is comparable with the gardens of Egypt, either for style or for antiquity. It might appear that Asiatic culture rivals the Egyptian in actual age; but the monuments of literature, and still more of pictorial art, begin at a much later date. The Babylonians arrived at their cultivation of Nature from a standpoint essentially different; the inhabitants of the Nile valley had been led to their kind of horticulture by the very abnormality of their country; but the Asiatics have the credit of being the real inventors of the park.

A park can only come into being in a well-wooded land, or at any rate where the ground is not so valuable that one cannot cover large tracts of it with plantations. In Egypt, of course, they had to consider too much the utility of their products, whereas in Western Asia the properties of mighty rulers would be set in the middle of forests. The Gilgamesh story, the oldest hero-epic of Babylonia, tells of such a woodland castle, and in this saga the wild creatures show unmistakable signs of a mixed ancestry, from demi-gods and from animals. Their homes have been the wood and the field; and, living so near to Nature, in their life and their joys they seem akin to the soil. In the story King Gilgamesh sallies forth with his friend Engidu to slay the Elamite tyrant Humbaba, the guardian of the cedar wood, whose business it is to terrify all human beings. His breath is like the roaring of the storm in the forest. They seek him in his lair on a high mountain with woods all around. When the heroes arrive, they

> . . . Stand and look upon the wood:
> They behold the high tops of the cedar,
> The entrance to the wood,
> Where Humbaba goes in on lofty tread.
> The ways are straight, and the path is wrought fair.
> They see the cedar mount, the dwellings of gods, the sanctuary of the Irnini.
> Before the hill stands the cedar, abundant and tall,
> Her good shadow is full of rejoicing,
> It covers the thorn-bush, covers the dark-hued sloe,
> And beneath the cedar the sweet-smelling plants.

Here we are not talking of a park at all in our modern sense. There is no enclosure with wall or fence, which differentiates park from wood; besides, the word which is used (*qistu*) means wood, not park or garden. That would be expressed by the word *kiru*, which implies a regular arrangement plain to see, a place laid out and planted by the hand of man. All the same, this cedar wood of Humbaba's, so realistically described, with its straight, cared-for paths (the keeper of the wood is mentioned), and its bushy undergrowths and sweet-smelling plants, is a forerunner, a kind of starting-point, for the park of history. Deep-rooted in the people is their veneration for lofty trees, for shadows "full of rejoicing." The same admiration that we see in the Gilgamesh Epic is remarked on in very late days by the Greeks as a leading characteristic of Asiatic nations.

Besides the park for trees, we may also assume in the most ancient times the existence of vineyards and other useful plantations. In the earliest account of a war between Egypt and Asia, about 2500 B.C., we are told that "the army returned in good order to Egypt after it had cut down their (the enemy's) fig-trees and vines." This was an act of revenge, which constantly occurs in the East. We find a picture in the palace at Kuyundjik, belonging to the latest days of Assyria, which shows a hostile army at the siege of a town busily hewing down palm-trees. The account of every war with Egypt takes us to the west coast, near the ancient Byblos.

Again a little later, about 2340 B.C., there is a mention of vineyards that the Sumerian King Gudea planted, and also of fish-ponds which he bordered with reeds, a matter of pride for the later kings. But the first person to boast of a park is Tiglath-Pileser I., about the year 1100. "Cedars and Urkarinu (box), Allakanu-wood have I carried off from the countries I conquered, trees that none of the kings, my forefathers, have possessed, these trees have I taken, and planted them in mine own country, in the parks of Assyria have I planted them." There were parks, of course, in the days of his forefathers, but he likes to be the first to acclimatise foreign plants in his own land, as the Egyptian queen liked to be the first in hers some five hundred years before.

FIG. 27. PRIMITIVE DRAWINGS OF TREES

The parks were the chief ornament of the country, and therefore the first to be exposed to the enemy's depredations. Not only were these useful and valuable woods and trees highly esteemed, but the hunting grounds were greatly valued. Tiglath-Pileser boasts that he brought to his capital Assur young wild oxen, and stags, and wild goats, even young elephants, and let them grow up like flocks of sheep in his park; he also imported dromedaries; and foreign kings sent him as presents "beasts from the great sea," for which he had large fish-ponds made.

In dealing with this period we feel the want of pictures. In the reliefs right on to the ninth century there are practically no backgrounds of country scenery. Every now and then we get a badly drawn tree (Fig. 27): among these trees the palm is often recognisable, also the vine. Together with the attempts at realistic drawings, sacred plants are also represented allegorically, but they are purely conventional, and therefore must not be taken to show any sentiment in these people towards the trees themselves (Fig. 28). It is only after the second half of the eighth century that the younger Assyrian monarchs begin not only to boast in inscriptions of their park sites, but to picture them in reliefs on the palace walls. And it is now that we find that the interest of the artist is not concentrated on man alone, for in many cases there are landscapes as a background. In the customary scenes—hunting, war, palaces, banquets—they now take great pains to show landscape, with carefully-drawn trees and plants, although this precision of detail is at the cost of proportion and perspective.

The first of these pictures comes from the palace built by the enterprising King Sargon. He founded above Nineveh a city called Dur-Sharrukin (Sargonburg) "by command of the gods and the desire of my heart." He set round it a great wall, and added a park, "like the Amanus mountains, wherein all flowers from the Hittite land and herbs from the hill are planted together." The ruins of the palace have been found among the villages of Khorsabad, and the park was probably close to it, adjoining the north wall. The Amanus, which plays so great a part in this and in all subsequent park-inscriptions, lies on the extreme north-east coast of the Mediterranean, and extends to the Cilician Taurus.

In all the royal inscriptions of later date, the words "like the Amanus" occur as though it is an "epitheton ornans" in the praise of a park. Far too often are words transferred in an almost meaningless way from one monument to another; and the only question is, what was the original similarity (or imitation) of the Amanus? The Amanus is the hill country of the Hittites, and there is no doubt that a certain influence was exercised by the Hittites upon Egyptian architecture. In the inscriptions it is said in a formal way about certain palaces, "I built a Chilani in the manner of a Hittite palace." If one assumes that by this Chilani is meant an open hall supported on pillars, such as is constantly found in or near a park, there is no evidence that it was from the Hittites that the style of a park-site was either borrowed or influenced. It is more likely, and more in accordance with the wording of the Sargon inscription, that the beautiful flora of this fertile mountain, full of trees and well-watered, was acclimatised in the Babylonian plain, just as Queen Hatshepsut brought a "Punt" to her temple garden.

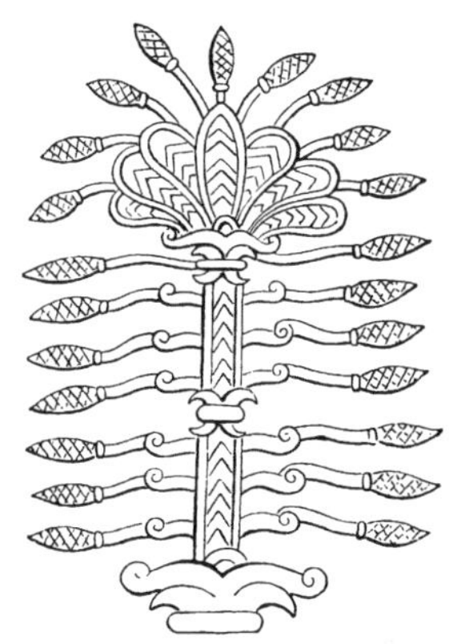

FIG. 28. A PRIMITIVE DRAWING OF A SACRED TREE

We must entirely reject the view that the Assyrian kings meant by the phrase, "like the Amanus mountains," to suggest any imitation of form or of landscape, thereby pointing to a picturesque and artificial style such as is meant by the "Chinese-English garden." These Western Asiatic gardens were laid out formally, as is shown by the very old Chaldean word *kiru* (see p. 29) and by the fact that Xenophon, who greatly admired the regularity of the rows of trees, bears witness at a much later time to the same thing. The monuments that have been preserved are all to the same effect.

The Assyrians had a great fancy for artificial hills and terraces; their palaces stand on enormous heaps of piled-up earth, and chapels and temples are erected on little hills in the parks (Fig. 29). These mostly have small open pillared corridors, near to running water, on rising ground planted formally with trees, such as pines and cypresses; and the hill is often crowned with a little altar with the hall near by. Whether these temples are built in the "Chilani style" is not yet known. It is only by looking far ahead that one can discern the hill made (many centuries later) for the sake of a view, called a spiral mount (German *Schneckenberg*), of which the Hill of Cypresses in the Villa Medici gives a classical example.

Sargon's son and successor, Sennacherib, gives us far more information about the nature of his park. He says he built "a palace that knew no equal," on a gigantic sub-structure of terraces. In his magnificent rooms there are numerous reliefs, which tell of the countless hosts of men by whose work the house was made, and in all the really important events the monarch appears in person. By the side of the palace was a park, "like the Amanus mountains," wherein grew the products of Chaldea and of the hills, both spices and trees. "I have given and distributed," he says, "to the men of Nineveh the lands outside the city, 3 Pt. in size, to be laid out in plantations." [EDITOR'S

FIG. 29. A TEMPLE AND ARTIFICIAL HILL IN A PARK AT KHORSABAD

NOTE: If the abbreviation 3 Pt. means three plethra, the area would be about 30,000 square feet.]

In another place he says:

> I made gardens in the upper and in the lower town, with the earth's produce from the mountains and the countries round about, all the spices from the land of the Hittites, myrrh (which grows better in my gardens than in its native land), vines from the hills, fruits from every country; spices and Sirdu-trees have I planted for my subjects. Moreover, I have cut down and levelled mountain and field from the land about the town of Kisiri unto the country near Nineveh, so that the plants may thrive there, and I have made a canal; one and a half hour's journey from the Chusur river have I brought water to flow in my canal, and between my plantations for their good watering. I have set a pond in the garden to keep water there, and in it I have planted reeds. . . . By the grace of the gods the gardens prospered, vines and fruit, Sirdu-wood and spices. They grew tall and flourished greatly, trees, and reeds also . . . palms, cypresses, and the fruits of trees . . . the reeds in the pond I cut down, and used them for divers purposes in my lordly palace.

Sennacherib is evidently particularly proud of his water system, and an abundant distribution of water was a matter of life and death for the whole country. Like the popes of the Renaissance in Rome later on, the king brought the water first into his own garden, and only afterwards into the city. It was not in Sennacherib's nature to shrink from difficult undertakings, so he built a festival house for the god Assur on the naked rock. Traces have been discovered by excavations of a large garden which he ordered to be made round the house of the god (Fig. 30). In front of the building there is a walled-in canal, and running parallel to it there are smaller water courses between the plants and trees, of which the old pits are still visible, quite round and dug about one and a half metres deep into the rock; the traces of them extend to the amazing area of 16,000 square metres; and it is evident that this mighty ruler used whole nations, that he had taken prisoner, to carry out his works.

Here also were huge reservoirs, three of which have been found, and we are able to think of them (judging by the various inscriptions) as all alive with fish and encircled with reeds and rushes, like the ponds that King Gudea made in his own park. Many reliefs show us what they looked like, and the reeds and sedges were so high that a sow

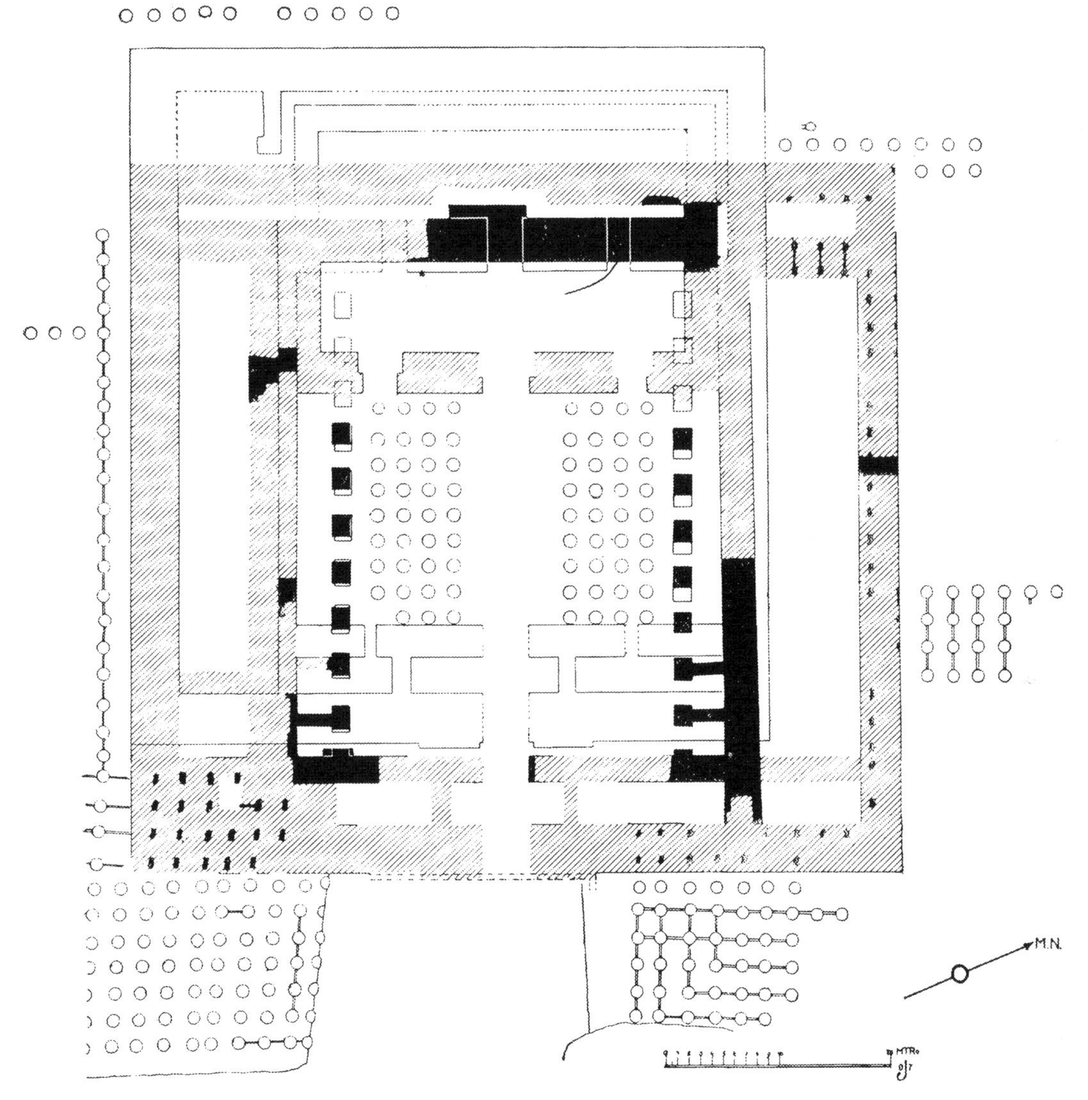

FIG. 30. THE BANQUETING HALL OF SENNACHERIB AT ASSUR

with her litter could hide in them, or stags and roes—even, if we may trust the pictures, a horse and his rider. In this large park stood in deep shade the festival house of the god Assur, and there he received and fêted all the other gods each New Year's Day.

Some light is thrown on the system of irrigation by a fragment of a tablet which is unfortunately nearly all destroyed. There is one wide canal, with little ducts branching out of it, and there are straight rows of various kinds of trees. It almost looks as though the different kinds were set in separate gardens and divided by low walls, as in Egypt, but perhaps these are only meant for paths (Fig. 31). In this picture we see many fishes and crabs, a kind of gondola, and on one side a man being let down by a rope, apparently to fetch water. Generally the edges of the pond were planted with rows of trees on the inner, or garden, side; but here there are straight rows outside, along the course of the water on the level ground, whereas there are irregular groups on the mountains, which are seldom drawn in a formal way. In backgrounds with many trees it is usual to represent the deeds of warriors, or even an actual battle.

FIG. 31. A GARDEN SURROUNDED BY WATER

Another fragment of a tablet was found that has to do with a garden: on it is depicted a row of pillars, supporting a very solid roof, on which there are several trees, planted regularly, probably indicating a terrace with a thick stratum of earth under it: in fact, part of a hanging garden (Fig. 32). Literary sources teach us that the Babylonians and Assyrians were the inventors of the hanging garden; but, however that may be, this tiny piece of a garden gives no suggestion of the astounding size of the so-called Gardens of Semiramis, of which Greek authors have so much to tell.

The erection of large buildings has always been connected with the name of Semiramis, whose actual life on earth (for those who do not share the belief that she was entirely mythical) belongs to the ninth century, but even in ancient days her invention of the hanging garden was argued and denied. Diodorus Siculus, whose account of Babylon is founded on Ctesias and Clitarchus, says emphatically that they were not due to her, but to an Assyrian king, who devised them for one of the palace ladies who came from Persia as a reminder of her old home. Berosus is more precise, and ascribes them to Nebuchadnezzar, and says he made them for the love of his Median wife. Here the initial questions are so confusing that we cannot make out anything certain. One fact seems to be pretty sure, and that is that the two historians, Strabo and Diodorus, are drawing their information from different sources, and this seems to make good the assumption that the custom of having hanging gardens was general and widespread. In one point alone the two writers agree—namely, that the length of the side of the substructure was four plethra (480 metres), and that it was rectangular. Diodorus says that terraces like steps ascend from this floor, and get smaller as they get higher, "like a theatre," but that the way round the circumference is in the open air. Under the terraces the arches that support the whole are set up as fine royal chambers, getting their light from a skylight on the terrace above. There is a way through on each terrace by gaps ten feet broad which pierce the enormous supporting walls.

The top terrace of all, which according to Diodorus was above a hollow arch fifty yards in height, was the chief garden. Its flat floor was made of a very elaborate stratum, to protect from damp below, and to regulate it above; first there was a stone balcony, then a layer of reeds and asphalt mixed with brick and gypsum; next over this came a leaden roof, which was intended to support earth, of which there had to be a thick stratum to sustain the large trees planted in it. The whole construction, Diodorus says, is like a mountain. We may assume—and this Berosus supports—that the walks round the terraces were each planted; so no doubt the whole thing looked like a green mountain.

The architectural idea was not really foreign to Babylonian art; the towers in tiers, which were also square, and often were as many as seven in all, must have come about in imitation of similar designs in architecture. No ancient picture has been discovered which confirms the descriptions of Diodorus; it is found, however, in this part of the world, where tradition has mastered every change of fact and fashion for thousands of years, that in Persia there exist similar garden arrangements even to this day.

In the fortunate lands of Shiraz there is a terrace-garden on one of the steep acclivities, the Bagh-i-Takht (Garden of the Throne) (Fig. 33), which, though it does not stand unattached, does give an idea of what Diodorus calls the Mountain, because its terraces are formed in steps, and get smaller at the top.

More accessible to our sight—and perhaps partly affected by the description of Diodorus—is the artificial terraced hill of Isola Bella in Lake Maggiore. About the ascent Diodorus says nothing, but there were probably steps inside as at a theatre, and just as we find them in the garden at Shiraz.

FIG. 32. A HANGING GARDEN

Strabo's description differs from that of Diodorus in essential features. He says that the floor of the upper garden is supported on a series of arches that stand upon hollow cubes. These cubes are filled in with soil, so that they may hold the roots of the largest trees; accordingly it is possible to have a thinner layer of earth for the roof, as it will only have to nourish the roots of the smaller plants. This system is really the same in principle as in the garden of Sennacherib, where the pits for the plants are dug deep into the rock, to relieve the stratum of earth from the worst of the weight.

Strabo's account scarcely admits steps in the building, whereas Diodorus lays great stress on this feature. Strabo, however, may be describing briefly a relief from Kuyundjik, which dates from Sennacherib's time. In this there stands a wall on the top of pointed arches, and on the wall are lofty trees planted at regular distances, close beside one of the little temples such as we know, and an altar set on an artificial mound that is planted with trees (Fig. 34). However that may be, the arches seem also to have supported a water conduit, for a stream runs out and crosses the mound, branching off in several directions; and one may suppose that it passes over the terrace, and is distributed about the rest of the garden.

In any case, hanging gardens require special contrivances for irrigation, but in this matter also Strabo and Diodorus are at variance. Diodorus says that in the uppermost cavity there is a water arrangement, which is fed from the Euphrates, and cannot be seen from outside; whereas Strabo affirms that the water is brought up by spiral pumps at the side of the winding stairs, and that special men are always there to draw the water up from the river. The writers both say the Euphrates is close beside the garden, and

FIG. 33. BAGH-I-TAKHT—THE GARDEN OF THE THRONE AT SHIRAZ

Diodorus goes on to assert that the Acropolis is close by: by this it is thought that he meant the palace on the right side of the Euphrates, in whose park Alexander the Great died: no trace remains of any building that suggests a hanging garden.

In all these elaborate accounts of hanging gardens we are not told one word about the way they were planted, nor do the park inscriptions give any clue to their plan or arrangement. On the whole we may take it that these people felt a greater interest, and at an earlier date, in the care and cultivation of trees than of flowers. Just as in the dim past the heroes of the Gilgamesh Epic had stood in amazement before the magnificent growth of the cedars, so did Xerxes, as later authorities tell, stand before a wonderful plane-tree which he came upon on his way to Sardis in Lydia. The beauty of this tree so affected him that he gave it presents, as a lover might to his beloved, winding gold chains and armlets round the branches, and setting a guard to stay behind and watch it. In all the inscriptions there is talk of trees, of different woods—cedar, plane, box—and of all sorts of fruit-trees (unfortunately not often identified), but we scarcely ever hear of flowers.

When Assurbanipal completed the house he made for the women, and adorned it with all manner of colours, he added "a large grove with many kinds of trees." There exists a little picture on a cylinder of jasper (Fig. 35), which shows a women's garden, wherein women are plucking ripe fruits, and beside the tall trees are seen low-growing plants, one of them looking like a dwarf palm, and the middle one perhaps meant for a flower. There is no doubt that such civilised people would have grown flowers at an early date, although flowers cannot have played the same part with them as they did with the Egyptians, for it is clear that they did not use flowers either in honour of the dead, or in wreaths for breast or head. On the other hand, in their earliest pictures one often finds winged genii holding either branches or actual flowers in one hand; and the old ornamentation of coins with plant designs indicates a certain cultivation of flowers. The late reliefs from Kuyundjik show long processions of people carrying all things needed at a feast or a sacrifice; and many of them hold in their hands jugs or vases, in which bunches of flowers have been put, or whole branches of blossom.

FIG. 34. A TEMPLE AND HANGING GARDEN AT KUYUNDJIK

Where flowers were to be cultivated, hedged gardens were necessary, and they would be near the house, or they might be actually inside it, as in the pleasure-palace of Sennacherib. Also a special site and special protection would be necessary for the women's garden. In the same way the great hunting-parks with their supply of wild animals must have required a separate division and strong protection. The inscriptions say nothing of this, and the reliefs seem to make a point of ignoring all boundaries. This appears in an unusually pleasing picture of the time of Assurbanipal (seventh century), when art at its best showed a decidedly realistic tendency in the delineation of individual objects. By that

FIG. 35. WOMEN'S GARDEN, FROM A PICTURE ON A CYLINDER OF JASPER

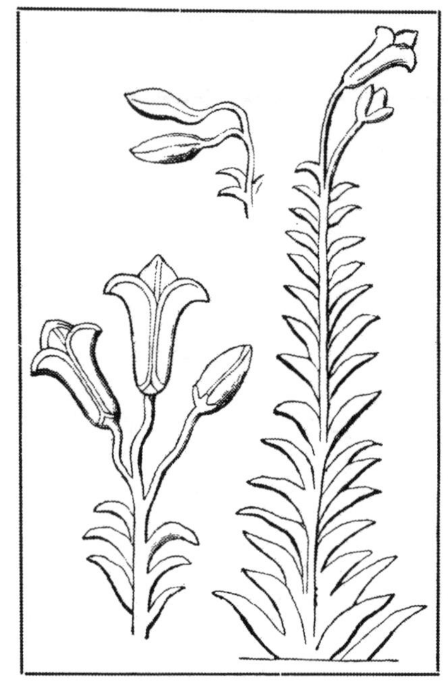

FIG. 36. LILIES FROM THE NORTHERN PALACE AT KUYUNDJIK

time the artist had learned how to represent flowers in a thoroughly natural way, and in the relief from the northern palace at Kuyundjik the lily is depicted (Fig. 36), sometimes in bud, sometimes in full bloom, growing on its slender stalk; and these attractive flowers are in a grove of pines and palms, placed between the trunks of the trees, which are planted in regular alternation.

Another tablet in the same relief has flowers, but the particular kind is not easily determined. The trees show that the same park is meant, but beside the flowers and under the trees there are two lions (Fig. 37). Huntsmen show us that this is the royal park, for on the other side they are dashing through the wood with a couple of dogs. Overhead the vine, delicately and accurately drawn, is twining from trunk to trunk.

If we may trust the memorial records we possess, the vineyard underwent a change in the time of this prince. The earlier monuments always show the vine creeping on the ground, and these date from Sennacherib's time. In addition there are only figures of individual plants without any leaning towards naturalistic treatment, except that here and there we find several vines grouped close together. But in Assurbanipal's time the vine is seen climbing from tree to tree, which implies some sort of training, such as one finds in Italy to-day, and very often in the Orient. At the king's palace the park shows no essentially different character. On another carving of the same series the king is shown having a meal in the open with his wife (Fig. 38). He reposes on a lordly couch; and on a seat,

FIG. 37. AN ASSYRIAN HUNTING-PARK WITH VINE-BRANCHES AND LILIES, KUYUNDJIK

as grandly decorated as his own, his wife sits at his feet; both are holding goblets in their right hands, and lifting them to their lips, while in their left hands they hold flowers —which reminds one of an Egyptian banquet. Beside the couches stands a fine table with food on it, and behind the master are slaves fanning him. Above their heads there is the arch of a vine arbour, but without visible support from stakes or columns; at the side it extends farther among the trees, which alternate regularly between a date palm in fruit and a fir-tree with lower bushes between. A procession of servants, followed by musicians, come carrying dishes, and in the tree-tops birds are hopping about. But even this peace-lover among Assyrian princes cannot in a scene of tranquillity forget his conquered foe. On one of the branches hangs the head of his last antagonist. This sight, promising rest and security, doubtless adds a peculiar spice to the meal!

Thus from special scenes and fragments we have to piece together the picture of an Assyrian-Babylonian park. These nations do not feel the Egyptian's love for telling a

FIG. 38. ASSURBANIPAL AT A MEAL IN THE PARK OF KUYUNDJIK

story realistically. The chief impression is produced by their rows of splendid trees, with (in the later times) vines growing between them, and low bushes and flowers filling in the gaps. The water of canals or ponds, and little pleasure-houses set up on a hill or a terrace, and scattered about the park, enliven the scene. This park is in its first intention the spectacular scene for a great man's hunting-ground, but is secondarily the place where feasts are given, and where assemblies, and audiences granted to great princes, take place.

The inheritors of this civilisation, the Medes and Persians, continued to develop the garden, like everything else, on exactly the same lines, and they brought with them their love and their veneration for trees. In their religion the cult of trees plays an important part, and with them, as with the Assyrians, the symbol of eternal life was a tree with a stream at its roots. Another object of veneration was the sacred miracle tree, which within itself contained the seeds of all. Among the Persians tree-planting was a sacred occupation, and, as Strabo says, was part of their education: boys received instruction in this art in the evenings. And so it came about that this reverence was seated deep in the souls of even the lowest stratum of the people, the common soldiers.

A story is told by Plutarch of Artaxerxes, on his campaign against the Cadusii, once halting in a barren treeless plain close to a royal estate, where there were gardens, very extensive and well kept. As it was mid-winter, and terribly cold, the king gave permission for the soldiers to cut down wood from the park, and not even to spare the best trees, cedars or cypresses. But the soldiers could not make up their minds to fell trees the size and beauty of which they admired so much. Thereupon the king himself seized an axe, and began to cut down the tree that he thought the largest and loveliest, and then the soldiers were no longer afraid to fell the trunks they needed, and to kindle fires therewith so that they might endure the cold.

What true reverence and awe for trees on the part of a whole army speaks to us in this simple Greek story! It indicates, too, the size and extent of a park which could thus receive great bodies of troops and accommodate them for a night. Moreover, we here have no exception; on the contrary, we continually hear of the mustering of armies in parks, where in all probability the necessary protection and shade are both supplied. In the same way the younger Cyrus owned a great park at Celænæ, which extended above the town on both banks of the Mæander, and there he used to hunt on horseback for the sake of exercise. The place was so full of wild creatures, and so large, that he was able to hold his review there of the 130,000 Greeks. Great public festivities were also held by the kings in the same parks. We read in the Book of Esther how King Ahasuerus kept a great feast in the garden of his palace for the nobles and for all the people, and that it lasted 180 days.

In countries such as these, where the same manners and customs may last for thousands of years, the old parks even now maintain their size and importance. So recently as 1908 the Shah of Persia mustered troops in his park near Teheran against the revolutionaries, just as Artaxerxes and Cyrus had done before him. In a castle that was built for pleasure in the middle of a great park, the insurrectionary parliament passed the day and held their tumultuous meetings.

The joy and pride that the Persians felt for their parks, they took for granted in other nations, and when the satrap Tissaphernes wanted to pay special honour to Alcibiades, he bestowed his friend's name on a park of royal splendour, which he had set out with fountains and beautiful lawns. And all the races oppressed by a Persian tyrant knew very well that, when a rebellion was at hand, they could not do a greater hurt or better express their desire for revenge than by laying waste his park. The Phœnicians began their hostile attack (which proved fatal to themselves) on the Persian oppressor by laying waste the park that the Persian king had made for a retreat at Sidon, and cutting down its trees.

The Greeks found fine parks all over Media and Persia, and their size and beauty gave them such a romantic appearance that there was a fancy for saying they had been founded by Semiramis. One of these was the great park in the Valley of Baghistan, under the rocky wall adorned with inscriptions about Darius; and again another park in Chaucon round a cliff, on which stood a pleasure-castle so high up that it commanded a view of the whole country round. To the mind of the owners these parks were so distinctly the leading feature that the house proper, or palace, faded before them.

Xenophon in his progress through Asia saw many of these, and admired them greatly. "Everywhere," he makes Socrates say to his pupils, "the Persian king is zealously cared for, so that he may find gardens wherever he goes; their name is Paradise, and they are full of all things fair and good that the earth can bring forth. It is here that he spends the greatest part of his time, except when the season forbids." Xenophon is the first writer to use the appellation "paradise" in a Greek narrative in the sense of a Persian garden, and in the Persian inscriptions the word does not appear, but only in the *Avesta* in the form Pairadæza. In Hellenistic times the Greek word also appears in the Bible.

To Xenophon also we owe the best account of the absolutely regular arrangement of the parks in the Orient. In a further conversation Socrates says that, when Lysander brought gifts from the allies, Cyrus himself showed him the paradise of his palace at Sardis. Lysander marvelled at the beauty of the trees, at the evenness of their plantation, at the regular rows, at the neat way the corners were made, at the prettiness of it all, and moreover at the many sweet odours which followed their footsteps. "All these things," he said, "do I admire; I admire the beauty of the whole, but far more, O Cyrus, do I praise the mind that has designed and ordered it." Cyrus was much flattered by his praise, and informed Lysander that he himself had been the only artist, and had even planted some of the trees with his own hand. This story becomes all the more vivid, if we bear in mind that as part of their education the young boys were instructed in horticulture, being told off in line, and headed by some royal prince or satrap.

FIG. 39. DARIUS HUNTING IN A GROVE OF PALMS

We find from all accounts and from monuments also that these paradises were first and foremost hunting-parks, with fruit-trees grown for food, just as in the Babylonian-Assyrian sites. There is one rock-crystal, very finely cut, which represents King Darius in a grove of palms where he is hunting, the trees being made all exactly alike, as in the Assyrian carvings (Fig. 39). The Persians were also familiar with the chase in the open country. A grand hunting-ground was given to the young Cyrus by his grandfather, in the hope that it would keep him at home, but he despised it, and, fired with longing, summoned his companions and went off, for in this park there were so many animals that he felt as though he was only shooting captive creatures.

Just as in the case of the Assyrians, monuments are wanting by which to judge the garden, which perhaps formed the part immediately adjoining the house. The mighty ruins of the palaces, built in the age of Darius in the very heart of the nation at Persepolis, certainly cry aloud for impressive garden surroundings; but when excavations were being made, the question of gardens was carelessly ignored, though the palaces were raised above three immense terraces. A careful eye can detect a certain intention in the large open stairway that leads to the first terrace, and then (turning at right angles) to the second. Still, the separate palaces are scattered apparently according to no rule, so that it is not an

FIG. 40. PERSEPOLIS—THE SIDE WALL OF THE GREAT OPEN STAIRWAY

easy matter to reconstruct (as in a measure it was possible to do in the Egyptian temple at Deir-el-Bakhari) the great regular garden-terraces. But in any case the terraces with large garden sites were indubitably here.

The approaches in Egyptian temples cut through the walls in a very gradual sloping path, but here—and in antiquity this example is of some importance—there is a great open stairway bordered by high walls. The decoration of the steps themselves shows that we may assume regular garden sites both at the side and above; on the uppermost tier there is a strip of earth with cypresses rising at regular intervals (Fig. 40), so that we may imagine the same thing in front as well. It is only at a much later time that we can get a real view of the evolution of the Persian garden, when poetry is able to tell us of the wonderful love felt by this people for their flowers and their rose-gardens.

The groves about their tombs were very important to the ancient Persians. The grave of Cyrus was enclosed by a grove, and his son Cambyses entrusted the care of it as a hereditary office to a family of Magi; when Alexander the Great saw it, it had grown high but had been neglected. The spot is now identified, and is an erection similar to a temple, raised very high on steps, in the neighbourhood of Pasargadæ (Fig. 41). Herodotus gives an account of the tomb of Alyattes, the father of Crœsus, saying that it is the greatest in the world with the exception of Egyptian and Babylonian buildings; but as he gives the circumference as 6 stadia, 2 plethra, that is about 1·1 kilometres, he must be including the whole estate with a large park on it, and this is further suggested by the presence of the lake that is near the monument, called the lake of Gyges, and supposed to be inexhaustible. The tomb had a foundation of freestone blocks piled up with earth—a structure which by analogy with Roman custom would mean plantation.

FIG. 41. THE TOMB OF CYRUS AT PASARGADÆ

The successors of Cyrus, the Achæmenides, are buried in strong rock-tombs near Persepolis, where they lived, and it is only in modern days that we have been able to see the tombs with our own eyes. The graves of Hafiz and Saadi, the two great poets of the Middle Ages, are at Shiraz, the latter alone in a valley. Within the marble enclosure stands the cenotaph in the middle of a garden of cypresses, poplars, flowering shrubs, and rose-trees (Fig. 42). The grave of Hafiz, nearer to the city, is still a favourite retreat for the townsmen, and the cheerful poet does not rest there alone even in death, for a number of other graves are within the precincts, and it is considered a great honour to find one's resting-place there.

In the same fashion the garden was used by the Israelites, and Abraham bought himself a double grave, that was in a tree-garden. Such was the custom among Jewish kings, who were at first buried at Jerusalem, where the gardens lay in the City of David, but from Manasseh onward they found their resting-place at Ussæ, also in a garden close to the royal house. But even in later times this custom was carried on, and the body of Jesus was buried in a private garden belonging to Joseph of Arimathea.

The Jews took as the beginning of all things the creation of the Garden of Eden; and this garden of Genesis is exactly like the oriental tree-park. For Jehovah makes all kinds of trees to grow forth from the earth, the fruit good to eat and fair to look upon, and in the middle of the garden the tree of the knowledge of good and evil, and the tree

FIG. 42. THE TOMB OF SAADI AT SHIRAZ

of life. Later accounts have added the water that would be needed, the stream which had its source in Eden, and from which four other rivers flowed. In this garden the beasts lived together, just as in the pictures of an Assyrian park the lions lie peacefully under the palm-trees and have their place among the lilies. This garden gave much refreshment both with its fruits and with its cooling shade; and God Himself, says the naïve anthropomorphic narrative, walked in the garden in the cool of the evening.

The Jews, who from patriarchal times had found a home in these parks, set up many sanctuaries and altars as well as graves, and it seems as though they adopted the custom from the heathen people around them. Abraham certainly planted a park, in which he sacrificed to God, but later on there was much opposition to the practice, especially by the zealous prophets, for it was feared that the Jews might be led further into superstition by the heathenish custom of sacrifices in the open air.

The sessions of justice were held out of doors, in rich men's parks. Joakim, the man of property, whose wife was the lovely and chaste Susannah, lived at Babylon, at the time of the Dispersion, and owned a fine park beside his house, and there the two justices held their court. Every day after the people left, they saw the fair woman walking in the garden and were inflamed with love for her. There was a large bath out in the open, just as in the Babylonian royal park where Alexander died, an ornament to the garden. Oaks and mastic-trees grew there, for the false judges betrayed themselves when they said they had witnessed wrong-doing, since each mentioned a different tree, one the oak and the other the mastic. [EDITOR'S NOTE: The mastic-tree of Southern Europe is *Pistacia Lentiscus*, an evergreen.]

Strabo talks about a palm grove in the plain of Jericho, saying it was well watered and contained many buildings, and also of a royal palace as well as a famous garden of balsams, from which precious spices were extracted. Josephus describes Solomon's palace elaborately enough, but says little about his gardens, except that they were fair to look upon. One of the chambers was decorated as a garden-room. The ornament was all regularly set out, the first three rooms in priceless stonework. In the fourth was a wondrous show: here one beheld trees and many kinds of plants, their branches and leaves hanging down and throwing their shade below. The trees and plants covered the stone, which was below, and their leaves were delineated with such marvellous delicacy that they almost seemed to be moving. It was clearly plastic art with naturalistic painting most impressively carried out, with the trees overhanging the walls and casting their shadow. The scheme of decoration must have arisen from the same desire that prompted Sennofri in Egypt to make his grave like a vine arbour; what they wanted was to enjoy the beauty of a garden in a private room.

At Jerusalem itself there were great gardens round the walls outside, both for trees and for vegetables, but they could not be put inside the holy city, for no manure, nothing unclean, might be brought within her sacred walls, and only rose-gardens were permitted. We may assume that the Jews loved and tended flowers, because their poetry, e.g., the Song of Solomon, so often used flowers in similes. There has been very little change in the main either in their religious or their daily use. The gardens outside the town were enclosed by walls with small turrets, in which guards were placed. From all we can learn of them, it seems that Jews' gardens were very like those of their neighbours.

When we include India in our group of the gardens of Western Asia, we do more than merely overstep geographical boundaries, for there is also much in the civilisation of its wonderful people that binds them closely to the countries of Eastern Asia; and more than anything else we must attend to the course of the great religious movement, Buddhism, which here takes its beginning. Almost everything we know about the early Indian gardens is closely connected with this religion. If then, in spite of it all, we separate India from the Eastern civilisations of China and Japan, and draw her into the circle of the West, we are driven to take this course by all the information we can get about her gardens. There is nothing that can distantly suggest that India adopted a picturesque style, as was the case with China and Japan. And neither to the foreign invader, like Alexander and his successors, nor to travellers arriving from the West, does there appear to be anything different here from their own parks at home, whereas in China every traveller is struck by the difference. Unhappily very few sculptures are preserved, and of these not one relates to the main continent. So much the more will the sacred texts have to tell us of the love felt by this people towards their gardens.

The worship of trees was a very ancient custom among the Hindus, just as it was in Western Asia, and there are innumerable Buddhist pictures of a scene that is always the same. There is a tree standing in an enclosure, the hedge made either of woven osiers or wooden stakes, or possibly of stone with a decorated balustrade. There are people offering sacrifices, generally a man, a woman, and some boys carrying the gifts, sometimes a whole number of believers, or again sometimes only animals, who reverently take their place—lions or gazelles, or perhaps whole herds of creatures. This scene occurs at least sixty-three times on the gates of Sanchi.

To each Buddha a different tree was sacred, and under it he received his Illumination. It was also under a tree that Queen Mâyâ brought forth the infant Gautama, whose destiny it was as the most holy Buddha to enlighten the world. The garden where this tree grew was called the Lumbini garden, and it had been laid out for the mother of Mâyâ-Divi as a present from her husband. When Mâyâ felt that her hour was come, she wandered in the garden from bed to bed, looking at the trees, until she came to the plaksha-tree, which bent before her, greeting her as the chosen woman. The garden was one of those pleasure parks which Indian grandees owned in front of the city, for Mâyâ betook herself to a carriage that was waiting on the beautiful road.

Pleasure gardens with their flower-beds and lotus ponds, shady avenues and leafy pippala- and sal-trees, play an important part in the story of Gautama's life. [EDITOR'S NOTE: The pippala-tree would probably be the peepul- or bo-tree, the sacred fig (*Ficus religiosa*) of India. The sal might be the hard-wooded *Shorea robusta.*] The prince was on his way for a pleasure excursion to the gardens of his parents when there came to him the fourfold Illumination of the transitoriness of this life, which caused him to flee from the home of his fathers.

But it was not only the kings but the community as well who owned great recreation parks near the city. These were accessible to everybody, and contained buildings of all sorts made for the convenience and pleasure of the townsmen. Such parks appear to have served from the first as headquarters for the Indian anchorites, and all the pictures of Buddha on his road to Illumination show him visiting the holy penitents, who lived in huts made of bamboo or in stone grottoes in a palm grove or some sort of park. When the lure of Buddha had become powerful, men vied with each other in giving to him and his pupils parks where they might take repose during the rainy season. Their scriptures always describe them in the same formula: "Not too near the town, and not too far away, well provided with entrances, easily reached by the people who like to come, not too noisy by day, perfectly quiet by night, removed from disturbance and crowds, a place of retreat and lonely contemplation."

One of the most famous parks was the Dshetavana, given to the Buddha by one of his very open-handed votaries, the great merchant Anathapinta. In the fifth century A.D. the Chinese pilgrim Fa Hian thus describes it: "The clear water in the pond, the tall greenery, and the countless flowers of many colours combine to make the picture that is called the Vihara of Chi-un (Dshetavana)." The park had once been a royal seat, and Anathapinta was only able to get it after long negotiations with Prince Dsheta, by offering to give him as much gold as would cover the whole of the ground. His reward was that the holy

FIG. 43. ENTRANCE TO THE PLACE OF THE HOLY TREE, CEYLON

Buddha chose it as his favourite home. Not only this merchant, but kings who had become Buddhists, presented such properties as this to the yellow-clad monks, for in that way they could keep them attached for a longer time to their own cities, and every fresh conversion made by the envoys of Buddha after his death ended with the presentation of a sacred garden.

Rather more than two centuries after Buddha had entered Nirvāna, Ceylon was converted to his creed. To this there is not only the testimony of the great Pali Chronicle, the Mahāwamsa, which throws light on the marvellous royal parks that grew up there in the course of the centuries, but we also have ruins, long concealed under ancient forest growths, but rediscovered in the last decades by English excavators. The accounts of the Buddhist writers, obscured by the network that covers a history written in verse, are confirmed by these discoveries, and are found to be accurate, and topographically remarkably trustworthy.

The first act of King Tissa after his conversion was the presentation of a wonderful park. The Apostle Mahinda had first of all preached in the garden of Nandana, which means the garden of Heaven, and the king wanted to present this place to the monks; but they scorned the gift because it was too near the city of Anuradhapura, and quite close to the king's palace. So then Tissa offered them the Mahamigha garden, which his father had laid out, "not too near the town, and not too far off, a valuable estate, well provided with water and shade." The very next day, with great pomp and ceremony, the boundaries of the place were removed, the king in person ploughing around it with a golden plough. Then they built temples and shrines to hold relics; but the chief treasure of all was a branch from a fig-tree, under which Gautama had once received Illumination —a miraculous branch, which in a few minutes grew to be a tree. Round it there are still preserved fragments and ruins of mighty forts, barriers, halls and gates (Fig. 43). Saplings from it are planted round the sacred grove.

Henry Cave thus describes the building by which the tree used to be enclosed:

> A wall of great blocks of granite, the projections and sloping parts inlaid with chunam, which resembles ivory, encloses the marble-paved court. Four entrances of high architectural pretension admit to this court, each covered with a metal roof like a baldachin, which rests on twenty stone pillars, each cut out of a single block. The plinth is made of huge granite plates with steep projections. A number of steps lead up to the lower part, with fine carving on them. The lowest step, which projects in a semi-circle, is wonderfully carved, and has side-posts of stone, ornamented with bas-relief pictures. Inside the court there are beautifully decorated halls, in which there are set up portraits of Buddha, sometimes carved in stone, sometimes made of a precious metal, and an inner wall with three stone terraces encloses the sacred tree itself. This tree and its surroundings are the most remarkable sanctuary in the whole island.

The Mahāwamsa gives an account of an imitation that was made of this tree, out of very costly materials. In the second century before Christ, King Dutthagamini had it put up in the great Ruanweli or Gold Dust Dagoba, of which we have the ruins to-day on a great hill overgrown with trees.

> In the middle of this shrine made for relics, the king had a fig-tree set up wrought in precious metals; the height of its trunk was eighteen yards, the roots were made of coral, and were fixed in an emerald floor. The trunk was of pure silver, and the leaves set with precious stones. The fading leaves were of gold, the fruits and the young foliage of coral. On the trunk were pictures of the eight lucky signs . . . above the tree spread a handsome baldachin . . . at the foot were fine vases arranged in rows, in which flowers shone that were made entirely of precious stones, and in the vases were the four kinds of sweet waters, which wafted their pleasing scents around.

This is certainly one of the earliest descriptions of an artificial tree. The religious character, the imitation of so great and holy an object as Buddha's tree, entitles us to discern in this piece the original source of that remarkable work of art, which—handed down continuously from age to age—eventually found its way as a fine garden decoration into the western and northern countries of Europe.

The excavations, which naturally cannot give any picture of the garden as a whole, have at any rate revealed traces of its beauty, in water-basins and baths, called pokuñas, of which the chronicle has so much to tell. The whole cultivation of ancient Ceylon depended on its excellent artificial irrigation, and the ability of the chief men displays itself, not only in their important religious buildings, but particularly in the way they provided immense, artificial tanks, so that the water could be taken all over the island in the time of drought. A good water-supply was one of the first requirements of a Buddhist grove, and many of the baths now dug up prove to be more than forty-five metres in length, over eighteen metres broad, and upwards of seven metres deep. The floors are paved with marble, and there are often marble pillars round them; also there are marble steps on each side with good balustrades. At one of these there is in addition a smaller bath, and a platform; this made a separate place, and was roofed over with a canopy on pillars. This inner bath opens on a room whose walls are made of wrought freestone (Fig. 44).

The Mahāwamsa has more to tell after the story of the capital city Anaradhapura, for we come at a later date to the period when under the great King Parakama in the twelfth century the city of Polonnaruwa rose to an unforeseen height of glory. Scarcely

FIG. 44. POKUÑA—BATH POND IN CEYLON

had the strong hand of the king secured peace when he began to build lordly temples and palaces, and by the side of his palace there arose a garden, which he called Nandana, in memory of the place where the Buddhist apostle first preached when he came to Ceylon.

This park is carefully described:

> Jessamine climbed about the trees . . . swarms of bees fed on the honey of various flowers; there were fruit-trees, some brought from abroad; . . . fruit-blossom and other flowers were there to delight the eye . . . there were many ponds, their borders decked with beauty, lovely with the abundance of lotus and of lilies . . . and there was the trumpeting cry of the crane (in India called the Cry of Sara).

The palace was girt round with pillars and rows of carvings in ivory, but the best decoration of all was a bath, blinding to the eye of a spectator: in it spouts of water, conveyed thither in pipes and machines, made the place appear as though clouds were incessantly pouring down drops of rain. There was a great array of different baths that delighted masters and men. We hear also of other parks and other baths, and even of gardens that the rich made for the recreation of the poor. The stories in the Mahāwamsa stretch back farther than the times we know about, and much of what we read suggests that there may have been some European influence. In its leading features, however, the garden seems to differ but little from those of the monks in the fifth century at Anaradhapura.

Our knowledge of Indian parks is helped by very few carvings, and it is to the tenth century that the great strip in relief belongs, of the Bôrô-Budur temple in Java. The life of Buddha is depicted here, and quite in accordance with legend, it is all played on the background of a park. The trees are drawn with extreme delicacy, but too conventionally to show what kinds they are (Fig. 45), and they remind us in their arrangement of similar

Assyrian pictures. Basins are rarely absent, and they are generally rectangular, with stone walls and floating lotuses, and in between waterfowl roving about, and in most examples also fish. The religious feeling of India provides that animals should play a great part, and we always find birds sitting on the bushes, and squirrels or monkeys playing about. In grottoes or rocks, which are conventionally drawn, antelopes are often found lying down, and lions too, as well as other beasts.

Flower-beds we do not often find here, but Indian stories and legends have much to tell of flowers as votive offerings. In the Mahāwamsa every festival procession has an abundance of them. Any person out of the numerous participators who also went up to the shrine of relics would lay his flowers there as alms, till altar and steps were literally

FIG. 45. RELIEF FROM THE TEMPLE OF BÔRÔ-BUDUR, JAVA

buried under them; and on days of high festival the lofty domes were wreathed from top to bottom. Whenever the chronicle takes us into a garden, we are at once surrounded by an intoxicating mixture of perfumes; and the eyes rest admiringly on the beds, and on the climbing plants in flower, which twine between tree and tree. So wandered Mâyâ, the Buddha's mother, admiringly from bed to bed in the royal garden.

The picture of Indian Buddhist gardens grows out of many individual examples into a whole. It harmonises with the great development of Western Asia, but not with that of Eastern Asia, where we shall find a completely different feeling for art.

Indian monks were the first to choose the garden as the proper setting for their lives, which were devoted to the contemplation of the divine; but with a prophetic eye we may see that the garden will often be dedicated in a like manner: at a later time Greek philosophers, and monks in early Christian days, will retire into their gardens for united, yet silent, contemplation.

CHAPTER III

ANCIENT GREECE

CHAPTER III

ANCIENT GREECE

THE Greeks were filled with sheer amazement when they first beheld the magnificent parks of Eastern potentates. Their own civilisation had produced nothing to compare with the achievements of these mighty satraps and kings. In the best period, when the other arts in Greece were rapidly advancing to their highest point of development, we hear nothing about garden culture. Nor need we hope to get through excavations any satisfactory evidence of this most perishable of all the arts; moreover, the witnesses who give us pictures of Oriental and Egyptian gardens—poets, historians, and painters—are silent here; so it is only by listening for chance remarks in literature that we can trace any development of the Greek garden, so simple yet so important in its results.

The reason for this remarkable gap in Greek art of the best period is the constitution of the cities: a sort of frame enclosing the whole life, intellectual and political, of the Greek, allowing no space, even at the time of its greatest expansion, for the cultivation of private gardens. In all ages any important development of the art has been due to the educated class, politically powerful and artistically refined. But the growing democracy watched with a jealous eye lest any mental superiority should raise a man's family to high station. If in spite of this the Greek spirit in its results has proved important, we must seek for garden development in a quite different place.

Greece had known a time when the conditions for horticulture were favourable enough—the great time of Cretan and Mycenæan art. But unluckily no light from literature is thrown back so far as this. Every year new excavations have given proofs of the amazing importance of the epoch, and have revealed buildings of mighty strength and size, together with a manner of life refined to the utmost pitch of luxury. Palaces of every sort and size have been brought to light, but the indications of gardening are dubious and leave room for little more than guesswork. The castles on the mainland were for the most part fortresses, and had so small a space within, owing to the densely packed houses near the palace, that there must have been even less room for gardens than at the castles of later times.

In Crete, of which our present knowledge reveals the whole civilisation from its first beginnings to its highest development, the situation was very different. Here reigned royal families in profound peace, protected by the sea. Here there were no fortress walls to restrict severely the palace grounds. We can see their love of the plant world from the ornamentation of the vessels which they used, the painting on their vases, the frescoes that decorated their rooms: flowers and trees are portrayed with astonishingly artistic skill (Fig. 46). Cleverly choosing a site, protected in winter from rough winds, open in summer to cool airs, on the slope of the hill they built the wing to live in, three stories one above the other, at the palace at Knossos. From open pillared halls and terraces you look down upon the lovely green secluded valley (Fig. 47); steps at the side lead down to it.

Unfortunately the excavations do not take us any farther; but one must needs think that lovely pleasure-gardens adjoined, to gratify the æsthetic taste of a noble and wealthy lord. This page in the history of Cretan art, which its monuments are year by year disclosing, must for us unhappily remain a blank.

The heroic poetry of Homer throws a somewhat clearer light on the scene. Here we see mirrored the mode of life of a noble race who were perhaps inclined, in those intervals

FIG. 46. CRETAN FLOWER-PAINTINGS FROM HAGIA TRIADA

of leisure that wars allowed, to busy themselves with gardening. Homer's poetry gives good instances of this, though the ways of life were simpler than in the palace at Knossos.

It may have been through grief at the loss of his son Odysseus that Laertes was driven to his fields, there in a servant's dress to do menial work, but toiling with his hands in a garden was no new task for him. In early life he had trained Odysseus to it, giving the lad a row of fruit-trees and vines to stimulate his zeal. We might call the place a farm-garden, where Odysseus, returning from his travels, made himself known to his grief-stricken father: he found nicely arranged rows of fruit-trees and vine espaliers ringed by a fence of thorn-bush, and beside it the humble dwelling shared by the field servants and their master (Odyss. XXIV.). We may regard the property of Alcinous, with the lovely fruit-garden, a stone's-throw from the town and close to the grove of Pallas Athene, in a similar way. It was here that Odysseus had to wait till Nausicaa and her maidens reached her father's house.

But Alcinous owned another large and magnificent garden in the town of Scheria: it was close to the palace (Odyss. VII. 112 ff.), and the poet praises it as having every imaginable beauty and charm. Yet if we examine it closely, it is only a garden for utility, and is not very different from Laertes' farm-garden. Odysseus looks over it before he crosses the lofty threshold of the Phæacian palace: it is outside the court, near the door. This front court is enclosed with walls, and perhaps, like the palaces at Mycenæ, with buildings also. Round the chief entrance extends the large garden (of four hides), also walled in. Next to the court is the orchard, with apples, pears, figs, and olives, the same as

FIG. 47. PALACE OF KNOSSOS—THE GREAT HALL AND TERRACE

in Laertes' farm-garden, except that pomegranates are not mentioned there. Next comes the vineyard; and finally, in the last section, are the beds of vegetables, planted according to the different kinds. By this triple division, which excludes such ornamental plants as are not immediately useful, the Phæacian garden looks just the same as the others.

If the poet, describing the very best of his time, can see nothing more than this, we are justified in assuming that only the simplest kind of cultivation was known, and that in this respect the Greek could not vie with the Egyptian or with any Oriental.

If a poet would produce a picture of fairy-like charm within these simple limits, he must adopt quite another method. His garden is to remain the whole year through full of blossom and fruit, unhurt by any failure or any mishap. Pear after pear, fig after fig ripens, fanned by gentle zephyrs. Such descriptions are found everywhere, and especially about vineyards, where the grape is shown from the blossom to the winepress, and all at the same time. No doubt the rhapsodist has in his mind the mythical gardens of the Hesperides. The fairy-tale character is given by this timeless mating of flower and fruit; for always and everywhere there has naturally been a demand that in a good garden something or other shall bloom at every season. Now the favoured climate of Greece could more easily fulfil this demand than more northerly latitudes; yet Bacon himself claims it for a garden in England.

Of the proper care of flowers in this fairy garden we hear nothing, for the "fair ordered beds" are only vegetable beds. How else would the poet, who can paint in such rich colouring the pastures of the wild, be silent about the wonderful beauty in a garden? It seems clear that nobody had as yet hit upon the idea of transferring to the garden the lovely children of the meadow. Only at a later date was the vegetable plot embellished with flowers, so that from the garden made for use the garden made for beauty arose.

FIG. 48. SILVER VESSEL FROM MYCENÆ, WITH FLOWER ORNAMENTATION

The absence of floriculture in Homeric times appears again in the fact that flowers are not much used to adorn houses or clothes. It is true that very often Homer's goblets are described as adorned with flowers, but the word "flowery" may point only to floral ornamentation, for the goblets are not vases in which flowers are placed (in Egyptian and Oriental fashion), but sacrificial cups for holy waters, or possibly mixing vessels.

There is a beautiful example of the Homeric goblet in a silver cup from Mycenæ (Fig. 48). It is a vessel representing a flower-pot in three tiers, with a flower design on it. This kind of object may or may not have been due to Egyptian influence, for such pots were common in the houses; but the ornamentation by itself is not enough to prove that there was an extensive cultivation of flowers at Mycenæ, although some writers have held that it does prove this.

Odysseus standing at the threshold of the house looks over the whole garden. If we take the poet's words literally, there must have been rising ground in front of him; and this is consistent with the disposition of the water-supply. "There are two streams," we are told, "one of them traversing the whole garden, the other passing below the threshold of the court to the great house." From this stream the townspeople took their water.

In the Iliad Homer gives an explicit account of the ingenious water system used in gardens. We hear of a wonderful controller of the water whose function it was to conduct the streams over garden and plants. There were evidently canals where the water remained dammed up till it was wanted: then the overseer took away the obstructions and out poured the water with a rush, to stream over the land. This kind of water system, to which the Phæacian garden also owed its fertility, needs a rising ground. And so can be best explained the division into two streams. If one stream is dammed up, it cannot be used for purposes that need running water. So the second stream should lead straight through into the court, where it can be drawn off by the townspeople; and it must have an exit beyond the outer wall.

One further detail in Homer's description throws some doubt upon the situation. About the grapes in the vineyard we are told that some of them were getting withered in the sun on the level ground of the drying-floor. If this drying-floor is in the garden itself we shall have to imagine (since it belongs to the middle section) not only our rising ground, but terraces as well. We are also told in this fairy-tale account that the grapes are cut in one place and pressed in another, so the drying-floor may just as well have been where the winepress was, at the farm outhouses.

Because of its fertile fancy this description of the garden of Alcinous, supported by the poet's authority, has put into words the highest and noblest feelings of generations of men who belonged to the ancient world. And when the art had attained to heights unimagined by Homer, there remained the enchanting ideal for poets, floating before their minds like their dreams of the Isles of the Blest.

Though the garden of Alcinous was conspicuous for its prolific abundance, it is not likely that a palace garden was an exceptional thing. Probably every man of importance in Homer's time had a garden of this sort at his palace, in order to provide his tables with necessaries. In the home of Odysseus such a one is plainly indicated. Penelope says that as a wedding gift her father gave her the experienced Dolios, so that he might take charge of the fine tree-garden (Odyss. IV. 737). Dolios lived at this time on Laertes' land, but we must think of Penelope's garden as close to the palace, and under her own special supervision. Already in those days it had become the custom for the garden to be the particular property of the mistress of the house, as it had to provide fruits and vegetables for the kitchen. In similar social conditions we find exactly the same thing all through the Middle Ages.

In the open courts of the palaces, which were to play so striking a part in the future development of horticulture, there was no attempt at gardening either at Mycenæ or in the Homeric age. The front court was always paved; this was where the altar stood, and very often there was also a well beautifully encased in marble, an ornament to the place. Still, even in Homer's time there were some courts with plants, as is proved by the tale of Odysseus about the secret erection of his cunning bedstead. Round about a shady olive-tree in one of the lower courts Odysseus built himself a bedchamber. He used the trunk of the tree to support the bed, which he firmly attached to the ground.

It seems as though the courts in the women's quarters were planted with trees, but for the most part front courts were paved even at a later date than this. Lucian describes a House of Aphrodite: the front court, he says, was not paved "in the usual way," but adorned, as was suitable to the lady of the house, with noble shady trees, flowers, and arbours.

In Homer the sacred grove is mentioned much oftener than the garden. We must look upon it as the headquarters of religious worship, for the service in temples, though often spoken of in epic poetry, is far less important. Whether the grove was always enclosed and therefore a τέμενος, is uncertain, as it is generally called ἄλσος, but it is the fact that its altar was surrounded by shady trees, for the epithets given to the groves are "rich in trees" and "shadow-spreading." Generally a spring is mentioned, but the names of the trees are hardly ever given: in distinction from the garden they are always non-fruit-bearing forest trees.

Often a sacred tree stands between the spring and the altar, as for example the tall towering palm which comes into Odysseus' mind when he meets Nausicaa. And the Achæans also make sacrifice at an altar beside a spring which rises out of a beautiful lofty plane-tree. Most renowned of all was the oak at Dodona, from whose top Odysseus heard the voice of the Thunderer. Very common among Cretan and Mycenæan monuments are pictures of sacred trees in small precincts or enclosures of stone. Their reverence for trees the Greeks shared with the Orientals. It lasted longer than the Hellenistic period, and was inherited by the Romans.

There is no lack of records to prove that the spots sacred to the gods were carefully and skilfully adorned. There is the grove of Athene at Scheria on the road to Alcinous' estate; the altar stands in a meadow wherein flows a stream; shading it there are black poplar-trees. The sanctuary of nymphs at Ithaca is even more charmingly depicted. Odysseus is strolling up to the town with the swineherd when they come to a cunningly enclosed basin, which lies open to the cool streams foaming down from the rock. All round in a circle are planted the water-loving poplars, and high above stands the altar where travellers are wont to offer sacrifice to the nymphs. This picture was the work of three townsmen of Ithaca, whose names are explicitly stated by the poet, so attractive does the sanctuary seem to him. This, our first picture of a nymphæum, gives us a clear idea of the sites that the artists of a later antiquity, and still more those of Renaissance days, could adapt in the happiest manner to horticultural uses.

In all these pictures we are concerned with the works of man; but in the grotto of Calypso it is somewhat different, for here we have a pretty natural scene. There is no grove, but merely a wood with alders and poplars and sweet-smelling cypresses, in whose boughs are nesting hawks, tree-owls, and loud-cawing sea-crows that know the trade of the waters. A vine, heavy with grapes, stands at the entrance of the grotto. Four streams rise near by, and meander hither and thither about a meadow-land teeming with violets and wild parsley. This is a picture that even an immortal stays lingering to behold.

Thus we see in the Homeric age germs many and diverse of the garden of the future; they will develop very differently in different social conditions. We know that until the fifth century the Greek lived happily on his land, and that not even the devastations of the Persian wars prevented him from rebuilding his home in the time of peace that followed. But the land was now split up into small properties proportionally divided. The rich had accumulated fine collections of furnishings for the house, but in their gardens they still had only useful produce, just as in the days of Odysseus. When Cimon had the boundaries of his garden removed, so that everyone should have free access to it, he was not abandoning a pleasure-ground, it was only that now foodstuff was procurable by all. It is possible—unluckily we lack information about this period—that people had by now begun to grow flowers, for the fashion of wearing wreaths, quite unknown to Homer, begins from the sixth century to be more and more prevalent. Every religious ceremony was performed by persons who were crowned with wreaths, every victim was wreathed, and though a symposium began with rites of religion, the guests were also crowned.

But, however this may be, there were still only a few kinds of flowers known to the Greeks; and the old folk-song that Athenæus has preserved, "Where shall I find the rose, the violet, and the lovely parsley?" gives nearly all the flowers known in these early days, if we add poppy, lily, crocus, and hyacinth. Perhaps the rose was known as a garden flower soon after Homer's time. At the foot of the Bermion Hills in Macedonia there lay in the days of Herodotus the fruitful garden of Midas, the son of Gordias, where sixty-petalled roses grew with surpassingly sweet scent, and maybe the Greeks owed their rose-culture to the Macedonians. It increased rapidly: Demosthenes knew rose-gardens where many different kinds were grown.

As to the peculiarities of these gardens, Greek literature leaves us in the dark. The author of the pseudo-Platonic dialogue *Minos* only excites curiosity when he mentions "Writings about Gardens, compiled by Gardeners," of which no trace remains. And the odd word περίκηπτοι for beds is some sort of argument for the existence of gardens with beds of flowers in them. When the demand for flowers increased, it brought about a special trade for gardeners; but for the needs of a house no doubt the ordinary men-servants attended to the flowers and vegetables together.

FIG. 49. ADONIS GARDEN

How far the culture with a view to medicinal use had proceeded in Greece (which is so plain to see in the Middle Ages), is not made clear from our sources of knowledge. Aristophanes calls the gardens sweet-smelling; and they liked to have odorous plants on graves, because their sweet smell signified the purification of the dead. And since graves form so often a part of the garden, we must needs find that herein was a stimulus to the culture of flowers. They are also grown in the groves of female divinities. At the mouth of the Alphæus there are flowery groves sacred to Artemis and the nymphs. Aphrodite above all others was a patroness of flowers and gardens: she was called "violet-crowned."

In the cult of Aphrodite at a later time—already naturalised in Greece in the sixth century and perhaps derived from Syria with the cult of Adonis—we believe that we may find the germs of a later garden-craft in the so-called Adonis gardens. At the Festival of Adonis, celebrated at midsummer by the Athenian women, who sang dirges over the death of Aphrodite's lover, they used to set up on the roof a figure of Adonis. Round this they placed earthen pots filled with soil wherein were sown fennel and lettuce, and also wheat and barley. The plants sprang up soon, and withered as quickly, and this signified the fate of all vegetation, which after its great beauty in springtime fades early, dried up in the hot summer of a Southern land. This is symbolised in the mourning for the early violent death of the beautiful youth Adonis.

Women kept the festival (which was not officially recognised) and grew attached to it as an old popular custom, such as is our own May Day. And so the lesser art of vase-painting took kindly to this subject. An Aryballos at Karlsruhe depicts the scene (Fig. 49) with a graceful naïveté: Aphrodite herself appears, and being a goddess is naked. She stands on the lowest rung of a ladder, and this shows that she is climbing up to the roof to carry there a pot for flowers (one of the halves of a jar that has been broken in two) which Eros is handing to her. The other half of the jar and another pot are still standing on the floor. To the right and left are Athenian women making gestures of surprise at this apparition. On another vase we see Aphrodite and Eros pouring the water from the jars before the festival, to make the seed germinate quickly (Fig. 50). Later on, in Alexandrine times, these shows were grander and had a more official character.

FIG. 50. ADONIS GARDEN ON A VASE

When the wife of Ptolemy II. celebrated the festival, Aphrodite and Adonis were carried out in a pompous procession, seated upon a silver couch under a flowery canopy decked with fruits and blossoms. But by now the old simple custom was far away in the background and had become a children's game. Boys sowed quick-growing seeds in great pots, delighted when the green began to show. The reason why Adonis gardens are so often mentioned, even by Plato, is that the name came to be applied to things of small importance that produced only short-lived pleasures. But in this cult and in the childish games we do get the beginnings of gardening in pots. Everybody who had to do without a garden and yet wanted to adorn the home, found a substitute in these pots, and Theophrastus bears witness to the fact that in his day pot-gardening was carried on for other purposes.

The Adonis garden was known even in imperial Rome. When Apollonius of Tyana, the worker of miracles, was the guest of Domitian at the Palatine in Rome, he found the emperor in the court of Adonis: "This court was adorned with flowers just as the Assyrians plant them on the roofs in honour of Adonis." The narrator had accurate knowledge of the cult, all the more because he had gone to Greece at one time from Syria, the native home of the Adonis cult. In the palace of Domitian tubs were set as a decoration all round the roof of the pillared court, and later on we shall find traces of a similar custom at Pompeii.

At first it was only in the short period of a festival that the Greek women adorned their flat roofs with flower-pots; but later on they kept this pretty custom the whole year through, till at last there came about the decoration of balconies and roofs in Rome at all seasons. Pliny's account of the Adonis gardens is somewhat exaggerated when he finds them like the gardens of the Hesperides, the garden of Alcinous, and the hanging gardens of Semiramis, grouping them all together as exceedingly wonderful.

At the beginning of the Peloponnesian War there came about a great change in the homes of the Attic town-dwellers. In consequence of the persuasive eloquence of Pericles and partly through dire necessity, those who lived on the land, peasants and gentry alike, moved into the town with all their portable property. The flat country was laid waste by Aristodemus. With much murmuring and with many delays they gave up their pleasant country life and left their fair estates to grow wild. But necessity compelled, and the days of old returned to Attic lands no more. Soon men were looking back as on a happy past upon those days when Attica was so safe that her lands were covered with country houses fairer than any in a town.

It was only the people of Elis who "according to old custom still lived on the land: they loved their country life, so that there were well-to-do families among them who for generations had never come into the town." Their early conditions were never destroyed, for the peace of God protected their lands; ay, they could even enjoy their own jurisdiction —an ideal which for long enough floated before the eyes of Attic gentlemen, even to the days of Theophrastus, as we can see by his jeers at the Aristocrat.

It was the ever-growing, hated democracy, far more than the disturbance of wars, that forced the country gentlemen of Athens to abandon their lands for ever. It is obvious that such a change would be far from propitious to the development of the art of private gardening.

The effect was of course restricted to the mother country. In Asia Minor the close connection with the East may well have brought about a garden culture adapted to an Eastern country, though we are not in a position to prove that it did. In the gardens of Macedonia, already mentioned, we can perhaps see an early direct influence of Asia Minor at work. With favourable social conditions a love of parks and gardens could extend on Greek soil where certainly Oriental influence had not been present. The powerful tyrant Gelon owned in the land of the Bruttii in the year 500 B.C. a park which excelled in beauty and was splendidly watered: in it was a site called the Horn of Amalthea, after the goat that gave milk to the child Zeus.

We can only take this to be a nymphæum such as Homer described, one of those sanctuaries that included well-arranged trees, artfully enclosed water, and perhaps a grotto as well. What kind of water system was in Gelon's park we can only conjecture, but possibly the name *Horn of Amalthea* points to a fine waterfall, or possibly it only refers to the unusual fecundity of the place. At a later time these nymph sanctuaries

FIG. 51. A NYMPH GROTTO. RELIEF

were very common in gardens. There is a relief in the Lateran Museum (Fig. 51) which (though clearly Hellenistic) relates to the earlier Amalthæum, and gives a charming picture of that type of grotto. Similarly, too, the sanctuaries to the Muses are found in a great many places. In Sicily the elder Dionysius owned a garden wherein he planted plane-trees, of which he was very proud, although they did not attain to a great height. In the time of Theophrastus these trees adorned a gymnasium which was established on the same territory.

But in Greece proper, the very same democratic conditions that robbed the private person of the means of making gardens on a large scale were clearly favourable to the cultivation of public grounds. Indeed it is here that we must seek the path that leads us to an art of very wide scope.

The sacred groves, which in Homer's time had still for the most part only a wall round the altar, grew to be more and more akin to the garden-like surroundings of a temple, and the grove became as important as the temple itself, and remained so. At Delphi was a laurel hedge close beside the temple of the god, probably over the treasury, which in the south must have been above the deep subterranean cellars of the temple. Euripides in the *Ion* makes Hermes at the end of his prologue step aside into the laurel grove. The fact that at the sanctuary of Æsculapius in Epidaurus there was a grove, is proved by an inscription wherein it is said that curious persons climb up a tree to see the people who are asleep in the sanctuary. Even excavators are beginning to glance at the groves in the temple precincts. At Miletus there is round the temple of Apollo a bare tract of earth sixty to a hundred metres in circumference, and this is rightly supposed to point to a sacred grove.

There were some holy places without groves, and the latest authorities would have us be prudent in dealing with the evidence that trees supply. A curious example shows the risk we run in accepting an old tradition: when Strabo went to the temple of Poseidon at Onchestos, he had been led by Homer's account to expect to find a grove of wonderful beauty, praised in the Iliad and also in the Hymn to Hermes. Instead of this he found a place bare of trees and any sort of greenery, and in his vexation he blamed the poet who calls every sacred place a grove just for the sake of embellishing his verse. About 180 years later, however, Pausanias goes the same way and finds a grove, for a century and a half had been a long enough time for the trees to grow to a fine height, and possibly Strabo's annoyance had given occasion for a fresh plantation. . . . Pausanias was greatly surprised, and formed the opinion that this was the same grove that Homer had praised so warmly in the past.

It was only at a later date—and then probably for the benefit of the servants of the temple—that fruit groves were planted round temples and nymph sanctuaries, delightful in the springtime with sweet-smelling blossom. But for the most part the Greek was deeply concerned that his sacred trees, dedicated to the gods, should never be touched. As in Mycenæan times, so on Greek vessels in the sixth and fifth centuries, we get pictures of sacred trees with votive tablets hanging on them; and more in number as well as in closer detail are the pictures on Hellenistic reliefs, where at the side of the semicircular shrine which encloses the tree, there is also a door and very often a pillar: above is a tent-like roof, which betokens a resting-place, and in the surrounding space, open at one side, there may have been a seat (Fig. 52).

In addition to these sacred groves and trees dedicated to the high gods, we have in post-Homeric times hero sanctuaries. The grave of the ancestor of the race, of the founder of a town, demanded a cult to itself, and was nearly always surrounded by trees, which were deemed so sacred and inviolable that the Athenians punished with death anyone who cut down even the smallest of them. Most towns possessed such a hero, venerated with awe. Ancient oaks, olive-trees whose fruit had never once been gathered, cypresses called *maidens* because of their slender growth—everywhere, even up to a late period, these shed the profound shadow of very old trees over the grave of the ancestor. Terribly, moreover, did a hero punish any crime committed at his grove, as is shown by the revenge Anagyrus took upon the countryman who cut down his sacred trees.

FIG. 52. A YOUTH BEFORE A TEMPLE WITH A SACRED TREE

By the side of this hero cult we find the Greek games: originally, as known to Homer, a kind of wake, they were held at the annual feasts in honour of the heroes. "We find the original root of the important system of games, which brought out the peculiar characteristics and the individualism of Greek life, in the older cult of heroes." But it was in honour not so much of the heroes as of the gods that the most important games were held in historical times, though even these appear to have aimed at the further exaltation of Heroic games by thus raising their status. With a view to uniting the whole of Greece, a great god came to the fore instead of the local hero, and this god granted to the games festival all rights of hospitality on his territory. For the games did take place in sacred precincts. On the Isthmus was Poseidon's sacred grove, one part planted with lofty fig-trees in great numbers and in regular order, the other part, where the games of the athletes took place, adorned with statues of the victors. Even chariot-races could be run in the τέμενος of a god: in the sanctuary of Pagasæan Apollo, called specifically a sacred grove by the poet, Heracles conquers Cycnus, the son of Ares.

So from the very start the holy places of heroes and gods were the arena for the games. Pindar's third Olympian Ode gives an easy explanation of what happened, but he turns it the other way about. He sings of the introduction of chariot-races into the hippodrome, ascribing it to Heracles. The hero looks upon the empty fields, a "naked garden," wherein is no green tree, burnt dry by the heat of the sun; he hastens to the Hyperboreans, admires their lovely, fruitful lands, fetches away an olive-branch and plants its shoots "there where the chariots twelve times in their courses thunder to the goal." And thus it came about that a shady grove enclosed the place.

So sings Pindar, but even in this tale of his we find the bond between the sites of the early games and the sacred groves. At the same time, moreover, the poem points to the practical need in a Southern land to have some soothing shade, both for the spectators and for the competitors in the games. That the planting was in regular order we know, not only from chance observations, but also because of the practical necessities of such places: the actual arena had to be quite free and open with rows of trees all about it.

As the Greek games became more and more important to their country, the need grew greater of having special places established where the young men could practise, as otherwise this first adventure of a gymnasium might be only temporary. It went hand in hand with a further development of gymnastic exercises in the education of young Greeks, but even here it was still connected with a cult, with a hero shrine. It is the easiest plan to follow the origin and constant growth of the one most famous Greek gymnasium, the Academy. The name comes from a hero called Academus, whose sanctuary was given up. This Heroon, where games were played to honour the memory of a hero, grew to be the famous home of Greek education.

Other sanctuaries were linked to it, first and foremost the famous τέμενος of Athene with the twelve sacred olive-trees, of which one was reputed to be an offshoot from that very olive which Athene herself planted near the Erechtheum. Other gods, too—Zeus, Prometheus, Hephæstus—were honoured there; also Heracles and Hermes, the proper patrons of all gymnastic sport; and there was an altar sacred to Eros himself. Even from the days of the tyrants this gymnasium was in existence, but only under Cimon's governance did it become very much admired. He took pains to supply the Academy with water, for it had suffered badly from drought, and so turned it from a place that had only grown unfruitful trees into a beautiful fertile park which was renowned far and wide. On the higher parts grew poplars, elms and planes, whose wonderful height and strength roused the admiration of Pliny. Cimon also made wide roads and shady walks.

It was in Cimon's day that people first began to take a wider interest in making towns beautiful with gardens. Cimon had the Agora at Athens planted with trees. Especially at the time when towns and town life began to flourish were pretty, charming grounds laid out. There were imitations in other towns. The Agora at Anthedon, near Thebes, had a double colonnade, the inner half of which was flanked with trees. In Sparta there was a place for exercise in the middle of the town, quite round, and like an island, encircled by a "Euripos." Two bridges crossed it, one ornamented with a statue of Heracles, the other with a statue of Lycurgus. The place was planted with fine plane-trees and so got the name of "Platanistas." And at Corinth there was the old gymnasium near the theatre, and beside it a lovely spring called Lerna, round which were cloisters and seats for pedestrians in the summer-time, and within the portico there were garden grounds.

The gymnasiums in the towns were naturally smaller ones. The four large gymnasiums at Athens were outside, and were all finely ornamented with park grounds. In the third century the three gymnasiums—the fourth did not yet exist—are all described as flourishing gardens, full of trees, ornamented with grass-plots, a place for philosophers of many schools, a happy resting-place for the mind. A generation later than Cimon, Aristophanes in the *Clouds* describes the Academy as a park with many trees: the spirit of Justice invites the youth to leave his silly affairs, for then he can stroll under the olives at the Academy crowned with white reeds, and run races with his friends. There, beneath the sweet-smelling yews and the trembling leaves of the aspen, he will be happy in the springtime, while elm- and plane-tree whisper to each other.

From the beginning gymnasiums were looked upon as public places of resort: people met there, partly to look on at the exercises, and partly for the sake of intellectual conversation. In the dialogues of Plato and Xenophon we generally find that the gymnasium is the background of the scene.

In Plato's time the gymnasium and the park were so closely connected that the philosopher wanted to have gymnasiums in such places only as were well-watered and specially favoured by Nature. In Rhegium such a one was in existence in the Paradise of the elder Dionysius, and in Elis at an earlier date there was wild woodland round the gymnasium instead of garden. But we may be quite sure that gymnasiums were attached to hero sanctuaries, for the clear light of history gives us the origin of one that was set up in Syracuse: the town raised a tomb in the Agora to Timoleon the liberator, and founded a Heroon with annual games. Hence arose later on a magnificent gymnasium with halls, exercise grounds, garden grounds, and an Odeon for musical performances.

In Roman times Vitruvius is the first to describe the appearance of a building erected with a view to gymnastic sports. He calls it expressly Greek, not Roman, and palæstra, not gymnasium. He is mainly interested in the actual building, and it is only when he cannot help himself that he even mentions the garden sites. He first marks out a peristyle, and round this are the different rooms which the philosophers used for their conversations. Here we get a wrestling school; a colonnade where in winter-time the games can be carried on in bad weather. It is called a "Xystos," and beside it there are walks under the open sky, which the Greeks call side-paths. The Romans, however, give the name Xysta to these walks, where in the bright days of winter the athletes can go out and take their exercise. Their Xysta are a kind of thicket or plantation of plane-trees between two porticoes, and under the trees are seats to rest on: the stadion is generally at the end of the Xysta. Vitruvius cannot say enough of his astonishment that the Greek and Roman use of this word should be so different. But the Romans must have adopted the word at a time when the Greeks still applied it to the open spaces. In the old gymnasium at Elis, where the practice for the Olympian games took place, the open space, walled in and planted with tall planes, was certainly called a Xystos. The name may perhaps have been given because Heracles, who practised in the place, used to cut away the thick acanthus growths every morning. (The supposed etymology cannot be proved, notwithstanding the acanthus, but we may compare the use of the German word *Rasen*.)

By other writers these walks in the open are often called Xystos paths, and sometimes Xystos gardens. When the gymnasiums became more and more educational, covered halls were needed, so that the instruction should not be interfered with by every freak of

FIG. 53. GENERAL VIEW OF THE GYMNASIUM AT DELPHI

FIG. 54. BATH BASIN AT THE GYMNASIUM, DELPHI

the weather. These in time became the chief places, the Xystoi proper, while the gardens in the open, now a sort of parterre by the side of the portico, were still only looked upon as side-paths.

Vitruvius calls his structure, as we said before, palæstra, not gymnasium; but no doubt in the later days the two words were used synonymously. All the same, we must bear in mind that what Vitruvius is talking about is only those parts that were of immediate use for gymnastic instruction, whereas in classical times the gymnasiums embraced not only the exercise grounds, but the palæstra proper, and also sanctuaries enclosed within gardens that were continually being made finer and better watered. There were temples, altars and chapels adorned with a great number of statues: there were also swimming-baths in the open air, just as we can see them at Delphi. Moreover, the baths were not for the most part closely connected with the palæstra proper.

Though at first it was quite a modest affair—in Priene no warm baths have come to light—the bath became so important at a later date that the words "gymnasium" and "thermæ" are often interchangeable. This we assume when in Miletus the buildings of the gymnasium are explicitly mentioned. And the ruins of the earliest gymnasiums strengthen our opinion.

The gymnasium at Delphi (Fig. 53) is constructed on two terraces, for the steep slope gives no possibility of making it on the flat. The oldest parts, as for example the fine polygonal walls of the upper terrace and the peristyle below built in a court with rooms and passages round it (called in a Delphic account the place for playing ball), date from the sixth century. Above on the upper terrace is the wide open court for exercise, the paradromos. The covered winter hall, at the back touching the hillside, belongs to the fourth century. It is, moreover, not improbable that here in the oldest times there was, as in Elis, a place like a Xystos under the open sky before a covered court was built.

At the same period the lower terrace was widened on the western side, and a fine circular swimming-bath was added, the first one made in the open air. Steps led down into the bath, and from the walls behind water streamed into stone basins (Fig. 54). Round this part there were no doubt garden plots on the smooth grass of the terrace. On the upper terrace a little canal cut through above the bathing-place, and this canal served as a reservoir for the bath. Here we must also conjecture thickets, and other garden land.

The above-mentioned Delphic account, however, describes, among the kinds of work done, the digging up and smoothing of the Xystos with its adjoining land, and the care of the peristyle and the place where ball-games were played. The Xystos was covered with white earth, and the ball-place with black.

Unfortunately the excavations of other classical gymnasiums do not help us, since only the buildings of the peristyle have received much attention, and this, according to Vitruvius, is but one part of the gymnasium. The wonderful peristyle at Epidaurus, with its grand propylon and huge colonnade, was built in the fourth century; but since the racecourse has not been dug up, we can do no more than guess at the existence of gardens in the great middle court. The diggings at Olympia are unsatisfactory in the same way. Perhaps further excavations will show that the colonnade of the so-called Great Gymnasium which adjoins the peristyle on the north is really a Xystos. But the lower gymnasium at Priene, far better excavated, has not only garden grounds in the peristyle but all round about in parts not yet dug over there are park sites and sanctuaries.

There is one and only one gymnasium (of the later period) whose whole imposing extent is open to our view, namely, that at Pergamon (Fig. 55). The ground rises in three great terraces, with a difference in height of twelve to fourteen metres. Walls of colossal size, and under them niches that had probably been filled with votive statues, above them huge porticoes—these enclosed the whole estate. All the three terraces have buildings on them, which are really parts of the gymnasium, and the middle one seems specially suitable for fine garden grounds; here we have a temple among several other edifices, probably dedicated to one of the gods of sport, perhaps to Heracles. But only the

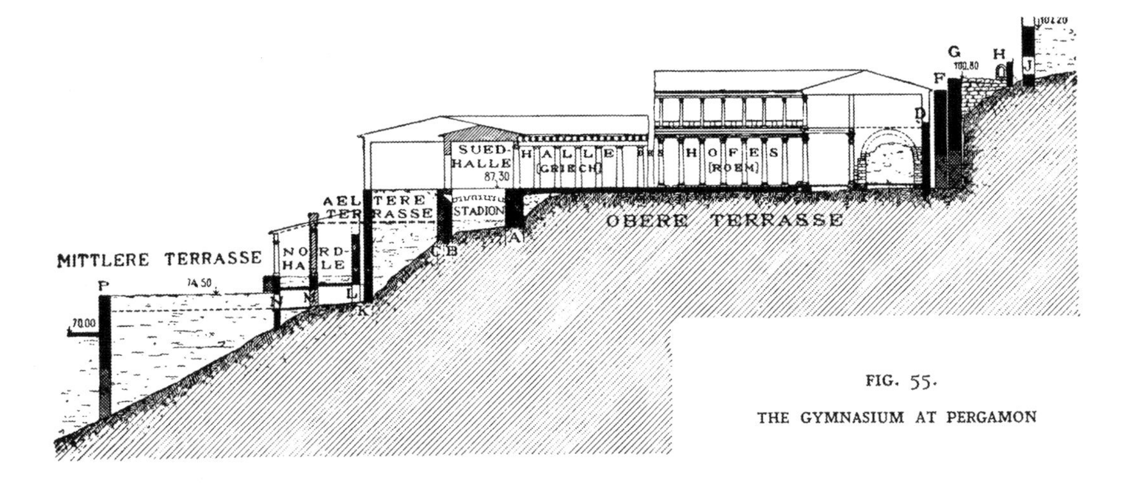

FIG. 55.

THE GYMNASIUM AT PERGAMON

uppermost terrace has an important peristyle, complete, as Vitruvius wishes it to be, with living-rooms and bedchambers built round it. Here also the inner part of this peristyle will have been made beautiful with a garden; and there were votive statues almost everywhere. We may imagine a great park all round the place.

It is a curious thing that, considering the fancy the Pergamenes must have had for terrace sites, which induced them to cut these out of the solid rock, a task for giants—in spite of this, the steps, the approach from terrace to terrace, are not treated as an important feature of the whole. They are dealt with quite separately, as a covered-in, winding stairway, leading up to the highest terrace, but they are of unequal width and almost hidden from view. This observation is of striking importance with regard to the evolution of the villa garden of the ancients in its relation to the Italian Renaissance.

This gymnasium is certainly one of the most imposing of all antiquity. The powerful work and the "pathetic" art of the great Pergamene style here finds its noblest expression. It appears in striking contrast with the little strictly measured-out plans of the Delphic gymnasium; and chance has so willed it that a very early example of a terrace site has been preserved, whereas up to this time we have no satisfactory specimen of the large flat gymnasiums of the Classical Age.

In Greece we also find for the first time one of the public gardens that arise from a democratic rule. A site was selected with the sole purpose of providing people with shade, cool springs and nicely kept paths and seats, partly for the good of their health, partly (as we learn from the Platonic dialogues) in order that in taking their walks they might meet for conversation, and then sit down comfortably on a seat. All the practically useful functions of a garden were excluded, though usefulness had been the leading aim in Egypt and also in the temple grounds; and there was no idea of a hunting-field, such as was essential to the Oriental paradise. So here we can first behold a pleasure-garden in the modern sense of the word.

Soon, however, people were not satisfied with the public gymnasium. Precisely at the time of that great change in the life of Attic citizens, that strengthening of their municipal life, we are told that members of the higher class were no longer content to hold their practices and to meet one another in the public gymnasiums, or to use the public baths. So they began to separate from the Demos, and to set up their own private gymnasiums and baths. The philosophers were in the forefront of this movement, perhaps from the comprehensible reason that public places were too noisy for their meetings and too liable to be disturbed by crowds.

We hear first about Plato, whose name is closely connected with the Academy, where to begin with he used to teach, that he moved his school into his own garden. He possessed a piece of ground quite close to the gymnasium, and made it over to the school; it was in their possession for centuries, until in A.D. 529 the Emperor Justinian annexed it. It was called the Academy just the same after this change, and so was the gymnasium: thus the name applies to the school as well. The garden must have included several buildings: there is talk of an Exedra and a Museum. Plato himself and several leaders of schools lived there entirely: under one of them, Ptolemon, the pupils made for themselves little arbours or huts so as to live quite close to the Exedra and the Museum.

One generation later Epicurus bought a large piece of ground for eighty minas within the walls of the city close to the lately built Dipylon gate, with a view to making a very fine garden for himself and his scholars. In lofty contempt for outsiders he withdrew here with all his pupils, men and maidens. Pliny in a sort of way ascribes to him the invention of the town garden; but in a polemic against a despised sect he makes use of the fact that it was within the walls that Epicurus had his garden. He desires to impress on the Romans, who in his day under the name of gardens owned large pleasure-parks, that they were taking as example this Master of Leisure (*magister otii*), because, he continues, "hitherto it has not been the custom to bring the country into the town." Unfortunately we have no information about the garden of Epicurus.

At the same period Theophrastus made a garden, which he bequeathed to his school, "for friends who will meet there and discuss philosophy." The garden was close to the Lyceum, and Theophrastus annexed this after Aristotle (who once taught there) had left, naming him as his successor. From the will we learn that in this place there was a sanctuary of the Muses, and a hall containing maps for teaching geography, and also a statue of Aristotle; further, that a second hall just as fine was to be built, and that Praxiteles had been commissioned to make the statue of Nicomachus which was to be put there. He entrusted the care of the houses, the garden and its walks, and the memorial that he desired to have erected, to his philosophical slave Pomphylus.

Gradually it became the custom that every head of a school which seemed important enough should have his own garden. King Attalus I. presented to Lacydes at the Academy a garden afterwards named Lacydæum. Lucian, who found all the philosophers mindful of their gardens, said it was a shame that Socrates should go short; so he makes out that in the Isles of the Blest he is rewarded for his services with a garden, which he calls a Paradise. This garden Socrates calls the Academy of the Dead, and there he is wont to parley with his friends.

How much it became the fashion in the time of Theophrastus for rich men to make playing-grounds, or a palæstra, is shown by a joke of his about the "Pleasure-seeking man"; this person offers to the philosophers his little palæstra, and then at their meetings he walks in last, so that everyone may say to his neighbour, "There is the owner."

It is true that ever since the Peloponnesian War, when the oligarchy was more strongly opposed to the democracy, as its unruliness increased, there had been a falling off of enthusiasm for gymnastic sports. At any rate this was so among the rich and important citizens, whether they carried on the sports in their private places and avoided the public ones, or whether in Greece the "Games man" was retreating more and more into the background. In either case we understand the accusation brought by Andocides, the orator and leader of the prosecution, against Alcibiades, that he was a corrupter of youth and a destroyer of the gymnasium.

Here also in these gymnasiums with their sanctuaries, their tombs, their walks well supplied with seats and statues, their stadiums, and their hippodromes with avenues of trees, we ought to look for the beginning of garden craft in Greece. It is true we can only grasp its import by passing beyond Hellenistic conditions to the garden's highest development in the splendour of the Roman villa. Only there, where beside the immediate cult of the Hellenistic garden in the literary circle that centred round Cicero in the classical period, do we find a conscious harking-back to the garden of a Greek philosopher—only there can we see with perfect clarity the real Greek elements of a later horticulture.

But sometimes a mighty impulse came from outside. The raids of Alexander the

Great opened to the Greeks the whole of Asia at one blow, and all its elaborate garden culture. Before then individual persons had extolled the marvels of Oriental parks, and the Greeks of Asia Minor, living so near to Persians and other Oriental peoples, were far ahead of the Mother Country in the love and care of gardens. Darius I. had praised a satrap for transplanting fruits into Asia Minor from the other side of the Euphrates. After the Persian wars there arose a lively interest in parks, even in Greece itself. If it is true (as Plutarch alleges) that the satrap Tissaphernes laid out a very lovely park in honour of Alcibiades and gave it his name, he could obviously get no result from his courtly flattery unless he could assume that Alcibiades would feel an interest in the park and an affection for it. Xenophon, too, describes the Persian paradises that he saw on his march, with the confessed intention that his fellow-countrymen should have a pattern before them which they could and must imitate.

Alexander himself brought his passionate love of hunting from his wild forest-home to the countries he conquered. In India he found groves and woods shut in by walls, well watered, with plenty of big game, but he shot his lions from a raised platform. The paradise of Bazisda was so rich in game that four thousand head could be brought down to serve as a feast for the whole army. On his return from India Alexander crowned his men with ivy, and as this plant did not grow in the parks of Babylon, he ordered it to be transplanted. Harpalus, to whom he had entrusted the care of the park, had no success with ivy, though he did wonderfully well with the acclimatisation of Greek plants, making the lime, and after much trouble even the box, at home there.

Alexander throughout took pains that the parks should be well looked after, and when he found at Pasargadæ on his way back that the tomb of Cyrus in the royal paradise had been neglected and robbed, he punished severely the Magi, the hereditary guardians. With the wish to see the gardens of Semiramis, at the foot of the mountain of Bahgistan, he made a detour on his march from Celonæ to the Crack Willows of Nyasa. The king loved, as Oriental princes did, to do his business in the paradise; and to receive his generals there: he sat on a golden throne, his counsellors on silver-footed stools. And when he came sick to Babylon, he had himself carried to the great park on the far side of the Euphrates that he might die there.

The successors of Alexander always felt that they were the heirs of great princes of the East, and loved to cultivate the luxury of gardens and fine parks; and they tried for the joys of acclimatising foreign plants on a large scale. The Seleucids planted the cinnamon in Syria, the anemone and spikenard in Arabia. Finally Egypt, being now just as accessible to Greece as Western Asia was, contributed a new splendour to the wide parks and gardens with her own experience of flowers, now some thousands of years old. The Greeks were astonished to find fresh flowers growing the whole year through, and under the Ptolemies cultivation was encouraged by the extravagant demand for flowers at their festivities. No longer were people content to have them for head or breast, but the hall itself had to be dressed out with flowers in the Egyptian style. At a great feast which Ptolemy Philadelphus gave in the winter he had flowers spread over the floor of his tent in such glorious abundance that it really looked like a heavenly meadow.

Besides the Oriental influence, there was another thing that encouraged the gardening art—the development of large towns in the Hellenic world. It was difficult for the inhabitants, with the ground walled round and thickly built over, to get to any open land;

and so, as they became richer, they were impelled to seek Nature, who had become more and more of a stranger, by bringing her into the towns and making town gardens handsome and large. It was not really a new thing for the Greeks; but it was carried out now on a large scale.

In Alexandria the public gardens and royal gardens together occupied one-fourth of the whole area of the town; they were all connected with each other, and situated both inside and outside the walls. Especially do we hear of the garden of the Museum, where scholars were wont to walk, of the Great Gymnasium, and of the grove of the Dicasterion which was in the middle of the town. South of it was the Paneion, a curious place: there was an artificial hill with corkscrew paths to go up by, and from the top one enjoyed a panorama of the town. Nothing is said about trees there, but the pleasant walk is an argument for them, since similar places of a later time and even in our own day are always provided with shade on the slope whereby we climb to see the view.

In the suburbs and the outer parts of the towns abundance of the most beautiful gardens lay. Strabo saw near the Necropolis many gardens in which family tombs were to be found.

In another way also the Oriental fashion affected and extended Greek usage. At one time the Greeks set up a grove round the specially honoured graves only, in order to protect a sacred spot from desecration, and in the classical period such graves were in the gymnasiums or the parks, or other places where the Greek games were held. Also the resting-place of Athenian victors, the Ceramicos, which Cimon moved from the Dipylon to the Academy, was planted with avenues of trees. Although it is stated that Herodes Atticus lies buried in the stadium, we must not look for his grave in the racecourse itself, but in the park grounds on the hill which belonged to the stadium, inasmuch as people named the whole estate after the game-place, as in the case of gymnasiums; and in the Academy at Athens a monument was set up to Philiscus the sophist. It is even uncertain whether Plato is buried in the gymnasium: it is far more likely that by *Academy*, where the grave was found, Diogenes Laertius meant Plato's own garden. The pupils who later paid Heroic honours to their great master will have kept the grave in the garden of the school. About the grave of Theophrastus this is quite certain, for (as was mentioned before) he enjoined on his friends to keep a suitable place in the garden for his own tomb. Thus we see how the philosophers set an example for family graves.

No wonder that in Hellenistic times the rich devoted some specially-cared-for bit of their lovely parks to the purpose of a family burying-ground, or else made themselves graves in their own separate plot. Later on—but there is no proof of it till the fifth century A.D. —a word was coined for graves that had gardens round them, the word Cerotaphion.

Antioch was even more renowned for lovely gardens than Alexandria. The town had a wonderful site. The main street was a long continuous portico, with houses on one side, and on the other side gardens that extended right to the foot of the mountain: they contained all manner of summer-houses, baths and fountains, which, however, did not appear on the hill, "so as not to destroy the impression of regularity." A second street reaches to the Orontes. In front of the West Gate the road passes through a wonderful suburb called Heraclea; farther on one gets to the famous park called Daphne after $7\frac{1}{2}$ kilometres of vines and roses on the south of Antioch. From this park the whole great estate took the extra name of Epidaphne. This too was originally a sacred grove which enclosed the

Τέμενος, the asylum with the temple of Apollo and Artemis. The great shady park had a circumference of eighty stadia. The springs here were more abundant than the earth had ever beheld, and the wonderful cypresses, three hundred in number, were, according to tradition, planted by the Seleucids.

Later generations could not say enough of the beauty of the baths, the portico, the places of amusement. The crowds found excellent hostels where vines grew even in the rooms, while in the gardens were wafted to them the aromatic odours of the flower-beds. Daphne is praised as the fairest spot on earth.

Syria is reported to have developed the art of gardening with special success. Pliny, too, reports that the Syrian gardens are the most perfect. Next to Daphne the park most admired was that at Batnæ: the Emperor Julian visited it, but found only a useful fruit-garden, and in the middle of it flower-beds and vegetables. Round about there were other trees planted in regular order, and the emperor specially praises a fine cypress-grove full of well-formed trees.

But it was not only the Greeks in Asia who thought it important to adorn their towns with gardens: in the much smaller places in the Mother Country there was an attempt to follow their example. In descriptions of Greek cities which we get from Heracleides in the third century much is said of gardens, and he especially praises Thebes, which is the best of all. In summer-time the quantity of water and the greenness of the hills made a wonderful impression.

In Hellenistic times the private house seems to have been a very modest affair in comparison with public places. It is true that Demosthenes grumbles that private buildings are better than public, but this is only in relation to past times. There may no doubt have been certain individual gardens on a larger scale, such as those of Plato and Epicurus: in the fourth century a man of means made himself a garden when his neighbour gave up his ground, but this would be an exception. One need only look at the plan of a town like Priene, that was newly founded in the fourth century, to see how incredibly small were the Greek dwelling-houses; but all the same it is to these that we must look for the chance of developing fine court gardens at a later time.

The Greek house is really a court-house, i.e. the living-rooms face on the open court, on one side of which is a hall supported by pillars. Out of this originated the peristyle found later on in all the good houses, a court with a portico all round it. In most private dwellings that have as yet been dug up this court is paved, so that it cannot have been planted with flowers, although the Homeric Age did know garden courts. But what was possible for a royal palace was for a long time to come beyond the possibilities of the modest dwellings of private citizens. In Priene itself all the courts are paved. The statues discovered in the courts of certain well-to-do houses, especially in that house in the Theatre street where a bearded Hermes and a lion in terra-cotta were found in what seems to be their old place, lead us to believe that there were also pot plants between the statues, put there to make the court a pleasant place in which to stay. Another house at Priene may possibly have had a strip of garden, for there is a pretty portico opening on to a narrow strip which most likely was planted as a garden terrace. And the so-called House of the Priest at Olympia distinctly proves that in rather larger places the court was planted, for here traces are shown by excavation. It is obvious that as the affluence of the upper class and their love of private life increased, the peristyle (expansible at need

FIG. 56. THE FARNESE BULL—A PARK GROUP

as it had no roof like that of the Roman atrium to limit its size) would be made use of for plants.

We can see this development and also its contrast with the Italian atrium house which entirely forbids the possibility of such gardens, and can really understand it when we study Pompeii; but this is to stray too far into the days of the Roman Empire. Still, as we might expect, Hellenism had already converted the peristyle into a fine garden court, and what excavations cannot yet tell us we must get from literature.

In the time of the Diadochi, when the great influence of the East was again making itself felt, these court gardens on which the state rooms opened were set up very luxuriously. Even a temporary erection (apparently put up just for a feast), as for example the Great Tent of Ptolemy Philadelphus, erected by him outside the city of Alexandria, was supplemented with fine garden courts. The huge dining-room, arranged with all imaginable expense for three hundred guests, opened out on the peristyle. The walls of the portico were hung round with rugs and choice skins, but the open space in the middle was planted with box and myrtle and other shrubs, and flowers bloomed gaily between. Up to the tent marched a procession, with various chariots representing the triumph of Bacchus. On one of these Semele the mother of the god was carried. She lay in a wonderfully devised grotto overgrown with ivy and other creepers, with doves fluttering around her. In the actual grotto were two streams flowing with milk and honey. Then above the colossal statue of Bacchus was a canopy of vine and ivy and autumn fruits.

Grottoes of every sort and kind were made. Pliny tells of a plane, a tree which liked to have its roots near a spring, that it grew in Lycia to such a huge circumference that a grotto was made in the hollow trunk, decked out with pumice and moss, so large that eighteen people could dine in it. And in a park such artificial grottoes were often made. Even in a modest park (described by Alciphron) in the neighbourhood of a Greek town, possibly Corinth, there stood an ivy-clad rock grown over with laurel and planes and encompassed with myrtle. At intervals nymphs were stationed, and behind them stood Pan gazing at the nymphs. With all this ornament one may safely guess at grottoes inside the rock. The rest of the garden arrangements at this villa are quite modest —only groups of cypress, myrtle and a small flower-garden.

To what an imposing park may we fly in fancy if we would behold a grotto-hill in the lands of a wealthy Rhodian, on whose summit there stands a work of art like the Farnese Bull (Fig. 56)! Only when one tries to imagine it thus is it possible to see how this work (which in the Museum appears in spite of its huge size to be capricious, unrestful and even bewildering) acquires life and significance, for here the grotto-like plinth is shown by the artist as it emerges from the greenery around it. All the outside detail—herdsmen, animals—can be understood in the proper natural surroundings, where, though the eye sees them not, the central group is strikingly exalted beneath the open sky.

Such gardens, of which this is only one type, were used by the rich as a background and theatre for their gluttonous feasts as well as for their daily life. They had become indispensable to Hiero II., who probably links up with an unbroken tradition of princely gardens in Sicily. Like most Eastern monarchs, he conducted all State business in the garden, nor would he part with it even on the water. He had had built a gigantic pleasure-ship that was to exhibit to the eye both his magnificence and his glory as a warrior, and on this ship he put a garden. The vessel was built by Archias of Corinth, and Archimedes contributed very remarkable devices to it. On the upper deck a gymnasium and walks proportionate to the dimensions of the ship were arranged. Here were beds containing all sorts of plants confined by strips of lead. The walks were shaded by ivy and vines, their roots getting nourishment from pots of earth, which were watered in the same way as the beds. In addition to all this there was an Aphrodision, its floor interlaid with precious stones and agate. It was profusely decorated with pictures and statues, but it had only room for three chairs. Another equally elegant garden house had a small select library. Where the two basins, of which we are also told, were placed, cannot be determined. One that was filled with sea-water and contained fish must be supposed to be open and on the upper deck, probably in connection with the garden.

It is quite evident that Hiero made in the scanty space of a ship's deck a little garden of costly workmanship. The Greek architect has made use of all the constituents of the garden as we know it. And incomparably the greatest are gymnasiums and promenade paths. These "peripatoi" again show the connection, not of the paths alone, but of all that portion of the garden designed for pleasure-walks. The beds in their leaden frames can have held only flowers and such plants as have short roots; ivy and vines, whose roots strike deeper, were put into tall pots. The pergolas made by these give the shade required, which otherwise, in the lack of lofty trees, the ship could not have enjoyed. Two fine pavilions and a fish-pond complete the whole, showing once more the intimate alliance between gymnasium and pleasure-garden. The first must be thought of as a small place, perhaps surrounded by a light portico and lying among the flower-beds.

In this manner we obtain a picture of the way in which the private gardens and parks of princes and other great men in the wide tracts of country now open to them were laid out. Unfortunately we have no definite description of a villa garden extant at that time; here, too, we have to reach our conclusions by working backward from the days of Roman emperors, where the streams flow more abundantly, and so we arrive at the Greek culture which was the mistress of Rome.

It is only in very late Greek authors that we find in a natural return to the past a description of gardens, that is to say in the love-stories of the rhetorical sophists and romancers, who flourished from about the second to the sixth century A.D. But though these tales come from so late a time, they are really imitations of a kind of poetry that began in the Alexandrine Age with the Idyl. Just as the idyl takes the open country for its background, so does the romancer love to take the garden. The constant repetitions and stereotyped phrases of such descriptions point to prototypes from a much earlier art, and from another side lead us to look at later imitations in the Byzantine epoch, indicating that garden culture, with its unbroken tradition from the antique world, has been further preserved in the Byzantine Middle Age. To be sure, the garden in Greek love-stories presents an unusual scene. It is not the public park, neither is it the seat of a magnificent prince; it is the rural farm-garden, and only now and then can we see the rare traces of a larger scope of interest. Such farm-gardens did as a fact always exist in Greece: the classical example being the garden of Alcinous.

As conditions must always be determined by reasons of utility, there has been little change at any period. Longus does certainly call that garden a paradise which Daphnis made with the aid of his foster-father Lamon, watering it and getting it ready with all possible care and pains for the visit of their landlord; and the word paradise generally implies at this date an ornamental garden rather than a garden for useful products —in other words a park. In the story it is "after the fashion of kings," one stadion in length, with a lovely view over the plain, whence one can look down on travellers and on the sea and the ships as they go by, and so the view is really a feature of the garden.

Longus knows very well that such a site as this is demanded in a Hellenistic garden, and even in Homer the garden of the Phæacians stands high or else on the slope of the hill. In the middle rises a temple, and beside it an ivy-clad altar to Dionysus, and this gives an artistic unity to the picture which is absent from the garden of Alcinous. The greater part of it is naturally occupied with fruit-trees, in kinds not much more numerous than Homer's, and between the trees are vines. The garden is enclosed by rows of cypresses, laurels, firs and planes, all planted at equal distances and entangling their higher branches one with another, with ivy also growing like the vines between the other trees. Flowers, too, are not forgotten: there are always the same kinds: roses, violets, hyacinths, narcissi. Like Homer, Longus requires of his garden that it shall bring forth flowers at every time of the year. In spite of the insistence on park-like surroundings, this is distinctly a garden for use of the same nature as that other one which Longus thus describes: "I have called my own a garden, wherein I may rest in the leisure of old age; it yields a harvest according to the season of the year: in spring roses, lilies, hyacinths, and both kinds of violets; in summer pears, and all the apples; and now the vine, the fig, the pomegranate, and the green myrtle. One might fancy it was a sacred grove."

And just as it appears in the story, so is it described by the theoretic writers of that day, whose works are collected in the Geoponica. Close by the house, we learn from this, is the proper place for a garden, not only for the enjoyment of its beauty, but that its sweet smell may fill the air and make the home healthy; for, as we learn elsewhere, the neighbourhood of a garden is most necessary for health, as it helps a convalescent to recover.

Plants must not be mixed up or put in without order as though the differences added to their beauty, as some say, but the individual ones must be carefully separated from one another, in order that the larger kinds may not rob the smaller of their nourishment. Between trees may be set lilies, roses, violets and crocuses; these by their look, their scent, and their uses, are most pleasing; moreover, they increase the revenues of the place and supply food for bees. So order in arrangement is the first requisite, that same order displayed by Egyptian gardens thousands of years ago, and in Homer's sterner pictures. The garden of Alcinous is the pattern for the romances; with passion they cry, "Blest were the Phæacians, not because they sprang from gods, but in that before all else they prized their gardens." Thus begins one description, and the phrase, "I feel as if I were in the garden of Alcinous," often occurs.

Most of the stories show a kind of architectural unanimity in laying special emphasis on the middle part. In Longus this is a temple and an altar, and in most of the others there is water set up in different ways. Sometimes it is a fountain splashing into a square basin of clever design appearing among the flowers. This water acts like a mirror, showing us two gardens, the real one and its reflection. A portico runs all round the garden, and between the flowering and shady trees are set ivy and vines trained on sticks: the flowers, here again the same kinds, in their loveliness vie with the songs of the birds.

Though the gardens are described as very much alike, the water shows a pleasing variety: first we get a foaming brook, swifter than the wind, then perhaps a crystal bath wherein the lovers bathe; forth from this dance from shell to shell tiny cascades with flower-petals floating on their waters—a constant scene of beauty.

From this type of romance, which for centuries to come was to reach deep down into the Byzantine Middle Ages, there is nothing to be gained in the way of garden tradition. We learn nothing new, only that the Byzantine gardens developed in the same direction as the Hellenistic. And in particular features we may trace the recurring strong influence on Byzantium of the East, and particularly of Persia and Mesopotamia. In all the earlier accounts we notice simplicity and moderation, something of the imperishable Greek feeling for the golden mean, μηδὲν ἄγαν, appearing in climes far distant. But after that the love for the glories of Asia presses to the front ever more insistently, to find its satisfaction in the multiplication of ideas, the glittering pomp of costly possessions and the creation of elaborate masterpieces.

CHAPTER IV

THE ROMAN EMPIRE

CHAPTER IV

THE ROMAN EMPIRE

THE Romans inherited the villa gardens which had been developed in Hellenistic times; and they came into possession just when country life —which in an earlier day was cut off from the Greeks—was the dominant idea all over Italy. Country life—*res rustica*—was the distinguishing, if not the one, cry of the Roman gentleman. Only in passing did the Romans visit their town houses; and after Rome had moved in an utterly different direction, and they had almost completely forgotten how their town started as a market-place for country farmers, they were yet unwilling to abandon their old allegiance to the soil, their piety towards the land of their fathers.

With what pride does Cicero, the busy lawyer, insist that the villa where he stays at Arpinum is his real home! Certainly the Tullian family had lived there from remote times; his sacred places, his own people, and many traces of his forefathers were there. So, when his friend ventured to object that he did as a fact live in Rome, it was with a full consciousness of the nature and growth of the town that he answered: "True, I and every citizen, we all have two homes: one comes by Nature, the other by the State: if Cato is a native of Tusculum and adopted by the Roman State, he is by nature and family Tusculan, but as regards the State a Roman, and he has one home in locality, another in law."

The original farm properties were not called "Villa," but "Hortus," in the Twelve Tables of the Law; and it was on the revenues from arable lands and enclosed gardens that the wealth of the Roman depended; it was in fruit and in vegetables. And so in earlier times it was for the sake of the State that good cultivation of country property was insisted upon. We have it on Cato's authority that people were punished for neglecting their farms, or letting them get dirty, and for not taking proper care of their tree-gardens. Buildings were unimportant except as giving cover for fruits, animals and men; and Cato's dictum is always the same, "First plant, then build." There is very little known about these early villa sites; but the aged Cato plainly shows his disapproval of the strong tendency to luxury in his day, when he insists that his own villa must be quite simple and unadorned, and must not even have the walls plastered.

But the villa of the man who first adopted the pure Attic style, and through whom its light spread into his own land—the country house of Scipio—was always kept in the fashion of his fathers, showing no trace of the Hellenistic manner of life. Seneca paid a visit to Liternum near Cumæ, where the great man spent his last years, and found his house built of freestone; it must have given the effect of a knightly castle of the Middle Ages, with its ramparts and towers. The park was enclosed by a high wall, and beyond the buildings and the court there was a very large pond, big enough to serve a whole army; but the bath was narrow and dark in the old style, "for our forefathers thought that if a place was to be warm, it must be dark."

If we want a picture of the outside of Scipio's country house, we must turn in thought to the early Italian villas which lie on the borderland between the Middle Ages and the Renaissance, and we shall find that the Medici Villa at Caracci, before it was first rebuilt, and still had a battlemented roof, and lofty towers also battlemented, and bulwarks, and a garden with a high wall round it, was not unlike the villa of the Scipio family at Liternum. Seneca also says of the villas of Marius, Pompey, and Cæsar, which were on the high ground above Baiæ, that they looked less like villas than castles or even watch-towers. These seemed, moreover, by their warlike aspect, to have lost Scipio's idea of simplicity; at any rate the villas of Marius near Misenum were considered too magnificent, too foolishly grand, to befit the warrior who owned them.

Marius was really of the generation whose manner of life was denounced by Seneca when he compared it with Scipio's. "Who could endure to-day to wash as Scipio washed? He would feel poor and squalid unless his walls sparkled with precious stones, if he had no Alexandrine marbles or Numidian inlays, if his windows were not glazed. The baths of to-day are called bat-holes, unless there are wide windows to let in the full daylight, so that bathers may get a combined view of sea and land."

In the very last days of the republic the luxury of the villa life was much greater and more common. In Cicero's family, for example, the generation before his had broken with the tradition of the dark and narrow homestead. Cicero, when he first expressly describes his villas, gives credit to his father's love of building, in that the family house at Arpinum was now a comfortable bright home, whereas in his grandfather's day it had been a small villa such as their ancestors had lived in, and such as could still be seen in the villa of Curius in the Sabine hills. But Cicero himself supplies the best example of that passion for building and buying which had taken hold of the Romans.

One early result had been the separation at this period of the Villa Rustica (farm) and the Villa Urbana (gentleman's house). The villa urbana indicates not so much the place in the town or the neighbourhood as an elegant residence, and the term is best translated by the French name, *maison de plaisance*. Cicero's villa at Arpinum was originally nothing but a villa rustica, and only acquired a more urban character from certain alterations that he and his brother made.

Vitruvius passes over the plans of the villa urbana in a very superficial way, not so much because there were very few of them, as because he gained all his knowledge from textbooks, which had not yet treated of any particular style. But about the villa rustica we have the most precise information, not only from Vitruvius but from all the authors who, like Varro or Columella, wrote about farming, wherein are included farm buildings, kitchen-gardens, and places for vines, fruit, and vegetables. It is true that very little is said about the ground-plan for a kitchen-garden, but there are careful descriptions of the kind of fencing used. It is easy to see that for practical reasons the garden must be next to the villa, and that it is important to have a well or a stream for irrigation. The planting of trees, shrubs, and vines in quincunx formation seems to have been a specifically Roman plan, at any rate Greek writers never mention it. Xenophon too, in the passage where he admires and extols the regular order of planting in an Oriental park, is only talking of rows; and Cicero naturally interprets this as the quincunx order, as he has no other in sight. Pliny describes it as "universal and necessary."

Not only in the time of transition, but also later on, a mixture of the two types was very common; and Columella says that a complete estate requires a third department, the "villa fructuaria," in order to include the different kinds of storehouses. But we shall find from the description of Laurentinum by the younger Pliny that these useful granaries were occasionally to be found associated with elegant villas, and close to the living-rooms.

A striking feature about Roman property is the way in which a man's belongings were scattered about. Cicero, in addition to what he inherited, bought in his lifetime seventeen other estates, which were partly made by uniting smaller garden grounds into one whole; and in this he was by no means exceptional. We may find different reasons for it: the chief estate of a Roman, originally his only estate, was still the one place of importance; and to statesmen it seemed so fundamentally important for the welfare of the people, that when it was not adequately cared for by the owner it had to be protected by a decree of the senate. Rich men bought such scattered properties partly so that their revenues might not be affected by bad weather, partly to get for each product the place that suited it best. But there was another reason, and not least in importance—that they might enjoy the various beauties of mountain, undulating lands, and sea. They wanted, moreover, to be sure (when they were travelling) of finding at certain stages a roof of their own under which they could spend a night. It came about, that with all this scattered property the owner could not himself look after the farms, still less work them; and so the whole management of the villa rustica was entrusted to a villicus, or bailiff, and the owner could live at the villa urbana undisturbed.

The Romans had seen from the beginning, and with a sore conscience, that this turn of affairs could not last long; the feeling was very deep-rooted in the best of them, that the dislike of agricultural life was the beginning of the end for the Roman people. Quite endless are the warnings, the moral teachings, the satires, that are levelled against luxury in houses, gardens, and villas. It was not everyone who could boast, as Cato could, of personal renunciation; but everyone was eager to put forward that view which Pliny condensed in the famous phrase, "Large estates have ruined Italy." Whether he was a writer about farming, like Varro and Columella, or a naturalist like Pliny, or a poet like Horace, or a moralist like Seneca, every man was a *laudator temporis acti,* and all preached on the common theme, "Greatness and Simplicity dwell together."

We learn, however, from this contemporary picture of superabundant luxury, that horticulture, from the later republic onwards, advanced with giant strides, and once more the name of Cicero is associated with the first stages of this development. We do not indeed possess a detailed description of any particular villa of his, but many of his dialogues have for background his own or a friend's villa with the garden, and in this way a mental picture may be arrived at. A garden such as Cicero's was quite different from the kitchen-garden of the villa rustica; and what we find is an ornamental site and park land, where he and his friends stroll about, plunged in philosophical discourse, as in the days of old. For Cicero, who was so eager to bring back a philosophical Renaissance in the Greek style, and to prove himself the immediate heir of the great thinkers, betrayed more conspicuously than any other writer that what we have to do with here is the conversion of a Greek gymnasium site into a Roman villa-garden.

It is not only from Pliny that we learn that Cicero had a villa at Puteoli which was specially famous for its portico and park, and which he named Academy after the Athenian model; Cicero himself is continually talking of places so named, which evidently were to be found on all of his estates. In his place at Tusculum there was one part called Academy, and another higher up called Lyceum. At the home of Crassus also, where the greater part of the scene of *De Oratore* is laid—that is, in his Tusculan villa —the gardens are still more plainly copied from Greek models. Even when the friend is made to express a doubt as to whether (in their day) time, place, and human understanding show any real leaning towards philosophy, he has to admit that this portico, these palæstras, and all the other adjuncts of a gymnasium, are bound to incite to disputations; and on another occasion Cicero, strolling in the grove of planes that belongs to Crassus, congratulates himself on the fact that Socrates too had enjoyed the like cooling shade, but that his friend's well-kept place provides him with comfortable seats to rest on, whereas the Greek philosopher had to sit down on the grass. That these gymnasiums carried no arrangements for athletic games is clear from what Cicero himself says; and Vitruvius expressly emphasises the fact that Greek palæstras were not customary in Italy. The Hellenistic philosophers had been the first to convert the park gymnasiums into private gardens; and now the Roman philosophers (centred round Cicero) were consciously copying this later plan.

We also learn much of the ornamentation of these gardens, for their purchase plays a large part in the correspondence between Cicero and Atticus. Indeed, they invented a new adjective, and Cicero begs his friend to send him "ornamenta γυμνασιώδη, if you can find any suitable to the place, which you know so well." In these gardens, following a Greek example, there was a great deal of sculpture. We find Cicero impatiently awaiting a consignment Atticus has promised for the villa at Tusculum; and when a Hermathena arrives, he is enchanted, and it is so well set up that the whole place seems to exist for this statue alone. This villa he seems to have adorned very sumptuously with works of art, for he asks Atticus to send him pictures also, to hang on the walls of the ante-room. Unfortunately the spot where the villa stood has not been identified; and whether the place that goes by the name of Cicero's Villa really belonged to him, is quite as uncertain as whether the so-called amphitheatre beyond is to be taken as part of a garden construction. One might more easily suppose that a place behind the theatre, the purpose of which is not clear, belonged to a garden. It is a large basin with a straight end on the side next the hill, and a round end in front, with several grottoes in its walls. Here one would expect to see a nymphæum, so often mentioned in Hellenistic gardens; and it may be that the amaltheum, which Cicero made in imitation of one that he had greatly admired in the garden of Atticus, was really a nymphæum.

The amaltheum of Atticus was in the park of his country house in Epirus; it was shaded with plane-trees, and apparently had its water from the Thyamis. Cicero asked his friend to send him any literature which he possessed on the subject. It is as a sort of literary interest that Cicero has his amaltheum made, or, in the fashion of his circle and their Greek traditions, it is an imitation made as far as possible similar to a place known by description, or maybe still in existence at that time, which was a sanctuary for the nymphs who nourished Zeus. It is clear from the amaltheum of Gelon that places of that name must have existed for a long time past. As Cicero says nothing more about it, one can only guess that it was a grotto with water running through it, and shaded by trees. What such grottoes looked like in Roman gardens we can see from a fresco at Boscoreale (Pompeii) (Fig. 57), where a grotto is depicted made out of unhewn blocks

FIG. 57. A GROTTO—A FRESCO FROM BOSCOREALE

that are piled below rose-covered pergolas; there is a stream foaming into a much decorated fountain, and falling down into a bright basin; there is green ivy all round the grotto, and there are birds skipping about in the plashing water.

The two grottoes which Seneca saw in the villa at Vatia were of a larger kind, "as striking as two large halls," which may imply that they were lighted from above, as we find them to-day in Italian gardens. The interior was coated with tufa, or pumice-stone, and shells; and the floor, especially the part near the water, was covered with soft moss. The same sort of moss grew on the borders of the basin, in the central tank, and in the houses. For protection against the broiling sun red curtains were stretched across.

At first the sanctuaries of the Muses were similarly arranged, for the Muses were also honoured in grottoes; but when the philosophers assembled for learned discourses, halls were made instead. Thus always have the museums of a philosopher's garden been metamorphosed, though with the nymphæums the case is different. We learn from poetry that they kept up their old character for solitude, seclusion, and cool shade, which had shown itself in Hellenistic times and always remained the same. In Horace's lovely ode, "O fons Bandusiæ," we have a description in the Latin tongue of one of these beautiful nymph sanctuaries.

One great attraction of Cicero's Arpinum was the fine supply of natural water, which he no doubt procured for the sake of the nymphæum. The rushing stream of the Fibrenus flows round a small island and forms a delta before it empties itself into the River Liris. The villa probably stood below the island, but in any case the island belonged to the garden. Cicero had made a palæstra here, as we may now assume, a garden with shady trees and comfortable seats, perhaps like the Platanistas at Sparta around which the Euripus flows. That Cicero had a place here for games is not at all probable, for this was his favourite resort from the orator's point of view, where he retired when he wanted to think or read by himself, and where he brought his friends for the dialogue about the Laws. With this abundance of water Cicero might well laugh and tease his friends when they, who had far grander villas, spoke of an artificial canal as "like the Nile," or "like Euripus." But, when it was needed, Cicero was most anxious to get artificial irrigation, and at the Fufidian farm, a place belonging to his brother Quintus, he wanted both a piscina and a fountain: at another of his villas Quintus had himself given the name of Nile to his canal, but he had no grove and no palæstra, and Cicero advised him to put these in. He paid a visit of inspection to the place when Quintus was absent in the field, and in a detailed letter sends him a report, mentioning various important particulars. The porticoes have a prominent place: they often contain statues, and open on palæstras, or xysta, and other features of a park, which Vitruvius always insists upon. Since people were fond of having the villa at the foot of a hill, and the dwelling-house as a rule somewhat higher up, it was natural to make the garden in the form of terraces; and it is clear that this was already done in Cicero's time from his description of Tusculanum, where he not only talks of the upper and lower part, but also "walks down" into the Academy.

Cicero's words show that it was not a mere tradition—already fading after the time of Hellenism—that bequeathed to this Roman group of educated men the sort of garden that owed some of its characteristics to the Greek gymnasiums, but that they themselves felt a wish to have their gardens just like those of the philosophers in the most flourishing days of Greece. It is quite likely that these men tried to pave the way for the adoption in Italy of the Greek garden style more than other people did; at any rate, none of the other gardens and later villas show so lively a sense of relationship. Be that as it may, Columella reproves the fashion of wanting so many departments in a place, great pillared halls, immense bath-rooms, and almost everything "that the Greeks had in their gymnasiums," also libraries, museums, towers for fine views, ponds, fountains, and waterfalls. So here, too, we have evidence of Greek influence, and after what we have seen in the Ciceronian villas we may assume that garden sites were of the first importance.

The style of Cicero's villa and that of his brother shows the Greek renaissance spirit in a comparatively simple way. The great love of display in Roman gardens was still in the future, but before Cicero's time we hear of individual cases of extreme luxury. Varro says that the villas of Lucullus and Metellus vie with public buildings, and the edifices built by Lucullus in particular are regarded partly with astonishment and partly with dislike. He owned many villas near Tusculum, and these had watch-towers commanding very distant views, countless lovely avenues and pavilions, and he looked on all this as a contrivance (as he himself declares) for flying off any day to another house, "like a crane or a stork." There is even more to tell of the great ponds at another villa, where he introduced a dam, to keep the sea-water for his fish.

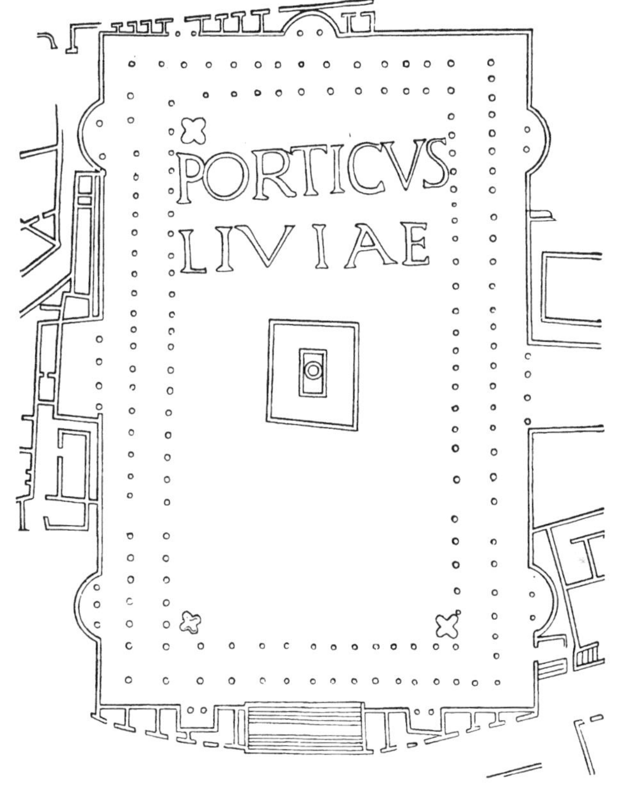

FIG. 58. PORTICUS LIVIÆ, TOWN PLAN OF ROME

The passion for the chase, which came from Greece, had hold of the Romans after the time of the younger Scipio, and hence there came about the construction of parks of a size hitherto unheard of. In Varro's time Quintus Hortensius had already made a park of fifty yokes of land, and enclosed it with a wall, and on this estate he had set up on the higher ground a shooting-box, where he entertained his friends in a peculiar way. He had a slave dressed like Orpheus who sang before them, and then sounded a horn, whereupon a whole crowd of stags and boars and other quadrupeds came up, so that to him who told the tale the spectacle seemed more delightful than the hunt itself.

The more money people had to spend on buildings after the peace of the first empire, the more they strove to produce extravagant homes in the villa centres of the Latin and Neapolitan coast, and on the hills near Rome and in Tuscany. There came a time, deplored by Horace, when the great villas cramped the farms more and more, and instead of elms and olives were planted planes and myrtles, laurels and beds of flowers, enclosed with fine shady porticoes. And although we must not forget that Horace's poem belongs to the rhetorical-poetic type, as the allusion to Cato shows, we still have quite enough evidence of the great increase in the number of ostentatious gardens attached to imperial and private places.

Chief of all, Rome herself in the early days of the Empire was not only surrounded by a circle of flowering gardens and villas, but also within her walls looked like a city of gardens with so many public and private grounds. It sounds poverty-stricken and fussy when Vitruvius asks, in his book, on behalf of the health of the town, that garden sites should be made near theatres and other places of assembly; for one has to remember what an abundance of public gardens the first emperors had already made, enclosing the town as in one lovely girdle. Augustus in particular, who was naturally prudent and disliked luxurious private villas, had a very extravagant villa destroyed that belonged to his grand-daughter Julia; he did not care solely for the rich and great, but also, and far more, for the welfare of the people. When Vedius Pollio bequeathed to him his great showy house, he had this razed to the ground, and in its place, north of where later on were the Baths of Trajan, he made the beautiful Portico of Livia for a public garden.

In the town plan (Fig. 58) there is still to be seen the outline of this famous place: there was a sunk lawn, 115 metres by 75, into which a stairway 20 metres wide led down on the narrower side. In the encircling walls there were niches, alternately rect-angular and semicircular, and in the middle of the space was a great tank, and a portico round the lawn. This simple ground-plan was laid out with plantations of different kinds of trees, and with flower-beds; the paths were shaded with pergolas; Pliny mentions one marvellous vine, which, growing in the open, serves as a shady pergola.

In Strabo's description of Rome he makes a great deal of the fine works of art, and all the sites which he tells of seem to be very much alike. The whole form of the place follows pure Greek examples; in the same way the town markets were adorned, also theatres and gymnasiums. In Hellenistic towns they no doubt had developed in the same fashion, and with a similar object—to conduce to public enjoyment and recreation. Here the people bustled about, and here the men of the city had their rendezvous. Ovid says that one of the most favoured walks for gilded youth was the Portico of Vipsanius, near the Portico of Livia, which had been made by Agrippa's sister in the Via Lata, now the Corso, inside the great Campus, which had been designed for gymnastic exercises. Two others, loved by the rich people, were the Portico of Octavia and the Pompeian Portico, to the south-east of the Campus Martius. But there were grounds laid out in the temple precincts also.

Beside the colonnade, Ovid says, is the sacred grove of Apollo, which Augustus had set on the Palatine, and which was also used as a rendezvous of no very holy kind. During his lifetime he had his own tomb built on the north-west of the town, between the Via Flaminia, northern continuation of the Via Lata, and the Tiber. The under-structure was of white stone, and the inner part of the wall is still to be seen (Fig. 59). Perhaps the grave-chambers were here. Above rose a mound, which ran up to a point, and on the top supported the standing figure of the emperor. As this mound was grown over with evergreen trees, it must have had a wall round it, and have been made either with steps or in a spiral, for only so could it support the statue from below. Round the tomb Augustus made a fine park with attractive walks. The place where his corpse was burned was an artificial mound, in the same way surrounded with white marble and cut off by an iron grating. But during his lifetime the whole place was devoted to the people as a public park.

To such public places as these, rivals could be found in the private gardens of the emperors and of rich men, who were for ever adding to the unsurpassed beauties of Rome. The term "Hortus" was applied only to the villas of the town and its suburbs. The garden was and remained avowedly the chief feature, but in all gardens there was at least one palace to live in, so constructed that even the emperor and his suite could stay there for a time.

If one came by sea, and approached by the Via Portuensis, one came upon Cæsar's gardens. There in the time of his dictatorship he once entertained Cleopatra when the young Egyptian queen visited him in Rome, an episode mentioned by Cicero with some bitterness. When he was emperor, Cæsar had these gardens turned into a public park, and left them by will to the people. From valuable remains, we see how wonderful the place was with its statues and its marbles. Judging by a sanctuary to Hercules that has been dug up, we may guess at an arena for gymnastic games. Close by were the gardens of Antony, and the whole tract of land on the western bank of the Tiber, dominated by the Janiculan Hill, must have been especially full of gardens on the north.

FIG. 59. THE TOMB OF AUGUSTUS IN THE TIME OF THE RENAISSANCE

Of first importance was a public place given by Augustus. In the year 2 B.C. he proposed to present to the people the spectacle of a sea-fight, and with this object had a wide oval basin dug out, 250 metres long and 350 metres broad, with an island in the middle. This remarkable Naumachia site had gardens round it (*nemus Cæsarum*) enclosing the mirror of water. The mimic sea-fight was transferred later from the Campus Martius to the region between the Villa Lante and the Lungara.

Not far to the north is the Vatican territory, and here Caligula's mother, the elder Agrippina, had gardens reaching to the Tiber, and there was a xystus separated from the river by a colonnade. Caligula had a circus made where St. Peter's now stands, and an Egyptian obelisk was set up, which stood in its place till the year 1586, south of the old church of St. Peter, when it was moved into the middle of the new Piazza di San Pietro. This circus became terribly notorious in Nero's time, for it was the theatre of the first Christian persecution, which Tacitus describes so realistically, when Christians were butchered as the supposed instigators of the great fire. Nero liked to practise his music here, and made use of the living victims as torches on his nightly excursions.

If we stroll towards the north-east along the river, we arrive at the gardens of the Domitian family, which also were imperial property, and had some connection with those of Agrippina. Later on Hadrian built his tomb here, much in the same style as that of Augustus on the other side of the river, though there does not seem to have been a mound with cypresses on the top. The great building must have made a strong impression, with its marble and bronze surround, 120 metres in length, and all about it avenues, old trees, colonnades and other garden works. The palace was probably a very fine one, for it was Aurelian's favourite residence. These gardens were still in existence in the fifth century, possibly because they were over the imperial tomb.

On the other side of the river was the Field of Mars with its innumerable gardens and buildings. In the Villa Pubblica, where the State Bailiff lived, and where foreign ambassadors were received, a garden was kept up later on, in memory of past times. Here Pompey chose magnificent sites for public gardens. The first real theatre that he built had pillared halls, no doubt for people to wait in. He laid out gardens. But he also owned private gardens, and the southern part of the Field of Mars in the last years of the Republic was a quarter for elegant villas; the elder Scipio had one, and the Gens Æmilia had extensive possessions. Had not Cæsar's vast building plans been put a stop to by his death, this part of Rome would have been entirely metamorphosed by diverting the course of the Tiber. But Agrippa had at least a certain part of his plans carried out, and the buildings of the Pantheon, in particular the baths, were not without beautiful gardens. There was a large pond in a grove, and Nero would row about on it, after his disgraceful orgies, on a boat made of ebony and gold. Once more we hear of a Euripus! On the other side of the Via Lata we find the Campus before mentioned, with its colonnades, and between them thickets of laurel.

The garden hill, which went by the name of Collis Hortorum all through the time of antiquity, lay at the north of the town, and the good air and fine view naturally enticed rich people to build. Lucullus had one of his famous gardens there, which later on passed into the possession of Valerius Asiaticus. But their beauty attracted the greed of Messalina, and by her wiles she persuaded her weak husband, the Emperor Claudius, to condemn Valerius to death. The lot decided that he was to kill himself, so he went back into his garden and opened his veins; but while he was dying he tried to find a place for his funeral pile, so that the flames should not spoil any of his old trees. Messalina did not enjoy the property very long, for she was put to death by order of Claudius.

These gardens seem to have covered most of the southern shoulder of the hill. On the north-west were the gardens and tomb of the Domitian family, where the ashes of Nero were laid beside those of his mistress Acte and some of his freedmen. Still farther north, probably where we now find the Passeggiata Pubblica and the Villa Medici, were the gardens of the Acilians. It is not certain whether the little octangular building, hidden under the hill of cypresses at the Villa Medici, is part of this garden or not; and there is the same uncertainty about an imposing exedra, of which there are traces found in the garden of the cloister at Sacré Cœur; both erections showed important remains even in the sixteenth century, as Bufalino's plan demonstrates. Perhaps the exedra is the end of a colonnade, which here overlooked the valley. It was only at the end of the fourth century that the Gens Pincia possessed this hill, and gave it the name it still bears, Monte Pincio.

The gardens that were laid out by Sallust the historian in the valley between the Pincio and the Quirinal near the Porta Salaria, paid for with the treasure that he had extorted in Numidia, must have been wonderfully large and extensive. Cæsar seems to have owned a place previously which Sallust bought and greatly enlarged, spending much money on adorning it. It remained in his family till the year 20 A.D. and then was added to the emperor's possessions. With Vespasian this place was a special favourite. Nerva died there. Aurelian preferred it to the Palatine to stay in. He either built or decorated a colonnade there, which was a thousand feet in length. This served as a riding-course for the emperor, and every day he rode on it till he and his horse were both worn out. Probably these colonnades, being the centre of the valley gardens, were planted with avenues, such as we shall find later in places of the kind. Terraces, perhaps with buildings on them, mounted to the top of the Quirinal, and an octagonal one with a dome has been brought to light. On the north side also there would be the same sort of terraces with pillars and houses in groups; one obelisk which stood here is now set up in the *piazza* in front of the church called Santissima Trinità del Monte.

On the Quirinal Atticus had a house described as beautiful "not so much for its gardens as for the way they were placed." South of the Quirinal, we come to the great Esquiline estates, which were probably covered with villas and gardens, especially alongside, and on the other side of, the Servian Wall. Mæcenas was one of the first to build here, attracted by the view from the hill towards where now is Santa Maria Maggiore. Horace praises this high, airy spot, the fine views of the Sabine and Alban mountains, and the walks on the high Servian Wall. There was also a tall building, reaching, it seemed, to the skies—possibly that very tower where Nero, according to Suetonius, stood to watch Rome burning, declaiming the Fall of Troy—and a large swimming-basin, of which traces can still be seen.

There is another building which may or may not have been part of these gardens, called the Auditorium of Mæcenas, which was unearthed south of the Porta Esquilina: it is one of those half-underground garden-rooms of which we shall learn more later on. On the narrow side to the north there is a semicircular exedra, with seven marble steps. The walls are ornamented in their lower part with pictures of gardens. People have been inclined to think it was a forcing-house, but it was one of the cool rooms for use in the hottest hours of the summer.

In the time of Mæcenas the Esquiline was the poets' hill; in the shade of their great patron, Virgil and Propertius also built, and later on Martial, and the younger Pliny. Mæcenas left his house and garden to Augustus, and Tiberius lived there after his return from banishment. Nero, moreover, made of this estate the boundary point in his monstrous building scheme, wherein he would unite the Palatine with these gardens; even before the fire he had entertained the idea, and begun new buildings, making use of old sites, and calling the whole complicated creation his *Domus transitoria*. But this was still all in the air, and it was only after the great catastrophe which laid this quarter of Rome in dust and ashes that his ideas really began to take shape. With the aid of his two architects, Celer and Severus, he proposed to cover with gardens and buildings a space of about fifty hectares. Concerning the luxury of this place the later writers are wont to speak with a sort of inward shudder. The chief palace, known as the Golden House of Nero, faced towards the Forum. Where the ruins of the Venus Temple and the Temple of Rome now are, there stood a large three-storied colonnade called *Miliaria,* and in the middle of it (perhaps in an exedra) the colossal statue of Nero; and again magnificent showrooms.

The gardens extended down the hillside to the valley of the Colosseum. Where the amphitheatre is, there was an immense oval pond, in which were reflected other palace buildings, so that it all looked like an actual town. One main wing, excavated under the Baths of Trajan, has a colonnade in the middle in a trapezium shape, in which there was probably a xystus, with pillared courts to right and left containing chambers which were completely cut off at the back, ending in a covered portico that was carved out of the solid hill. Fishponds and the remains of other constructions are found scattered over the whole neighbourhood. But brilliant as this was, "the real luxury was not in gold and jewels, which were common by this time, but in having country meadows and parks in the middle of the town, where one found, in complete seclusion, first shady thicket, then open lawns, vineyards, pasture lands, or hunting grounds."

This place of Nero's, which with the utmost refinement of skill brought the country into the heart of Rome, must certainly have stretched from above the Baths of Trajan, south of the Portico of Livia, to the gardens of Mæcenas. It was no doubt the crazy want of proportion in Nero that made him design this monstrous estate, for it placed a barrier across the town, and so checked trade. He himself said he could now "at last live like a man," and told his architects that what Nature denied, Art must compel. No wonder that Nero did not long enjoy this pride in its completeness, and the later emperors thought it was better to "give Rome back to herself." The Flavian Amphitheatre was built over the pond, and Titus put up his baths on the site of the Eastern Palace.

Nero was only fulfilling in an exaggerated way the insatiable desire of all Romans, who were for ever striving after some new attraction. To them a palace in the town would seem a very cramped dwelling, if it did not cover more than four acres. Martial describes a villa belonging to his patron and friend Petilius, on the Janiculan Hill, and calls it a regal estate. The house was at the very top, and so enjoyed the best view; the real country had been brought into the town; there was a better harvest of grapes than on the Falernian Hill; there was space enough to take drives within the walls; it was absolutely quiet, so that a sleeper was never disturbed by noises from the town. Private villas like this competed on a smaller scale with the emperor's, but they lay on the outskirts and did not press into the densest parts of the traffic.

Nero could not be content with the proper home of the emperors, which had been on the Palatine since the time of Augustus, for that was in the moderate taste of the first of his line. Although the Temple of Apollo had, as we have seen, fine garden surroundings, the actual palace had only small gardens. Ovid praises a pleasant shrubbery with laurels in front of the gates. The extensive garden which stretched to the westward from the palace, with its large lofty exedra, was one of the great outside erections put up by Domitian.

South of the Capitol lay the Plebeian Quarter, so crowded that no garden ground was possible. It was only at a later time that a fine place appeared there, the Baths of Caracalla. Directly to the south and actually on the circumference there is yet another villa quarter to be found. At any rate, the Servilian Gardens must have been on the Via

Ostiensis, and these were adorned with works of art in Vespasian's time. Suetonius says that Nero went there in the course of his last flight to Ostia.

Sallust says that not only did they try to make country houses in the town, but the real country villas were also like small towns; and this remark points (with some disapproval of their vast size) to the fact that all the Roman villas with their scattered buildings had a townish look if seen from a distance. The Roman loved to change his home continually, and he shows this not only by keeping up a great many villas in different localities, but also by moving about, here and there, in the same one, according to the time of year, and even to the time of day. And so Lucullus was able to fly "like a crane" to another home. The dwelling-house proper, the villa urbana, was mapped out into different kinds of separated pavilions that could be lived in. We cannot be certain how far the Roman villa was taking the Hellenistic as a model, and this can only be positively known when the whole of some Hellenistic villa has been excavated. Still one may say with some confidence that the Romans adopted a style that was already fully formed.

FIG. 60. A ROMAN VILLA—FROM A WALL-PAINTING AT POMPEII

We saw in the Greek gymnasiums and the philosophers' gardens the scattered buildings, i.e. the very different sorts of buildings and houses collected into one garden scheme. If we next look at special instances of villas (Fig. 60), as shown in the Pompeian frescoes, we find the same front view constantly recurring, the colonnade, generally with three wings, and sometimes with conspicuous pavilions at each corner. We can tell from the dominance of the peristyle that the inside atrium was rather old-fashioned even in the first century, and later on practically a thing of the past. At Pompeii we shall return to the difference between Roman Atrium houses and Greek Peristyle houses. But here in Rome the pure Greek peristyle—for one may consider the three-winged colonnade to be an open peristyle—comes immediately to the fore; and the conclusion is forced upon us that not only the garden but the whole plan was adopted from a Hellenistic source, as a perfected product of Greek design.

One must not, however, be deceived into supposing that frescoes depicting villa types were first known in Roman times, though wall-paintings of the Hellenistic age are so few that the material gives no opportunity for comparison. When Pliny relates that in the time of Augustus a certain Tatius or Scudius was the very first to paint pictures of villas on the walls, with shelters, *opera topiaria*, groves, woods, hills, fishponds, Euripus rivers, the sea-beach, and all manner of varieties of the garden or seaside villa, the remark only holds good of Roman times. This artist, a man of note, had no doubt painted

FIG. 61. SANCTUARY, CHAPEL, AND TREE—FROM A WALL-PAINTING AT POMPEII

the imperial villas of Augustus, but these pictures are not instructive about the style of architecture, and rather show the design of the gardens, and still more their decoration. At that time undoubtedly Egyptian taste prevailed, and the art of gardening had been furthered by the incorporation of Egypt as a Roman province. Anyhow the name "Nilus" was used for an artificial canal in Cicero's time, and that is only explicable if something of an Egyptian look had been given to its surroundings. This is supplied by the accessories shown in the wall-paintings: the Nile is strewn with the lotus, encircled with palms, and made lively with pygmies and crocodiles; round about there are Egyptian sanctuaries.

Many such scenes have already occurred in Hellenistic carvings, and the whole art was no doubt much affected by Hellenistic influence. We see it, for example, in the treatment of the symbolic tree in frescoes at Pompeii (Fig. 61). To be sure, the veneration for trees is very old among the Romans, and in Caligula's time there stood on the Palatine the sacred cherry-tree out of which sprouted the spear of Romulus. Lucan has a nice sketch of a dead oak, whose bare trunk casts a shade, and holds the place of honour in the midst of green woodland.

The landscapes were mostly painted as views seen between the buildings, or as a background to a small garden; in both cases the idea was to make the room look larger. It was not enough to have a view right through, but the actual walls must look like a garden, and in every case the illusion was encouraged of being on the green hill-top. Similar attempts

FIG. 62. GARDEN-ROOM AT THE VILLA OF LIVIA—A WALL-PAINTING

were made in ancient Egypt, and the overseer of the gardens of Amon had desired to rest in death under his arbour of vines, and had made his tomb accordingly. There are not enough monuments to demonstrate the direct influence of Egypt on Roman taste, but it is very probable that what had been granted to the dead would in later days be granted also to the living.

We have already looked at one of these garden-rooms in the so-called Auditorium of Mæcenas, but the best of all is found in the Villa of Livia by Porta Prima in Rome (Fig. 62): here there is a subterranean room almost three metres below the old ground floor. Romans were glad to escape from the hot sun into a cellar like this, and they decorated it, so that any pampered soul might be able to find enjoyment there. The room was probably lighted from the roof, but in the Villa of Hadrian at Tivoli there are dark subterranean rooms quite as grandly set out as if they were living-rooms with artificial lighting. The confessed aim of the Empress Livia living at the gates of Rome was to transfer her garden into this cool room out of the summer heat. All four walls, not interrupted by windows, are painted as a green garden full of flowers, and anyone who comes in is in the middle of it. The illusion is helped by the arrangement of the room: round it runs a wooden fence, broken here and there, then a broad green path, and on the other side a second fence, more ornate and complicated, showing three patterns, which are separated at fixed intervals. There is a semicircle with high trees; among the trunks we have acanthus, and in some cases ivy winding round. Close to the fence there runs to right and left, as a border to the path, a belt of flowers with very pretty white blossoms, and at equal distances between them a small cactus or an ivy-plant tied up. Above the outer fencing there is visible a medley of trees in the garden beyond; and yet again one

FIG. 63. FRIGIDARIUM OF THE STABIAN BATHS, POMPEII

can recognise the trees and shrubs, all planted in regular order—oranges with flower and fruit, and in between them splendid colours in a fine flower-bed appear over the barrier. Farther back there are cypresses, palms, and other shady trees. In the branches a number of many-coloured birds are flitting about, enjoying the flowers, and their freedom also, for only one of them is shut up; his golden cage is standing on the balustrade.

Another room with much the same intention as this of Livia's, and just as attractive, is the little Frigidarium of the Stabian Baths at Pompeii (Fig. 63). It is a circular room, letting in the light overhead: the blue sky above mixes its own colour with the roof, which has stars painted on a blue ground. The walls and the four semicircular alcoves that are meant to enlarge the room, are also covered with a garden fresco, of trees over a red border. There are many fountains, where doves are seen quenching their thirst; while others are fluttering about among the trees, which are planted at regular distances (Fig. 64).

The water starts from one small recess, opposite the entrance, and from its round basin it encircles the whole place in a small narrow stream.

In everything we have so far looked at we have described individual instances only, whether by pictures or verbally. We first get the description of his own villa from Pliny's two letters, which are the only business-like accounts of much value; and we find all features united in an intelligible fashion. No attempt shall be made here to reconstruct these villas in precise detail, for our knowledge is incomplete, and must remain so until the sites have been found and dug over; any description would be rhetorical and obviously full of gaps. It is only when buildings by their arrangement actually imply the existence

FIG. 64. A WALL-PAINTING OF A GARDEN AT THE FRIGIDARIUM OF THE STABIAN BATHS, POMPEII

of gardens that they must be taken into consideration (Fig. 65). The landscape is definitely described by Pliny, with his villa, called Tusci, in the middle of it, and it is a good example of the villa scenery of Tuscany. It lies at the foot of the hill, and by a gentle ascent one reaches a point where there is a fine view of the valley of the Arno with its ring of mountains resembling an amphitheatre. It has been possible to determine with tolerable certainty the site of this uniform slope, so well suited for a villa of the kind.

This place is clearly a villa urbana, for the farm buildings are so completely ignored that we are compelled to think they were in some separate group at a distance. However that may be, Pliny has only described the important features of the place, but fortunately he lays special stress on the gardens. His account shows three marked groups: (1) first in importance the main building and its accessories, this section ending with the words, "this is the façade, this is accessible from the front"; (2) next a group of buildings of which we can only say from their position that they were at the side high up on the hill; and (3) the hippodrome, the park, which also lies on one side only of the house, and adjoins the triclinium wing of the chief façade.

What do we see from the front? First, a wide colonnade with two wings, in which are both the important rooms—a dining-room (*triclinium*) and a large living-room, one on each side. In the middle of the colonnade towards the back is a little court, visible from outside through a double row of pillars: this is planted with plane-trees, and has a fountain in the middle. Round this court are three rooms, one of them described as a particularly charming garden-room. It has a marble foundation, above which are walls painted with trees, and with birds in the branches; a little stream ripples through the narrow pipes, scattering its cool water. In short, this is one of those garden-rooms in which the walls are hidden under the pictured garden, just as we have seen them at Rome and at Pompeii, and the fountain helps the illusion.

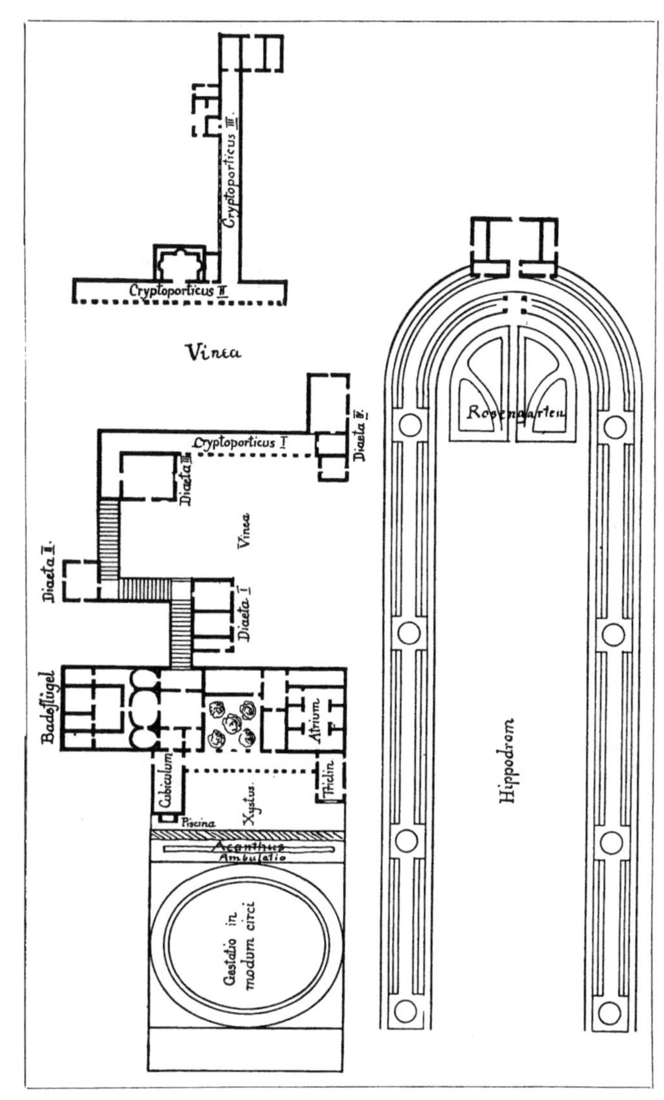

FIG. 65. PLAN OF THE YOUNGER PLINY'S VILLA TUSCI

In front of the three-winged façade is the ornamental garden, or xystus. In the time of Pliny this and other technical words were probably so incorporated in common speech that there was no sense of their being derived from the Greek gymnasium, though this feeling was strong in Cicero's time. But as in the Greek gymnasium, so in the Roman xystus, the pillared hall is indispensable in front of the garden, which has now become a highly decorated place, and well cared for. We hear from Pliny nothing of those thickets that Vitruvius demands, for now it is a front garden that serves the double purpose of adorning the façade and keeping the view clear and open. It follows the gentle incline in terrace steps, and round it there is a wall.

On the uppermost terrace, the xystus proper, there is a lawn cut out with flower-beds edged with box. Before the window of one of the wings there is a fountain fixed in a marble basin. Pliny says nothing about the disposition of the beds, but the Romans liked variety, and we hear of round and rectangular beds, and of wavy borders. Outside this box edging there was a favourite trellis border (Fig. 66), and the pictures also show

FIG. 66. A GARDEN AT POMPEII—FROM A WALL-PAINTING

fountains between the beds (Fig. 67). From the lawn downward there is more plantation, leading to the next terrace; and raised box borders are cut into various shapes of animals facing one another. Pliny speaks now and then of this clipped box at the villa, but we have no picture of it in antique times. The lower terrace is planted with the softer-leaved acanthus, which here takes the place of our mown grass. On this terrace of acanthus there is a path bordered with clipped greenery, the so-called *opus topiarium.* Again, following on this, we find a wider terrace which is circular, and enclosed with box cut in many shapes, and with dwarf trees. Box cut in the form of steps conceals the outside walls, so as to obliterate the appearance of a separation of garden and open country. For it is always an object of desire that the front of the villa shall have a view over the landscape, "where Nature seems as good as a garden made by art." Thus, then, has the xystus developed from an addition to the colonnade of a gymnasium into a beautiful ordered work of art, which is destined later on to be maintained in the world of horticulture, and which leads our minds onward to see what the future will be.

The cutting of evergreen plants was important for attaining this end. It is difficult to ascertain the date when a device first appeared that was so great a step towards the art of formal gardening. Pliny ascribes the invention to a friend of Augustus, named Cnæus Martius, a Roman knight; and although rightly somewhat incredulous over the tendency of the ancients to give definite names to such inventions, we still have to admit that before the days of the emperors there is no mention to be found of this very peculiar treatment of plants. Cicero once uses the name *Topiarius* for the gardener who had the special care of the ornamental garden, but this was giving the appellation to the service first, and it was only afterwards that the term *opus topiarii* was applied to that clipping wherein the gardener could exhibit his greatest skill. It is noticeable that neither Cato, nor Varro, nor Columella, ever mentions this special pride of the garden, when they are inveighing against the luxury of villa life. In the time of the elder Pliny they must have been considerably advanced in an art that is none too easy, for they had whole scenes cut out, such as fleets of ships, hunting scenes, and so on, all in box and cypress.

At the west side of the colonnade is the bath-place, whose situation is not quite accurately determined. Steps lead to an upper terrace, next to three small living-rooms, one of which looks down on the plane-tree court, the second towards a meadow that lies far to the westward, and the third on the vineyard, which climbs the hill in a terrace formation. Behind, apparently on a similar terrace, is a cryptoporticus (covered portico), and under it a dining-room looking out on the hippodrome, the vineyard, and the mountains; lastly there comes a fourth living-room, connecting the villa and the hippodrome.

FIG. 67. A GARDEN PICTURE IN THE PERISTYLE OF A POMPEIAN HOUSE

The second group of buildings lies in the vineyard, and seems to have a different approach. Here, too, the place is marked by two other colonnades and several more rooms, and also two pavilions, with no garden, but only looking out on the vineyard. Apparently one would feel entirely in the country here: in the Laurentinum villa we shall see how completely the kitchen-garden was adapted for the pleasure of the occupants of these best rooms.

"All these places, and their amenities, are surpassed by the hippodrome"—thus Pliny begins his description of the third part, the park. This place, too, takes its name from the old athletic games; and much as it has lost of its former significance, it still keeps the rectangular form with one side rounded, and the chief features of its plantation. The hippodrome lies to the east of the villa, which fronts southward, and it seems to be on a lower level, for the windows of the eastern triclinium look down on the tops of the trees; but the whole plan can only come out clearly when excavations have been made on the spot.

Pliny's account is so pleasant and easy to follow that it shall be given in his own words:

In the front of these agreeable buildings lies a very spacious hippodrome, entirely open in the middle, by which means the eye, upon first entrance, takes in its whole extent at one view. It is encompassed on every side with plane-trees covered with ivy, so that while their heads flourish with their own green, their bodies enjoy a borrowed verdure; and the ivy twining round the trunk and branches spreads from tree to tree, and links them together. Between the plane-trees are planted box-trees, and behind these laurels, which blend their shade with that of the planes. The raised path around the hippodrome, which here runs straight, bends at the farther end into a semicircle and takes on a new aspect, being embowered in cypress-trees and obscured by their denser and more gloomy shade; while the inward circular alleys (for there are several) enjoy the full sun. Farther on there are roses too along the path, and the cool shade is pleasantly alternated with sunshine.

Having passed through these manifold winding alleys, the path resumes a straight course, and at the same time divides into several tracks, separated by box hedges. In one place you have a little meadow; in another the box is interposed in groups, and cut into a thousand different forms; sometimes into letters expressing the name of the master, or again that of the artificer; whilst here and there little obelisks rise intermixed alternately with apple-trees, when on a sudden, in the midst of this elegant regularity, you are surprised with an imitation of the negligent beauties of rural nature; in the centre of which lies a spot surrounded with a knot of dwarf plane-trees. Beyond these are interspersed clumps of the smooth and twining acanthus; then come a variety of figures and names cut in box.

At the upper end is a semicircular bench of white marble, shaded with a vine which is trained upon four small pillars of Carystian marble. Water, gushing through several little pipes from under this bench, as if it were pressed out by the weight of the persons who repose themselves upon it, falls into a stone cistern underneath, from whence it is received into a fine polished marble basin, so artfully contrived that it is always full without ever overflowing. When I sup here, the tray of whets and the larger dishes are placed round the margin, while the smaller ones swim about in the form of little ships and water-fowl. Opposite this is a fountain which is incessantly emptying and filling, for the water which it throws up to a great height falling back again into it, is by means of connected openings returned as fast as it is received.

Fronting the bench stands a chamber of lustrous marble, whose doors project and open upon a lawn; from its upper and lower windows the eye ranges upward or downward over other spaces of verdure. . . . In different quarters are disposed several marble seats, which serve as so many reliefs after one is wearied with walking. Next each seat is a little fountain; and throughout the whole hippodrome small rills conveyed through pipes run murmuring along, wheresoever the hand of art has seen proper to conduct them; watering here and there different spots of verdure, and in their progress bathing the whole.

This is the clearest and most intelligible account of a garden ground that antiquity has bequeathed to us. The plan is fixed by the shape of the hippodrome, and its original planting, and is very simple. The chief feature is as usual several avenues, and they no doubt are wide enough for the small vehicles used in those days to drive in. This plan is easily to be seen from the middle space that makes one whole out of the various parts.

It is planted with acanthus, like the racecourse of Hercules at Elis, and a visitor on entering would look over the rose-garden, whose scent would reach him, as it is open on the south side; behind him stand the slender vine-clad pillars of the stibadium (semicircular seat) with a fountain, and still farther back the pretty pavilion. The whole picture is framed by a dense wall of cypresses, and to right and left there are avenues. The massing of foliage is cleverly distributed: on the outside great shadowy laurels and planes; farther in, the clipped box and dwarf planes; and quite inside, the acanthus plants. All the separate groups, which have paths round them, skirted with box, are treated in striking variety. No gardens except the French can boast that they ever surpassed this picture of skilful inventions in the way of special grouping. Behind the pavilion, at the highest point on the north, rises the spring, which first serves the fountain, then the waterworks of the stibadium, and lastly the many canals that intersect the whole of the garden.

Pliny was the very man to enjoy in his innermost soul the peace and beauty of villa life. What a joyful longing speaks in the questions he asks his friend Caninius Rufus in his short note:

> How stands Como, that favourite scene of yours and mine? What becomes of the pleasant villa, the ever-vernal portico, the shady plane-tree grove, the crystal canal so agreeably winding along its flowery banks, together with the charming lake below, that serves at once the purposes of use and beauty? What have you to tell me of the firm yet springy avenue, the bath exposed on all sides to full sunshine, the public saloon, the private dining-room, and all the elegant apartments for repose both at noon and night? . . . Leave, my friend (for it is high time), the low and sordid pursuits of life to others, and in this safe and snug retreat emancipate yourself for your studies.

Pliny's description has one great defect, for not a word is said of plastic decoration, and this is not through forgetfulness or with any rhetorical intention, but simply from want of taste. In his letters he gives in elaborate detail what he holds to be the highest culture, but to pictorial art he is entirely indifferent. He once had a statue left to him, and the best he could do was to give it away! He makes no mention of statuary in any of his own villas, although since the time of Cicero there had been a great advance, if not in knowledge, at any rate in the fashion for plastic art, and consequently for the joys of possession. Although Cicero bitterly reproaches Verres for the luxury shown in the works of art in his garden, that is nothing compared with Pliny's own account of the riches of Domitius Tullius, written shortly after his death. Pliny says that Domitius had hosts of art treasures, and could adorn immense gardens, the very day he bought them, with quantities of statues. Tastes of this showy inartistic kind indicate that it was a common custom to decorate gardens with as much statuary as possible. The best evidence of this comes from excavations, and, since the days of the Renaissance, gardens have been the chief hunting-ground for statues, partly in the villa urbana outside and still more in Rome itself; for the best pieces, the original Greek statues, are mostly there, and both Pliny and Martial despise the specimens outside in comparison with those in Rome.

A plan has been preserved of a garden hippodrome belonging to the place, formerly called Stadium, beside the Palace of Domitian on the Palatine (Fig. 68): its form is familiar from Pliny. Domitian, who first laid out the garden, set up a wall with alcoves in it, and the colonnade which now stands there belongs to a later date. Quite late is the oval enclosure, but the two fountains on either side may perhaps belong to the earlier date. The original planting was in avenues with varied groups of trees, and probably there were flower-beds round the crescent-shaped fountains. It is plain that there was an open space in the middle, which (like the niches in the wall and afterwards the colonnade) was adorned with statues; for this is shown to be so by excavations. On the south side, where the oval rounds off, there is a building that stands higher than the hippodrome by some two metres, and has steps leading down from it. It would answer to Pliny's pavilion but that it has more of a town character and is not so hidden in green. It is not known whether or no the great exedra, on the eastern of the two long sides, belongs to the Flavian building group; but we do know that an exedra was much liked in a garden, and perhaps was copied from gymnasiums, for we are told that Plato's own private garden had one. In this hippodrome on the Palatine it is an organic part of the ground-plan, rising steeply right in the middle of the garden, and so providing for the imperial family a good view over the place itself and also far beyond. The grouping of house and garden here is much less precise than at Pliny's place; the garden is all on one side of the palace, and can only be reached by a little gate. There is no talk among the Romans of unity or cohesion between the house and garden, and it seems as though there cannot be much connection between these loosely-arranged groups. The utmost they attain is the one composite whole made by the relation of each individual part to the xystus.

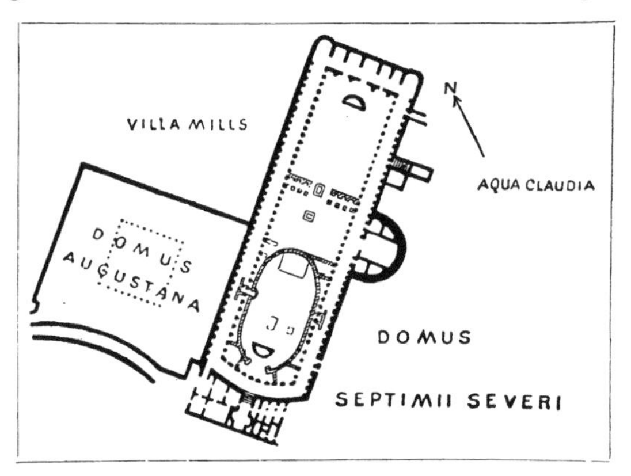

FIG. 68. GROUND-PLAN OF THE HIPPODROME GARDENS ON THE PALATINE

We shall presently find a similar garden design at Hadrian's Villa. It is clear that hippodromes, as well as other places for games, were still used for their original purpose, because it was just at this period that the Romans were particularly fond of Greek competitive sports. So Martial, after speaking in praise of plane-trees, laurels, and rushing water, goes on to tell of a hippodrome where "the flying hoof rings, scattering the dust." Pliny, too, speaks of a room in his second villa, the Laurentinum, which he calls "the gymnasium of my household." It is in a corner in front of the chief façade of the house which faces the sea, and is protected by certain parts of the buildings that jut out: it cannot be anything but a playground or garden site.

Laurentinum is very different from Pliny's Tuscan villa. It is on the sea, and so near Rome that the owner can get to it in the evening after his day's work in the city. It is a mixture of villa rustica and villa urbana, the buildings so arranged that everything needed for farm use is on the land side, but the house to live in is next the sea. Pliny says the place is very convenient from the point of view of utility, and not expensive to keep up; and he had no large park or important pleasure-garden here. A xystus he certainly had, but this he speaks of as "scented with violets," so it was probably a mere lawn with flowers, enclosed on the sea side by a colonnade. But the colonnade and xystus are not necessarily joined together; and there were no borders of box, for Pliny says that only the tree-garden behind the colonnade can have box borders, because they are protected by buildings, but that where there is exposure to the wind and the sea foam, box must be replaced by rosemary. Perhaps the beds were enclosed with low interrupted stone

FIG. 69. A VILLA ON THE SEA—FROM A WALL-PAINTING AT POMPEII

walls, or with woven osiers, such as we see used very effectively on some of the frescoes at Pompeii. The tree-garden, lying on the other side of the colonnade away from the sea, produces in particular mulberries and figs, which grow tall here, but here only. Round about are lovely walks, first a path of vines laid with fine gravel, so soft that one can walk on it barefooted. The other wider paths for driving are, as was said, bordered with box or rosemary.

At the end of the xystus and colonnade is a summer-house, where Pliny can flee from the noisy Saturnalia to a beloved spot, made by himself, and so enjoy peace and quiet. There are several rooms in it—one a living-room with a charming veranda, which gives a fine view of the sea, the shore by his villa, and the woods between. The colonnade and immense tree-garden are between this summer-house and the villa fructuaria (the barn, and the threshing-floor), which includes very attractive rooms, and enjoys a view over the tree-garden or perhaps a kitchen-garden of fruits and vegetables adjacent to the

FIG. 70. VILLA ON THE SEA OR ON A LAKE—A WALL-PAINTING AT POMPEII

main entrance on the land side. For the chief dwelling-house has no separate garden of its own, its front being close on the sea. There are several courts, one behind the other, the

FIG. 71. VILLA WITH XYSTUS—A WALL-PAINTING AT POMPEII

first set with the slaves' dwellings round them, so pretty and neat that they can be used for guest-rooms. The best thing of all at the villa is the view of the sea enjoyed from the state-room at the great house. As there is no fine view outside, on account of the flatness of the country round, the outlook at the back through the courts is especially famous. The chief room in the house is a large dining-hall, in the middle of the façade, jutting out like a balcony. The most distant view is gained from either of two towers, one flanking the main building, the other beside the villa fructuaria.

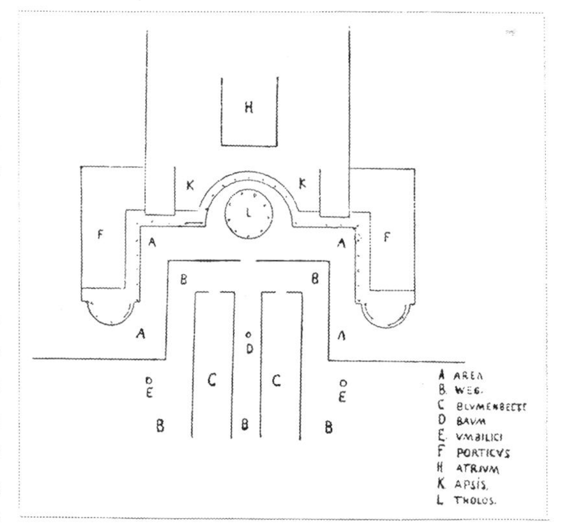

FIG. 72. GROUND-PLAN OF A VILLA WITH XYSTUS
(*see also Fig.* 71)

The appearance of Laurentinum outside we can easily arrive at, helped by a long series of villa pictures found on the walls of Pompeii (Fig. 69). With variations here and there, the pictures all show villas beside the sea with towers, and as a rule with statues on the banks, but of statues Pliny says not a word (Fig. 70). Wherever the second type of mountain villa occurs, there is nearly always a three-winged colonnade before the house, and a xystus in front of it, and behind various diætæ (summer-houses), or towers set up on higher terraces. The picture (Fig. 71) from Pompeii (with a ground-plan after Rostovtzeff. Fig. 72) may be taken, just as it is, as giving a good idea of the chief group at a Tuscan villa: the three-winged colonnade runs round the xystus with its pretty bordered flower-beds, and in the middle, behind the buildings, there is what seems to be the roof of the atrium, while farther back we observe, reaching to the top of the picture, more diætæ and more colonnades.

The great difference from the terrace buildings of the Renaissance lies in the absence here of any consistent architectural idea, such as is expressed in symmetrical building (an "axial lay-out"), and one regular scheme for the whole: as yet there was no con-

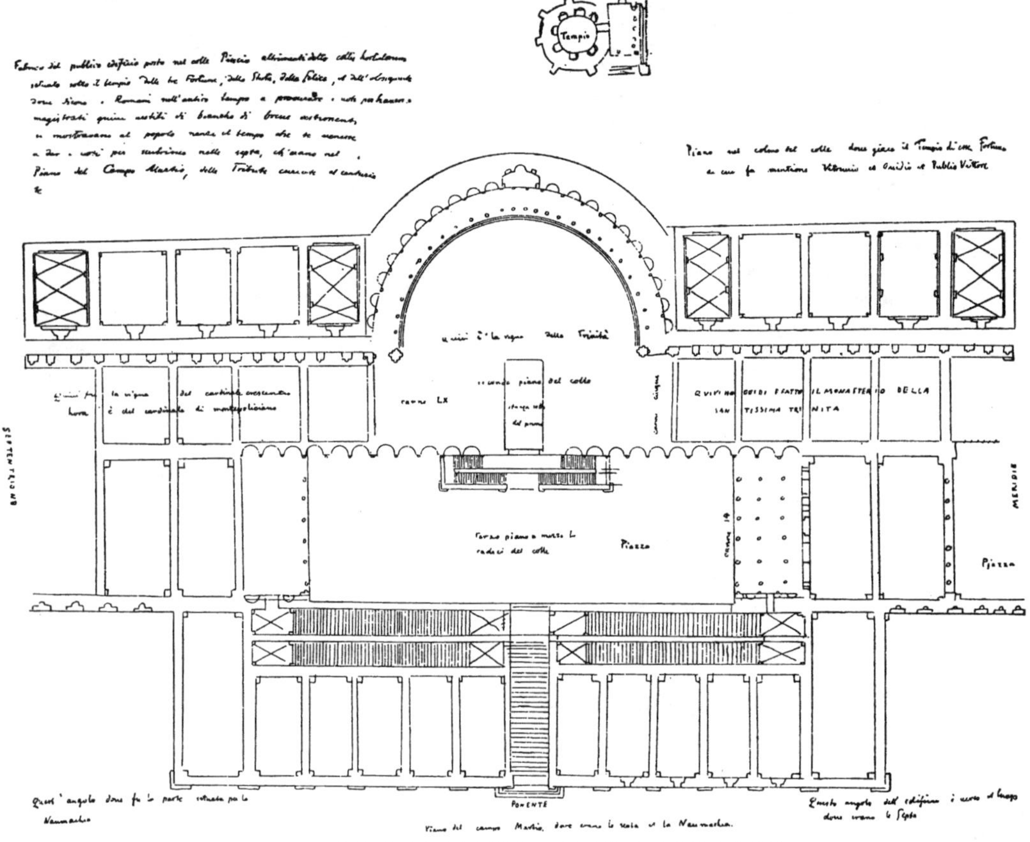

FIG. 73. DESIGN FOR A STAIRWAY APPROACH TO MONTE PINCIO

nected plan at all, though generally the chief house was below, and the other lesser buildings above, the hill. But not always: Statius describes the villa of Pollius Felix on the Punta della Calcarella which covered the whole territory between Marina di Popolo and Sorrento; below, on the brink of the sea, a warm bath was put up with two towers, a temple to Neptune, and a temple to Hercules. A colonnade, apparently zigzag, climbed the rocky cliff, and the villa stood on a plateau at the top. According to Statius, this had meant a great deal of digging, and the colonnade "creeping upward" must have had an unusual, and for those days a strikingly picturesque, effect. A little picture from Pompeii, unfortunately ill-preserved, shows this kind of colonnade "creeping upward." Pliny gives yet another description of the two types of villa: they are on the Lake of Como—one actually on its border, on level ground, like Laurentinum. There is a very large xystus on a slightly curving inlet, and the villa behind it takes all its charm from the lake. The other one, on the top of the spur, enjoys a lovely view which includes a straight avenue passing along the back of the mountain. Pliny calls this (in his witty, antithesis-loving style) *Tragedy*, because it is poised on the cothurnus, and the other one on the strand is *Comedy*, because it is on the soccus, nearer to the ground. To him these two represent the two types of villa such as were built at Baiæ, the great fashionable bathing-place for the plutocrats of Rome.

In all these descriptions of villas one feature is wanting, which will prove of great importance in the architectural forms of the Renaissance—the stairway. Of course there were stairs of some sort out of doors, and Pliny mentions them at his Tusci, but they appear to be only the means of getting to a higher part of the villa; and if there was any architectural treatment of stairs, they were entirely isolated, like the fine covered stairway of the Hellenistic Pergamon. It was only at the time of the Renaissance that the possibilities were realised of connecting links, such as the sloping terrace, acting as a rib in the whole wall of terraces. Accordingly we may justly look with suspicion on the drawings of Ligorio, the Renaissance architect, who represents staircases in antique villas. One of his sketches (Fig. 73) has been far too credulously accepted. It purports to show the ascent from the Field of Mars to the present Villa Medici and Monte Trinità, where formerly were the Acilian Gardens with the round temple already mentioned, and the remains of the great hemicyclium (semicircular seat). It would almost appear that the hemicyclium is the one and only centre and starting-point for the whole of Ligorio's imposing plan (Fig. 74), because it is not confirmed either by literature or by monuments.

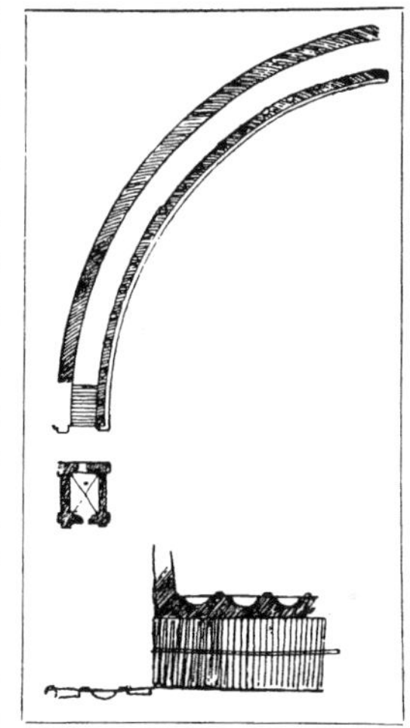

FIG. 74. HEMICYCLIUM IN THE GARDEN OF THE SACRÉ CŒUR, ON MONTE PINCIO

These sketches, like so many thrown off by this great artist, must be regarded as misrepresenting the architecture of the ancients. Further, the great villa built by the emperors in four terraces above the Alban Lake was far from presenting the appearance of an enclosed place from the Renaissance point of view. At a distance the chief building looked like a castle. The remains are so scanty that we cannot form a mental picture of the whole enormous place. But what we do get from ruins is the sense of the ever-growing leadership and dominion of the emperors in the first century, until Domitian, at the same time as he made his residence on the Palatine, gave to such places their final form.

The palaces with their gardens, the colonnades and towers, enclosed a great stretch of the lake, which spread out below them like a huge Naumachia. Certain parts of a *cryptoporticus*, leading to the lake, have been preserved. There are also traces of a theatre and an amphitheatre; and the Emperor Domitian encouraged games of every kind and instituted Greek games, calling them Capitoline. The slope of the ground made irrigation an easy matter, and the reservoir may well belong to the ancient fishponds: one of these, resting on 36 pillars, could hold 20,000 cubic metres of water, and a second one, a little smaller, 16,000. The water was conveyed by a system of canals to every part of the villa. Connected with these reservoirs are two nymphæums or grottoes, one of them still recog-

FIG. 75. NYMPHÆUM AT THE IMPERIAL VILLA, ALBANO

nisable by its decoration (Fig. 75), and at the time of Piranesi still giving a good impression of the magnificence of the emperors. In both cases natural grottoes were altered, and turned into elegant rooms, with alcoves in the walls, and with stucco or mosaic floors. The place was kept cool by fountains and waterfalls, and adorned with statues and other carved work; from the opening the bright green garden seems to look into the cave-like darkness. The villa at Tusculum was not much smaller, but has left fewer traces. This villa was set up by Domitian on the ground where Lucullus had lived, and now a great part of the town of Frascati stands on the same territory. We know the names of many other villas of Domitian, at Antium, Gaeta, Baiæ, and elsewhere, from what is said by Martial.

Although the individual remains of Domitian's villas are so impressive, we cannot get a connected view of them. A very different idea is given by the villa, built several decades later by Hadrian, at the foot of the Tivoli hill. The whole extent of it is not yet fully known, although it has been explored most studiously from the days of the Renaissance to the present time. The ground stretches from north to south on a rise of about forty metres, starting from the foot of Monte Arcese, at the top of which is Tivoli (Fig. 76). The chief approach was no doubt from the north, for the great road from Rome was on that side, as were also the façades of the palace. The place was not really unsuitable for a grand entrance with terraces and stairways, but the object most in favour was to attain the chief point of view by a very gradual ascent, and there to build the main palace. The emperor must be able to get to every part of his villa by easy ways, and probably they led by beautiful gardens: of these no trace is preserved, and instead of an explicit account, there is nothing but what Spartianus has to say, namely, that the emperor

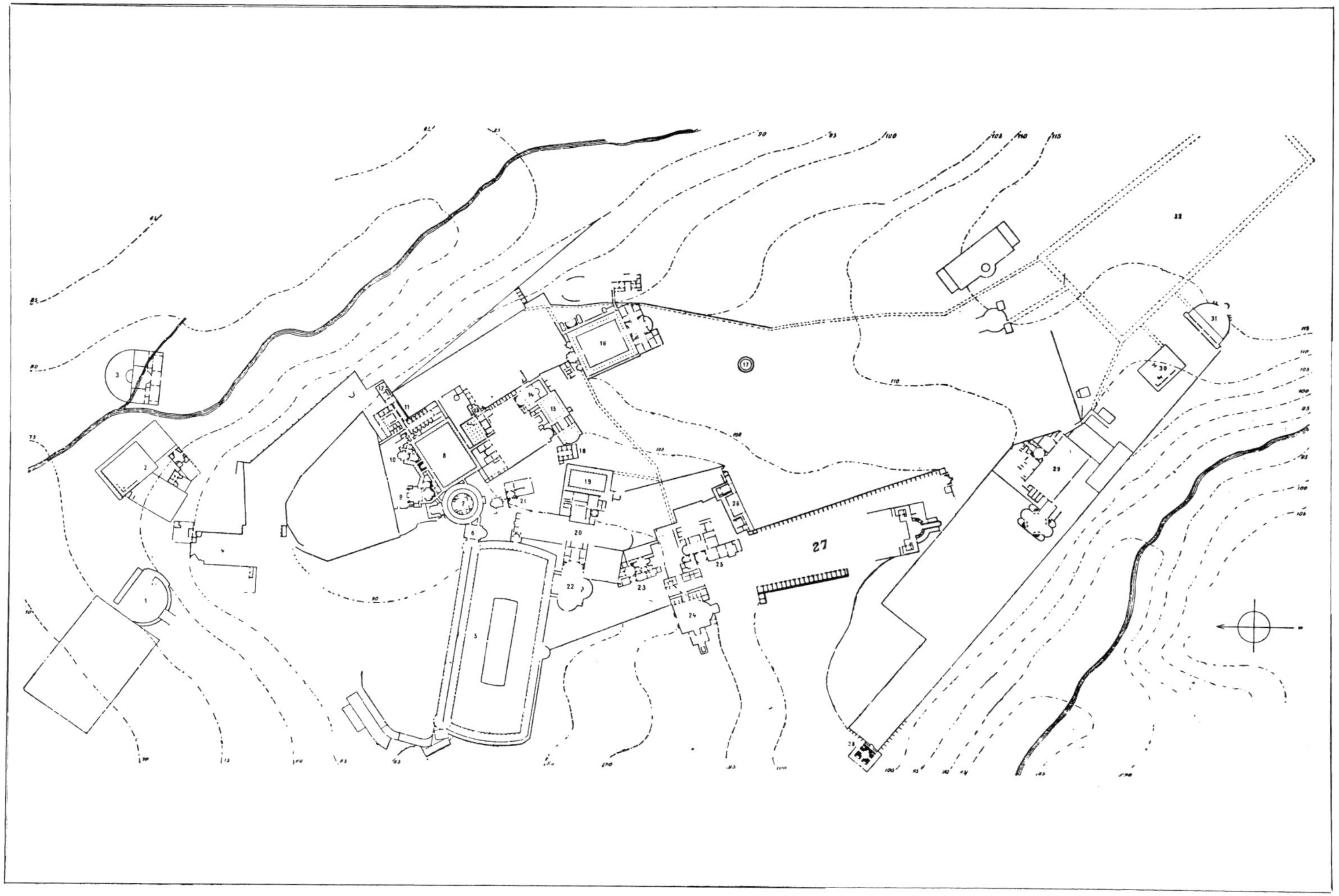

FIG. 76. GENERAL PLAN OF HADRIAN'S VILLA AT TIVOLI

had made certain parts of his villa in memory of his own travels, and called them after important spots, such as Lyceum, Academy, Prytaneum, Canopus, Poikile, Tempe, and had even made an Under-world. All this was to serve him as a symbol, in his dream of re-establishing the glory of Greece.

These varied tales, casually thrown together by Spartianus, were destined to act as a bone of contention among later inquirers. The first person to take action was Ligorio, and he was quick to bestow these names on what was left of the ruins; they remained, and we have only advanced to-day in so far as doubt has been thrown upon all Ligorio's design. Garden sites are attached to most of these places, but by reason of the destruction

FIG. 77. HADRIAN'S VILLA—THE WALL OF THE SO-CALLED POIKILE

and looting that went on for hundreds of years, their traces have utterly vanished at this villa, and the hope of finding any light later on is now very weak.

To a visitor who enters from the north, the first imposing sight (if he has passed over the theatre and the various colonnades) is no doubt the portico, two hundred metres long, that Ligorio calls Poikile (No. 5 of the plan, Fig. 76). It is above a xystus, extremely well preserved, and forms the north side of a great double hall, whose huge dividing wall is still standing at its whole height (Fig. 77). On the south this wall forms the long side of the great hippodrome (5), which has its usual form with similar colonnades on all four sides. It must have taken a great deal of digging and undermining to make this ground level on the south-west. In the middle of this hippodrome, which is rounded off on both sides, there is a basin that suggests an extension; and on the eastern narrow side there is a charming summer-house, and in the middle of the long side on the south an exedra.

The ground-plan is similar to that of the Palatine, and we should think of the plantation as like that of the hippodrome at Pliny's villa. But it is hard to imagine the immensely high halls joined on to the wall—which is all we see—and mentally to cover it with its own proper decoration. It is not easy for us moderns, whose buildings, however imposing, are in their first intention meant for utility, to grasp the idea of these monumental erections of antiquity; in the later days of the empire great and luxurious buildings were erected for purely decorative purposes. One door only unites north and south halls, and these pillared

FIG. 78. HADRIAN'S VILLA—THE SO-CALLED NATATORIUM

halls were evidently two stories high: the rounded corners of the walls, and the circular shape, show that people could drive round in light carriages. But one question remains unanswered: how were these gigantic walls covered? It is quite possible that they had frescoes on them, and in that case the name Poikile would not be so ill applied. But there is no doubt that the so-called Canopus was a garden site. This name was used for it even in Renaissance days, because a number of Egyptian statues were found in it; at the same time it is less like the fine bathing-place near Alexandria (which is commonly compared with Baiæ) than the Poikile is like its Athenian model.

On the south, embedded rather deep in the ground of the valley, on both sides, there is a very large exedra, larger than any of the others of which the emperor was so fond. It has a great many alcoves, and a fountain in each. A narrow bath evidently led from a reservoir, and there were several rooms on each side. A small terrace passes to the bottom of the valley, enclosed with corridors that have now disappeared, and which, according to Piranesi, descended by steps. The condition of the excavation does not permit us to tell whether or no there was a large basin here. If there was, the name Canopus might be more defensible, as it would remind one of rowing on the lake at the bathing-place.

A curious building on the north-east of the hippodrome seems to be some kind of garden architecture, although it does not look very promising for vegetation—the so-called Natatorium (Fig. 78). It is a circular edifice (7, Fig. 76), surrounded by a strong wall, inside which is a portico supported on pillars, and within it is a canal with a draw-bridge. This goes round a building that consists of little rooms, grouped about an atrium,

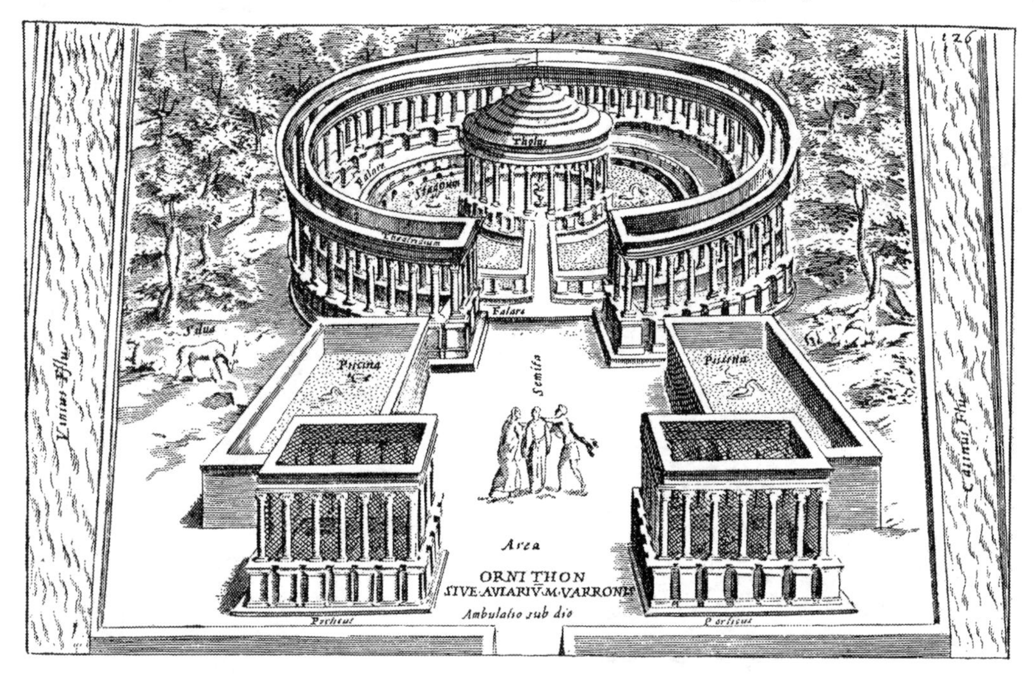

FIG. 79. VARRO'S ORNITHON

which stands on a pillared foundation, following a wavy line. In front of the north side is a hall, whose portico opens out on a garden terrace. This puzzling building cannot be a miniature house, nor can it be an imitation of some famous island; very probably it is a pleasure-house, erected for some special purpose.

Varro describes a place which does not completely explain our building, but does throw more light on it. Varro's place is an aviary, put up to amuse him at his villa: it has often been imitated. The whole place, enclosed by a wall, consists of two main parts, one square and one round. The first has at the entrance two porticoes covered with nets to serve as bird-cages, then two piscinas, and between them a way leading to the Tholos (round building). This circular part begins with a colonnade that goes round it, its outer pillars made of stone, and the inner ones of wood; these are also used for cages. A narrow stone path divides the colonnade from the canal which runs round inside, and is connected with the piscinas. Its waters encircle a covered house on pillars, that encloses a kind of triclinium. At the time of the Renaissance an exact reconstruction was attempted after this description (Fig. 79). If one compares the picture with Hadrian's round building, the similarity is obvious: in both we find colonnade and canal, with house in the middle. But the place at Hadrian's villa cannot have been used for the same purpose as Varro's aviary. The outer colonnade with enclosing wall would be quite unsuited for bird-cages, and the inner house has a great deal of masonry work on the walls; the peculiar little rooms might be supposed to hold birds and other creatures, if the floors were strewn with sand.

There are many examples in Renaissance times of masonry used in aviaries, such as those at the Villa Borghese and the Farnese Gardens; and very often drawbridges were made, crossing a canal, not only for the look of the thing, but to give a free run to the waterfowl. It is no objection that the masonry was ornamented inside; perhaps this is the vivarium of the emperor's villa. We know how specially beloved such rooms were, and that (as Varro says) they were used as pleasure-houses.

This building, which stands alone, turns a little out of the axial line in a south-westerly direction towards a great group of buildings, which is broken up into several disconnected sets, rightly believed to be the main home of the emperor. For many years he avoided Rome, and conducted the world's business from this villa. The great desire of the ruler, weary of wandering, was to make his own home beautiful and bright with gardens and pleasant waters. The south-eastern group of the palace must have been a masterpiece among garden villas, the so-called Piazza d'Oro: on a pillared court there opened a large oddly curved hall with a dome, in the middle of which stood a basin, pointing to the fact that there was an aperture in the roof, as in the Pantheon of the same date at Rome. This room with its alcoves for fountains was a real home of the waters, which is further suggested by a frieze of turkeys riding on the backs of fabulous sea-beasts. A series of rooms in various groups open into this large hall. The court has a double corridor all round it, and to right and left is flanked with a covered way. This peristyle was arranged as a pleasure-garden with flowers and shrubs. A tank ran the whole length, with Hermes statues on either side, and at the north end was a curving vestibule, with niches and fountains, and two small rooms at the side: this corresponded with the domed room at the other end. These are doubtless private apartments, not the emperor's entertainment halls and gardens, where gay festivals took place.

On the east (although not actually attached) there is an informal garden terrace, over one side of the so-called Vale of Tempe. By unimposing steps at one side we reach a suite of rooms; on the other side is a colonnade with a great exedra. On the lower terrace there remain traces of a huge basin, but without any connection, though the view into the valley is peculiarly fine and lovely. But this place must have been joined on with some stairway, if the owners had any use for the garden. The view into the valley can be had from a pavilion also (12), which is set up at the north end of the terrace that lies in front of the two groups of buildings.

For the purpose of the present book, it would be going too far to describe all the separate groups; and in their similar characteristics we always find the same picture: a suite of rooms round a more or less large court, and each group independent of its neighbour. The exedra, with niches for fountains or statues, is conspicuous; occasionally there is dominant instead of it a rectangular room with alcoves, as in the case of the unusually small site of the north group (beside the triclinium, 13): the room with alcoves opens on an attractive peristyle with a basin and flower-beds. This is above a *crypto-porticus*, and is dark; for it gets no light except from above; but it has such fine traces of grotto work and mosaic that it must have been a place to stay in when the weather was very hot, like the garden-room at Livia's house. It was probably the same at the "*crypto-porticus* subterranea" in Pliny's villa, Tusci, which, he says, "in the midst of summer heats retains its pent-up chilliness, and, enjoying its own atmosphere, neither admits nor wants the refreshment of external breezes."

Wherever we look, there is this great regard for water, in gardens and indoor rooms —even in halls we find it in fish-tanks, wells, and fountains. There is a specially artistic

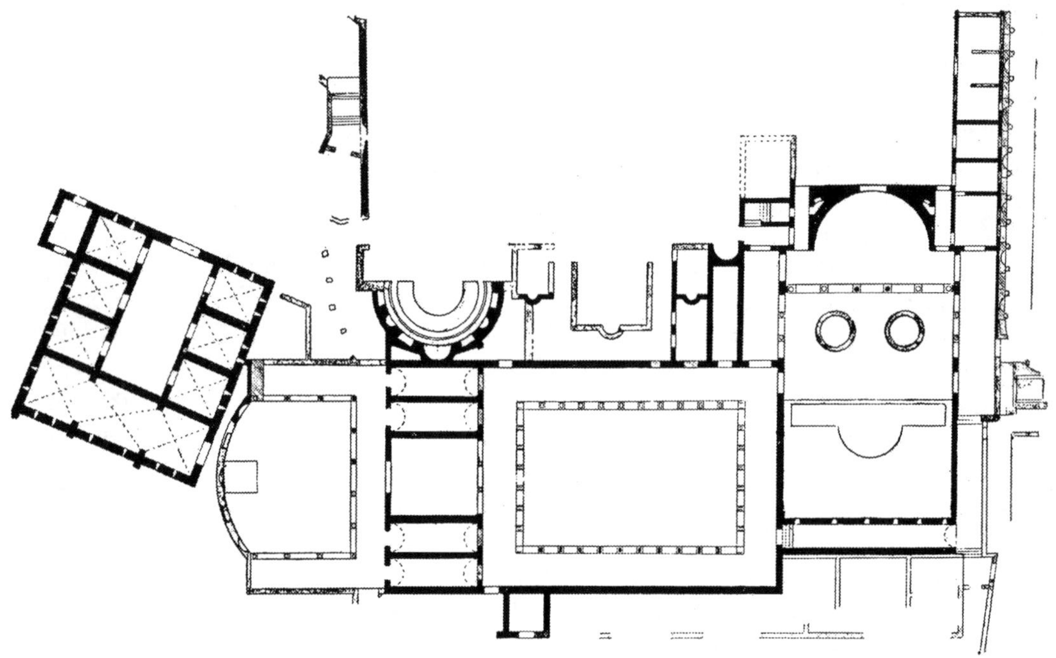

FIG. 80. HADRIAN'S VILLA—A PLAN OF THE NYMPHÆUM

water arrangement in the northern group of the chief palace, beside one of the large state-rooms (Fig. 80). Here again is an exedra, with a pillared hall in front used as an entrance; the pillars were probably in two stories, to allow of the height. In front of the hall lies a flat piece of ground with two round basins, no doubt complete with fountains and flower-beds. On the opposite side this ground rises in steps like an amphitheatre; and very probably this was a waterway, and the stream flowed down into a basin that extends the whole way across. To anyone who entered by way of the exedra, this rushing waterfall, with fountain, flowers, and greenery, must have shown a very charming picture. Ligorio calls the place a nymphæum, in the fashion of the Renaissance, when every room with water running in it was so called: whereas it is a covered—though perhaps only partially covered—inner room. But in order to grasp the plan of these marvellous garden-courts, we must again turn our attention to their beginnings.

The old Italian house had not really any convenience for the provision of gardens inside. The atrium, which was the heart of the whole, and around which the various buildings were grouped, was a roofed and floored court with a good light above the piscina—the impluvium—meant to let in the rain-water. A proper garden was out of the question, but flower-boxes were used, either round the impluvium, or forming a kind of wall, as is seen in the atrium at an African house (Timgad, Fig. 81). The boxes forming the wall are set in a curving line, and are filled with soil and flowers and creeping plants. But this atrium is only one example; moreover, it stands on the border-line between peristyle and atrium. At an Italian house they made, as we can see in some of the older houses at Pompeii, a small garden, for they kept a little strip of ground for plants

FIG. 81. FLOWER-BOXES IN THE ATRIUM OF A HOUSE AT TIMGAD

against the back wall, out of which a portico opened. At the house of Sallust at Pompeii, where later a little peristyle house was added to the atrium house, the garden was granted two strips of ground next to the high wall at the back. An almost equally wide portico opened on this garden, which was only planted in two narrow borders. It was made to look wider by paintings on the wall, in the fashion we know. The corner was shut off by a delightful arbour, a triclinium under a pergola (Fig. 82), which mingled its living creepers with the paintings, so that the whole produced a pleasing picture (Fig. 83), in spite of the drawback of the small space. Of course this does not suit an atrium house. We find much the same at Priene, the Greek colony in Asia Minor. It was only through the influence of Greek peristyle houses in Italy, with open court and unlimited power of extension, that the garden was able to penetrate into the town house as an attractive spot with shrubs, flower-beds, and other beauties and luxuries.

At Pompeii the eruption of Vesuvius preserved a town where it is easy to follow garden development with the utmost clearness. For here, in this early Hellenised town, the old Italian house is closely connected with the Greek peristyle, especially in the way that the atrium remained as a paved court, and the peristyle behind was turned into a garden, about which the living-rooms were grouped. Sometimes these peristyle houses

FIG. 82. GARDEN IN THE HOUSE OF SALLUST AT POMPEII

are actual extensions, which in Italy joined on to neighbouring houses that had been pulled down. In the grander places there might be two peristyles, and the second one was the larger; there were only a few living-rooms in it, so that a great piece was won for the garden.

Unfortunately the excavations of gardens have been carelessly made, so that there is seldom anything to say about their sites. On the whole, then, we shall do well to call in the help of the peristyles that lie around living-rooms to be our examples of flower-

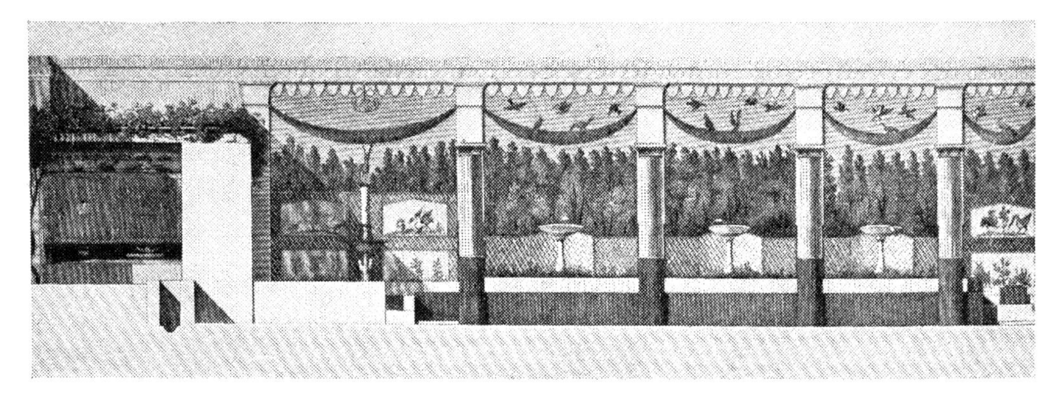

FIG. 83. A WALL-PAINTING IN THE HOUSE OF SALLUST AT POMPEII

gardens, seeing that the larger places (seldom to be found in the narrow spaces of the town) were probably used for the practical purposes of a vegetable-garden. This was certainly the case in the house of Epidius Rufus (Fig. 84), and probably in that of Pansa and of the Silver Wedding (Fig. 85). At the first of these there is a small terrace, which was obviously laid out prettily with flowers, behind the kitchen-garden. These little raised terraces were liked as a suitable ending, and in addition to the flower-beds, fountains were set up with mosaic, and also statues. Most of them were so made that one could not quite reach up to them.

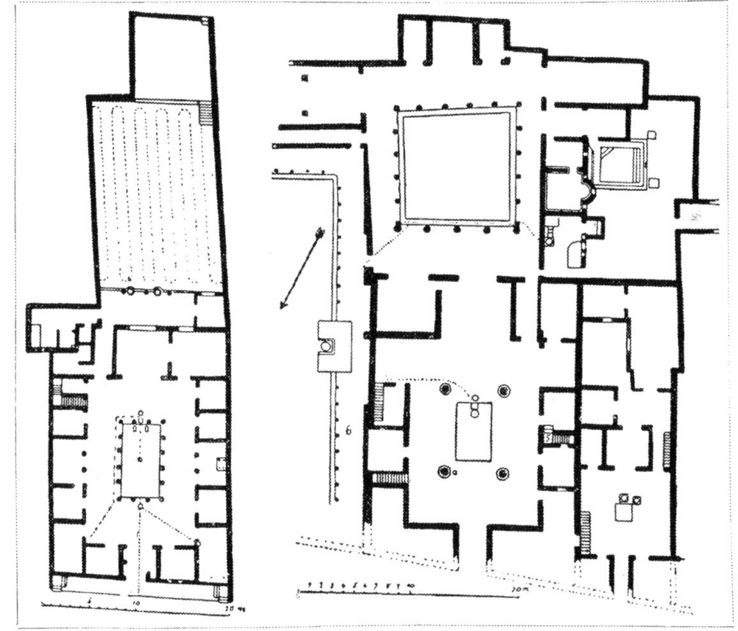

FIG. 84. THE HOUSE OF EPIDIUS RUFUS, POMPEII—GROUND-PLAN

FIG. 85. THE HOUSE OF THE SILVER WEDDING, POMPEII—GROUND-PLAN

The garden of Lucretius (Fig. 86) at Pompeii is peculiar; it ascends a little towards an alcove for flowers, wherein stands a Silenus, pouring water out of a vessel over a little stairway: the water is caught in a channel or gutter, and flows thence into a small round fountain. Standing around are little animals in marble, a small row of other statuettes, and four figures of Hermes; the whole thing, now that flowers have been replanted there, makes a pleasing picture, if somewhat bizarre. Peristyles generally ought to be regarded as part of the dwelling-house, simply rooms in the open air. The furniture and decoration are not intrinsically different, for indoors one walks through rooms that (by pictures) have been converted into gardens. We find this at Pompeii, whichever way we turn, for walls are all painted with garden pictures—atrium, peristyle, and terrace background alike. From these we can easily make out what the real gardens were like, the delicate climbing roses of the pergola (Fig. 87), the pretty lattice, the lovely fountains, the ivy tied up in little knots.

FIG. 86. A SMALL GARDEN TERRACE AT THE HOUSE OF MARIUS LUCRETIUS, POMPEII

The peristyle in the house of the Vettii (Fig. 88) which has the ground-plan preserved in a sketch, and is now restored and planted out, belongs to the last times of Pompeii, and gives a striking example of an antique domestic garden in the first century A.D. The open

FIG. 87. FLOWER-GARDEN—A WALL-PAINTING AT POMPEII

space is bordered by eighteen white pillars with coloured capitals, but the size of the whole is only eighteen by ten metres. On the edges of eight basins, four-sided but with rounded corners, stand twelve statues pouring out water, and in the middle of the garden

FIG. 88. PERISTYLE IN THE HOUSE OF THE VETTII, POMPEII

there are two more basins with fountains. There are also marble tables, shells, and little pillars with a Hermes on them. The wavy flower-beds, fashionable at the time, are bordered with box; and there are knots of ivy, bushes and flowers, thus repeating the ideas that are depicted on the walls of the surrounding portico.

One remarkable type remains in the only suburban villa found at Pompeii, that of Diomedes (Fig. 89). This villa suburbana is something between a house in town and a villa urbana in the country. Vitruvius insists, in his short notes appended to his account of the town house, that in the villa suburbana one must not go straight into the atrium, for the peristyle has to be next to the entrance. The atrium comes second, and then the portico, which will look out on the garden. In the villa of Diomedes we do as a fact step straight into a great peristyle from the street, and this actually has one corner on the street front. The ground-plan, a group, connected as in a town house, forms an isosceles triangle, with the street for hypotenuse. From the entrance a stairway leads to the peristyle, which may be pictured as a small garden with a basin in the middle. Beside one of the sides of the triangle there is a garden on the level of the street, unfortunately not excavated, into which a bedroom in the Cyzicene style protrudes at an acute angle. Rooms like this were a Greek invention, and Vitruvius says they were not used in Italy; in this age, however, they belong to the general luxury of Roman houses, and Pliny mentions them in his letters. The garden would probably be a tree-garden, which would give shade and quietude to the bedroom. At the other side of the triangle the ground drops sharply; so there is another story, where perhaps the slaves lived. One steps down into a lower garden with a colonnade round it, the roof of which can be reached from the top of the terrace. In the garden remains of trees were dug up, but unhappily these were set aside without examination. Opposite the terrace is a raised pergola, and a large basin in front with

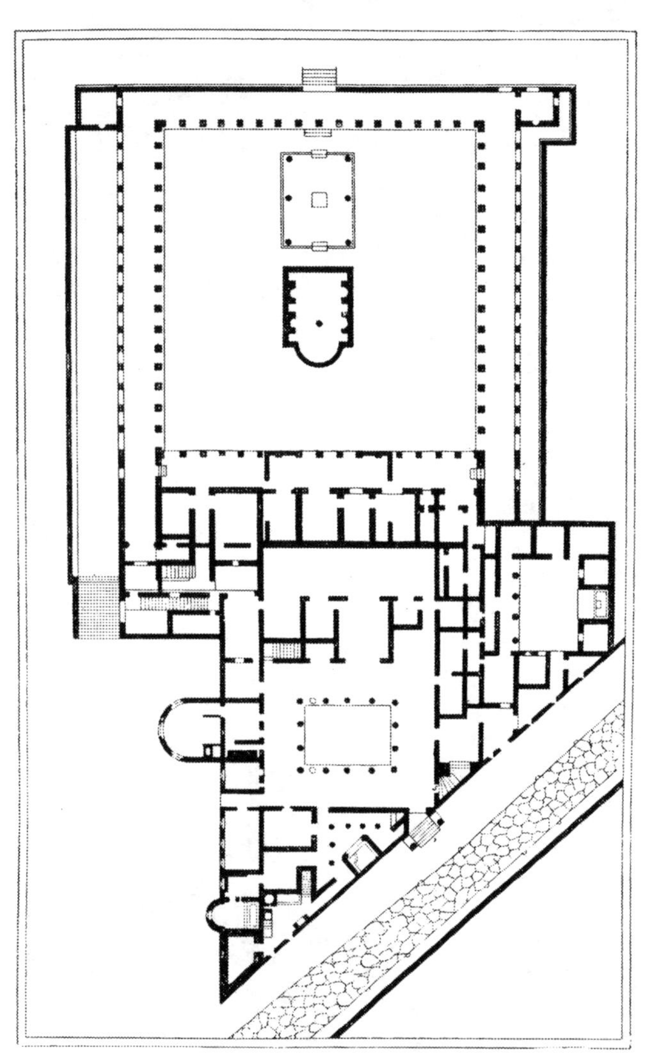

FIG. 89. PLAN OF THE VILLA OF DIOMEDES

a fine brim and a fountain. Before the left side of the portico is a broad terrace which also opened on a garden not yet dug up. At the end of each portico is a room, perhaps made for the view, or as a pergola, such as we see on the pictures at Boscoreale and in many Pompeian frescoes. The atriums, demanded by Vitruvius for a villa suburbana, are not here, and at that time they were very old-fashioned. Indeed the small tasteful villa at Boscoreale, which is before all things a villa rustica with rooms for the owners, has no atrium, but the rooms are grouped round a great peristyle, which, judging by its paintings and the four little fountains in the corners, will have also had a bright garden. Pliny speaks of an atrium in his Tusci as "in the fashion of our fathers." In the

FIG. 90. A VILLA WITH GARDEN DECORATION—A FRESCO FROM BOSCOREALE

suburbs it was really a half-open room built for pleasure, like the one at the African villa at Uthina, where a very large peristyle had five atriums, each the centre of a group of rooms. At that time far more light was needed than an atrium could give, and the first intention of an atrium, to supply water, was nothing now to the Romans; for wherever there were villa centres in imperial times, irrigation was one of the chief features, and the gardens were enlivened with fountains, water-stairways, and all sorts of devices. How far water-organs, so beloved in Renaissance days, were used to adorn and enliven the gardens in ancient times, we cannot say, but that they were known and very much liked we can tell from a host of authorities.

In imperial times the great contrast between divers ways of living was always becoming more marked. While the circle of suburban villas was getting more and more extended, and the public places grander, the homes inside the town were contracted and narrow. The barracks, of a height that has never been exceeded except in the modern towns of America, were surrounded by streets on all four sides, and the courts must have dwindled into mere shafts of light. Even in the town palaces a court with plants seems to have been a rare thing. Crassus had a court in his, and its six lotus-trees [Editor's Note: Probably *Diospyros Lotus,* the date plum] were very famous, and only perished in the fire in Nero's reign; to the older Pliny this possession inside the city seemed a marvellous luxury. But when it was wellnigh impossible to secure a patch of garden ground, there was indoor gardening, such as we have nowadays in large towns. Little flower-gardens were made in front of the windows, very probably on wide balconies, which were attached to each story, "so that every day the eyes might feast on this copy of a garden, as though it were the work of nature." One of the frescoes at Boscoreale shows a house with balcony and open galleries with flowers (Fig. 90). In the smaller towns they very likely used the roofs of a peristyle in a similar way; and at Pompeii, at the house of Sallust, one can climb up on the roof of a peristyle, laid out on two sides as a real garden on the house-top. Such balcony-gardens must have been of a fair size, for Martial, no doubt in fun, acknowledges the receipt of a bit of land, with thanks to the donor for a place no bigger than the garden in front of his window. Naturally these balconies were a fine opportunity for thieves; and Pliny says that daring burglaries compelled many people to do away with their balcony-gardens. The flat roofs of the large town houses as well as their balconies and colonnade roofs were used in the same way, and had an under-structure made to protect them from damp. On the top, boxes were laid, or fixed, so that tall plants could grow there, flowers, shrubs, vines, and even trees. Seneca says that on "high towers they planted fruit-trees and shrubberies, with roots where their tops ought to be." Even fish-ponds were there, and one might climb up to a balcony like this by a stairway of two hundred steps.

In later days Rome exhibited great art and luxury in the culture of flowers, and with the aid of artificial heating both roses and lilies were made to bloom in winter. No importations from Egypt were wanted now, for Rome had surpassed the Nile with her winter roses. In the forcing-houses, protected by mica in the windows, very fine bunches of grapes were artificially produced. Still more important, there were great nurseries in the Campagna, especially for growing flowers to be sold in the town. At Pompeii a small nursery was discovered with a whole array of painted pots, presumably for raising seed or cuttings.

Wonderful gardens were flowering on town roofs, the markets were supplying fine flowers in winter, and so it is not surprising that the emperors were envious of Hellenistic display, and tried to force gardens to grow on the poor ground by the sea. Like Hiero, Caligula built a ship with ten rows of oars, one above the other, and studded with precious stones. From the masts waved sails of divers colours, on the deck were pillared halls and dining-rooms, and many vines and fruit-trees. Under this canopy lay the monarch, borne along the coast of Campania to the strains of music.

The art of gardening reached in the first centuries of imperial rule in Rome a height which it could never surpass. To the master of the world-kingdom nothing seemed unattainable. Naturally this feeling was not confined to the limits of Italy; on the contrary, the farther the borders of Roman territory extended, and the more rich men had to live during one part at least of their lives outside Italy or in the provinces, so much the more did their luxurious nature increase and develop at their magnificent country seats. It was precisely in the provinces, where they could not indulge the desire to spend the best part of the year in Rome, and where they were compelled to live almost exclusively in the country, that the Romans sought compensation in the size and magnificence of their surroundings. The riches that were piled up in private hands grew to be something

FIG. 91. AN AFRICAN-ROMAN VILLA—A MOSAIC FROM THE BATHS OF POMPEIANUS

enormous. It is said that in Northern Africa half the Roman territory was in the ownership of six men, and it is no wonder if such a state of things excited the envy and greed of powerful persons at home. We hear of a certain Julius Calidus, whose name was put on the proscription list on account of his vast possessions in Africa.

It is easy to understand that such mighty lords built themselves very extravagant country houses. First in their esteem was the large park for hunting, and after the chase came the training of horses, and racing, but of all this we know very little. Up to the present time no villa has been entirely excavated, and our knowledge of country life among Romans in Africa is, as a fact, limited to what we get from a series of mosaics. Their favourite subject is the chase, and the villa is often represented as in the middle of the park, as, for instance, in the great hunting mosaic found at the Baths of Pompeianus (Fig. 91). The house shown in that piece looks more like a villa of the early Renaissance than a Roman villa with its scattered buildings: there is a great group of trees flanked by lofty towers, and the park is indicated by trees of various kinds. As was said before, we have to do here with the different relations of the parts. In Africa the villa is the centre of a large piece of land, and must often have served as a refuge for the visitor and his retinue against the wild and savage Berber races around.

FIG 92. AN AFRICAN-ROMAN VILLA WITH GARDENS—A MOSAIC

In houses like these, which are enclosed from without, and often provided with battlements and towers, one is obliged to think of the different parts and their relations, just as they were once brought to completion in Italy, at Scipio's Villa. No doubt the wealth and display became greater indoors, and we have the cheerful colonnade with its accompanying xystus, as in Italy. The arrangement of the garden is the same in all the other mosaics, which in African villas take the place of the frescoes at Pompeii. As a rule, they show a pleasure-garden in front of the façade of the colonnade, which generally has side-wings. Behind the villa is the kitchen-garden with vines tied up (Fig. 92), and also fruit-trees planted in rows. We have another mosaic scene that is curious and comes also from the Baths of Pompeianus; among trees grow flowers and low shrubs, and on the left stands an erection which was perhaps meant for a door or window into the garden-

FIG. 93. A LADY IN A GARDEN—A MOSAIC FROM THE BATHS OF POMPEIANUS

house. As accessories to this scene there is a lady with a fan, sitting down, and standing in front a man holding a sunshade over her with his left hand, and with his right leading a small dog on a string: he may be a slave, or possibly a lover (Fig. 93). Above their heads is written "FILOSO FILOLOCUS," and perhaps the inscription makes some distant allusion to philosophers' gardens, or possibly the owner uses the diminutive out of modesty.

The great bath sites, in which many of the mosaics were found, show what a rich man the owner was. This bath is for the time left exposed and quite isolated, with no connection with a villa. There is a whole array of living-rooms close by the actual bathing-rooms, which themselves could be used to live in; so it is possible that this is a small ordinary house belonging to a great landowner, who would live there at a favourable time of year. Here, as at Hadrian's Villa, we notice a fancy for the exedra, and for semicircular sites. Especially attractive is the so-called atrium—an uncovered, or perhaps half-covered space, which has one of its ends made in a half-circle, and has short colonnades running down either side. At the other end there are three steep steps leading to a large exedra, within which is a basin. Possibly these steps were to let the water down, as in the so-called nymphæum at Hadrian's Villa. Round the exedra is a semicircular space, filled in with a swimming-bath, which can also be reached by steps from the spaces at the side. Outside, again, there is a large semicircular colonnade, and here we may suppose there were gardens, as there is plenty of room according to the plan. South of this bathing-place, which is ornamented with fine mosaics, there are the living-rooms. A dining-room opens, through a passage with columns, on a front room probably unroofed, which opens into a garden again on the opposite side, also through pillars, that stand on a little plinth, and this garden lies alongside the set of rooms.

Very different are the villas that Roman colonists built in the northern provinces, wherever they set foot. No doubt the province of Gaul had at an early date been influenced by the purely Hellenistic style of villa, especially in the districts near the Mediterranean, when Marseilles was Greek. The nobles, inclined as they were to culture and civilisation, and very sensitive to the refinements of manners and customs, had certainly not held aloof from Greek and Roman life, whenever their homes were safe enough to allow of it. But unfortunately it is just in these territories that excavations have given so little that bears upon our study. It is not in these early times, but only much later, actually on the threshold of the Middle Ages, that we get indications of any important development. It is easy to see how rapidly the regions about the Moselle and the Rhine were covered with fine villas when the Romans were masters. Ausonius, as an eyewitness, is the first to describe (in his poem *Mosella*) the state of things that existed towards the end of the fourth century. But he, like other writers of the time, employs the rhetorical expressions of a far earlier age: he praises the lovely banks of the Moselle, the heights covered with villas and high towers, enjoying the fine views of river and valley, and enclosed in their own gardens, thickets and meadows. But careful excavation has proved that a proud and wealthy race settled here, and founded villas in no way inferior to those of their great Mother Country, so far as their ground-plan shows the extent of their properties, and the beauty of the sites they selected. As a fact, these plans are a help towards understanding much that is said in the literature of classical times. It is, for instance, now known when a number of great villas—unearthed where Metz is to-day—were originally built. The plans are so reminiscent of Italy in the early days of the empire—a period which here also enjoyed comparative peace—that (unless other proofs were at hand) it might be imagined that they were built in the first and second centuries after Christ.

Nearly all the villas found both here and farther up the Rhine are constructed with three wings, the middle one for the guest-rooms, one of the sides for private living-rooms, the other for baths. Whenever it was possible, a hill was chosen that faced south; and it appears, from a row of separated buildings which are often placed round the chief one, that in the north there was no objection to the scattered plan. But there is a preference for having the villa rustica at a distance from the master's house. One example of this, the villa at Ruhling (Fig. 94), shows its colonnade opening to the south-east of the middle building, which is bounded by bath-rooms and living-rooms placed at an obtuse angle on either side, and no doubt there was a xystus in front. These main buildings are greater in breadth than in depth, and grouped around them are several dependencies, which are naturally connected by park grounds, just as they are in Pliny's Laurentinum.

The villa at Ulrich has a peculiar ground-plan, and the features remind one rather of a villa suburbana than of a real country house. According to excavations, it is apparent that the chief entrance to the villa (which stretches from east to west on a flat part of the hill) is on the eastern side, beside a court that has no colonnade. But one gets the best view of the place from the northern valley, where there was a long colonnade on the flat ground, whence there is an enjoyable prospect. The bath-wing is on a terrace lower down. Outside the court, behind this colonnade, there is a large peristyle which we may suppose to be a pleasure-garden, and in the house on the mountain side is a pillared hall. The gardens proper will have extended east and west of peristyle and court. In its

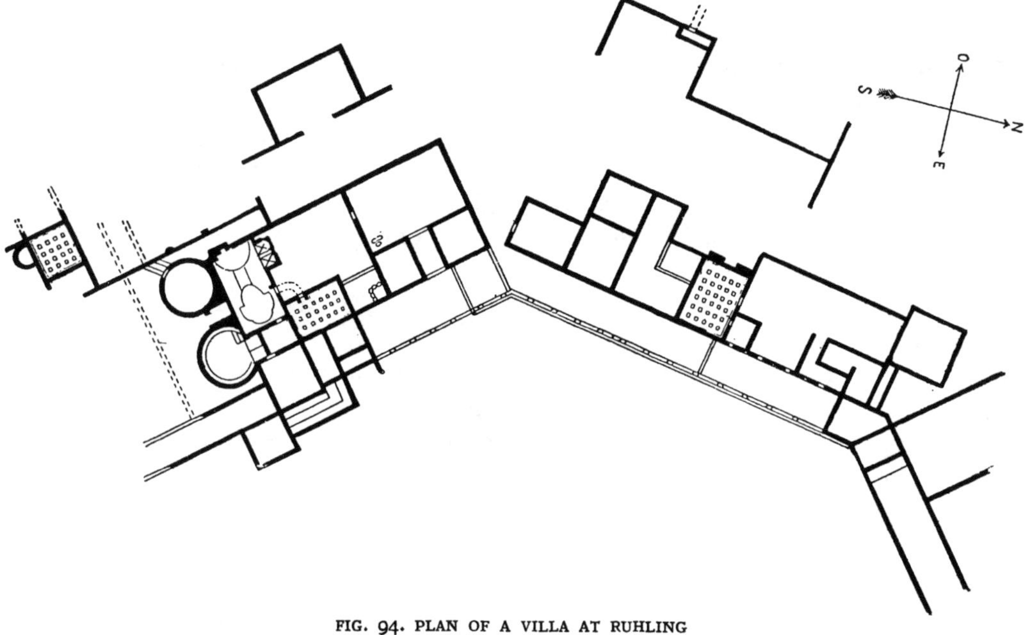

FIG. 94. PLAN OF A VILLA AT RUHLING

whole plan this villa shows some resemblance to the Villa Madama on Monte Mario, near Rome. An unattached building, probably the villa rustica, has also been found here.

But the most remarkable of the villas in the Moselle district is the one found at Teting (Fig. 95). It is one of the largest on this side of the Alps, certainly a seat of which none of the greatest men in Rome need have been ashamed. The imposing façade with its three wings is towards the south-east, and in a splendid situation. In front is a wide court, eighty-eight metres in length by sixty metres in depth; and from the house there would be a full view of its beauty, planted as a fine garden, whose extent it is not possible to determine. The central building forms a large broad exedra (forty-four metres), a size which would be imposing even for Italy: on both sides this is flanked by two semicircular rooms, in the so-called Cyzicene style. The exedra was formed by a pillared hall, in front of which lay a kind of terrace two and a half metres broad, probably a double colonnade, and farther back a large room with an apse. It would appear from excavations up to date that there are only a few rooms behind the exedra, and the living-rooms proper are in the east wing, which also has a pillared hall extending to the narrow end of this wing. In the western part are the baths; and quite disconnected to the north, there is a diæta which contains several rooms; stretching still farther is a colonnade ending in the form of a cross. Turning off at a right angle there is again a colonnade, fifty-three metres long, with another small diæta at the end of it. At once we call to mind the descriptions of Pliny, especially of the colonnade in his Laurentinum, with the garden in front smelling of violets, and the charming summer-house at the end. So far the excavators at this sort of villa have had no time or inclination for exploring the garden land round the buildings,

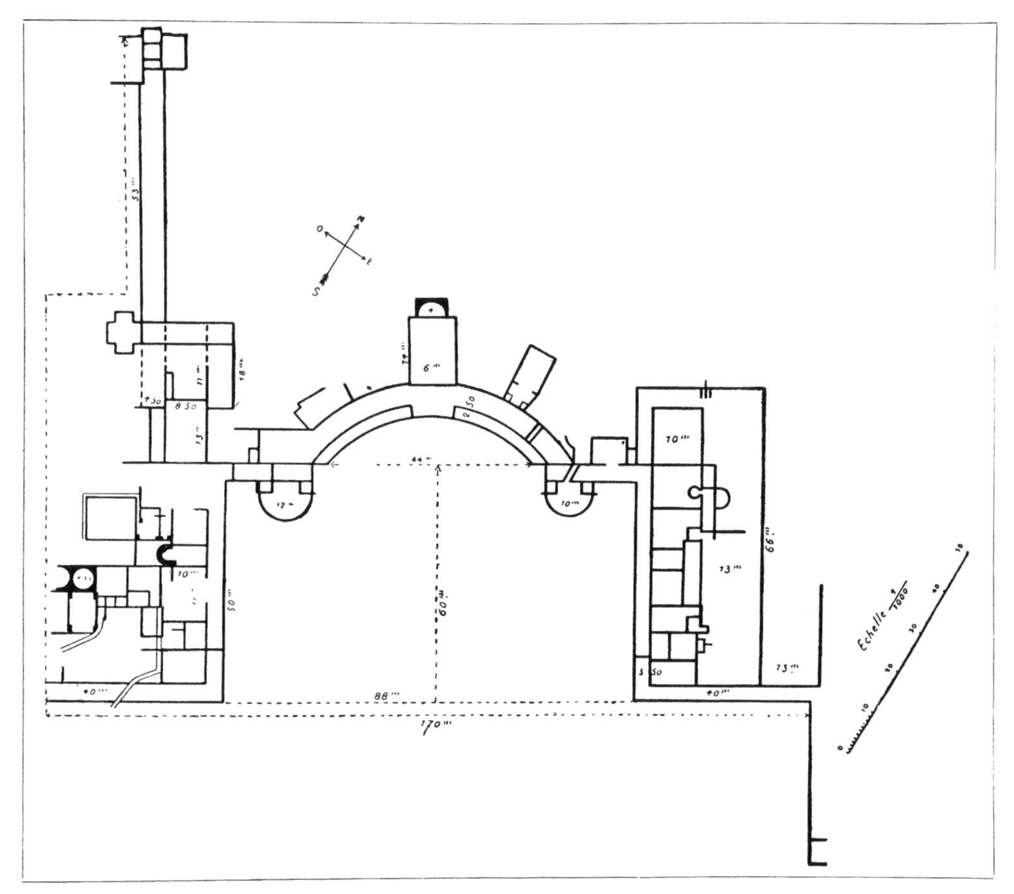

FIG. 95. GROUND-PLAN OF A VILLA AT TETING

but the ground-plan of this particular large villa may serve as a means whereby with the help of Pliny's words we can reconstruct the beauties of the original garden.

None of the villas dug up on the Rhine is so large as this one. There is, however, one lately excavated at Wittlich, not three-winged as they mostly are, but with a light colonnade following the hillside and answering to a second one running parallel to it (Fig. 96). These two colonnades unite the three familiar features of baths, reception rooms, and living-rooms, which are grouped around two intermediate courts, whose water arrangements point to some kind of plantation. The river has now shifted its bed, and runs close up to the façade, and this makes it impossible to know whether there was a xystus or not.

The Romans carried their habit of building villas even into Britain. Faustinus, a Roman

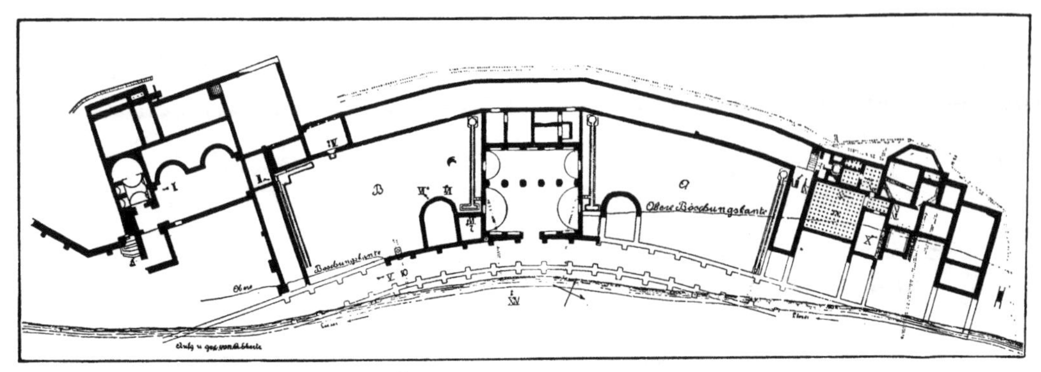

FIG. 96. PLAN OF A VILLA AT WITTLICH

who owned a villa at Naples, built a castle in Suffolk, and this shows the three-winged type. A good example is a villa at Spoonley, near Winchcombe, unearthed about 1890. The threefold colonnade has become a veranda; and the fourth side, generally open, is shut in by a high wall. A paved way leads from the entrance gate to the middle building, and on both sides of the path there would be a front garden. The open Roman villa has developed into the secluded home of Northern lands, suited to a Northern climate.

A lucky chance has preserved a piece of real garden architecture in the great pond at Welschbillig, one of the villas on the Rhine (Fig. 97). This piscina is fifty-eight metres long, and eighteen metres wide, and lies in a park, which, from the elegance of this single feature, must have been an important one. The pond was meant for boating, and a wall with fountains at each end (of course partly covered by the water) shows the extent of the course. On the edge of the oblong basin there are six bays built out, and the whole tank is bordered with a graceful balustrade. But for its chief decoration there stood at equal distances 112 Hermes pillars, set up so as to face the water: they should be looked at from that side, and would show up well against the green hedge, which would no doubt be planted at the back.

Everything that the Romans did in these provinces, in the way of gardening and cultivation, was destined to perish when they were compelled to remove their troops under the German onslaughts. The open villas and gardens were the first victims, and these Northern lands had to wait for hundreds of years before they could make a humble beginning with a new garden art. In the North the thread was broken, and only after wild, stormy years can we discover a new trace, again starting from the South. Of the original homes of ancient culture we can scarcely say in the same sense that all continuity was checked and wiped out, but in particular districts, and especially in Western Italy, where so many people had crowded back home, there does seem to have been complete darkness for a long stretch of time. These countries have no history, not only as to the art of gardening, but as to every kind of art.

It is not only Byzantium that maintained the tradition of garden and villa, till other nations seized upon it, but in Spain, the south of France, and Italy we can trace in individual cases a continuation of the old style—sometimes weaker, sometimes stronger—right on into the sixth century. Even the conquerors, the German invaders, did something to preserve the buildings and gardens; for they liked taking possession of the villas of their Roman predecessors. Procopius relates that the Vandals, when they had grown effeminate in North Africa, discovered that they could do very well in the fine, shady, well-watered paradises that they found ready to hand at Carthage. We are told by Luxorius that these places looked much the same as in classical times, when they were in their glory. In two of his epigrams he writes admiringly of the views from the towers, the clear rushing springs, the fine summer-houses, and of how he enjoys the scent of flowers and the song of the birds.

By that time, and even sooner, both taste and expression were corrupted and perverse. We see this particularly in a description given by Apollinaris Sidonius of his Villa Avitiacum in the Auvergne. His attempt to rival Pliny is obvious, and, quite in accordance with his model, the pleasantness of the southern aspect and the fine view are commended. He lingers over the baths, especially a piscina, which without boasting he feels he can compare with a public one. Water is supplied from canals and flows from lions' heads. There is a long colonnade, which he cannot call hippodrome, but does call *cryptoporticus*. One can see that he has no notion of what was meant. He goes on to describe a green court (*area*) as a delightful grove (*nemus*) This "nemus" lies at the exit of the colonnade towards the lake, which with its island in the middle is the joy and pride of its possessor. In the "area" two lofty lime-trees mingle their tops to make thick shade. This is the substance of a very wordy description: the ancient clothes sit loosely on him.

The open villa of the old time was threatened by the flood of nomadic peoples, more and ever more, in the place and time at which Sidonius lived. The great uncertainty of life demanded stronger dwellings; and Sidonius shows us, in an invaluable description, how the transition took place between the villa and the castle of the Middle Ages. He writes of the castle (*burgus*) of Pontius Leontius, his friend, in a poem. It stands on several terraces upon the slope of a high mountain, close to where the Garonne and Dordogne

FIG. 97. A ROMAN BASIN AT WELSCHBILLIG, RECONSTRUCTED

mix their streams, and is visible from afar: round it are walls and watch-towers, which serve both for ornament and protection. A ditch, following the line of the fortification, he calls *Thermos*, perhaps to suggest that it can be used for cold baths, as he also mentions the warm bath indoors. Separate colonnades divide the parts of the building, and look towards different points of the compass. Two of them are oval in outline, and are crossed by crooked paths; so we may suppose that there was the usual plantation outside the walls. There is a summer and a winter colonnade, also a winter house for the gentlefolk, a women's quarter, and a so-called chapel. The granary has to be inside a castle of this kind.

The picture is completed with a burying-place, which is on the highest ground, with a great balcony above it for the sake of its fine view. Beside one of the double oval colonnades there is a canal of flowing water called Euripus, filled from a lake that lay higher on the hill, and no doubt falling into a basin below which is not mentioned, but would be on a terrace. There is of course a laurel grove, where pedestrians find shade, and also a pretty nymphæum. There is a fountain, not artistically made in handsome marble, but in a porous tufa-stone with grotto work all round it. Outside the walls the whole castle hill is planted with vines. Thus we find all the characteristics of the villa, and at the same time we recognise the different parts of a knight's castle of a later day, which are really descended from villa buildings. Everything is more compressed, so as to save space, and also to give the appearance from outside of a defended stronghold. In these homes of the proud Gallic nobles with their retinues of vassals, there was coming to birth the whole feudal system of the Middle Ages. The proof of this development is given in lively fashion by the works and letters of Sidonius.

In Italy itself, on the western side—that great road for the Germanic nations as they passed through Tuscany on the way to Rome—the devastations had been so many that little can have been preserved of garden culture until after the fifth century. But when the emperors perceived that a safer place of refuge was ready for them on the east coast, and took up their residence at Ravenna, circumstances were more propitious, and 150 years of peace were secured for the civilisation of the ancient world. The coast of Ravenna had already been covered with villas and castles built for pleasure. Nero had killed his father's sister Domitia, because he wanted her fine house, and he had himself built a great place here, that was still admired in the days of Dion Cassius. Martial, too, praises the glorious sea with its villas, saying he would like to end his days there. The pinetum was renowned far and wide: it was a wood of stone pines a mile in width, whose foliage yielded a cooling shade in summer, whilst in spring its floor was covered with countless flowers. After Ravenna had become the imperial city, everybody of importance migrated here from their unsafe villas in Rome. But Theodoric the Goth, who considered himself the heir of a Roman line, gave the order to his architect: "Our works must only differ from the old in being new," and he not only spared the old, but used it as much as he could as a model for his new creations. There is nothing at all left of his famous palace; but a mosaic in the church of San Apollinare Nuovo bears the inscription *Palatium*, which probably means the chief part of the building. Here again there is a colonnade flanked with a tower, and behind it a peristyle with other buildings. Between the pillars there are traces of statues, which have had to give place to the disfiguring addition of curtains.

Another memorial is the tomb of Theodoric. At the present day this building is in a thickly-wooded park quite outside the town, but at first there was a flourishing suburb. Just as Augustus built his own tomb in his lifetime, set a park round it, and then presented it to the people, so Theodoric erected this proud Rotunda; and we may assume that there were walks round about and shrubberies. It may be that the great time of Ravenna passed away with Theodoric, but the town remained as the headquarters of Byzantine art. And we owe it to this that Italy was not stripped of those models by which the monks, who carried on the culture and civilisation of the Middle Ages, were enabled to learn the art of gardening, and (though the threads were thin and few) to carry forward the traditions of antiquity.

CHAPTER V

BYZANTINE GARDENS AND THE COUNTRIES OF ISLAM

CHAPTER V

BYZANTINE GARDENS AND THE COUNTRIES OF ISLAM

THE Romans attempted to preserve their spiritual existence in a nursery of Greek culture in the East, on the threshold of Europe and Asia, hoping that their own traditions as masters of the world might be grafted on another stem. When Constantine, after his decisive battle with Maxentius, removed his Residence to the Bosphorus, and called it Constantinople, he found Byzantium a small but prosperous town. It had certainly never played an independent part, but its position and natural advantages had often made it a bone of contention for warring nations. Even in antiquity Byzantium had had the reputation of being a sacred city, by reason of its many temples and shrines. Both inside and outside the city there were stadia, gymnasiums, and racecourses near the Temple of Poseidon, called new by Dion. It was a Greek-Hellenistic town, such as we are now quite familiar with.

Constantine extended the boundaries of the Residence, which was already of commanding size, and in the year 330 laid the foundation of his imperial palace, the size and magnificence of which developed to such a point in the course of eight hundred years that travellers in the Middle Ages marvelled at it even in its decline. Unfortunately the accounts in literature are not yet confirmed by excavations, and the information about gardens is very scanty. The picture which we have shows what is found in all Byzantine civilisation —a combination of Græco-Roman and Asiatic elements. Thus, if ground-plan and site, courts and gardens, do not appear to differ much from those of the palaces we know as Roman, we must never forget that in Roman villas, at any rate, the Hellenistic style had a great influence. The imperial palace at Byzantium may have shown more variety than the Palatine, due to its long architectural history and the influence of the Church, but it was more shut in than Hadrian's Villa, and necessarily so, because of the restricted area of a town.

The palace reached on one side to the sea, and on the other to the great hippodrome made by the Emperor Severus. But even a palace built so late as the Triconchos, the work of Theophilus in the first half of the ninth century, is in some parts reminiscent of Hadrian's Villa, though it of course betrays Asiatic influence as well. This Triconchos is like a triple shell, as its name implies, a hall with three large alcoves; the front room is entered by three doors, and is called Sigma (Σ) from the shape of the Greek letter. It is oblong, and open in front, the fore part supported on fifteen pillars. The Sigma was a kind of theatre-box for spectators, from which they looked down on the court below when dances and games were going on.

The emperor's own box, with his throne, was a baldachin of an unusual kind, supported on four pillars; from it there were steps leading to the court, and these were used as extra seats for the imperial retinue. On the middle of these steps was a kind of arbour on pillars, which was probably a favourite seat. Perhaps we may see in the steps a late development of the commanding staircases which produce so great an effect in the Cretan palaces at Knossos and Phæstos.

The peculiar way that a palace theatre changes its style is noticeable, as the theatre in a town remains the same as ever. The staircase leads down to the court, which is really a peristyle, having a shell fountain in the middle made of brass, with a border of silver, and at the top a golden pineapple. This basin was filled at specially festive occasions with all procurable fruits, including pistachios and almonds, and the sweet juice of the pineapple ran out. This was called the Mystic Fountain of the Triconchos—a name which has never been explained, but which perhaps arose from some custom at the feast.

On the east side of the Sigma there were two brass lions, which spouted out water, and filled the whole empty part of the room with it, "which caused no little pleasure." But as the lower room, a hall with mysterious echoes, was also called Sigma, this may be

FIG. 98. A TERRACE-GARDEN IN A BYZANTINE MINIATURE

FIG. 99. A GARDEN AND PINEAPPLE FOUNTAIN

the one meant: it is on the same level as the peristyle, which has an apse, or a hemicyclium, on the far side. Nothing is said of plants in this court, but Theophilus was very fond of the place, doing his ordinary business and taking daily walks there, and he had five gardens put round it in the style of the palace, but what this means is unfortunately never explained.

It need not be said that all the other palaces had gardens round them; and of some of these we have a weak and very incomplete picture, as for example of the Anadendradion of the Magnaura, the reals how-palace. As the name tells us, it must have been rich in trees; one could pass into it directly from the bridal chamber; there was water flowing through it, for it had a bridge over the canal leading to the bath at the end of the garden. At fêtes and on reception days the servants made the whole garden into a kind of large tent, with the help of Sidonian carpets that were hung up; and on both sides seats were erected for the chief parties concerned and for spectators. Also the court of justice, the Lausiacus, had a garden, called Mesokepion, where malefactors were chastised with rods. The Byzantine emperors were very fond of their balcony gardens, and we hear a great deal about them. A poem on Justinian I. has enthusiastic praise for a little house, open to the air, and yet protected from all winds, built in a balcony-garden with a lovely view over the sea. High terrace-gardens are often depicted in miniatures, in the prayer-books and gospels of the eleventh and twelfth centuries (Fig. 98).

The chief garden at the palace was laid out by the art-loving Emperor Basil I., the Macedonian: it was near the new church he built, and was called Mesokepion. On the east was a great oblong space with porticoes on two sides, planted like a garden, or so-called Paradise, with trees and flowers of all kinds, and water incessantly flowing. In the galleries round there were pictures of martyrdoms. The narrow side on the west contained the church, but on the other side the emperor erected the Tsycannesterion, a place for games, where Persian polo was played, a ball-game which was popular so early as the days of Theodosius' court. The ball was thrown down by a player on horseback, and knocked backwards and forwards between the two contending sides.

Of the site as a whole we have no information from ancient sources, but we must make use of what we know about the gardens of Roman hippodromes and colonnades; in its whole area, which is a longish oblong, we see a Roman colonnade-garden that has pillars round it in the usual way, and the cloister of the Middle Ages will presently show something very like it as an addition to churches. We know a great deal more about the history of fountains in Byzantine sites, and Oriental influence is perceptible in certain peculiar decorations. The most conspicuous trait is the splendour of the materials used, wherein we find a purely artistic purpose. In that pineapple fountain in the Sigma the original intention of supplying water has been lost sight of; but the precious stream that flowed from it suited well with its own fine receptacle, and contributed to the delight felt by spectators in the actual costliness of the materials. In like manner two fountains on the west of the new church poured out wine that was supplied from below. Even those that had only water in them were made of costly stone, and round the border of one (put up at the church) the artist wrought hens, stags, and rams in brass, which threw out streams from a good height as well as directly into the basin.

Another feature is the naturalistic sort of imitations. A pineapple sprinkled the water; and that form of fountain was very popular at the time, the oldest example we know being the one at the Giardino della Pigna at the Vatican. There is a pleasant garden that is merely symbolic, which is given in a miniature (Fig. 99). A fountain of classical simplicity stands between two overshadowing trees, which are to indicate an avenue, a clipped tree meaning a shrubbery, and separate flowers to represent a garden. A more important specimen than this is a fountain in mosaic at Daphne (Fig. 100), a cloister near Antioch, where the water is falling into two shells, the lower one made of variegated marble, and then down into an ornamental marble bath. In this example the garden is merely suggested by a few trees, though by the side there is an arbour prettily formed by the interlacing of the tree-tops.

The love stories of the period are much the same. They are closely related to their Hellenistic predecessors, for there is nothing new in their stereotyped descriptions of a garden background. The taste for glittering materials is apparent, and the fancy for elaboration shows how the strong feeling of the Greeks for moderation in all things has been supplanted by the Oriental love of pomp and display. The narrators go on describing baths and fountains, and about the eleventh century there is a masterpiece of this kind in a tale of Eustathius, wherein there appears a pillar of many-coloured marble standing in a shell of Thessalian stone, inlaid with black and white marble; and the water breaks into a thousand sparkles. On the pillar sits an eagle with wings outstretched as if he would bathe in the stream; water spouts from his beak, and falls down the pillar side, reflecting a thousand hues. On the edge of the round basin a goat rests her foot and drinks, behind is a shepherd milking her; a hare also supports a fore-paw on the brim, and all sorts of birds—swallows, doves, hens, peacocks—pour out water from their beaks; the apparatus moreover makes sounds that imitate their different voices. Round in a circle are marble seats and footstools, and the myrtles are cut to make a green roof overhead.

FIG. 100. FOUNTAIN AND ARBOUR—A MOSAIC FROM THE DAPHNE CLOISTER NEAR ANTIOCH

But the well by the Basilica of the Emperor Basil is a proof that, conventional as their tales appear, they yet described particular objects after the models they saw about them. The naturalistic imitation of animals, even by reproducing the voices mechanically, found its way to the grand imperial receptions. Bishop Liudprand, who several times visited the court of Constantinople as ambassador from Berengarius and Otho I., gives an account of the emperor's throne-room of the Magnaura: he saw a brass tree, gold-plated, that stood before the throne, and on its branches gilt birds of different sizes, which sang each with its proper notes. The throne was so placed that it could be raised or lowered, sometimes being quite high up. It was guarded by two lions of enormous size, whether of brass or of wood he does not say, but overlaid with gold. At an appointed time they beat on the ground with their tails, opened their jaws, wagged their tongues, and roared. This display did not impose on the bishop, but that the Byzantines themselves took it as a real expression of the emperor's greatness is clear from the similar account that appears in the *Book of Ceremonies* of Constantine Porphyrogenetos. Thus we have met again with the metal tree first mentioned in the Mahāwamsa, but the Byzantine court was really only copying the splendour of the Persian Great King, whose throne was shaded by a golden plane-tree. Of Arabian development we shall hear anon.

The tree idea found its way into the love romances also. Although one of these tales bears the classical name of *Achilleis*, this name is but a veil of the slightest, and once stripped off there is only the ordinary fashionable love story, with its usual descriptions of tourneys, fair maidens, palaces, and gardens. In one of these lovely gardens—alas, too lovely, for words always fail the narrator to describe it—there is the golden plane-tree once more, and in it are birds that begin to sing as the wind waves in the branches. At this period the artificial tree had entirely dropped any religious significance.

FIG. 101. A SILVER TREE FOUNTAIN

Later we shall see how the Crusaders brought news of it to Europe. In the thirteenth century a French goldsmith, commissioned by the Great Khan of Tartary, was clever enough to make a beautiful object of art (Fig. 101) that has been preserved; it is a silver tree with gold leaves, and dragons' heads on the branches, shooting out some liquid (which is no doubt wine) into four vessels guarded by little lions. In another romance there is a marvellous bath in a park: the vaulting of the roof is a tree studded with jewels for fruit, and a vine made of gold climbs up the face of the wall and mixes with the branches of the tree. Mirrors are set round about, so that anyone who comes in may see the garden repeated. (In the romance by Tatius the simple mirror of the lake had to reflect the garden picture.) As a fact, this Byzantine romance is made on a real pattern, and there is talk at Byzantium in very early days of a mirror bath. One room in the palace of the Chrysotriclinium was "turned into a flowering rose-garden, wherein different kinds of many-coloured stones imitated the forms of living trees, and round them twined climbing plants, making an extraordinary combined effect." A silver enclosure, like a ring, shut in the whole place, which gave much enjoyment to the beholders. A garden-room like this goes back to a Roman model, based on an earlier original. Of its development at the Italian Renaissance we shall speak later.

There is very little information about the gardens of those Byzantine villas that were outside the town walls. We hear of some names that stir our imagination, such as the Pearl, beloved of the Emperor Theophilus, and the Philopation (a park near the city), well known later on. The private pleasure-gardens were probably not very different from those of the palace. Among the Byzantines it was the same as with their neighbours in the East—they all took great delight in the chase and in keeping up extensive hunting parks. Their summer homes were generally in such parks, and often there were many

FIG. 102. PERSIAN CARPET OF A GARDEN DESIGN, ABOUT 1600

houses in the same park. In the year 1146 Odo de Diogilo accompanied Louis VII. to Constantinople, and thus describes the Park of Philopation: "In front of the town is a beautiful, spacious, enclosed place with all sorts of animals for game, also canals and ponds, and ditches and caves, so that, instead of woodland, the creatures have hiding-places. At this delightful spot there are shining palaces, built by the emperor for coolness in summer—all indescribably grand." The Philopation was destroyed in the German invasion; its site was lower than that of the palace, whence the devastation could be seen.

All this is very like the magnificent castle upon which the enemies of Byzantium, the last Sassanid rulers, had spent so much. Dastogard, so loved by Chosroes II., "the Conqueror" (under whose rule the power of the Persians came to an end), that he preferred it to all places to live in, "was surrounded by fine hedges, and there were gazelles and ostriches, wild asses and beasts of the forest, even lions and tigers. Here lived this art-loving prince, till he was conquered by the Byzantines, and his beloved Dastogard was destroyed." But this was only an anticipation of the last fatal destruction that the Persians were to suffer at the hands of a new and growing foe, the Arabians. Chosroes preferred to live in a castle of his own than at the Residence at Maidan (the Greek Ctesiphon). There are still to be seen ruins of that commanding edifice, the imperial palace, which show the size of the throne-room; it was 300 yards long and 120 broad. Wonderful gardens lay round the palace, where the Sassanid rulers held their proud feasts. Only when the rough winter forced them did they leave their garden, which even then they could not altogether forgo. The artistry of his people came to the help of their lord.

FIG. 103. A PERSIAN CARPET IN A GARDEN DESIGN

In the days of Chosroes I., the most important of the Sassanid princes (531–579), a marvellous carpet was woven, sixty yards square. It was made of the finest materials:

The ground represented a pleasure-garden, with streams and paths, trees and beautiful spring flowers. The wide border all round showed flower-beds of various colouring, the "flowers" being blue, red, yellow, or white stones. The ground was yellowish, to look like earth, and it was worked in gold. The edges of the streams were worked in stripes, and between them stones bright as crystal gave the illusion of water, the size of the pebbles being what pearls might be. The stalks and branches were gold or silver, the leaves of trees and flowers made of silk, like the rest of the plants; and the fruits were coloured stones.

This is the earliest extant account of a Persian carpet, and it was known either as the winter carpet, or as the spring garden, of King Chosroes. Several carpets of the same kind, but of a later date, have been preserved (Fig. 102). The one thing that never varies is the formal regularity of the design, which is generally enlivened by water-animals round a tank in the middle. Encircling the centre part are paths and narrow canals, with straight beds between. Sometimes the whole border is treated as a canal, and made bright with fish, or again there may be a small island in the middle with water branching off (Fig. 103). These designs must have been very common, for in Arabian carpets even now the border is called *Su* (=water), and the foundation or ground part is called both in Arabic and in old Persian by a name that means Earth. These words, coupled with the earliest descriptions of a Persian carpet, make it seem more than probable that this is the original idea of all carpet design. Rugs, the real Persian furniture, served the same purpose as wall-paintings in other countries to bring into the house the picture of the beloved garden of summer-time. Even the material used was much the same as for the paintings on the walls and for the stuff they sat on with crossed legs.

FIG. 104. A GARDEN IN A PERSIAN MINIATURE

The Persians were delighted with this illusion of a garden, helped out as it must have been by the sweet scents of their artificial perfumes. On these carpets we can realise a picture, though a conventional one, of the pleasure-gardens and those surrounding the houses. The miniature garden pictures look like carpets (Fig. 104). The love of trees is an inheritance for all Persians; as in ancient days, they still love to have little rooms fixed up among the branches, and steps to help them up into the tree (Fig. 105). In the flower-gardens they liked to have cypresses and flowering trees, and their parks had long rows of trees and hedges and beast-cages, and here was the great playing-ground for the national game of polo, which their neighbours learnt with enthusiasm. That gigantic carpet of King Chosroes which the Sassanid rulers used (even Jazdegerd, the last of them) was looted with innumerable treasures at the victorious invasion of the Arabs in 637, and we owe our knowledge of it to the bewildered admiration of their historians. Before the mighty attack of the Arabian foe the Persian empire, like the Byzantine, perished in the dust.

It is a characteristic of Asiatic art that, in spite of enormous changes and revolutions among the nations, it remained strong enough always to rise again and become re-established. Nations and nationalities might change, but the tradition of art was never really broken. The Arabian writer Ibn Chaldun (fourteenth century) is perfectly correct in what he says of the superiority of the East. "In the Orient," he writes, "the arts had time to strike deep root in a long succession of centuries under different rulers, Persians, Nabatæans, Copts, Israelites, Greeks, Romans, and others. All the habits and tastes of a sedentary people's life—habits of which art forms but one part—were completely established in these countries, and the traces they left behind will never be obliterated." Among all stay-at-home arts gardening stands supreme, and it was quick to receive the soothing effect of Eastern tradition. One nation stepped directly into another's footsteps, took a lesson, and added what was already its own. The love of a garden is clearly born in Orientals, as is abundantly shown by their history; and a great part of their life is spent there. The enormous riches collected in the hands of a few persons made it possible, with the aid of architect and gardener, to procure the satisfaction of their every wish.

FIG. 105. SEAT IN THE BRANCHES OF A TREE—A PERSIAN PAINTING

All this was learned by the Arabs, those wandering sons of the desert, when they came to the land of Eastern conquerors, who were so soon to be conquered. Scarcely ever has a people shown such aptitude for adopting foreign civilisation, and setting a common stamp on every art. In Damascus they were pupils of Byzantine art, in Bagdad of Persian. In Northern Africa and Spain they found the teaching of Rome and later Hellenism; and wherever they were pupils they quickly became lords and masters. From the very beginning they felt great interest in horticulture, and in the Residence towns, which grew up with amazing rapidity under every new dynasty, the caliphs took the greatest care from the first to increase the fertility of the soil by a water system on a large scale, and also to convert the country places round into gardens, and then to settle wealthy owners there.

We can trace this sort of activity very clearly in Samaria, that marvellous town whose whole time for building was only about forty years (836–876), and which in that short period grew up out of nothing at all, so that in the tenth century its size was amazing and incredible to strangers. Its founder was Mu'tasim (a son of Haroun-al-Rashid), whose first care was to conduct water from the Tigris by canals. He entrusted the cultivation of a piece of land to each of his prefects, and trees such as date palms, also vines and many other things, were imported from Basra and Bagdad. As there was so much water at hand, everything prospered on the virgin soil. The geographer Ja'kubi, writing about Samaria in the year 889, says: "The country places were converted by Mu'tasim into houses for the upper class. In every garden there had to be a villa, and therewith halls, ponds, and playgrounds, for riding and the game of ball." (The Persian polo was only lately known to the Arabs, but by this time no large house was complete without a play-

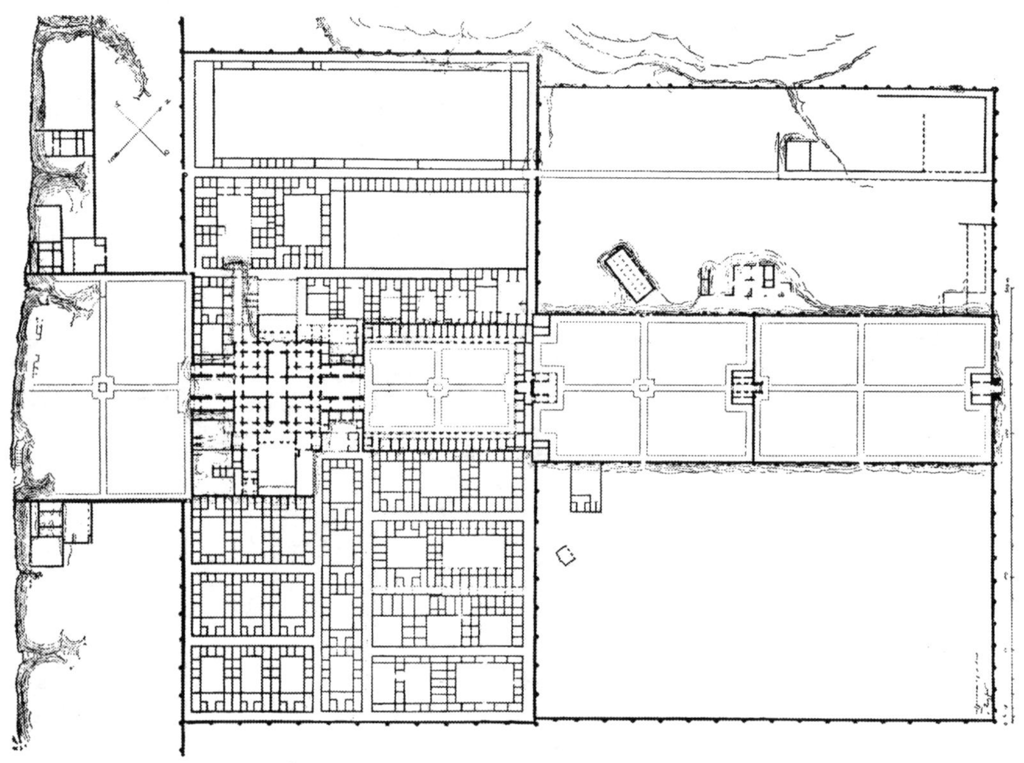

FIG. 106. GROUND-PLAN OF THE PALACE OF BALKUWARA, SAMARIA

ground.) The estates did extremely well, and rich people were eager to buy land, so that the price rose high.

A number of palaces sprang up, with a row of finely ornamented private houses near them; and then, to add to the marvellous, fairy-like character of the place, Mutawakkil, the son of Mu'tasim, made a second wonderful town to the north, which rose up in one year, was inhabited also for one year, and then, after the caliph had been murdered by his son, completely forsaken; and yet there was built in this place a palace whose area covered one and a third square kilometres—almost thirty-five times the size of Diocletian's palace at Spalato. A palace like this is entered by wide courts (Fig. 106), which are no doubt paved, and ornamented with flowering plants in pots in the ordinary Oriental fashion. These one would walk through before arriving at the state-rooms, and as a rule only after passing the master's own part; on both sides were the servants' apartments,

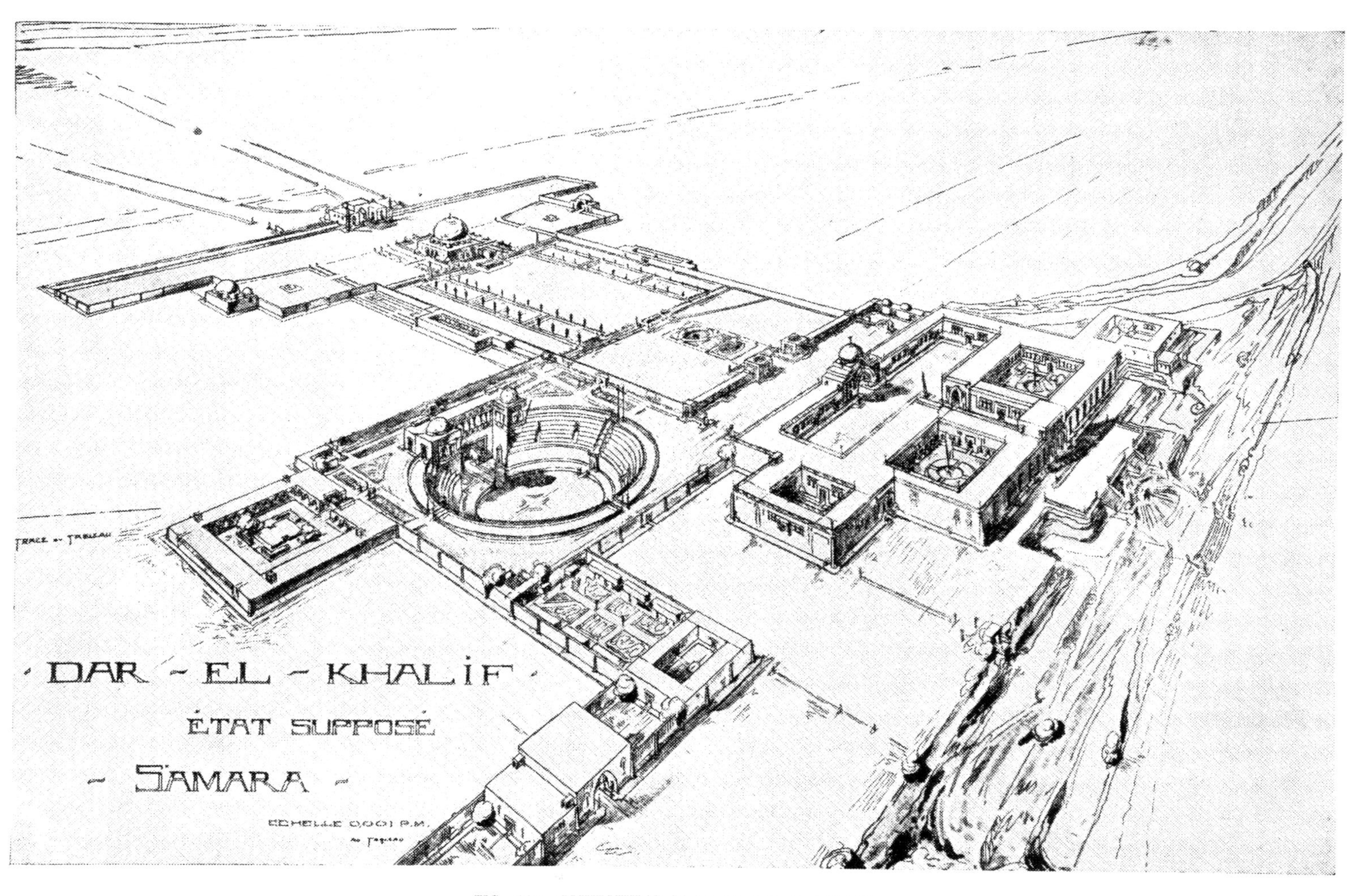

FIG. 107. CONJECTURAL PLAN OF DAR-EL-KHALIF

and the storerooms, and very often other large places, possibly tree-gardens and courts for games. The chief palace generally looked towards the river and often had large gardens in front. Such is the picture of the palace of the successors to the throne—the only one unearthed at present, and that imperfect—to the south of the town, the Balkuwara. Its most striking feature is the strict axial plan, which makes it possible to get a view on every side by help of the raised site of the chief palace. The garden on the river side is enclosed by a wall, with pillars that end on the bank with finely decorated pavilions. At the very extremity is a harbour for boats—an arrangement that calls to mind the palace of Tel-el-Amarna. In the middle is a kind of pond or large basin.

The same axial arrangement appears in the palace of the caliphs, which is in the centre, so far as the very superficial excavations allow us to judge (Fig. 107). Here, too, both the palace and its courts stand above the high river-bank as on a prominent platform, which may possibly be a garden. Farther inland one passes through an immense door into a great ornamental garden court, which is watered from a basin in the centre by means of a long canal. At the end there is a sort of grotto, with a basin in front of it, but its significance is obscure. Behind and crosswise to the main line is yet another enclosure, perhaps meant for the ball game—an arrangement that makes one think of the Byzantine Mesokepion. By the side of the garden is a large round place deeply dug, possibly for an arena, possibly for a large tank. All the pictures, and all the attempts at reconstruction (including the one given here), are purely hypothetical; but descriptions of other gardens are helpful for colour and form.

Even when there is a better supply of information from literary sources, as about the gardens of Bagdad, it is difficult to distinguish between the influence of Hellenistic and Sassanid art. The straightness of the main lines of plan and site is here, as at Byzantium, very like the Western style, whereas the ornament in plantation, water devices, and fountains, points to Asiatic influence. Bagdad was the residence of the Commander of the Faithful for nearly five hundred years, with the sole exception of the period of less than fifty years at Samaria. By reason of its favoured situation, it has maintained its importance, in spite of having been destroyed by Mongols in 1270, and notwithstanding many subsequent disasters. Immediately after the Residence had been removed back from Samaria to Bagdad, a great time of building began. The enormous array of the caliphs' palaces sprang up on the east bank of the river, and practically created a town of their own, with their many edifices and gardens, all included in an encircling wall.

A certain favourite of Haroun-al-Rashid, much addicted to wine and to poetry, made for his own use an extraordinarily handsome country place, at a little distance from the town, which later on he presented to the caliph's son, who added a maidan (racecourse), and a park for wild beasts. After the return of the caliphs from Samaria this place was greatly enlarged and built upon, so that certain Byzantine ambassadors, visiting the court of Bagdad in the year 917 (A.H. 305), reported that the number of palaces had increased to twenty-three. Among them the most conspicuous were the House of the Crown, called the Taj Palace, the Palace of Paradise, and the Hasani Villa, the oldest of all. They were all surrounded by one wall, and the Palace of the Pleiades was united to the chief palace by an underground passage three kilometres long.

Every individual palace had its own separate encircling wall. It very often happened that disagreeable near relations were shut up in a palace of this sort, as it might be in a kind of state prison. There they could enjoy the fine situation, and had any number of servants, but on pain of death they might not show themselves outside the walls. In such a way the Palace of the Tree was used for a long period. The gardens of the palaces reached down to the Tigris, and from its farther bank the old town looked down from the west with its domes and its palaces. The Byzantine ambassadors were amazed at the magnificence of what they saw. First they passed through marble halls and corridors at the so-called Riding House, beside which was a great animal park, with special houses. Various wild beasts were kept there: elephants, lions, giraffes, leopards. These were sufficiently tamed to eat out of a visitor's hand.

One of the most famous places of that day was built by the reigning prince Muk'tādōr, and named the House of the Tree. It stood in the midst of lovely gardens, and took its name from a tree made of gold and silver, standing in the centre of a great round pond in front of the large room of the palace, between the trees in its garden. This tree had eighteen boughs of gold and silver, and innumerable branches covered with all sorts of fruits that were really precious stones. On the branches sat birds which were made of gold and silver, and when a breeze passed through, they whistled and sighed in a wonderful way. At the side of the palace, right and left of the pond, were sculptures of fifteen knights on fifteen steeds, dressed in silks and brocades and girt with swords: in their hands they carried lances, and they were able to move forward in a line, so that it looked as though each knight was trying to hit his neighbour. Here we see how the fancy for mechanical imitations of nature has gone on as far as to include human beings. The golden tree and its singing birds were no longer a novelty to the Byzantine legates, and the caliph was only in his magnificent way extending a far-spread Asiatic custom. Wherever there was a fine court, there was bound to be an artificial tree. The Sicilian poet Ibn Hamdis describes a similar one made of precious metal and with singing birds.

After the House of the Tree, the thing that most interested the ambassadors was the new, or modern, so-called Kiosk. This word (*gausak* =kiosk) means a pleasure-house with its gardens. In certain cases the kiosk might be strong enough to serve as a defence at need for the caliphs, and for a considerable time. This "New Kiosk" had a fine garden round it. In the middle was a tank lined or filled with pewter, at that time more costly than silver, and a canal treated in the same material. The basin was thirty yards long, and twenty wide, and beside it were four grand summer-houses, their seats made of gold. All round were gardens, wherein grew 700 dwarf palms, each about eight feet high. The trunks of the trees were entirely covered with pieces of teak-wood, held in place by gilded copper rings. These palms bore large dates, and were so trained that ripe fruit was always there. In the beds were melons and other fruits. Palms set the stamp of the desert, so to speak, on all these gardens, since for the Arab they were the venerated trees of his true home, and gave him, all in one, his cool shade, his repose, and the beauties of the natural world. And so he took the palm wherever he went, and did a great deal towards naturalising it in the West. Apparently its stem was thought ugly, for we often hear that in rich men's gardens it was dressed in some costly covering.

There is a similar garden at the castle of the Tulunids at Cairo, laid out by the Sultan Rhumarawaih on a racecourse of his father's. Here too there were dwarf palms, whose fruit, mostly dates, could be reached and picked by anyone standing up, or even sitting. These trunks also were covered with gilded copper finely wrought. Between the copper and the trunk leaden pipes were introduced from which water was thrown upward. The concealed water appeared to come straight out of the palm branches, whence it was received in a fine basin and was then conducted by canals all through the garden. This may perhaps be the original model for artificial fountain trees, just as the other was for the little trees that were so popular later on all over Asia. There were sweet-smelling plants and all manner of roses in this garden, also oddly-grafted trees. The Arabs liked artificial culture: different fruits on one tree. Different grapes on one vine they thought specially pleasing. They liked to have flowers of unnatural colours, and to graft a rose upon an almond-tree.

All these things were to be seen in the Tulunid garden; moreover they planted saffron and other plants. Their gardeners cut plants into various figures, as well as the shapes of letters, and this had to be kept up regularly, lest a single leaf should stick out. We need not speak of the marvellous foreign importations that the sultan brought. A tower was made in open-work teak, to serve as a bird-cage, painted in many colours, with paved floors, and little streams purling. The garden, plants, and animals were all watered with well-sweeps, like the Egyptian ones used in the gardens of the Pharaohs. The birds, which filled this house with their sweet songs, not only found baths and food there, but also their nests, in pretty coloured pots prepared for them and let into the walls. There were also peacocks and fowls, and various wild creatures in great numbers running loose. Near this ornate bird-dwelling there was another house to live in, called the Gold House. Its walls were gilded, with inlay of lapis lazuli, and on them were curious figures, with realistic clothing and ornaments. In this living-room was a large tank, answering to the one at Bagdad that was filled with pewter, but this was filled with quicksilver, more valuable still; its rocking movement was meant to soothe and rest the sultan, who suffered from sleeplessness, and accordingly he had his bed put upon it.

Another North-African garden was unearthed near Tunis. It was of the seventeenth century, and had an immense four-cornered basin, into which water "poured down steep as a wall." At either end was a grand pavilion, standing on pillars of marble and mosaic. Other basins, kiosks, and tall shady trees made this dwelling the favourite haunt of the sultan.

In Arabian texts we get a great deal of detail; the lingering over curiosities and beautiful stuffs produces the feeling of romance very strongly, but there is no real grasp of the whole. To get an idea of colour and size, when there is such a poor supply of pictures to help the eye, we have to depend on examples of an earlier period—on Egyptian garden pictures, which give us straight avenues of palms, broken by basins and pavilions, or on the villas of Pliny with their xystus, figures and floral inscriptions and rows of dwarf trees, on Persian parks alive with beasts, and with little garden pavilions mirrored in water, on Byzantine marvels of irrigation, on the court-gardens of Pompeii; for it is out of all these things that the Arab garden has grown, and on all of them it throws, in return, some light of explanation.

The Roman-Hellenistic influence, always active, is evidenced anew in information about particular spots, preserved, one might say, by mere chance. In a story in the *Thousand and One Nights* a garden is described, shut in by a wall which is painted with all kinds of pictures; as for example two knights fighting, with pedestrians near by, and also birds. It is highly probable that we have in this story a place showing landscape and accessories as they are found in many Roman villas. It is certain that the tales are not mere inventions, from the confirmation given by the latest diggings at Samaria, for here have been found in private houses many frescoes that give human and bird pictures, and not only ornamental ones. There is unquestionably some Hellenistic influence, and yet it remains a surprising fact that Islam was not in every age hostile to pictorial art.

According to all we can learn as yet from excavations, the favourite kind of garden among the Arabs was the completely enclosed court-garden, either shut in by buildings or high walls very like them, and it is only at the great man's palace that the garden set high over the river bank gives the possibility of a fine view and seclusion at the same time. It is from this quiet scene of beauty found in the Arabian court-garden that their poetry takes its beginning.

The most simple place has at least one fine fountain; the paths are very often paved with costly marble or shaded with vines; sometimes the whole court is paved, and in that case the trees are planted in great boxes or in reserved corners where earth has been left. The beds are bordered with stone, and beside the paths are strong-scented plants, or clipped shrubs, salvia, myrtle, and bay hedges, and climbing plants hanging from tree to tree. The Caliph Kahir had at his castle, probably the Taj Palace in Bagdad, a garden laid out in a court. This was only a third of an acre in size, but it contained orange-trees brought from Basra, Omân, and India. On the regularly planted trees there gleamed yellow and red fruit, bright as the stars of heaven against the dusky foliage. Around grew all kinds of shrubs, sweet-smelling flowers, and plants. Many birds were there: turtle-doves, ouzels, and parrots, brought from foreign lands and distant towns. People said this garden was "the fairest one could see." The caliph, once so fierce and bloodthirsty, loved it above all things; here he drank his wine and held assemblies. When he was deposed and blinded by his nephew and successor, he would not let him have this last and dearest thing of all, but by cunning got it destroyed, giving out that there was a hidden treasure buried there.

The Arabs used flowers very extravagantly for display at their feasts, and had learned from an older world than theirs how to use scented oils and waters. Flowers decorated the table, and oils of roses and violets were indispensable for their toilet. Forcing-houses for cultivating plants were by this time also familiar to the Hellenistic and Roman garden-lovers. We hear marvellous tales about the display made by rich men, as for example the son of the eye-doctor who acted as physician in ordinary to Haroun-al-Rashid. This person, who drew an income of 800,000 fr., received guests in winter in a hot-house, that was laid out as a garden in the open; and behind the tapestry hangings on the walls his slaves kept up a constant fire. Similarly we know that at Hadrian's Villa there was a fancy for treating the rooms like gardens. Here beside the walls there were fountains, and there were water-cisterns in the middle; through the doors one had a glimpse of the flower-garden where antelopes were playing about, and in an aviary pigeons flew hither and thither: such was the reception-room of a mere private person in the time of Haroun-al-Rashid.

If we follow the Arabs into the West, we find the garden enlivened by many novel features. Their poetry, it is true, though incessantly dwelling on the subject, does not contribute much to our knowledge, for it mostly emphasises the feelings awakened in the singer by the beauty of the garden, and the poems often extol the renown of the wealthy owner. Such verses were often hung up as a decoration for rooms and halls. To this day we see laudatory poems in the inscriptions at the Alhambra. The Arabs came to Europe no longer as pupils; they brought with them their own native culture, and found a mighty inheritance to take over from Romans and from Visigoths. The founder of the Omiad Dynasty in Spain, Abd-ur-Rahman I., tried, in a way it is easy to follow, to ornament his new residence at Cordova as far as possible in the Eastern fashion. There was one villa, which he attempted to build on the model of a similar one at Damascus, and which he named Bussafa. In this he endeavoured to perfect the likeness to home surroundings by growing Syrian plants in his garden, in particular the well-beloved palm-tree, until that time as much a stranger in the West as was the exiled ruler himself. In one melancholy poem addressed to the favourite tree, he speaks of his yearning and his home-sickness:

> O lovely Palm, a stranger thou,
> Like me in foreign land,
> Here in the West dost languish now,
> Far from thy native strand.

Cordova and the whole of the Arabian domain in Southern Spain was in a flourishing condition during the comparatively peaceful period of the first Omiad Dynasty. In the time of Abd-ur-Rahman III., the banks of the Guadalquivir were covered with villas. This ruler in his poetry expresses a feeling about buildings common at the time of the Italian Renaissance:

> A prince should build who cares for fame
> And after death would keep his name.
> Behold the Pyramids how high,
> Though many monarchs in them lie!
> Where solid house and ground we find,
> We praise the Builder's noble mind.

After what we know, as a fact, about Samaria, it will not seem like a fairy tale when we hear how the wonderful town Az-Zahra first arose. In the neighbourhood of Cordova Abd-ur-Rahman built a town in honour of his beloved, after whom he named it: it was one household, arranged on a huge scale. The whole plan was of terrace formation. Lowest of all were extensive gardens, orchards, aviaries, and cages for animals, all confined within their own borders. On the middle terrace were the dwellings of the servants and attendants; on the third (perhaps on several) stood that world's wonder, the caliph's palace. The lovely view over the garden was particularly famous, the finest part of all being a golden hall and a vaulted pavilion; and here again there was a cistern of quicksilver. We may be quite sure that there was no lack of artificial ponds and fountains in the gardens or in the rooms. Very remarkable was the green cistern in the caliph's bedroom, brought hither from the East, and adorned with pictures wrought in gold of lions, gazelles, crocodiles, eagles, serpents, and any number of birds, with water gushing from their mouths. We are expressly told that Abd-ur-Rahman imported Byzantine artists, and in every sort of embellishment of his garden he certainly vied with Oriental architecture, at that time at the height of its glory in Bagdad.

The mountainous nature of Southern Spain was a strong reason for having the palaces built on the heights, though no doubt Roman examples also affected them; and when we hear praises of lofty terraces, fine views from pleasure-houses, colonnades, towers, it is just as though we were again listening to Statius. The great systems of irrigation, traversing halls and courts, answer quite as much to the needs of Orientals as to the fancy of the Romans. A hilly ground and terraces are wanted if the waters are to fall in a cascade "like the Milky Way.' Other features show clearly enough their Eastern origin, and in poetry the small pavilion plays a great part: it stands high and shines above the streams —an indispensable adjunct to the Asiatic garden. Here the master meets his friends, and gives them wine, and it is here that they practise the much-prized art of improvisation.

Wonderful things were related about the pavilion of Toledo's last ruler. In the middle of a pond he had a mechanical device set up that threw the water high into the air, so that it fell down again all over the sides of the pavilion. Thus the caliph was able to sit in a cool place with water all round him, even by lamplight. The same kind of pavilion was seen at a later time by travellers at Favara in Sicily.

Spain has preserved at least one wonderful legacy from the Arabs—the Alhambra at Granada. Much has been lost by actual destruction, neglect, and restoration, but the spirit of the ancient civilisation is alive and unconquerable, though we have now only the shadow of its former glory. Originally the place was (like Az-Zahra) one single and complete household; very likely the park occupied the lowest terrace—that is, the narrow valley between the real Alhambra hill and Monte Mauror, which still bears the name of Almadeo de la Alhambra, or Alhambra Park. The beautiful elms that are there now were probably planted by Wellington, though tradition would have it that this is a

FIG. 108. THE COURT OF THE MYRTLES, ALHAMBRA, GRANADA

FIG. 109. THE COURT OF THE LIONS, ALHAMBRA, GRANADA

burial-place of Nebrissian kings. Terrace-gardens climbed the slope of the Alhambra hill, and traces of them may yet be found in the "Jardin de los Arcades." The Renaissance gardens of Charles V. were there at one time with fine entrance and fountains.

The great Moorish castle stands on the broad spur of the hill. We get a picture of the old situation from the inside and from some of the courts: at the so-called Myrtle Court (Fig. 108) the original plan is very clearly shown. A canal, wide as a cistern, cuts through the whole length of this court, which has arcades round it. At each narrow side it ends in a fountain, and on both sides there are thick hedges of myrtle. We may suppose that originally the vegetation was tall and fine, the shrubs rich in sweet-scented blossom, and full of life with strange birds and other creatures. In the Court of the Lions (Fig. 109) there are no plants at all; but as there are great shell-fountains in all four squares there will have been plants too, either in pots or in separate beds set at intervals between the pavements. In Renaissance days the other courts were completely metamorphosed; but in spite of later changes the old state of things is preserved in the plantation of the little Patio de Daraxa (Linderaja, the Boudoir of the Sultana) (Fig. 110).

Fifty metres or so above the residence and the Alhambra hill, separated by a ravine, stands the charming pleasure-house of the Sultan of Granada, Generalife (the House of the Architect). The building is apparently of the second half of the thirteenth century, and therefore belongs to a late period in Arab history. In the year 1526 it was visited by Navagero, a Venetian noble; but at that time the supremacy of the Arabs had been dead

FIG. 110. COURT OF LINDERAJA (BOUDOIR OF THE SULTANA), ALHAMBRA, GRANADA

and buried for a lifetime, even in their last stronghold at Granada. Navagero found castles and gardens in ruins, but from his clear Italian style of description an informal account is to be obtained of their former condition—though it is wrapped in fantastic phraseology and is over-full of sentiment. Thus he writes:

One leaves the encompassing walls of the Alhambra by a door at the back, and walks into the lovely garden of a pleasure-house that stands a little higher. This, though not very large, is a striking building with wonderful gardens and waterworks, the finest I have seen in Spain. It has many courts, all abundantly supplied with water, but one in particular with a canal running through the middle, and full of fine orange-trees and myrtles. One gets a view outside from a loggia, and below it the myrtles grow so high that they almost reach to the balcony. The foliage is very thick, and the height so nearly the same that it all looks just like a green floor. There is water flowing through the whole palace, and even at will in the rooms, some of which are joined on to a grand summer-house.

Farther on he finds a court which is "full of greenery and wonderful trees," with a good conduit: if certain pipes are closed up, a person walking on the green lawn sees all of a sudden that there is water under his feet, and that everything threatens to be swamped, but he can turn the water off quite easily and without being observed.

There is another remarkable court, though not a large one, which has ivy growing so thick that the walls cannot be seen: this court stands on a rock, and has several balconies, from which one looks down into the deep valley where the Darro runs—a charming, ravishing view. In the middle of this court there is a fine fountain with a very large shell. The pipe in the middle shoots the water more than ten fathoms into the air; the amount is astonishing, and nothing could be more attractive than the appearance of the waters as they fall.

On the very highest part of the castle grounds, in one of the gardens, there is a wide stairway, leading up to a little terrace; from it there falls out of a rock the whole of the water that is distributed over the palace. There it is held back by a great number of taps, so that one can let it out at any moment, in any manner, and in any amount one pleases. At the present time the stairs are so made that after every few steps there comes a wider one, which has a hollow place in the middle for the water to collect in. The balustrade on either side of the stairs also has a depression in it like a small gutter. But above there are taps for each of these divisions, so that one can at pleasure turn on the water into the gutters of the balustrade, or the hollows of the wider steps. Also one can at will so increase the flow that it escapes all restraints and overflows the steps, wetting anybody who happens to be there: many little jokes may be played in this way.

In spite of all destruction, all restoration and rebuilding, we can reconstruct the past from Navagero's description, with its beauty and charm. The great court still shows its long canal cut right through the middle; and if the myrtles are not kept so well and are less tall, they still show the same plantation on both sides. From the loggia on the south we look down on the cool court, out of which we come to the chief garden-room, passing through a pillared double gate on the north side. This gate is the most attractive, owing to the vista which it gives, of all the relics that have been preserved of the architecture of the Arabs, who were unusually sensitive to the picturesque. On the other side lies the ivy-grown court that gives a view of the rock overlooking the Darro valley. Similar fine views are given from balconies on the west, and on the east we pass through the Canal Court, and other buildings put up later in the sixteenth century, into a beautiful little court, which is really a kind of water-meadow, with flower-beds planted at intervals.

Old cypresses tower above as a border to this charming picture (Fig. 111). Whether one of these ancient trees, called the Cypress of the Sultana, can really be six hundred years old, we must leave an open question. The many little streams, that appear in this place even more frequently than in the canal garden, flowing all over the cistern brims and the various

FIG. III. THE GENERALIFE, GRANADA—GARDEN COURT

paths, would make us suppose that here we see Navagero's so-called "conjuring mirror" (Figs. 112 and 113). A gate, belonging to a later age, leads out through the east wall to the terraces that ascend on the east side, and then to the water stairway described by Navagero —in these days to the Belvedere, which was built in the nineteenth century. The terraces proper are at the side of the steps, and were possibly always planted with shrubs and flowers.

The new feature is the water stairway, which we have never yet met with in an Arab garden; but as such features existed in Roman gardens, it is not impossible that the Arabs imitated them from Romanised Spain. It may well be, however, that Hellenistic gardens repeated this water scheme, like so much else, from Byzantine tradition, and the Arabs may likewise have known it in the East. A curious description lends colour to this view. An Arab geographer, who lived about the time (fourteenth century) in which the Generalife gardens were made, describes in his book on Egypt a water stairway as a work of great antiquity, saying that Joseph made it at God's command, and that at the time he was writing it was utterly destroyed. "One part of the estate of el-Menrah," says the author about an older place, "is the Park of Sahum. This is one of the wonders of the world, because of its marvellous waterworks; it has a water stairway between two towers; this is sixty feet long, and has streams above and in the middle, the upper one watered from the upper part, the middle one from the middle part, the lowest one from the lowest part, all in a strictly regulated amount." Here we first find fully described the practical joke of suddenly sousing an unsuspecting visitor, though this has appeared before in Roman gardens, and possibly the origin might be found in Hellenistic gardens, as artifices for water were very much admired. The practice in the Generalife is only a slight extension of a disposition shown in a thousand ways in Western Europe.

FIG. 112. THE GENERALIFE, GRANADA—A VIEW OF THE GARDEN COURT

The Generalife was only one of the pleasure-houses that the Nebrissian Dynasty built above their Alcazar (fortress), the Alhambra. Navagero says he climbed up to another and yet another pleasure-place, but that even then they were in ruins. No other country of those where the Arabs established their rule has such important relics to show. In Sicily everything is gone, and travellers in the fourteenth century, such as Alberti and Fazello, could find only poor remains, which were strung round the city of Palermo, like a "necklace on a fair lady's neck." Alberti does describe the villa La Zisa, which is still standing, but it is so completely rebuilt that one can scarcely find the court with the fountain that he admired so much. The whole of the house floor is traversed by a

FIG. 113. THE GENERALIFE, GRANADA—A VIEW OF THE GARDEN

stream, with a fine decorated hall above it of two stories and a vaulted roof. In front of the hall Alberti saw a wonderful fish-pond, into which streamed the fountain water, and in the middle of this was a good kiosk, attached by a bridge to the land.

Another Arab villa, which lay between Palermo and Monreale, is particularly interesting, because Boccaccio mentions it in the sixth tale of the fifth day, calling it Cuba, from the Arab *kubba* or domed pavilion. Traces of an important orchard, about two thousand feet long, have been preserved. "There was a splendid garden," says Fazello, following older accounts, "with all possible combinations of trees, and ever-flowing waters, and bushes of laurel and myrtle. From entrance to exit there ran a long colonnade with many vaulted pavilions for the king to take his pleasure in. One of these is still to be seen. In the middle of the garden is a large fish-pond, built of freestone, and beside it the lofty castle of the king."

Looking at these mighty erections of Arab princes, we are on the threshold of the great period for art in Italy, known as the Renaissance; but the movement in Western Europe goes to another measure, though now and then very insignificantly affected by Oriental influence, and has such different curves and lines, and such a different rate of progress, that this Arabian pocket of land in Western Europe can only be regarded as a little excrescence, outside the continuity of the great Asiatic culture. The centuries are linked together, and their cohesion prevents the possibility of any violent and vehement development such as happened in the West. However that may be, it behoves us to deal discreetly with the few relics that Western Asia has left us, spared from the storms and furies of the Middle Ages.

No country has suffered so badly as Persia. After the Arabs, host after host of barbarians raided the land; but though in the first fury of their attack they spared nothing, they yielded in the end to gentler customs, and the culture of the conquered never entirely perished. The most surprising thing is that such storms only raged above the tops of the mountains, and the regions that were sheltered and beyond their reach could continue to flourish, so that the "home" arts made progress—as Ibn Chaldun reports in the fourteenth century.

Persian ways and thoughts are liable to run into extravagances, and thus we find mystical sects, sometimes dreadful, such as the Assassins. Marco Polo, who travelled through Persia, did not himself penetrate into the secrets of the "Old Man of the Mountains," but he tells the wonderful story of the "Paradise" wherein he had his disciples carried in drunken sleep, though this is only on hearsay. The Paradise is a garden such as we know: the chief had laid out a most handsome place between two hills, and in this valley grew the sweetest flowers and the costliest fruits that one can possibly imagine. There were pavilions and palaces of every size and shape on terraces set one above the other and adorned with gold, paintings, and silken stuffs. Within were many fountains of fresh, clear water, and here there were streams flowing with wine, milk, and honey. Mandeville adds, though he admits he has it only from report, a characteristic that is purely Asiatic: "Over the head of Prester John there is arched," he says, "a bower of vines with growing grapes, white and red; the fruit is made of precious stones, the foliage of gold."

When, a hundred years later, Tamerlane (Timur) the Mongol fell upon the whole of Asia, he destroyed many nations with the sword; but even this bloodthirsty tyrant was a great patron of the arts, and tried to immortalise every fine feat of arms and every other joyful event by raising an architectural monument. The land of his fathers, Bokhara and Samarcand, flourished once more under his rule. Bokhara had in the ninth century under the Emir Ismael enjoyed a time when it was encircled with villas and palaces, whose orchards were praised by Arabian travellers. Tamerlane had eleven large canals made round the town, and it is said that on the banks of one of these there were two thousand pleasure-houses. Without neglecting the other towns, Tamerlane made Samar-

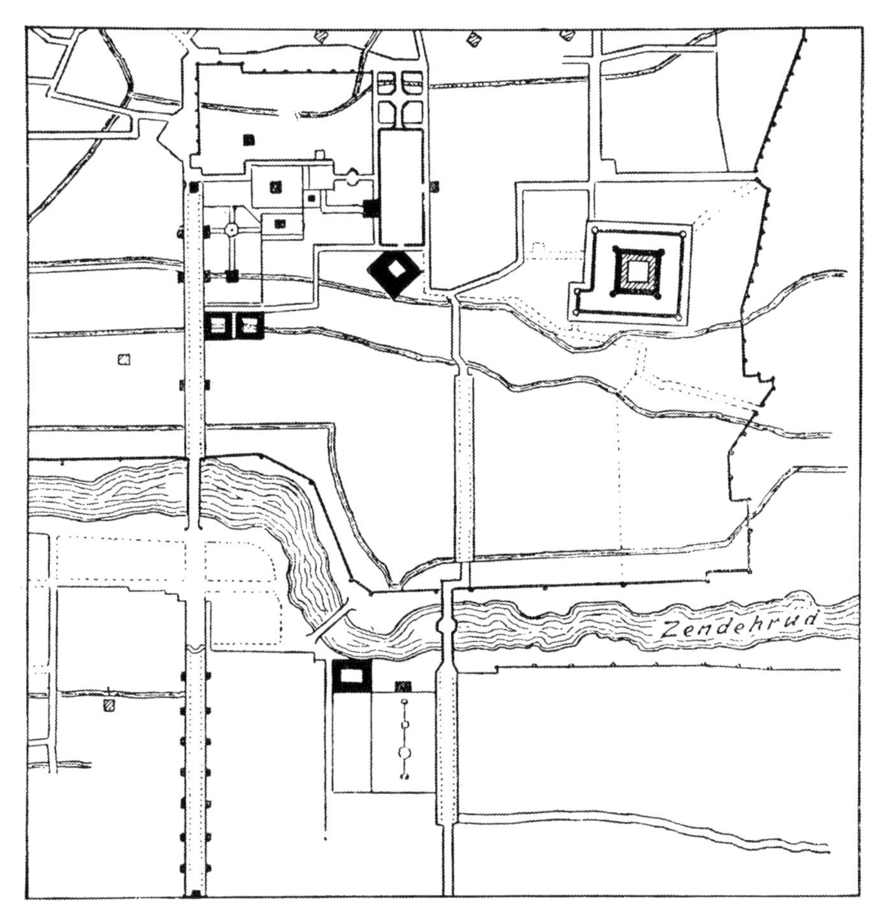

FIG. 114. ISPAHAN—PART OF THE TOWN PLAN

cand his capital. The gardens were one and a half to two square miles in area, and on the east side was the Bagh-i-Dilkusha, "the garden that cheers the heart," joined on to the town by a long avenue. But the favourite place of all, which its owner called his hermitage, was Bagh-i-Bhisht, "the Paradise Garden." Sheref-ed-Din, Tamerlane's biographer, describes it as made entirely of the famous white marble of Tabriz, built on a terrace having a deep trench round it, with two bridges leading over into the gardens, and a park for wild animals at the side. Another palace was famed for its mighty poplar avenues, and was called the Poplar Garden.

Of all the Asiatic races who were then in eternal conflict for the leadership of that part of the world, none is without descriptions of particular gardens, though no particular one is conspicuous by its individuality. The Turks, moreover, have wonderful places to tell of, and the advanced state of civilisation enjoyed under the rule of the Sultan Bayezid II., Tamerlane's great opponent, may be gauged by what we hear of the erection of a madhouse at Adrianople. In the midst of a garden, or paradise, the historian tells us, there was a very lofty and solid domed building, the top open like the dressing-room of a bath-house. Under the open roof was a cistern with streams and fountains splashing. Under the dome were winter and summer apartments, eight of each, and the winter ones seem to have been at the top. Their windows look alternately towards the cistern and

FIG. 115. TSHEHAR-BAGH, ISPAHAN

towards the garden. The summer rooms were open, with marble walls on three sides. The garden had long avenues of roses, vines, and fruit-trees, with many fountains, and in spring an abundance of flowers—jonquils, Greek musk, narcissus, wallflower, pinks, basil, tulips, hyacinths, and others. These flowers were especially valued for their fragrance, a healing property in sickness, particularly love-sickness.

A time of fine flowering came to Persia itself after many storms had been weathered, when in the sixteenth century the whole kingdom was united in the Sufi Dynasty. We have reached a time when our hunger need no longer be appeased with vague if boastful accounts. At this period Persia was much admired and much visited, and the wonders of the buildings were not only written about by travellers but actually surveyed, and thereby a great deal has been saved for our inspection which must otherwise have been lost for ever. This period of the final greatness of Persia brings to an end our inquiry into the story of West Asian gardens. The greatest among the rulers, Abbas the Great, who reigned from 1587 to 1629, removed his residence to Ispahan (Fig. 114).

He gave the town its first form, now much blurred. The chief thing in the town (and everything else was grouped round it) was called the Maidan, and it clearly shows how firmly established was this playing-ground, even if it had never been converted into the town market-place. It is rectangular, 386 by 140 metres, with avenues round, and an encircling ring of a two-storied colonnade. The lower floor was used for magazines, and the upper as open boxes for looking on at games and feasts in the square. On the west, Abbas made the great street called the Tshehar-Bagh, the "Street of the Four Gardens" (Fig. 114 and Fig. 115), more than three kilometres long, and crossed by a bridge

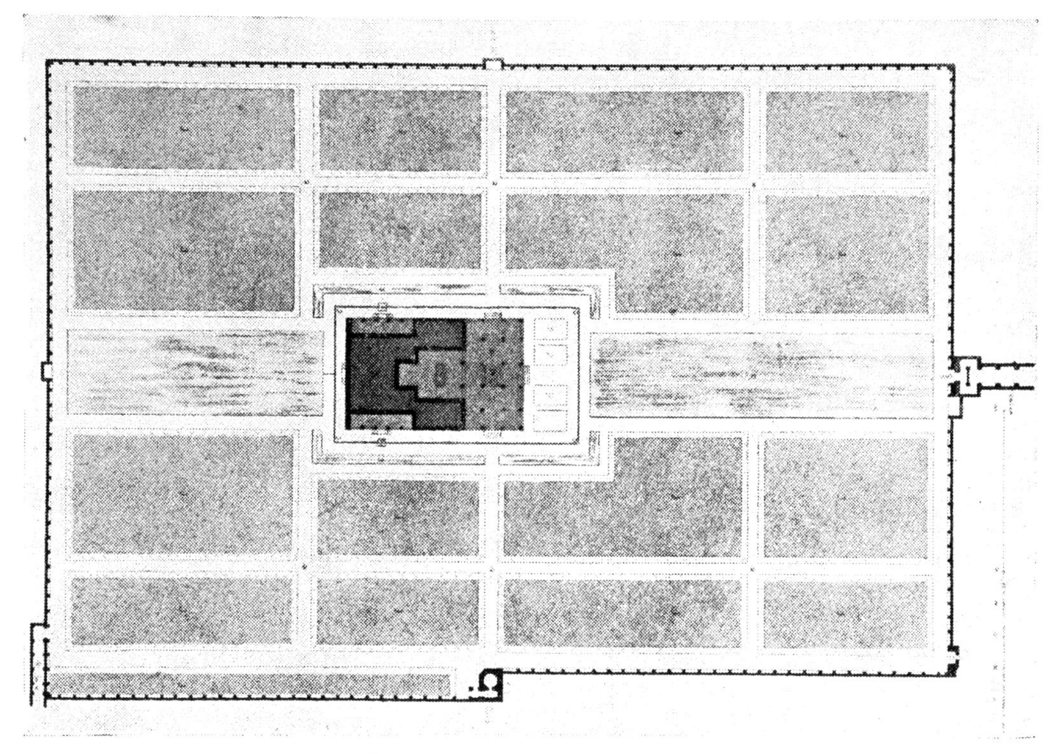

FIG. 116. TSHIHIL-SUTUN, ISPAHAN—GROUND-PLAN

of two stories. The street mounts up by low wide terraces, with a canal in the middle, which on the level of each terrace widens into a cistern.

Chardin, the French traveller of the seventeenth century, describes the whole place minutely: the cisterns and the canal are bordered with stone paving, so wide that two men could ride side by side. The tanks are of different sizes and shapes, but each has its own spring, and the water falls in cascades from terrace to terrace, giving a succession of pictures for anyone who is walking up. At each end of the avenue there are two pavilions put crossways to the street to form its end. Similar ones serve as entrance gates to the gardens, which lie along the sides of the street. Each of these gardens has a second pavilion in the middle, nearly all gilded over, and of the same size, though of different shapes. "Between the Maidan and the Tshehar-Bagh is the wide square of the palace precincts, embracing various pavilions with gardens round them, among which the most noteworthy is the Tshihil-sutun with its forty pillars (Fig. 116). The original building was raised anew by Abbas the Great after a fire, at the end of the seventeenth century, but the garden kept to its old ground-plan: inside a rectangular wall stands the pavilion, almost in the middle; at its narrow side in front is a fore-hall with a heavy wooden roof supported on three rows of pillars, six to a row." A narrow canal flows round the whole of the pavilion, cutting it at both ends; issuing from the terrace of buildings, a very wide canal passes through the whole garden, which is mapped out in regular beds, with

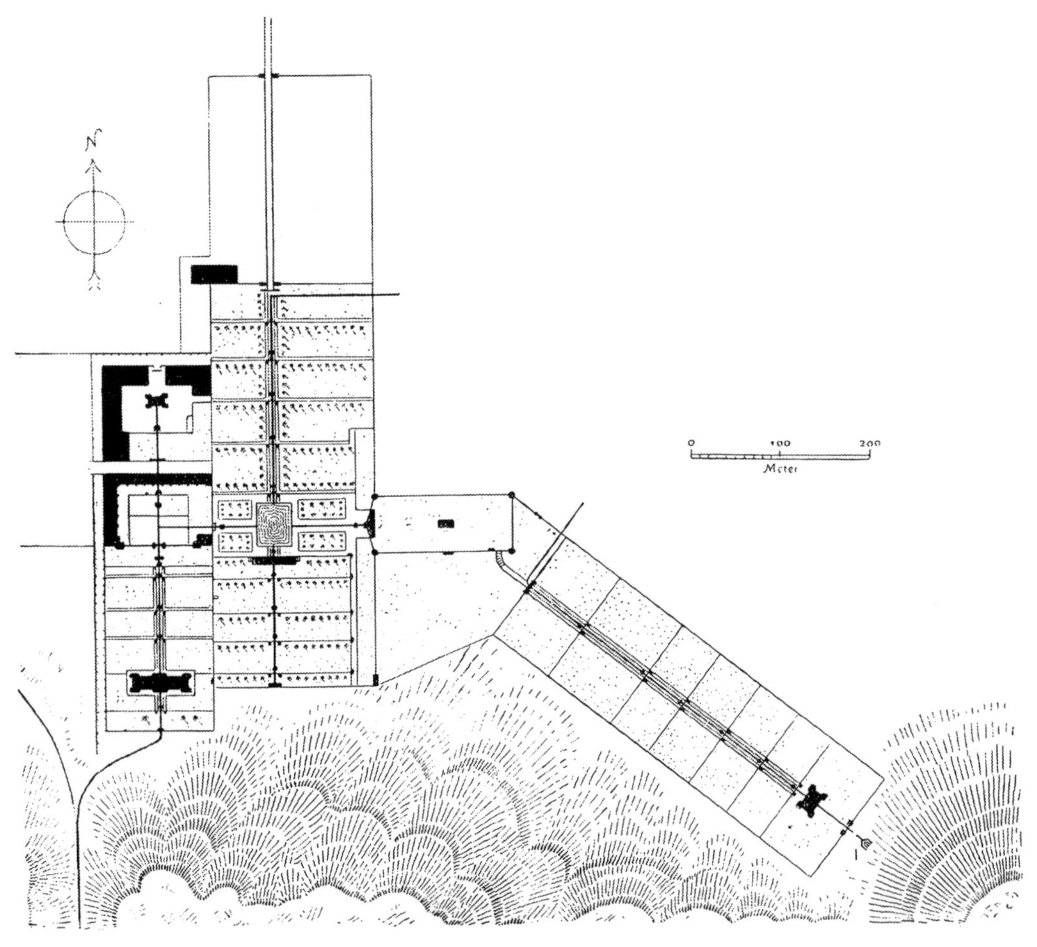

FIG. 117. ASHRAF, PERSIA—GROUND-PLAN OF THE GARDENS

avenues between. One of the rooms inside has an alcove adorned with decorations that recall the Byzantine Golden Triclinium. "Over the whole field of ornamentation there is a smooth raised surface, kept in place by strongly-rooted trees, on which are leaves and bunches of flowers. In the branches sit bright birds of many colours and curious appearance." Very likely most of the pavilions were the same.

Chardin also describes the palace precincts outside the town, the pavilions, and the great quantity of water, saying that "so much water flowing in these grand palaces makes one feel it is fairyland." The gardens are arranged in long avenues, and have beds filled with flowers. He considers an eight-sided pavilion of two stories very wonderful, for water falls down from it over the terraces, "and if one stretches a hand out of the window, it is at once covered with water." The pavilion of the caliph at Toledo was much the same as this.

The large rectangular water cisterns on whose glittering mirrors the Persian looked down with such delight, were called by him Deriatcheh, that is, little sea.

Fortunately there has been preserved, far from the town, still recognisable in its ruins, the site of a real villa urbana. This is at Ashraf, a few kilometres from Astrabad, on the slope of the Elbruz mountains, and dating from the time of Abbas the Great.

FIG. 118. PAVILION IN THE SHAH'S GARDEN, ASHRAF

"There are seven perfectly regular, rectangular gardens, not arranged to correspond to one another according to any main design, but simply placed side by side as would be most convenient on the land they occupied." So does Sarre describe this place (Fig. 117), and the travellers of the seventeenth century have complained, just as he did, of the want of unity in the plan, and this without reflecting that it was the late Italian Renaissance which first conceived the notion of that unity which they desire here.

As always in the East, these Persian gardens are separately walled in, each with its chief buildings arranged as terraces, dropping towards the north and north-west. Through a large fore-court one arrives at the head garden, called Bagh-i-Shah, the Shah's garden. This is the largest, just short of 450 by 200 metres, and it rises in ten terraces. A wide canal between walls cuts through from one terrace to another, falling down the middle in cascades, and streaming through the pavilion (Fig. 118) which overlooks the open pillared

FIG. 119. A POND IN THE SHAH'S GARDEN, ASHRAF

hall on the fifth terrace. On this, wider than the others, the canal broadens out into a rectangular pond (Fig. 119), enclosed by four flower-beds, with an arm of the canal separating them crosswise. Most of the other gardens are laid out in the same way, including the one at the top with its domed pavilion, and the Bagh-i-Sahib-Zeman, i.e. the "Garden of the Lord of Age," whose grand pavilion is built like a palace. The harem has its own "Home Garden," with an unusually high wall round it. Further there are high terraces to the east of the main garden, approachable from this place by steps, and serving as a temporary abode or reception-rooms for the ladies. One striking ornament of this garden—which has now become a sort of romantic overgrown wilderness—is still to be seen in the cypresses, bordering the pond and canals with their magnificent old trunks.

The ancient garden traditions of Asia have remained unbroken to this day, and what the Sufi Dynasty did for Ispahan, the shahs of the Kahar Dynasty did for Shiraz, when at the end of the eighteenth century they moved their residence to its fortunate fields. Its gardens of great beauty earned incomparable fame from the earliest times, and even now all travellers who make the difficult journey hither from the north take great delight in them. Shiraz seems to be much the same even in a period of revival: not only is the grave of the poet Firdousi cared for—it was neglected for a long time—but late years have seen the restoration of the wonderful steep terrace-gardens, which the first shah of the Kahars, Agha Mohamed, set up about the year 1789 in place of an older garden. This bears the significant name Bagh-i-Takht, the "Garden of the Throne," or the "Throne of the Kahar-Shah." In many narrow terraces, getting smaller as they ascend, it mounts to a wide pond, whose enclosing walls are faced with marble, and adorned with alternate pillars and arches; there are covered stairs within the walls that have no architectural connection, and there is a group of pavilions at the top; in the restoration there is also a corner pavilion on the lowest terrace. Each of these was, no doubt, planted with trees and flower-beds, and outside them is a high garden behind the pavilion on the side of the mountain.

The picture of the restored place (see Fig. 33, Chapter ii.) gives a better impression of the garden than the later one. It was described by Jackson in 1906 as having terrace above terrace, fountains, canal, a stream flowing in cascades over marble slabs into cisterns edged with stone, walls to enclose the watercourses, cypresses and orange-trees by the side of the paths. The similarity of the whole scheme to the hanging gardens described by Diodorus has been observed already, although this garden leans slightly towards the steep hill on one side, and so loses something in symmetry of design. The gardens on the plain show, however, the simple lines that are familiar from earlier times; no garden is without a canal cutting through its middle (Fig. 120) or small bright cascades interrupted by basins; very often there are cypresses in avenues, with simple beds between. In the axial line the pavilion generally starts with an open pillared hall or veranda, very often with a fountain basin on it. And in Persia, wherever fertilising water is found, there appears an oasis with its fine gardens, such as Fin-nahē at Kashan, where the springs come down from the neighbouring hills (Fig. 121).

Though flowers and many kinds of trees and shrubs have in the course of centuries given a different aspect to these gardens of Western Asia, there is but little change in the general scheme of the picture. The civilisation of other races seems only like a new

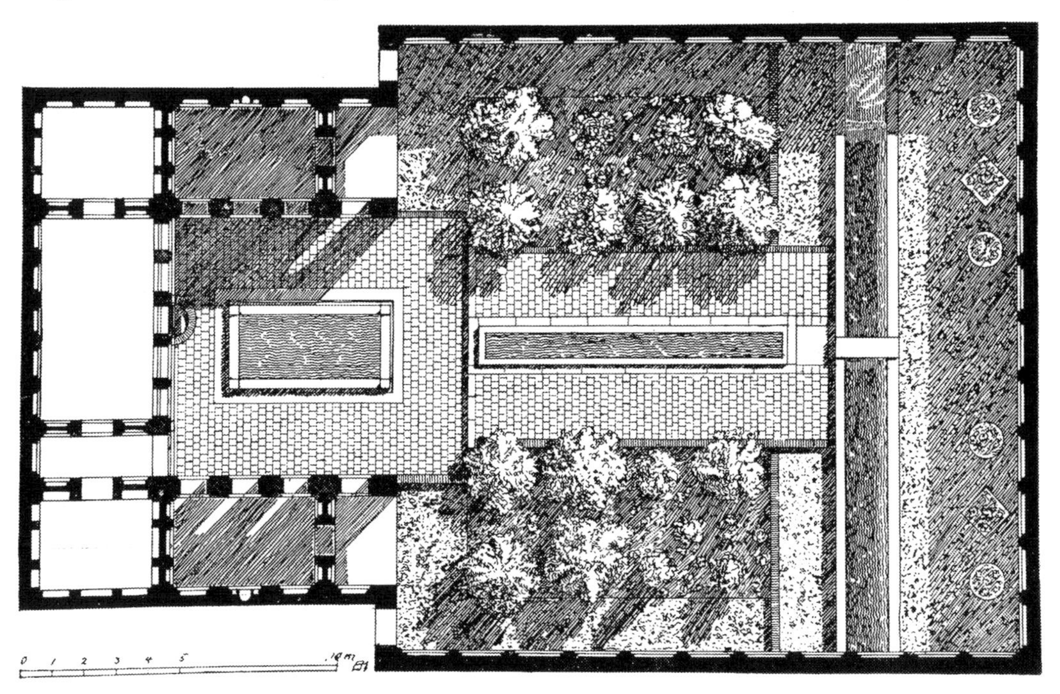

FIG. 120. A GARDEN HOUSE AT SHIRAZ

FIG. 121. GARDENS AT FIN-NAHE

graft on an old stem, and the scion has grown in obedience to the impulse communicated by the old sap.

An entirely different picture opens before us when we now turn our eyes to the West. The old civilised nations were far more fundamentally demolished by German hordes than ever before by Asiatic marauders. Yet one advantage has been gained from this utter destruction—the roots of the new development are more open to investigation, in so far as they draw their nourishment from ground that had been fertilised by the valuable legacies of former civilisations. The awakening of the new world was now in the hands of a religion likewise new, in which lay the seeds, still unseen, of all that culture which the lands of the West were to unfold. Christian monks were destined to prove the guardians and promoters of the art of gardening in the West.

CHAPTER VI

THE MIDDLE AGES IN THE WEST

CHAPTER VI

THE MIDDLE AGES IN THE WEST

THE history of the beginning of Christian monasteries is somewhat obscure. We only know that in the East men were attracted by the hermit's life, and even after they had taken to living in the cloister as a community, they preferred separate isolated cells. These monks lived in the early days (as so often later) on alms that came from outside; and therefore we cannot assume that they cultivated gardens in any regular way. For monks, like other people, only began to busy themselves in the garden when they were driven by the necessity of working for their own livelihood.

The development in the West seems to have been very different, even from the start. We find quite early that like-minded brethren form a community, where they live together with no personal property and the same religious practices. Saint Augustine assembled his company about him in this spirit. He tells the tale of the foundation at Hippo: "I assembled, in a garden that Valerius had given me, certain brethren of like intentions with my own, who possessed nothing, even as I possessed nothing, and who followed after me." Just as in another age Plato, Epicurus, and Theophrastus took their pupils into their gardens; and just as, still earlier, the Buddhist saints withdrew into the garden given to them by their Indian co-religionists, so these Christian brothers with their great teacher annexed a garden—perhaps in slightly conscious imitation of their ancient predecessors. Later than this Augustine built his church (with its cloister site corresponding to museum, exedra, and portico in the old philosopher's garden); and from the account given it is obvious that in this particular instance the habit of meeting and living together in a garden preceded the foundation of church and monastery. The buildings of a monastery were grouped round a peristyle quite as invariably in the West, where the monks had the strict discipline of a life in common, as in the East, where they were allowed more personal freedom; for even the isolated cells in an Oriental cloister were mostly grouped round a central court.

The old Christian monasteries have so many forms, ranging from the small town house to the magnificent villa urbana, and yet are so persistently of one original type, that it is hard to point to any immediate predecessor. A place of this kind certainly satisfied the needs that Southern people would feel; it gave the monks two things which they required before all others—a common centre and complete seclusion. No doubt cloisters have often been made from forsaken villas, rebuilt and utilised, and we also hear of estates and great houses being presented by rich fellow-Christians. The "hortus" that Valerius gave to Augustine will have had some building attached, for the word "hortus" nearly always means villa.

On the other hand, we have already found many examples of the portico in connection with gardens; and seeing that there was generally a portico, always planted as a garden, adjoining the ancient basilicas, one is reminded of the fact that the portico of the monks, the cloister, was always close to the Christian basilica, the church. But most of all do the ancient cloisters of the Forum, and the house and temple of the Vestal Virgins, show a likeness to the Christian places, in the grouping of their parts; and so it may be that the problem of every Order, desiring to live together, and apart from the rest of mankind,

FIG. 122. A PARADISE AT THE MONASTERY OF ST. MARIA LAACH

has never found a better solution, either in antiquity or in the Middle Ages, than this of the cloistered life, although the actual manner of life of the famous priestesses and that of the poor monks seem poles apart.

The front court of the early Christian basilica is adapted from the antique portico, which was sometimes called Paradise, sometimes Atrium. It appears in very different forms, either covered in—even with an upper story—or an open pillared court. Here we may see a change of meaning analogous to that which we noticed before in the Greek word "xystos." We saw how clearly it is established that the name Paradise was given to the Roman portico-garden in Byzantine times, after the great hunting-grounds of Persia; and in precisely the same fashion was the same name Paradise applied to the open pillared court of a Christian basilica, where sinners had to wait before they gained admittance "in ecclesiam." Later on, the covered-in front court, which fulfilled the same purpose, and also was a sanctuary, received this name of Paradise. The so-called "Expulsion of Adam" at Halberstadt seems to be connected with this last meaning given to the word: it was a popular fête, representing the expulsion of the first man from the Garden of Eden, and it took place in the paradise that stood on the west front of the church. The Latin name "atrium" was applied to the open court. The tradition which the Bollandist Fathers believed in may have arisen from the contiguity of these two places; they held that the Greeks were the first to plant trees in the atrium.

FIG. 123. THE CROSSWAY AT S. PAOLO FUORI LA MURA, ROME

These open courts were either paved all over (like the atrium with pineapple fountain in the old basilica of St. Peter's at Rome), or—in the earlier period this was more usual—they were planted over. In the church of St. Felix, which was built by Paulinus of Nola in the year 900, an inscription is put over the door that leads into the "pomarium," and bears the words, "Exitus in Paradisum." And even to-day it is possible to see a paradise with plants, as for example before the west front of St. Maria Laach in the Eifel (Fig. 122).

Unfortunately we do not know what the paradise looked like which was built "in Roman fashion" after the rebuilding of Monte Cassino in 1070; we only know that its whole south side was occupied by a great well. There always had to be a well, and it was needed partly for religious purposes. The pineapple fountain in the open atrium at St. Peter's was probably one of the oldest: but there may have been the same sort of thing even in the ancient world, and certainly this one has served as a model for many others, especially in the East. The idea spread in the West also, and the pineapple appears in the front court of the minster at Aix. The spaces between the crossways that marked the centre of the cloister were planted like the paradises of an earlier time, as at S. Paolo fuori in Rome (Fig. 123).

In the eighth century Pope Hadrian had this place "most beautifully restored"; by that time, as a fact, it was a complete ruin, and cattle and horses browsed contentedly on the vegetables. These courts were not often planted as kitchen-gardens, perhaps only when there was no larger place to turn to: more often they were used as churchyards, generally full of trees and flower-beds, with a well—a place where the brothers could meditate.

We often hear of their beauty at a later period. William Rufus once in his headstrong fashion made his way into the abbey at Romsey, where Matilda (Henry's future wife) was staying for a time, and the abbess, afraid the child might come to some harm, dressed her in a nun's habit. The king stepped out into the cloister, as though he "only wanted to admire the roses and the other flowering plants," and let the child go by unmolested among the nuns. Still earlier, in the tenth century, a pious singer describes a cloister at the monastery at Reichenau: "Before St. Mary's house on the far side of the threshold is the garden, well-nursed, well-watered, and lovely. About it there are walls, boughs swinging every way; it glows under the light, like an earthly Paradise."

From an early date the garden was not limited to the chief court. The founder of the Benedictines, more than any man, inspired the cloister life of the sixth century in Western Europe; and he at once ordered that "all the necessaries" for the support of monks should be supplied within the walls, and among these "necessaries" water and gardens stood in the first rank: of course these gardens were for herbs and vegetables. We can only guess how far the establishments founded by St. Benedict himself, and especially the mother-cloister, were able to comply with his demands. The mention of a tower, and a portico, where St. Benedict lived with his pupils, makes us think of pictures of a Roman villa; but in any case the Benedictines, whose rule enjoined work in the garden, were the men who handed down the practice of horticulture right through the Middle Ages.

Those Orders which were not influenced by the Bènedictine Rule, and forbade the monks to do farm work, still seem to have thought a garden indispensable. The Spaniard Isidorus in his Rule makes a special point of having a garden within the cloister, attached to the wall and entered by the back door, so that the monks should be able to work there and not have occasion to go outside. There was a certain tradition in the old Roman provinces about the cultivation of the choicer kinds of fruit, and it is hard to say how long it survived the storms of the Middle Ages, whether the monks are to be connected with this tradition, or if they started afresh on their own account, as is doubtless what did actually happen in the case of the German nations farther east. It is well worth noting that in Norway even to this day none but the finest and choicest fruit-trees are found on the site of an old monastery.

There is a very clear picture of a great monastery (according to the Rule of St. Benedict) in the plan preserved at the library of St. Gall (Fig. 124). It was sent to the abbot of St. Gall in the year 900 as a model plan. The whole design, which really includes all the "necessaries" within its walls, falls into three divisions: first, the church with the buildings for Regulars in the centre; secondly, on the north-east, schools, hospitals, and guest-houses; and, thirdly, to the south-west, stables and farm-buildings. Not only the monastery proper and the church, but almost all the other buildings are grouped round a centre space, an open court with a portico. The cloister for the monks has four doors, and paths leading across the square. In the very centre there is a little circle drawn, with the word "savina" (which perhaps means stoup for holy-water) written on it, surrounded by flowers and grass; also the four squares left blank were intended to be used for flowers. Perhaps this is indicated by the ornament in the design, although it may mean to suggest a mosaic pavement, seeing that for some strange reason there is no sign of water in this plan, which is otherwise so complete. For such a place there would have to be plenty of water; moreover, the founder had included water among the necessaries. Wells may have been omitted from the plan because the idea was to distribute the water in suitable proportions over all the various parts of the ground.

The school and the hospital have similar peristyles; and other buildings, the guest-house, a second school, etc., seem to have a covered atrium as their central place. Though we learn so little about the planting of the peristyle, the artist is most explicit as to the special gardens at the school and hospital. South of the hospital is the physician's house, so arranged that the very sick patient could be taken there. On the west side is the physic-garden, a small quadrangle, divided into sixteen straight beds. This may have been following Roman tradition, but the raised beds, straight and square, point to the ever-present need of careful tending. Poets in the later Middle Ages have thought a garden like this looked like a chess-board with the pieces moving about; and the parterre no doubt originated from this arrangement of beds.

The medicinal herbs which grow in our cloister gardens are plainly marked by name on the plan: first come roses and lilies, and then sage, rosemary, and other herbs that look pretty and are aromatic. Thus this little garden gave not merely healing medicines to the sick, but also a very charming view to the convalescent. The original idea, the germ, of all specialised flower-growing is here in this physic-garden; and so deep down was it in the heart of man that even when horticulture was highly advanced, in the days of the Renaissance, the place where flowers were grown was always called the garden of medicinal herbs.

To the north of the hospital is the burial-ground, which is also, in accordance with the practical good sense of monks, a garden regularly planted with fruit and other trees. In the middle lies a cross that may perhaps be a mosaic in many colours. Between the rows of trees, also set out regularly, are the monks' graves. The trees too are planted according to plan; but the designer was not much of a botanist; he has done his work very casually, simply copying from Charles the Great's *Capitulare de Villis*, without considering whether the kinds of fruit he names would be forthcoming in the neighbourhood of St. Gall. One must needs feel very doubtful about the fig, and pines and laurels would not flourish in that churchyard.

It is important for us to observe that a certain charm, almost architectural, is aimed at here, and that it was entirely absent from the physic-garden, and also from the vegetable-garden, which is to the north of the churchyard: the vegetable-garden is much the same as the physic-garden, but much bigger. "Here the fair plants grow green and tall," as the inscription says, in eighteen straight beds close together in a square. The vegetables cited are again copied out of the *Capitulare*. The care of this garden was extremely important for the monks, whose food was chiefly vegetable; and so a good house for the gardener and his assistants was set up close by. North of the kitchen-garden is the poultry, in two circular houses. Between the two there is a house for the poultry-keeper himself to live in.

The paradise shows one odd peculiarity of form; on both sides of the church there is a space unroofed and marked with the word "Paradise." These spaces are semicircular, and there is nothing to show whether they are meant for gardens or not. A still more curious thing is that next to the abbot's fine house, the only one not in the central group and with two verandas in front of it, there is scarcely any room for a garden, which the architect does not seem to have thought of. An abbot used to have a special garden of his own, as we shall see presently, and the abbots at St. Gall, who loved a garden, would not have given it up. But we must always bear in mind that we are not now dealing with an historical or real monastery that has been added to and enlarged, but with an ideal plan, happily conceived and beautifully executed.

The great number of gardens that appear on our plan is not justified by the real state of things; but on the other hand the love of flowers would certainly have demanded more space for their cultivation than the plan of St. Gall would allow—and especially would this be so at a time and place where ancient traditions were strongly maintained in monastic life. Although for a long time the plants of healing virtue lorded it over the other flowers, the Church had already taken care to give a symbolical meaning to flowers —first of all to roses and lilies, and then to violets. Roses and lilies, at first condemned as heathenish flowers by the early Christians, soon became the symbol of Mary and the reward of martyrdom, and it was not long before an elaborate symbolism was attached to their colours and scents.

In the touching letters written by the poet Venantius Fortunatus to his friend, St. Rhadegund, one can see the part that flowers played. This royal lady had established a nunnery near Poitiers as a retreat from the dissolute life of the Merovingian court, from which she fled. Fortunatus in poetical and friendly words often returns thanks for the food of many kinds that she sent to him, which was almost always accompanied by flowers. Once indeed Rhadegund had him to a meal where the tablecloth was strewn with roses, and the dishes wreathed; while garlands hung on the refectory walls in the fashion of the ancients.

In the sixth century we find traces of an older manner of life, even in this example of the intercourse of a truly pious nun with a poet-friend of equal rank; and such extravagance in flowers must have meant a large garden and skilful culture. There can be no doubt that with rougher customs the amenities of life were lost. But if there was no demand for the table, the service of God required a supply of flowers at every season. Therefore particular little plots were entrusted to the sacristan, who decorated the church on feast-days. In the ninth century a sacristan's garden was set up at Winchester Cathedral, and the custom was followed very generally in English monasteries, where an attractive flower-garden is almost always found, and is well kept up, as an ornament to the cathedral. Now and again the office of a special gardener is mentioned in quite early days, and in the plan of St. Gall he has a house of his own. Sometimes a learned monk took up the work, and according to tradition this was the case at the monastery of Ranshof at Salzburg right on to the time of its dissolution in the year 1811. At this place the Father had to go round the whole estate once a year, and instruct the peasants in the cultivation of

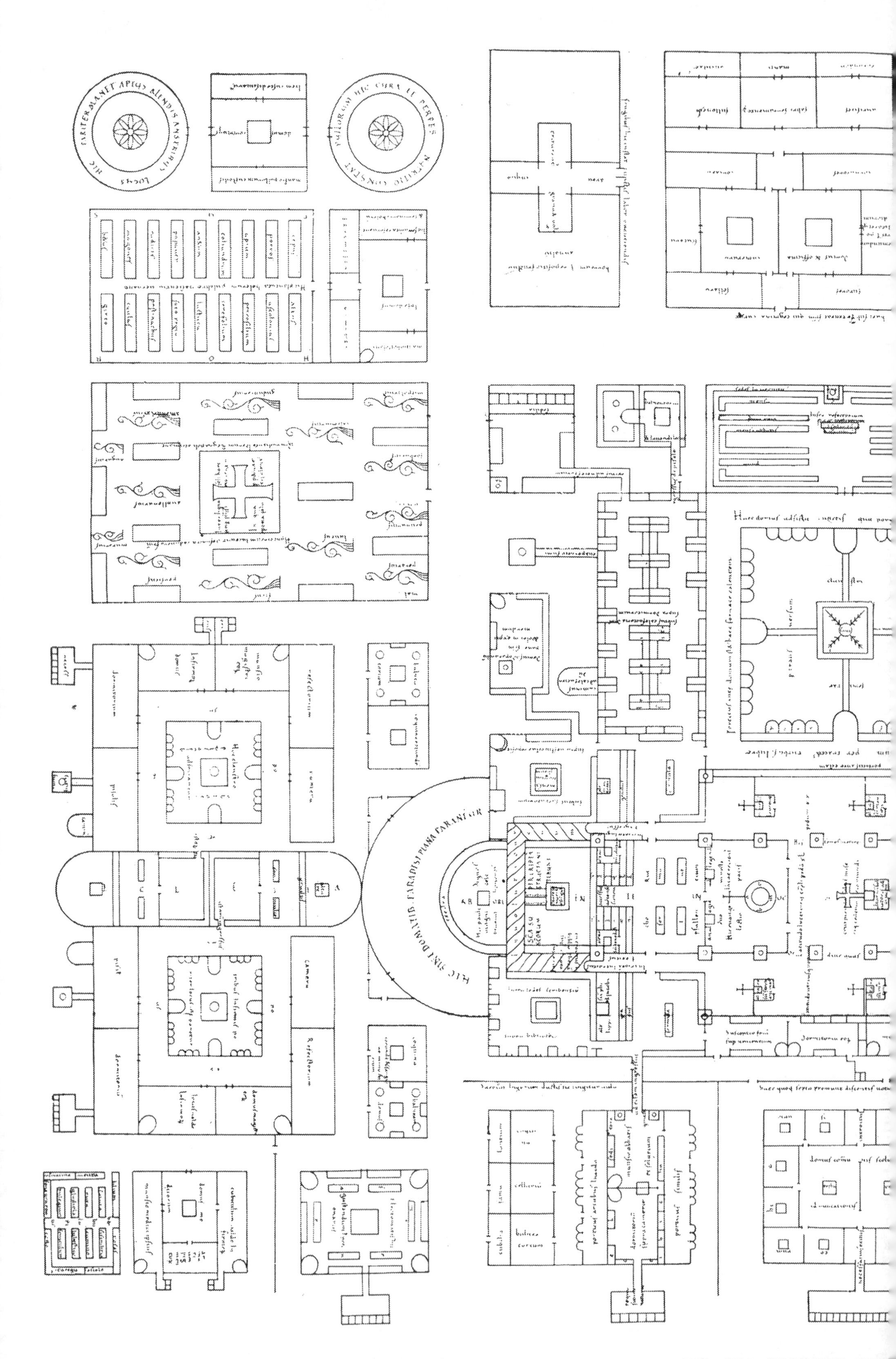

FIG. 124. PROPOSED PLA

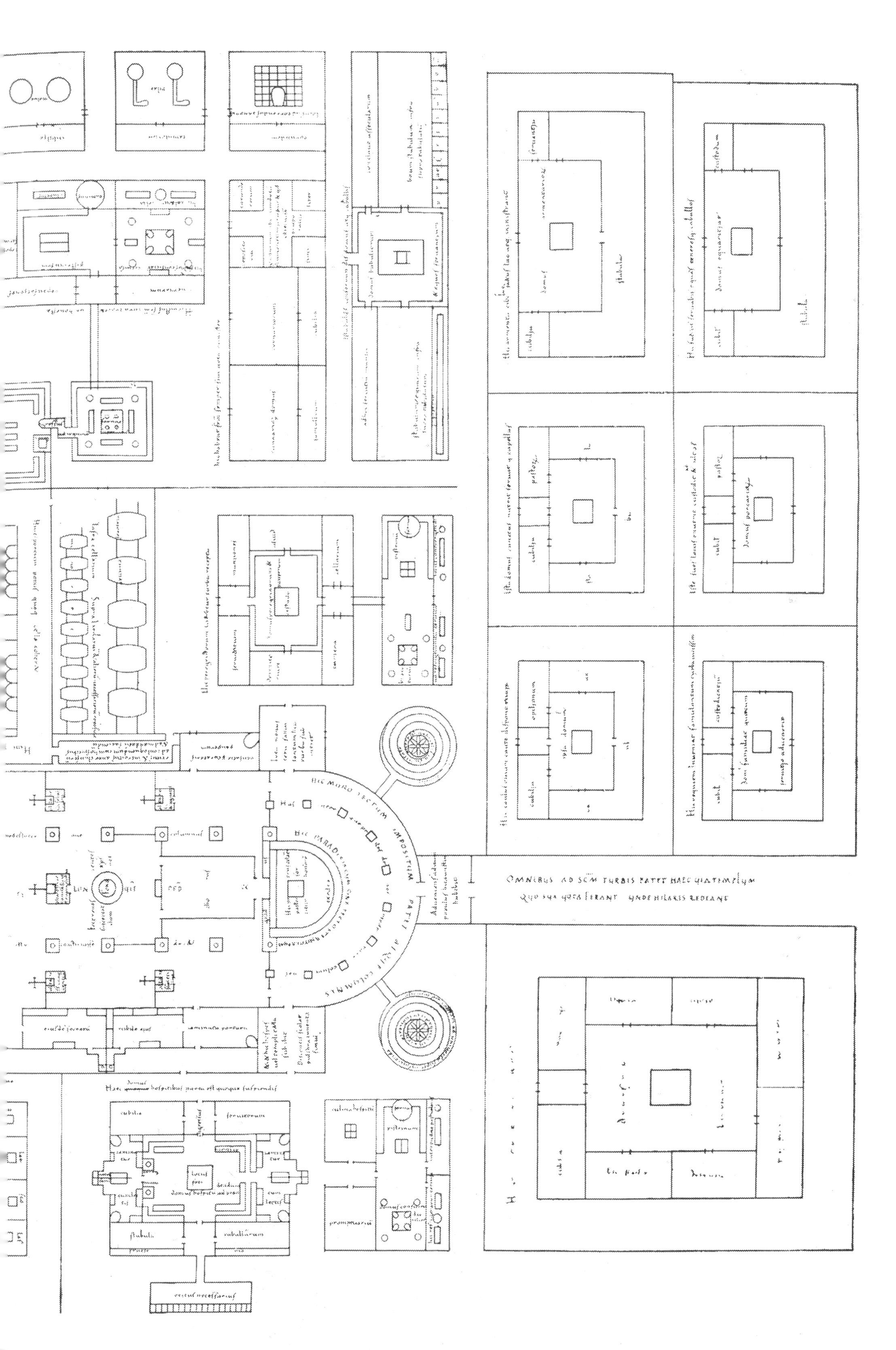

R A MONASTERY AT ST. GALL

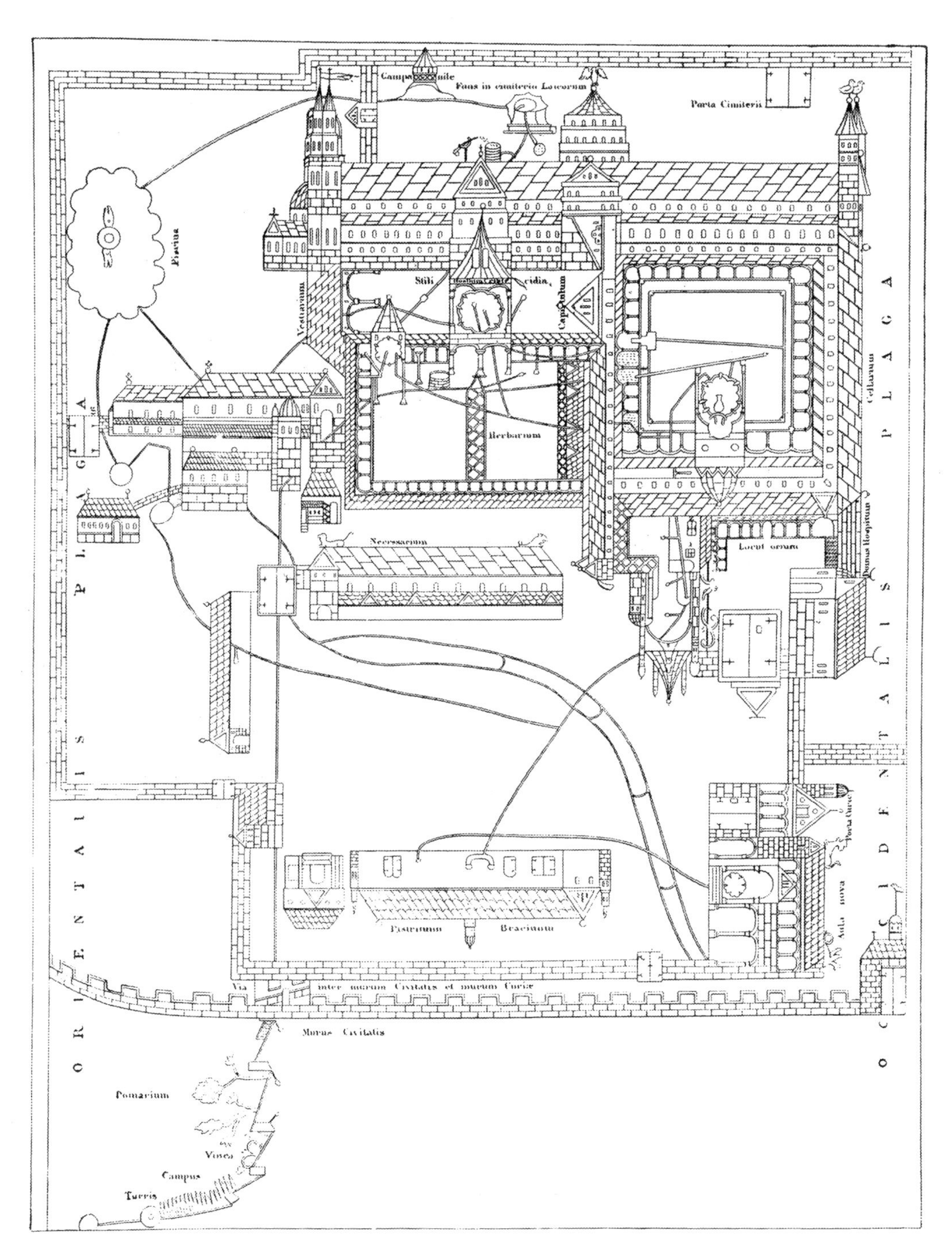

FIG. 125. PLAN OF THE ABBEY AT CANTERBURY

fruit and kitchen gardens. It is not unlikely that the poet who wrote *Meier Helmbrecht* in the thirteenth century, calling himself "Wernher the Gardener," was one of these men from Ranshof. A pleasing picture—the monks are teachers of other men, and also cultivate their own estates themselves.

There were many of these properties cultivated indirectly by the monks, and lying outside the walls; their own vineyards and their own fruit-gardens were also far too large to be kept within the precincts. King John of England made a present to Llanthony Abbey of twelve acres of fruit-garden, and this shows how the fashion had spread. Any vineyard in any country—and in the Middle Ages vine-culture extended far into the North of England and also to East Prussia—belonged indubitably to the estate of the monastery. At Cologne there is still to be found, amid the whirl and bustle of the great city, one quiet vineyard near St. Mauritius; and from its deep green seclusion there seems to come a living breath of the Middle Ages.

There is a plan of the cloisters at Canterbury (Fig. 125) which dates from the year 1165, and fruit-gardens and vineyards are shown outside the walls. This plan was probably made to explain the water arrangements, perhaps because repairs were needed: the thick black lines indicate this, running across the picture, which is partly a ground-plan and partly a bird's-eye view. The water system is conducted among the courts and gardens very liberally, for it was urgently required for plants. In one court beside the east wall there is a very large basin marked "Piscina"; it was intended for fish-breeding, a very important business in English monasteries. Then the churchyard for the laity, to the north but also inside the precincts, had its own special well. The infirmary looks out on a large court that contains the "Herbarium"; it is cut across the middle by a covered walk or pergola, and has its water-arrangement on one side, and on its other side the monks' own burying-ground. This plan is fragmentary, and, as we said, was made for business reasons. No doubt the close, with so many different divisions in it, will have contained other gardens than those marked on the plan; and we see, lightly sketched in, a vineyard and a fruit-garden, and two towers in a field outside the walls, perhaps meant for watch-towers.

A little earlier the Abbey of Clairvaux is described by a contemporary of St. Bernard. "Behind the abbey, and within the wall of the cloister, there is a wide level ground: here there is an orchard, with a great many different fruit-trees, quite like a small wood. It is close to the infirmary, and is very comforting to the brothers, providing a wide promenade for those who want to walk, and a pleasant resting-place for those who prefer to rest. Where the orchard leaves off, the garden begins, divided into several beds, or (still better) cut up by little canals, which, though standing water, do actually flow more or less. . . . The water fulfils the double purpose of nourishing the fish and watering the vegetables." This picture of a French monastery garden at the beginning of the twelfth century, with canals round it, is an early indication of the effect that the Renaissance will have on the gardens of France.

Very like this is William of Malmesbury's picture of Thorney Abbey, near Peterborough, situated in a marshy neighbourhood. He especially praises the tree-garden for being "level as the sea," the smooth stems of the fruit-trees, that stretch up toward the stars, and the luxuriant growths everywhere. "There is competition between nature and art, and what one fails in the other produces."

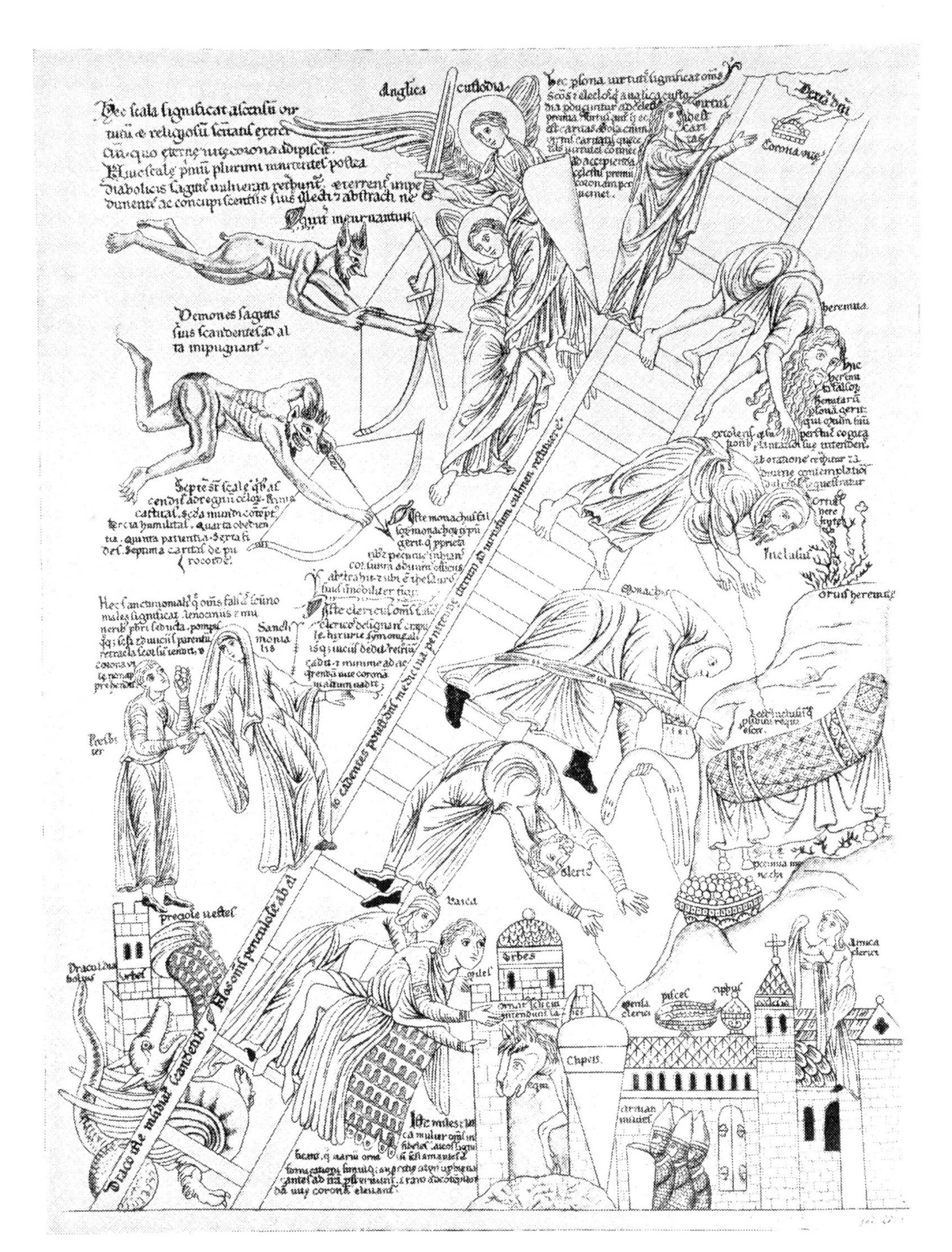

FIG. 126. HORTUS EREMITÆ

As early as the days of Charles the Great the cloister owned outside property, and just as at Canterbury we must conclude that the plan of St. Gall meant the orchards and vineyards to be outside. The whole time of Charles the Great—and the St. Gall plan may be supposed to belong to it—was of great importance for horticulture. Charles himself encouraged personally the cultivation of his own estates, and in his private garden he had a large number of new plants, chiefly of course for the kitchen and the workshop. The *Capitulare de Villis* (which contains among much else a very complete list of woods useful for building) had an unimagined influence on the arts of architecture and horticulture. The great ruler specially favoured the extension of monasteries, and he had them built wherever he could. Thus, in the full enjoyment of peace and protection, the monks busied themselves very diligently with the gentle art of gardening, and so reaped calm happiness and the useful fruits of the earth. There were some, to be sure, who detected a danger in this pleasure. In a work by Herrad of Landsperg, called *Hortus Deliciarum*, the author tells of a recluse who has climbed up to the very top rung of the ladder of Virtue, but then looks behind him. There he beholds his flowery garden, he is seized by a strong desire, and plunges headlong down among the beds, because he has preferred the earthly to the heavenly paradise (Fig. 126).

Happily the men and women of the cloister were not all so strict, but could see in garden culture the symbol of a peaceful quiet life; and they always had before them the example of the first garden, which God made for man; and this in every age was to serve as the ideal pattern for them to imitate on earth.

The word "paradise" in Anglo-Saxon and Old-Saxon texts is translated "meadow" or "pasture": Ottfried calls it the "field of bliss." From the beginning of a recognised art of gardening, the paradise was thought of as a decorative place, and Notker calls it "Ziergart" or "Zartkartin" [*zieren* =to adorn, to decorate]. Also the detailed descriptions given in the Old-German *Genesis* answer to the gardens of that day, with fruit, flower, and vegetable beds: they show a distinct likeness to early descriptions of real gardens.

The learned poetry of the monks ventured on the garden as a subject in the first half of the ninth century. A poem called *Hortulus* was written by the young Abbot Walafried Strabo, who secured in a very short time a far-reaching reputation for learning, at his monastery of Reichenau on the delightful islet on the lake. He describes his little private abbot's garden; and whereas in the plan of St. Gall the designer takes the west for choice, here it is on the east of the house, close in front of the door, so that the porch protects it from wind and rain, while a high wall on the south side keeps off the burning rays of the sun.

Strabo has twenty-three kinds of flowers, each one worthy to find a place there by virtue of its beauty and scent. He has not adhered so exclusively to the *Capitulare* as the architect did at St. Gall, and here and there an eloquent verse betrays his love for some particular flower. But the rose, as queen of flowers, and her twin-sister the lily, are praised as the noblest symbols of the Church—the blood of martyrs and the purity of faith. And what a lovely garden-picture we get in the final dedication! He sends his little book to the Abbot Grimaldus, his neighbour at St. Gall: "If you are sitting in the precincts," he says to his honoured friend, "in your green garden, under your shady apple-tree with its swelling fruits, where the peach-tree parts its foliage to cast a dappled shade, while the young lads from your happy school gather the grey downy fruit . . . then read my gift."

This is a charming picture of summer life, and we feel his love for it in what he writes, but of course the garden of the early Middle Ages is primarily a garden for useful plants and food. Where else than in the garden should men like Strabo attain to that life of security and freedom, which demands a state of civilisation where friends can meet together safely

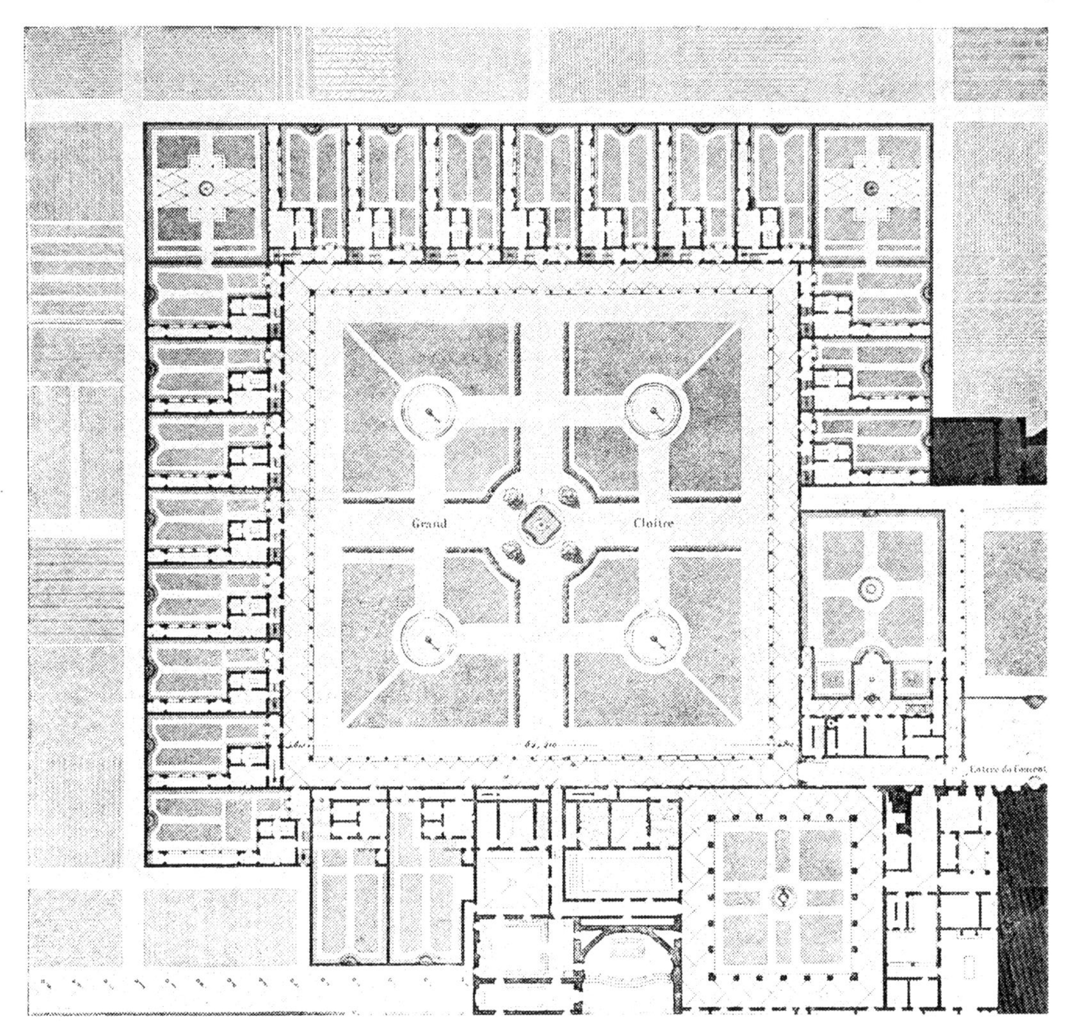

FIG. 127. THE CLOISTER COURT OF THE THERMÆ MUSEUM, ROME

in public and private gardens for their pleasure? First they had to learn once more, and by slow degrees, to be content in some measure with their gardens and their duties.

Another teacher of the same kind was the learned historian Rhabanus Maurus, who devotes one chapter of his great didactic work on the Universe to the Garden. The Bible is for him a kind of herbarium, and he must have every plant sanctioned by a passage from Holy Writ: still he has seen a great many things, and he describes them lovingly. "The Garden," he says at the beginning of the ninth chapter of the nineteenth book, "is so called because something is always growing there." It must not be supposed that in the cloister garden at Fulda there was something growing always; but we hear the old cry of Homer, and this remains the ideal through the ages. Then the writer compares the garden with Holy Church, which bears so many diverse fruits of the spirit, under the protection of God, and in which flows the sacred fount of healing. The garden signifies the innermost joys of Paradise.

The ancient rose-tree at Hildesheim may be regarded as a living sign of the love of flowers at that period. Whether or no it is true that it was planted a thousand years ago, cannot be proved, but at any rate it must have been renewed many times by now.

The foundation of the Carthusian Order marks a great advance in gardening among the monks. Their leading ideas take us back to the Oriental monasteries, concentration on the best education, and preparation for the life to come. As a logical consequence, the common life of monks, even in their dwelling-places, was as much as possible curtailed. No doubt the central cloister was still important as a common garden, but separate houses were grouped round it, with small gardens attached, each to provide for the monks a special duty and recreation.

FIG. 128. THE GARDEN DOOR—A MINIATURE

The plan of the cloister at Clermont gives the scheme of a Carthusian estate. On this simple plan what a profusion of ideas for the garden do we find! In the middle of the entrance court is a raised part, a large court with guest-rooms and farm-buildings round it, and here stands the abbot's house with a garden at the side and a rivulet flowing through; on the west is a tower with a dovecot.

The western section of the churchyard branches off from the cloister court; the rest is laid out as a garden, and round about this are the separate houses of the monks, and on the south-east the little private gardens. Exactly like this, the Certosa in the Val di Ema at Florence still keeps a mediaeval appearance. The monks enjoy a wonderful view from the little gardens by the cells, which are set round the large court like buttresses. Every visitor must be impressed by the court of the Thermæ Museum at Rome with its little concealed cubicle-gardens (Fig. 127); even though he only sees it first as disguised by Michael Angelo, the division of the gardens has remained the same. It is quite clear that, through stirring up the monks to a personal interest, a great impulse was given to the raising of flowers and in consequence to horticulture generally.

If we compare the cloister plans with the houses of the laity of the Middle Ages, we are struck by the want of space in these: the strongholds of the nobles and their garden grounds are crowded with buildings closely huddled together. We followed the course of development of the ancient villa into the castle on to the time of Sidonius. Then it was still a place of wonder and pride, not yet disturbed by the need for defences.

Even a hundred years later much of the joy in feast and garden appears in the poems of Fortunatus, for it was not only in the great monasteries that people cared for roses and other flowers; the Gallic kings also were concerned to lay out lovely gardens. Ultrogote, wife of Childeric I., was famous for the rose-garden she had planted by her palace; and the garden at the old royal castle at St. Germain had a great reputation.

But in the stormy days of the next centuries the haughty nobles had to relinquish many of their gentler manners and customs. They were compelled by the unrest and insecurity of those days to strengthen their places, and contract them into a smaller compass. Nothing remained of the fine buildings of Bishop Sidonius' time except the

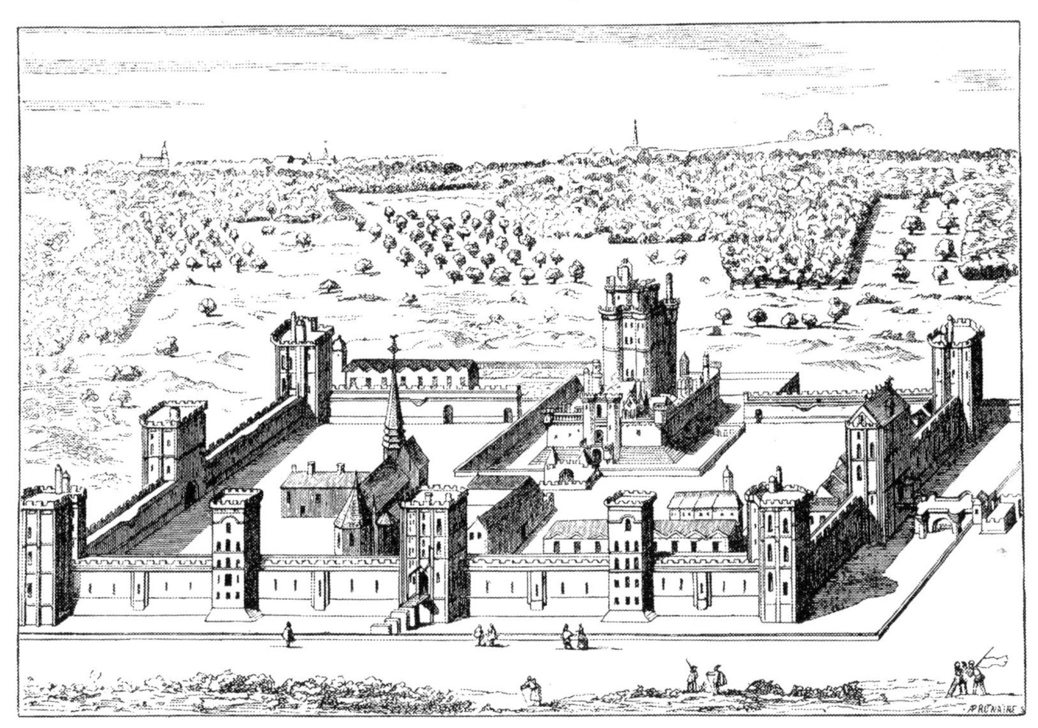

FIG. 129. CASTLE OF VINCENNES, PARIS

defensive parts of the walls and the towers of the keep, the stronghold with its dependent farm-buildings about the inner court. The noble owners were obliged to build their castles on almost inaccessible mountain-tops, where there was very little space, or else down in the plain with wide moats; and in neither case was there room to have a garden. Moreover there was not much inclination for peaceful cultivation of the ground, and the men brought in from the chase everything that was wanted in the kitchen.

In spite of all this, however, the garden was not entirely absent from the old castles. The ladies were the gardeners, for they had been taught by monks how to plant healing herbs among their vegetables, so that they not only got extra dishes and green food for the table, but were also able to help the sick and wounded in castle or village. In the season of flowers, they enjoyed the beauty of many colours, and the gay young people no doubt used to weave garlands for themselves and their companions. The garden was put near the windows of the women's quarters, so as to be under the eye of the lady of the

FIG. 130. THE CASTLE GARDEN AT THE BELFRY IN THE PICTURE OF "CIVITAS DEI"

castle. It was good to see it from above, for it lay like a many-coloured carpet, small and delightful to behold. It was not often possible to walk straight out from the women's rooms into the garden, which was generally set apart and enclosed, and they went out of the house by a "very narrow" door (Fig. 128). Quite as often the garden was close to the mansion, with a staircase leading down to it, as in the garden where the lady Iwein was wont to stroll. So too we find the charming garden of lilies and fountain in the picture of "Civitas Dei," enclosed near the belfry (Fig. 130). Places of the sort are also to be found in the French castles of the Middle Ages, like Vincennes (Fig. 129). Boccaccio (Chaucer following him) allows the fair Emily to wander around in the garden right under the eyes of the prisoners in the keep. Of course we have scarcely any information about the planting of castle gardens in the early Middle Ages, but about the year 1000 Notker tells of "flower-gardens where roses, marigolds, and violets grow."

FIG. 131. TURF SEAT BY A WALL; FROM "ROMAN DE LA ROSE"

At an early date tree-gardens are mentioned also, and we hear of their being enclosed by a wall—"round the tree-garden goes a very high wall," or "before the house a tree-garden lay, and round it a high hedge." The tree-garden which is described by Hartmann von der Aue in *Erec* has neither fence nor water, neither walls nor trenches, and is simply a garden of enchanted beauty, like the one in Laurin's *Rose-Garden*, which has only a silken cord instead of a wall for protection. (This pretty idea, however, seems to show that some wall was necessary.)

The garden of trees was the special "pleasaunce" of the Middle Ages, the pleasure-ground. Fruit-bearing trees came first, but we know, from Charlemagne's *Capitulare* and from the trees in the burial-ground at St. Gall, that it was not only fruit-trees that were planted, but a medley of trees desired for their shade and their beauty. The Latin fable, called "Ecbasis Captivi," describes a noble garden wherein grew an oak among flowers and herbs: it gave its shade to the sick king, and a pure clear stream flowed purling through the garden.

In *Parsifal* there are found in the Schlossberg garden, together with the fairest flowers, fig-trees, pomegranates, olives, and vines. And what a part does the lime-tree play, especially in Germany! It is not confined to the tree-garden, but is also the glory of the castle court with a lawn and a fountain. It is the proper tree for social life and parties, and it also stands out in the pasture-lands. Its branches are extended widely, supported on pillars, with a seat below; often there are benches actually in the boughs; sometimes the whole tree is surrounded by a barrier, as we still see it round the lime-tree at Michelstadt.

FIG. 132. WOMEN SITTING ON GRASS WEAVING WREATHS

FIG. 133. A GARDEN PICTURE FROM "ROMAN DE LA ROSE"

In the garden itself people liked best to sit on the smooth turf seats, which mostly ran along the walls and were either propped up with bricks or stood alone (Fig. 131). In the same way they sat about at games or in conversation, or for weaving wreaths. They sat on the grass (Fig. 132), for here the flowers were not set out in beds, but grew scattered about anywhere on the grass.

As early as the tenth century there is a Psalter which shows the meeting of the two Maries with Christ in a flower-meadow of this kind, and they tread on a fine carpet of growing flowers.

At a later time they were not content to have only one garden, but had several, separated from one another by lattices or gates. In the *Roman de la Rose* we have a pretty picture of separated gardens (Fig. 133). In the middle of the flowery part is the fountain, which keeps the lawn from getting dry and bare; out of the tank starts a narrower canal, which waters the rest of the garden, and flows out through a trellised gap in the wall. Behind an avenue of trees a trellis overgrown with roses is seen in the background. Yet another separates this garden from the next, which is full of flower-beds. These are mostly bordered by stone walls, as in the Brussels picture (Fig. 138); but in the David picture (Fig. 149) there is only a little wooden paling to border the earth. The paths between the beds were hard and sanded, very nicely and evenly kept. When indeed the garden was to be particularly ornamental, the beds had tiles round them, as we can see in the charming little garden beside the dwelling of St. Hieronymus in the Brussels picture (Fig. 138). In this miniature we also see that there is a clipped tree in one of the beds, and a little well in another. Very frequently these trees were set in the middle of flower-beds, and cut so as to make three wreaths (Fig. 134). This favourite form was used for the tree at the Festival of Spring on May Day, and artificial fruits were hung on the crowns as an attraction to the dancers (Fig. 135).

FIG. 134. GARDEN WITH CLIPPED TREES AND FLOWER-PLOTS

The arbour is of very great importance in the gardens of the Middle Ages. It was known to the ancients in the form of a pergola or trellis-work, covered with green, possibly supported on posts, and very attractive, but in gardens of that early date not so necessary; for the portico gave a convenient shelter against sun or bad weather in the larger gardens, and in the smaller ones at private houses there were generally buildings all round. But now the garden was mostly a thing set apart, and needed a real shelter in the open: roses and honeysuckle covered every sort of arbour and kept off the hot sun (Fig. 136); and a very slightly-built summer-house was enough in which to hide with a faithful lover. Also the walk leading to these bowers could be used to stroll about in (Fig. 139). A rose-tree was often grown "so broad and thick that it can give its shade to twelve knights together; wound round evenly and bent into a hoop, yet taller than a man; under the same thorny bush there is golden mullein and lovely grass."[1]

There were other and more substantial summer-houses like those known to the ancients, so that it was possible to eat and sleep in them, as in Pliny's little arbour at the hippodrome. The poet Lydgate seems to have been thinking of that kind as the scene of the adventure in *The Nightingale.* It was in a garden with flowers and grass all round, and shady trees. This garden was entered by the little narrow door mentioned before, and was placed in front of the knight's house, "where the air is better and fresher than elsewhere." There "the host has built a high arbour to sit in every day in summer time; when he eats, he finds the food tastes very good." Meals in the open (Fig. 137) were extremely popular. The Brussels picture shows a strongly made arbour, open to the air, and the scholar who is studying there can go for a change down into the front garden by a small flight of steps, and there sit awhile on the stone stool, and carry on his learned meditations (Fig. 138).

Another feature appears very early in the gardens of the period, and this, too, was meant for retreat or for domestic enjoyment—the maze or so-called labyrinth. When this first found its way into gardens is uncertain. The name carries us back to the palace of Minos at Crete: the story goes, that no one could find the way out of its numerous rooms without a guide, and in common speech the Greeks used the word in that sense. The symbol for it was a figure like a circle or a hexagon, within which were a great many lines crossing each other, and arriving at a point in the middle from which they led out again to the circumference. At Pompeii, for instance, there is a sign of this sort, and beneath it the words, "Hic habitat Minotaurus."

In the early Middle Ages the Christian churches adopted the same figure as a symbol, and it was marked in stone on the floor of a church and used by penitents. But we are far from sure as to the date when these mazes appeared in gardens. In one kind there were paths between hedges taller than a man, so that anyone wandering about and taking a wrong turn could not see over and set himself right.

FIG. 135. A MAY TREE WITH ARTIFICIAL FRUITS

We first hear of a labyrinth in England in connection with Henry II., who is said to have hidden the Fair Rosamond, his beloved, in the woodland retreat at Woodstock; but the earliest authorities of the fourteenth century only speak of a "House of Dædalus," where he kept her hidden away. But at this time the garden labyrinth cannot have been unknown. We learn that when the English envoy, Bedford, had the Hôtel des Tournelles in Paris replanted to make room for his huge elms, he "had to have the hedges of a labyrinth, called the House of Dædalus, taken up." Later on, no large garden was complete without its labyrinth, and in the design of any ground-plan the pattern of the old pre-Christian maze was for a long time preserved.

At an early date very large tree-gardens were laid out. In *Mai and Beaflore* the prince gives a banquet to a great number of people: "A little garden of trees full of white-rose bowers was under the castle. Many were sitting there on seats; poor and rich ate together."

This garden was often a park for animals as well: "Beneath the marble tower lies a wonderful garden of trees, with walls all round, and generally with wild animals." So writes the author of *Guillaume de Palerne*. The nobles soon began to make sure of the hunt, by keeping large areas as parks surrounded by high walls. Thus writes Hartmann von-der Aue in *Erec*: "The king took two acres or even more of woodland by the lake, and threw a wall around it." The whole description shows that this does not remain mere wild land: it was divided into three parts for different kinds of animals, and the king had a well-appointed shooting-box built, so that he could look on at the hunt with the ladies. Ragevin too describes one of these, built by Barbarossa about the year 1161, a red stone building of no little magnificence: "On the one side it had strong walls, on the other a fish-tank, rather like a lake, where fish and birds are kept, beautiful to look at and good to eat. The park adjoins it with plenty of stags and does."

FIG. 136. A ROSE ARBOUR

FIG. 137. AN ARBOUR USED AS A DINING-ROOM—A CARPET DESIGN

In the twelfth century the art of gardening took a most important step forward, for the whole spiritual life of the West was astonishingly uplifted. We are now at the time of the crusades: the Christian soldiery beheld, in the East, gardens of a splendour beyond their wildest dreams. The poets listened to their tales, and from now onward they sang of the East, which was the darling theatre for the adventurous knight-errant. Gladly do we try to depict these distant scenes, and the tale of Herzog Ernst is pleasant enough:

Near by they came to a valley in a garden near the hall; it was very roomy, and therein stood many cedar-trees rich in leafy bowers; they found two rivulets also that rose there and flowed through the grounds, winding as they pleased, and as the master had planned who made all by his skill. They found a bath too, bright and clean, wrought of green marble, well walled in and overhung with fifty high branches; the streams were brought therein by silver channels. Much water ran out in silvery course from the thicket, and flowed around the castle, in straight or curving line around the whole castle. All the paths were of white marble, all bridges made where men should walk.

FIG. 138. THE GARDEN HOUSE OF A LEARNED MAN

Such things as are here described modestly enough (and plainly based on observation of what was found at home with the mere extra ornament of cedar-trees, precious metals and marble) were as a matter of fact painted in far richer colours in the accounts that the envoys brought back as eyewitnesses of all the magnificence in the houses of the caliphs, hitherto anxiously concealed from strangers and foreigners. Of course, Oriental influence, at any rate Byzantine, had already left some traces, as e.g. in Charlemagne's work at Aix, where the pineapple fountain—which in the first instance was placed in front of the royal palace at Aix, spurting out water between its leaves—was in fact a direct imitation of the same thing at the king's palace at Byzantium. Charles owed to his friendly relations with the Caliph Haroun-al-Rashid the possession of certain curious plants, which he first acclimatised in his garden by the palace court, and then inserted into the *Capitulare* as an addition to the list.

The embassies sent to the Byzantine court were henceforth very numerous. In the year 968 the Bishop Liutprand of Cremona saw the imperial palace at Constantinople, and gave a detailed account (already mentioned) of the throne with the golden tree and the singing birds, which later, in the thirteenth century, made such a deep impression on the poetry of the West. In the year 1167 the whole German Embassy was received in audience by the sultan at Bagdad, and its members were only too sad that they were so quickly rushed through these scenes of magnificence. But still they did see the open courts with pillars, and walls adorned with fretwork, and golden garlands hanging between them; they did admire the costly mosaic of the floors, the splashing fountains in crystal basins, the animal-cages with strange foreign beasts, and the wonderful birds with brilliant feathers.

The domain of Frederick II., extending from Southern Italy to Germany, must have felt the effects of the East more than any other country. We hear that he had hanging gardens on the buttresses of his palace at Nuremberg, and also a park (destroyed in 1494); these things betray the Oriental influence. Frederick always encouraged horticulture, of which he was very fond, and he insisted that his procurators and counsellors for the Crown lands should supervise the gardens and the castles, and should see to the nurture of the foreign plants that he had acquired through his alliances with Saracens and Spaniards.

FIG. 139. SUSANNA AT THE BATH—A CARPET

Nothing seems to have interested the poets so much as the story of the golden tree, of which we have seen traces so often in Asiatic gardens. In *Wolfdietrich* there is a detailed description of this tree, which by now had become a linden, adorned with precious stones: "right through the trunk passed three golden pipes, which gave out the sweet voices of full many birds, and thereto two little trumpets made with all diligence." Again in *Titurel* the Castle of the Grail has a golden tree with singing birds; by its side stand four golden angels, making a fine gesture with one hand, and holding a sackbut in the other. Who could fail to recognise an Arabian picture here?

When artificial music has once made a start in mediaeval poetry, every single thing has to produce a sound—first the vine-shoots, then birds, which (in the *Rose-Garden*, by Laurin) are actually to be found astray in the helmet-crests of great enchanters. Even the lions of the imperial throne at Byzantium make their roaring heard in the German poet's woods, as e.g. in the *Krone* of Heinrich von dem Tuerlin. Laurin introduces other ornaments of Arabia into his *Rose-Garden*, crowning the roses with precious stones, which peer out from the green foliage. This was all only enchantment, which can scarcely have been imported from actually existing gardens; but on the other hand it was probably through Oriental influences that costly baths were set up in the gardens, found very often in the pictures of the time (Fig. 139). It was now a matter of necessity for ladies to refresh themselves by bathing their feet before a meal (Fig. 140).

In the first half of the thirteenth century garden description received a fresh impulse in France from the *Romaunt of the Rose* of Guillaume de Lorris, and with this story he gave the form of the love romance to the later Middle Ages. The theatre of the love-scene proper is the garden, just as it had been in old Greek romances. And when we remember how frequent was the interchange of Byzantine and Western literature, and how the Byzantine romance borrowed its subjects from the Western Hero-Sagas, and dressed them in its own raiment, it is surely more than natural that, *vice versâ*, the West should now become the theatre for love romances, for which the East has lent its traditional garden.

FIG. 140. BATH IN A CASTLE GARDEN

Even earlier than Guillaume de Lorris the scene of several Latin and French love tales was laid in a garden, but it was his *Romaunt of the Rose* that fixed and incorporated the traditional descriptions, in which successors scarcely ever made any alteration even in details. It matters but little that (in accordance with the feeling of the time in Western poetry) not only the individual persons who act a part in this dream-romance are allegorical figures, and must be regarded as personified qualities, but the different parts of the garden are intended to have an allegorical meaning. The description itself adopts the purely traditional form, but all the same it is realistic; and whether we are reading one of the ancient descriptions, or this, the earliest in the vulgar tongue, the type remains the same: the fountain has to be the centre of the piece, and it is as clear as crystal, its waters flowing over silver pebbles, ever fresh and glistening. And in Guillaume de Lorris's poem there is a well of peculiar importance (Fig. 141), that has on its edge two crystals, which show all the colours of the rainbow when the sun goes down. (In Eustathius it is the variety of colours in the marble columns which cause the reflection in the water.) Guillaume de Lorris says one of the crystals is a magic mirror, so that you can see half the garden in it, and if you look up, there is the same picture to be seen again. Achilles Tatius once described it in a far more realistic fashion, saying that the mirror of the square tank was so bright that you saw the whole garden in duplicate.

But Guillaume de Lorris, in spite of his touches of magic, is thoroughly realistic in other respects; and although the wall too is painted in the colours of allegory, it is still a real garden wall with a little entrance gate. It encloses a garden which is a perfect square, fine trees overlook it, and in them birds sing beautiful songs. The trees are set at exactly equal distances, they are tall, and their tops are interwoven so that no sun-rays can get through. Squirrels play about in the branches, and on the thick grass. Clear streams, and wells that have no frogs in them, are here, and the banks are made pretty by brightly-coloured grasses and lovely flowers. The special features are as true to type as ever, though it may be that now and again the towers for defence are planted round with roses (Fig. 142).

FIG. 141. WATER BASIN WITH CANAL

When the poet can sing no further praises, he tries to enhance the effect by exalting it all into the infinite, though he confesses that words fail him to describe the beauties of Paradise. But he will not have his account too meagre, and so borrows once more from ancient story, and gives an endless series of trees and flowers. Virgil supplies a long list, and Ovid too; and the later classical authors have a real passion for such lists: Seneca, Lucan, Statius, Claudian, all accept the same with hardly any changes, and clearly indicate the path whereby Guillaume de Lorris proceeds to re-spin the thread and give it to us again. So if we are familiar with one garden only, we find we know them all, whether they are in French imitations or come from England, where this story was most influential.

FIG. 142. A ROSE HEDGE AND A WATER TOWER

It is easy to see that Chaucer in his *Romaunt* as well as Lydgate in his *Love's Chess-(Cheker)board* (which is very like a translation) gives the garden-part word for word. So when Chaucer, in his *Assembly of Fowls,* steps into that wonderful garden, he begins with the same list of trees, quite unconcerned as to whether English soil can produce them, and goes on to the usual twists and turns of the old pattern. But still, now and then there are fresh features, independently observed. In *Troilus and Cressida,* for instance, he says that the garden was large, that all the paths were shaded by flowering branches, and that the turf seats were new, and all the walks strewn with sand (Fig. 143). (These seats must have had to be renewed often, for they were soon worn out.) Again in the delicious English poem of perhaps a later period, *The Flower and the Leaf,* there is an arbour like a room: the roof and walls are cut out of a hedge, as thick as a castle wall, and quite even, "as smooth as a plank"; the seats inside are newly turfed with thick, short, freshly cut grass.

FIG. 143. A HOUSE GARDEN WITH A TURF SEAT

These traditional descriptions did not immediately further the ends of horticulture, but indirectly they exercised an important influence, because of their illustrations. Artists could not attempt much from the ordinary general descriptions given in the books that they were to adorn with miniatures, and they could not seek refuge in the poets' subterfuge, that the beauty of their subject was beyond all words; so they took actually existing gardens as models. We owe our clearest conceptions of the gardens of this period to the miniatures painted for the *Romaunt of the Rose* and its imitations, and also for the prayer-books, which in the North maintained the traditional forms of the Middle Ages right through the whole of the fifteenth century. The love-gardens of the painter answer to the love-visions of the poet, and (without being exact illustrations of these) provide garden backgrounds for love scenes and for merry parties. In Orcagna's "Trionfo della Morte" in the Campo Santo at Pisa, the gentlemen and ladies—whose sturdy love of life is so very near to death—are seated on stone seats under a wall in the shade of lovely trees (Fig. 144).

Almost a hundred years later, near the middle of the fifteenth century, there is a copperplate engraving (one of the very earliest known) which represents a Garden of Love as in part an open pasture or perhaps a flowery meadow, in the middle of a park with a brook running through it. The people sitting on the grass have railings to lean against, and the centre piece is a hexagonal table with refreshments laid out (Fig. 145). There is another table like it in the beautiful garden where the Queen of Heaven is sitting, deep in her book, on a chair beside the table (Fig. 146), while high above at the back of her are lovely lilies, roses, and irises; her attendants are happy in the garden, some with wings and some without, playing with the child Jesus who sits on the flowery mead,

FIG. 144. THE GARDEN OF LOVE—FROM "TRIONFO DELLA MORTE"

or carrying fruits in woven baskets, or drawing water from a marble trough; others are discussing philosophy. A high white battlemented wall hides the heavenly company from profane eyes.

Another result of the close alliance with Eastern civilisation was the awakening of the desire for knowledge in the Christian world. Moorish Spain was now friendly with the lands of the North, because it had become customary for learned men to study medicine and mathematics at the Spanish universities. A good knowledge of plants was one of the first necessities for medicine, of course, and in the thirteenth century Albertus Magnus was the true pioneer, with a scientific work on the kingdom of plants; and though many fantastic errors are mixed in with some of the details, this study did lead him to the original sources of learning in the East. He had moreover seen a great many things, and observed them to good purpose, and had made all sorts of experiments in his own garden.

The most remarkable story about his gardening art is told by the chronicler de Beka in the middle of the fourteenth century. On 6 January, 1249, he says, King William II. of Holland (King of the Romans) was travelling through Cologne, and Albertus received him in his monastery and made him a great feast. He caused the utmost amazement among his guests, because in mid-winter he was able to show them in the cloister garden fruit-trees bearing ripe fruit, and other flowering growths, which he had brought out by the application of gentle heat.

FIG. 145. THE GREAT GARDEN OF LOVE

How far this tale rests on firm foundations it is hard to say after the lapse of centuries; but it is not impossible that Albertus had learned from the ancients (or still more likely from Oriental sources) how plants could be artificially trained, and had been fortunate in his experiments. At that time Cologne enjoyed a great reputation for skill with flowers and gardens; and cathedrals (as well as archbishops) owned magnificent gardens. Engelbert II. at his estate at Bonn kept lovely grounds, where he had not only a large cage for lions, but also many unusual plants. How great was the general interest of the burgher-folk in these, is shown by the following anecdote: Engelbert in 1258 had quarrelled with the city, and had to flee to Bonn, but his councillors were clever enough so to infatuate the foe with accounts of the magnificent plants in the Bonn gardens, that they were trapped by the offer of a peaceable inspection, and paid for the gratification of their fancy by the loss of their freedom.

In the towns it was by this time not only the great nobles who set store by beautiful gardens; the burghers began more and more to care about them. Every now and again there is heard some very early rumour bearing on this subject, as for example about the praise given to the city of Paris by the Emperor Julian when he was on his journey to the North. The particular commendation is due to the fact that Parisians used to keep vines and figs round their houses, and protected them through the winter with coverings of straw. But this is an exception, for it is not till quite late that we hear of any horti-

FIG. 146. A LITTLE GARDEN OF PARADISE—FROM THE HISTORICAL MUSEUM AT FRANKFORT

culture in the Northern towns. At first the gardens were just vegetable plots in front of the town walls, and the produce was sold in the town markets. In 1345 the private gardeners of the great nobles and gentlemen of London had a quarrel with the alderman, because the noise of the market at St. Paul's Churchyard annoyed the inhabitants and passers-by. Naturally there arose at an early date a trade guild of gardeners: a Roman document is known (1030) about a Gardeners' Company, and at that time the Northern towns were not much behind Italy. Fruit orchards and vineyards were carefully laid out and kept up, and any injury done to them was severely punished. Indeed, after the peace of 1187 anyone in the Kingdom of Germany who injured orchards or vines suffered the same penalty as for arson—proscription and excommunication.

Inside, the towns developed very slowly from want of space; and it was only when they were attached to the larger houses that one found more important gardens, taken from the space for building (Fig. 147). Sometimes people were able to have small gardens in the streets alongside the wall, or at the side of houses which faced towards the river, as so often in Paris with the more important houses. They had more freedom in the suburbs (Fig. 148). Thomas à Becket's biographer, William Fitz-Stephen, shows himself full of pride over the surroundings of London. On all sides, he says, outside the houses of the citizens who live in the suburbs, there are gardens planted with trees, very spacious and most pleasant to the eye.

Still earlier Landulf describes the beauty of Milan, and though, to be sure, this is rather a matter of imperial buildings, a castle, a play-house, and hot springs, he particularly praises the gardens, "green as God's Paradise, beautiful with lovely trees arranged in divers ways."

FIG. 147. HOUSE GARDEN WITH LATTICE FENCE

The gardens of village houses naturally developed in a finer way and at an earlier date than those in towns. In the songs of wandering minstrels we often hear of the peasant's garden; there was generally a plot of ground in front of the house, serving hospitable ends; the peasants met there as in an arbour to drink together, and the plot at the back was used for kitchen stuff.

It is a more important matter, that from the twelfth and thirteenth centuries onward the towns started public grounds, where people of leisure could meet and enjoy themselves. Near the end of the thirteenth century they were getting larger at Florence, and a statute of 1290 orders the extension of the "pratum commune," which plan was carried out four years later by the purchase of old houses which were then pulled down. The name has remained in the Via del Prato. At the same time it was quite the usual thing in Tuscany for the smaller towns also to have some of their streets laid out as promenades; and a statute ordering a Prato for Siena in 1309 shows what was aimed at in these public walks. It says that they are first of all meant to increase the beauty of the town, as a pleasure resort for the inhabitants and for foreigners: Siena ought to have places of the kind like other towns where they have been made and presented by the merchant class, in order that strangers coming from outside can enjoy the amenities of the city. The Prato before the Porta Camollia was laid out in the same spirit, and a place of this kind was bound to have a pretty large circumference, for annual markets were held there, and also reviews and horse fairs.

FIG. 148. SUBURBAN GARDEN AND SMALL FARMYARD

The name indicates that it was chiefly wide meadow flats traversed by shady avenues, but it must have also included other places of an ornamental kind, and in 1297 the Florentines decided to add a large artificial lake. Also the Prato della Valle at Padua shows by its size even now that it must have been a sort of common for the townspeople. Still the style of the central ornamentation which we see nowadays was only adopted in 1770, for in that neo-classical period it was made on purpose in the shape of a perfect oval, in reference to the Colosseum as model. The plantation in the middle, adorned with statues, makes one think of the plane-grove at Sparta, because of the moat that runs round it (Euripus), crossed by four bridges which are also decorated with statues.

Again in the Prado at Madrid and the Prater at Vienna, both name and style have preserved for us examples of the pleasure-grounds enjoyed by the townspeople in the Middle Ages. The first public garden in Paris was in the district of St. Germain des Prés, and bore the name of "Le Pré aux Clercs." The wealthy townsmen of Paris used to meet there in early days.

A great advance in the development of gardens for the people is due to establishing Brotherhoods, both lay and clerical. They were highly approved of by the great of both kinds, and were constantly remembered in rich legacies and gifts. One Florentine document of 1208 tells of a present of three pieces of land "nel corte di Ganglandi ai poveri," and from then onwards there are a great many documents of the sort. Though in these cases we are once more dealing almost exclusively with useful gardens whose produce was to be bestowed on the poor, it is still the fact that the Brotherhood of Archers (in Northern France and Belgium) were very helpful in the promotion of horticulture. The character of these Brotherhoods was half military and half religious: their patron was St. Sebastian, and naturally they were in high favour with the rich and great, for whom they trained good marksmen; and by them they were endowed with valuable privileges. They built fine club-houses with large gardens, and had their shooting-stands set up in them, and it soon came about that the rest of the townsmen assembled there for recreation and amusement.

FIG. 149. GARDEN IN THE DAVID PICTURE

At Boulogne a place of the kind with shooting stands was called "Courtils aux pauvres" —a similar expression to the one mentioned before in the Florentine document—and here we must take it for granted that there was a garden for the people. Not only for shooting, but for other sport, places were provided, and especially for ball games, which developed in the fifteenth century, chiefly in England; later they spread all over the Continent in the form of football, croquet, and lawn tennis, and when people had more room these games were introduced into private gardens with properly laid-out squares and courts.

Although in the thirteenth century it was as yet impossible to say that any one of the western European countries strikingly excelled the others in gardening, in the fourteenth the leadership is plainly Italy's, as Italian sources show. The towns now begin to show more and more a spiritual and political independence, and among them stands Florence, growing with marvellous speed. The historian Villani, fired by the splendour of the Jubilee Year 1300, which assembled round the Pope all the nations of Christendom, made up his mind to write the story of his own town and bring it up to date. This he did in the proud consciousness that his native Florence was a rising city, whereas Rome was declining. In the wonderful picture he gives of the flourishing condition of the town, country houses and gardens play no unimportant part. He says that there is scarcely one Florentine, gentle or simple, who has not a finer house in the suburbs than in the city itself; he adds, to be sure, that they are sometimes foolishly extravagant and expensive. The suburbs of Florence became so wonderful to look at that a stranger who had never lived there and was coming in from outside, might mistake the costly palaces, the towers, the walled gardens, for the actual city itself.

FIG. 150. LE JARDIN D'ALENÇON

In these rich men's villas the talk was no longer of vegetable plots with fine produce, but little by little there grew up the idea of a truly artistic garden fit for princes and nobles. The councillor and learned explorer Petrus Crescentius of Bologna was writing his great book about husbandry almost contemporaneously with Villani. Herein he draws a sharp distinction between the small gardens of from two to four yokes, and those of kings and other rich men. The first are merely for use, and their produce should not go beyond the Geoponica and other ancient writings: a hedge of sloes or red and white roses is to go round as a screen, and in front of this there should be a trench. There should be planted rows of apples and pears, and in the hotter parts palms and citrons, also mulberries, cherries, plums, and other trees, such as figs, nuts, and almonds, all separated from one another in carefully-made even rows, but between them vines may grow. The rest of the ground may be laid out as grass meadow, which will show after the rains a growth of grass that is tall but not good, and this can be cut twice a year, to keep it in better condition; also in certain parts they could set up pergolas to serve as a kind of pavilion.

He demands quite a different plan for the gardens of the rich, who may take twenty yokes or more—as much as they like—and enclose it with a high wall. Towards the north there should be a thicket of tall trees where wild beasts are kept. To the south they should build a very beautiful palace, where the king and queen can spend their time, whenever they want to escape from serious thoughts and to get refreshment and happiness. Crescentius, with a careful eye to the Italian climate, chooses this spot so that there may be pleasant shade close by, and so that the view into the garden is not spoiled by the blazing sun. He further asks for a fish-pond, and an aviary near the palace. In other parts of the garden more shrubberies can be put, and the tamer animals kept in them. But he is very emphatic in his injunction that avenues of high trees are not to be allowed to block the view from the palace to the thicket at the end of the garden, and so prevent people from seeing the animals that are kept there. He makes a longer business of the building of summer-houses, which have to be made simply of trees, with various passages and rooms, so that they can be used in damp weather as well as in dry. He recommends making the divisions with cherries or apples, or perhaps planting willows or elms, which one can train over stakes, poles, or rods, till roof and walls are completed; then dry wood can be used to finish the structure, and a vine trained thickly over. The garden can also be much beautified with evergreen trees, but here he must have a strict arrangement, and different kinds must be kept apart, so that they may be all flawless.

In another place Crescentius recommends the clipping of trees, and says that gardens and trenches may be encompassed by green trees shaped and cut out like walls, palisades, turrets, and embrasures. In spite of the fact that we have talk of plan and order, we detect an effort after picturesque form, which prepares the way for the early Renaissance. Crescentius wishes to be a teacher, and to give practical advice and leading; this object he attained not only in Italy, but also on the far side of the Alps, where his book was soon known and widely circulated, in Germany, France, and England.

Hand in hand with this scientific treatment of building on the land went the introduction of foreign plants. In Italy people soon began to lay out gardens of wonderful herbs with a view to their use in medicine; and this practice must have been prevalent in Petrarch's time, in the fourteenth century in Italy, for then the reputation of Italian plant-lore was so great that its professors were summoned abroad: and so it came about that Angelo, the Florentine, laid out a botanical garden for Charles IV. at Prague.

As is to be expected, Italy takes a prominent place, not only on the practical side of æsthetics in gardening, but also in poetical descriptions in the fourteenth century. Boccaccio, the poet who above all others is identified in every department of art with the immense stride from the Middle Ages to the Renaissance, stands on the threshold of both worlds. Second only to the *Romaunt of the Rose,* we have his description of a garden in *The Vision of Love,* which takes its place in direct line with the antique Byzantine love-romances. The poet enters the garden; over the walls flowering branches are beckoning, the beauty of the grassy lawn decked with a host of flowers overmasters his spirit, there are tiny rivulets, animals playing on the green grass, the comforting songs of the birds; in the midst is a fountain—and here his description soars above that of Eustathius, and takes two chapters in the telling.

We hear of a lovely basin of red marble, on whose four sides stand figures in different attitudes, also made of priceless marble. In the middle, on a very beautiful support, stands a pillar like diamonds to look at, with a golden capital. On this there are three female forms, their shoulders touching, each of a different-coloured marble. One of them is black, and the water flows like tears from her eyes; the second one is red as fire, and the water flows from her breasts; the third is almost white, and she flings the stream over her head. It then passes through shells and various heads of lion, bull, and man, and finally proceeds to water the garden.

The delight in these fantastic water arrangements has a further effect, for Rabelais in the court of his Abbey of Thelemites has a very similar fountain, which he describes in a humorous way. Almost all Boccaccio's accounts of gardens are in the conventional manner, and since he adopted the list of trees in his *Theseid*, the ancient tradition (as we said before) remained alive right on into Renaissance days.

But Boccaccio knows also how to keep his eyes open to actual fact, as no man before him had done. This is plain to see when he gives us a bit of the history of his own times, in his inimitable style, in the tales of the *Decameron*. As Villani in his history gives a general picture of the beauty in the surroundings of Florence, as Petrus Crescentius as a teacher exhibits a specimen of a lordly estate, so does Boccaccio's picture of the loveliest of villas produce a clear, intelligible view of some nobleman's seat of that day, standing on the heights of Fiesole over the city. Undoubtedly these villas did exist in truth and still exist in fiction, exceedingly like to one another: here the gentlemen and ladies of the *Decameron* spent their days. The round of ideas for the garden is still small, and it reverts to type, but what the poet sees and makes us see is what was really there: the house was on a hill, surrounded by meadows, and a smooth, pretty and almost untrodden path led up to it. A fine inner court was roofed with a flower-bedecked loggia, where the guests were received. The garden, where they went afterwards, was at the side and had good walls. Round by the walls and in the middle there were straight paths as broad as a street, with many arbours clad in vines, which in that year gave fairest promise of a rich harvest, and already from their blossoms wafted a delicious scent abroad. The sides of the paths were enclosed with red and white rose hedges, so that one could walk in the shade not only in the morning but also at any other time of the day.

How the other plants were arranged, the poet unfortunately had not time to tell. But in the middle there was a closely mown lawn, of such a dark green that it almost looked black, and it appeared to be painted with a thousand different flowers. Round about it was a border of green oranges and citrons, their flowers and fruit growing together. In the middle of the lawn is the white marble well, with beautiful sculptures on its border. A figure with a column or pipe inside throws a water-jet—whether this is from an artificial or natural spring the poet does not say—so strong that one could have driven a mill with it. The stream is conducted under the meadow and into canals for the watering of the garden. Naturally in this lovely spot men's hearts are rejoiced by the songs of birds, and they allow themselves to be pleasantly teased by the animals, deer and rabbits. On the lawn tables and chairs are put out, where people who are tired of singing and dancing can get rest and refreshment. What pleasure and what *joie de vivre* is shown in this account of life at the villa! We detect a new vigour in their gardening, and the whole scene becomes more living to us because we here have actual biographical details.

Petrarch seems to have wanted, wherever he went, to have a garden that he could care for himself. He possessed one at Parma, where vines grew between the trees, and here the famous meeting between him and Boccaccio took place. He said of his garden at Vaucluse that he had planted it with his own hands, and happily compared the divine activities of Bacchus and Minerva with his own, saying that he had not concerned himself with a shelter for his grandsons, but would walk in it himself.

And to this day, on the slopes of the Euganean hills, overlooking a heavenly lasndcape, there stands the little house which the poet kept as a retreat for his old age in the lovely village of Arqua, so renowned for its vines. From the village street one steps into the garden, where a thick laurel avenue leads up to the house. To-day, as then, flowers bloom to right and left in the simple beds. At the other side of the house is the open loggia, like the one in Boccaccio's garden, and from here one can both enjoy the lovely view, and look towards the gàrden at the back which is completely shut away. At the foot of the hill, near the church, stands the sarcophagus which holds the poet's bones; but above, green and sweet, grows the laurel, by him beyond all other men beloved and sung.

CHAPTER VII

ITALY IN THE TIME OF THE RENAISSANCE AND THE BAROQUE STYLE

CHAPTER VII

ITALY IN THE TIME OF THE RENAISSANCE AND THE BAROQUE STYLE

THE historian Villani, although proud of the beauty of his native town, shook his head gently when he told of the luxury of life at a villa. The poet Boccaccio shows his little company walking about, free, careless, happy, untroubled by the breath of pestilence that brooded over the town at their feet, while they were surrounded by elegance and beauty. Still it is not till a full century later that a new voice is heard in Italy praising with songs of delight the loveliness of a country life. When Leon Battista Alberti in his *Della Famiglia* describes the joys of a villa, he is proud of being on the side of Cicero, Horace and Pliny. He loves to breathe the pure air of Florence, where, in a bright clear atmosphere and cheerful surroundings and lovely views, everything is healthy and clean, with very little fog and no bad winds. In such a place he thinks men are bound to be good and just, "where there is nothing, as elsewhere, of fear or of treachery, and one is not cheated by debtors or attorneys. Nothing secret goes on here, nothing that may not be seen and known by all."

This theme is dwelt on pleasantly at some length in the dialogue. But "the great Alberti," as Burckhardt calls him, with whom the sun of the Renaissance arose and shone over Florence, has something more weighty to say on the question of villas and gardens, for he is an architect. He is the first in any field to try, with conscious intention, to knit together the glorious past of his people with the flowering time of the new Roman world. How far this union was purely literary is nowhere so clearly seen as in the architecture of villas and gardens. In Alberti's work *De Architectura* the villa appears in all its beauty and gaiety. Its open halls, disposed quite arbitrarily, "wherein the artist ought to keep his main lines in strict proportion and regularity, lest the pleasing harmony of the whole should be lost in the attraction of individual parts," seem to welcome with a smile the arrival of a guest. By gently rising paths he is to be led up to the house, without noticing the climb, but he is to be astonished at the view when he arrives. He is to wonder if he had better stay or, enticed onward by variety and splendour, go farther.

The garden which he sees around him is just as cheerful; because all that makes for melancholy is to be avoided, and dark shadows must be kept in the background. Porticoes and pergolas on the sides give protection from the hot sun, as also the cool grottoes of tufa, which Alberti will have introduced "after the fashion of the ancients." He also approves of their customs of using decorated pots for flowers, and of writing the master's name on the beds in box. All the paths are to be bordered with box and other evergreens. Bright streams of water must run through the garden, and above all must start up unexpectedly, their source a grotto with coloured shell-work. Cypresses with climbing ivy must be in the pleasure-garden, but fruit-trees, and even oaks, are relegated to the kitchen-garden. Comic statues may be tolerated, but nothing that is obscene is admissible. Circles and semicircles, considered beautiful in courts, he desires to see in the garden as well, and these he would have made with laurel, citron, and yew, with interwoven branches.

It is plain to see how strongly Alberti is leaning towards the two villas of Pliny, and we can really cap nearly every one of his demands with a passage from that writer. Indeed, his zeal for imitation goes so far that he is led astray, and thoughtlessly copies his model, when Pliny arranges to use box borders only where they are protected by houses. At Pliny's Laurentinum, rosemary had to be used instead of box in places exposed to the wind from the sea and to splashing foam, because under such conditions box would dry up and perish, and Laurentinum was close to the sea; but when Alberti recommends evergreen borders, and adds that it is better not to use box unless it is protected by walls from wind and foam, the advice sounds rather unnecessary, as he is first and foremost thinking of his own inland villa at Florence.

FIG. 150*a*. A ROOF GARDEN AT VERONA

In spite of this marked inclination towards his model, the place turned out to have a very different character. Although he seems a little vague about the garden as compared with the clearly designed house, both are certainly regarded as making one coherent whole. He expressly demands that the circular and semicircular parts that occur in halls shall be repeated in the garden, for a garden leads into the house, and the two meet at the threshold; therefore gardens also ought to exhibit the cheerfulness of the early Renaissance, which refuses to tolerate dark shadows except as a background—in opposition to the heavy and serious grandeur beloved by the later ancients, who desired to be free in their country seats from the many cares and burdens that oppressed them. The days were still far distant when Italian gardens could compete with the size and grandeur of antiquity.

But Alberti, so far as his theory goes, was ahead of his time, and it is only somewhat later that we can see his principles actually carried out. Indeed, when he himself is acting as a practical man, he, like the others, is very slow to relax the fetters of mediaeval tradition.

By a happy accident there has been preserved an exact description of a garden that with great plausibility may be set down to Alberti. This is at Villa Quaracchi, the country house of Giovanni Rucellai, a rich Florentine merchant for whom Alberti acted as architect and friend. Commissioned by Rucellai, he made the plans for the façade of Santa Maria Novella at Florence, the palace in the Via della Vigna (opposite the so-called Loggia dei Rucellai), which "outside Florence, on the right of the road that leads to Pistoia, is a great palace with trenches for water and beautiful gardens." As Alberti held staunchly to his main principle that an architect is only the designer and not the builder, and accordingly made nothing but plans, it is difficult to be quite certain about his works; but we are fully justified in saying that this villa and garden appear to be the true child of his genius.

We can see a portrait bust of Giovanni Rucellai in the Kaiser-Friedrich Museum, an intelligent, somewhat pensive head, showing a curious mixture of kindliness and strength. An unwelcome time of leisure, caused by the plague in 1459, was the occasion for his beginning a kind of diary. In this little book, "which contained something of everything," he set down important family events, mercantile news, philosophical and moral remarks, and poetry, and also talked of "all the beautiful and attractive parts of his garden." Owing to the superabundance of these, it is not very easy to get a picture of the complete plan. The pergolas are the most striking feature, three of them passing right through the garden, the length of which is a hundred ells (arms, *braccia*). These pergolas supply the necessary shade on the paths, and are sometimes of a barrel-like form, as in the chief avenue, and cut out of evergreen oaks, sometimes clipped to a point. The open paths have lattice-work on either side, with fine vines trained on it and white roses between, "which when they are in flower are so lovely that no pen can give the sense of joy and peace that the eye receives."

The chief pergola starts from the front door, and has a small loggia above it; it has sidewalks on either hand, and breast-high espaliers of box, over which the arms of the family and their relatives are placed as an ornament in a festoon. At the end of this pergola there is another door leading into a little walled-in garden which includes a small meadow: here we find terra-cotta vases and perhaps beds, with pots set all round, which contain Damascus violets, marjoram, basil, and many other sweet-smelling herbs. Here also there are hedges of box cut into divers shapes which we shall speak of later—especially one fine round bush made into steps, the peculiar joy and pride of a garden at that date. In the same line as the chief pergola there runs perfectly straight on the farther side of this *procinto* a continuation which takes the form of an avenue of lofty trees and wild vines reaching all the way to the Arno, a distance of 160 ells. The master, dining at his table in the chief hall, was able to see the vessels passing by.

The house, whose situation on level and healthy ground is insisted upon, would also (as Alberti, the designer, desired) be set on a gently rising eminence. The main garden, however, was mostly orchard, with a wide hedge round it, composed of laurel, plum, juniper, and various clipped shrubs. It had plenty of seats in it, and there was a pretty, neatly kept footpath. Inside this hedge there was a great variety of fruit-trees, some foreign and rare; among them was a sycamore of which Rucellai was particularly proud, and which he may have brought home from an expedition to Palestine. There was a rose-garden within the same hedge, and close to the pergola a tangle of rose and honeysuckle round a circular stone, one arbour of firs and laurels, and another of honeysuckle. The home for birds—indispensable feature of an Italian garden—was of course there, and also the spiral hill, covered with evergreens, with eight paths winding round to the top. Near the house, perhaps at the side of it, was a balustraded fishpond, shaded by evergreens.

Now all this magnificence was visible not only to the owner but to the passers-by.

FIG. 151. PAVILION WITH FOUNTAIN

Between garden and river was the Pistoia road; and on this road, just where the long avenue started from the gate to pass down to the river, there was a clump of trees (*alboreto*), and in it a small house for the ball games. This place, as Rucellai insists, was intended first and foremost for pedestrians—a spot where they were protected from the sun's heat and could enjoy the beauty of the garden when they had refreshed themselves, if they liked, in a little stream that was close to the edge of the trees. The inhabitants of San Piero appreciated this generosity, for at a church assembly held in 1480, the men of the small parish resolved that, seeing how the beauty and fame of the garden redounded to their own glory, and wishing to evince their gratitude for the many favours shown them by the House of Rucellai, two of their members should be appointed to keep up the garden in all its pride and beauty at the cost of the parishioners. These Florentine peasants have indeed shown here an example of their advanced state of horticulture.

In this account we hear little about water arrangements, and the house is surrounded, in the fashion of the Middle Ages, by a moat with a fishpond at the back. The stream, "as clear as amber" in the trees before the gate, does not appear to touch the chief garden at all, and the Arno is only useful for the view. The description is also silent on the subject of sculpture, and terra-cotta vases are all that it mentions. But stress is laid upon the clipped box, which apparently takes the place of statuary, as it does almost always in the Middle Ages. It appears in every imaginable form. In the flower-garden, which is best seen from the Pistoia road, the box cut into five steps is the central object: there we find giants and centaurs, ships, galleys, temples, arrows, men, women, popes, cardinals, dragons and all kinds of animals, and much else in a fine conglomeration. Sometimes they are cut out of the hedge, and sometimes they are stuck up as separate individuals. Alberti says nothing at all about this *opus topiarium*, and the other writers of the early Renaissance hardly mention it; the Italian garden of the High Renaissance period draws away more and more from this childish style of ornamentation, which of course exhibits gardener's skill rather than æsthetic meaning. We know how widespread the fashion for tree-clipping had become in the days of later Roman antiquity, but still the art had been so perfected (as we see it in the Quaracchi gardens) that it is impossible to think it is derived from faint indications in Pliny and other ancient writers; rather it must be due to long practice in the Middle Ages, and never since abandoned, although the threads are hard to follow.

FIG. 152. CLIPPED TREES

Rucellai's account of a garden actually in existence gives a solid foundation to two poetical works. One dates from the middle of the fifteenth century, much the same time as the Quaracchi garden. In the *Sogno di Polifilo* of the monk Colonna he describes in words and by pictures a round island in a garden, surrounded by myrtle hedges as well as by water. The garden itself forms the segments of the circle, the paths are shaded by pergolas, and most of them are overarched with roses and vines. Pavilions are set up where the paths meet. The separate parts of the garden are meadows, containing flower-beds or fruit-trees, the latter often clipped to make rings. The central piece would be a wonderful fountain, a pavilion (Fig. 151), or a very artificial shape in box, of the more important specimens of which he gives a set of illustrations (Fig. 152). Fantastic as the whole place may appear, especially in its ground-plan, even that (in its main lines) is clearly indicated by the paths which the pergolas mark out; and the individual parts are closed in. Even in this remarkable garden the absence of statuary is obvious, and in its place there appears a decided tendency towards the use of clipped trees in every kind of shape. The house is entirely suppressed, and instead there is this rigid formal plan which combines, in one whole, all the separate features of a garden.

The second poetical work is *De Hortis Hesperidum,* by Jovianus Pontanus. He presents a garden without reference to a house. He wrote the book in old age, about the year 1500, when he was enduring an unwelcome leisure near the end of his life. Thinking of the days when with his wife he had tended his Neapolitan garden of orange-trees at Vomero, he wrote an imitation of Virgil's Georgics, to advocate the culti-

FIG. 153. MEDICEAN VILLA AT CAREGGI

vation of this very profitable fruit, which in Northern Italy was not yet known. But even in his first book he deals in elegant hexameters with pleasure-gardens and the *opus topiarium.*

If you are not very much concerned about the produce of the garden, or the income it brings you, but only care to enjoy the beauty of the grove (*nemoris*) and a shady spot brightened with lovely things, choose before all some place where there is a spring or a gliding river to which a stream flows down, or where a fountain plays, so that the garden need never fade and wither in the heat, or your care of the fine wood (*silvæ*) be wasted by the bitter cold. Set walls all round it, against the storms to come, lay stones, dig trenches, arrange the place with care, and put up earthen banks. Plant young shoots, and arrange them in fixed rows (*trames*), support them with bast, that from the start they know what they have to do, each in its own proper place. When the tree, owing to the gardener's constant care and attention, begins to put out its branches and unfold its leaves, then choose the task for each, and make the formless mass into shapes of beauty. Let one climb to high tower or bulwark, another bend to spear or bow; let one make strong the trenches or the walls; one like a trumpet must wake men to arms and summon hosts to battle; another shall throw stones from slings of brass, storming the camp, sending the foe back to their ruined walls. In those ruins the hosts go forward, and stand at the open gates;

the conquering army presses into the town. Thus shall you by skill, time, native strength, and careful nurture, convert the tree into many new forms, even as a thread of wool is woven into divers figures and colours in a carpet.

If one ignores in all this the exaggerations and meanderings of a humanistic style, there is nothing at variance with Rucellai's garden: indeed it is quite likely that the poet had in his mind his garden at the Mergelina.

As compared with such detailed accounts, what we get from literature about the

FIG. 154. VILLA SALVIATI, FLORENCE

famous villa of the early Renaissance at Careggi (Fig. 153) is very scanty. The talk about trees and shrubs in a Latin poem from a humanistic source is quite worthless: after the conventional comparisons with the gardens of Semiramis and Alcinous there is nothing but the familiar rhetoric and repetitions. Instead, we have the villa preserved, and certain features of the garden also. The house itself is just as of old, to the roof that was put up on the battlements after a fire in 1517. It was built by Michelozzo for Cosimo de' Medici, a little earlier than Quaracchi.

There is very little to be found at Careggi of those Renaissance forms so fully adopted by Michelozzo in the Medicean palace—which was already built—in Florence itself, such as the delicate "rusticated" joints, graduated according to the relative heights of the stories, and the outspreading corona, and again the tastefully treated window-frames. So far as its outside is concerned, Careggi is still under the spell of the castle of the Middle Ages. The façade on the road widens at its lower story, as though it made a safeguard against an enemy's attack. The windows are small and unattractive, whereas the upper part shows those very picturesque battlements which are peculiar to Florentine palaces, and familiar to us in the Palazzo Vecchio at Florence.

Careggi was no exception to a general rule; indeed we must think of all the villas in Florence, perhaps in Italy everywhere, as built in this way in the middle of the fifteenth century. Caffagiolo, another Medicean villa, was built by Michelozzo even more like a castle, with moats and drawbridges. Most unfortunately, this house has entirely lost its character, owing to alterations made by the Princess Borghese in the later nineteenth century.

FIG. 155. FOUNTAIN: BOY WITH DOLPHIN; BY VERROCCHIO, ORIGINALLY AT VILLA CAREGGI

About the same time or a little later was built the Villa Salviati (Fig. 154), which in the year 1450 came into the possession of Allemanno Salviati, brother-in-law of Cosimo. In style it is so like the Medicean villas that Michelozzo was probably the architect. Quaracchi, like Caffagiolo, was something of a fortress with trenches round it; other villas, such as the Medicean one on the slopes of Fiesole, and Villa Rusciana, which Brunelleschi built for Luca Pitti, are too much changed for us to make sure of what they used to be like. The Belvedere, the first villa at Rome, then shyly rearing her head, built by Pope Nicholas V. on the Vatican Hill, was a castle with battlements; and the same character appears in the old part of the Villa Imperiale at Pesaro, built by Alexander Sforza, the Emperor Frederick III. laying the foundation stone in 1452. This villa, like the Palazzo Vecchio, had a tower on the façade. The garden, which according to descriptions of the sixteenth and seventeenth centuries adjoined the old villa at Pesaro, and was partly a field and partly an orange-garden, must without doubt have been part of the original plan.

At Careggi the garden has preserved its main features, though it has seen great alterations in plantation in the course of centuries. The principal garden stopped at the side façade, and was also shut in by a high battlemented wall. It is possible that the mosaic pavement which, near the house, has the same pattern as the flower-beds, was like this even in the fifteenth century. From the start the front parterre will have been separated from the large garden beyond by a little wall and a gate, and being a flower-garden will have been ornamented with terra-cotta vases as at Quaracchi. The planting of this main garden, to resemble Quaracchi, must be supposed to include pergolas. In the spare places it is filled up with fruit-trees, but it was certainly not lacking in clipped box, or indeed in arbours with benches and other places to sit on. Fortunately the prettiest centre-piece and ornament, the fountain of Verrocchio and its laughing boy, holding a dolphin under his arm (Fig. 155), has been preserved, not, it is true, in its old place, but in the Palazzo Vecchio at Florence. In the garden there is a pretty loggia, such as Boccaccio had already written about. In this particular one we get the most perfect architectural feature of Renaissance style, the handsome arches supported on pillars. Here Lorenzo was wont to collect his friends at a Plato-like academy, and from here they enjoyed the then charming view. The house, which, like Quaracchi, stood in the plain—for people in that day were

FIG. 156. POGGIO A CAJANO—VIEW OF THE VILLA IN THE DISTANCE

nervous of the keen air of the hills—was so well placed that here, only a short drive from Florence, one got a lovely panorama of the whole of flowery Tuscany. Duke Eberhard, who, in company with Reuchlin, visited Lorenzo de' Medici, saw a roof-garden in flower which was afterwards burned, roof and all. We cannot tell to which of the courts we ought to ascribe the labyrinth with a fountain that Duke Alexander had restored in 1530 after the fire.

Under the rule of Lorenzo, who as a connoisseur had important things to say about architecture, the Renaissance influence was completely accepted for the villa. In 1485 he commissioned Giuliano da San Gallo to build him a handsome pleasure-house at his favourite seat, Poggio a Cajano. This place had been between the years 1448 and 1479 in the ownership of Giovanni Rucellai, but he does not appear to have built there, and only received the rents of the estate, the inn, and a little house, to use for the expenses of putting up the façade of Santa Maria Novella. His interest was concentrated on the newly laid out Quaracchi. And so when the church was finished in 1470, he sold the place to Lorenzo, who with his friends took the utmost delight from the first in the pleasant spot and the fertility of the property (Fig. 156).

The stride taken from the battlemented towers of Cosimo to the cheerful building at Poggio a Cajano is an enormous one. There is nothing shut in here. The four pavilions at the corners, perhaps reminiscent of towers, are of intimate use and connection with the front part; the main entrance is emphasised with a loggia, which owes its style to

FIG. 157. POGGIO A CAJANO, WITH STEPS LEADING TO THE GARDEN

the antique temple. Round the top of the *pian nobile* runs a wide balustraded terrace resting on pillared arches, and giving a fine view over the open flowering valley, in the middle of which stands the villa enthroned on a gently rising hill. The great cloistered hall runs the whole breadth of the house, which in gigantic yet just proportions expresses the lofty and ambitious thought of this generation of wealthy Florentines. Steps of an elegant round shape lead down to the garden from this terrace—an idea often recurring in Florentine villas, and lending them grace and dignity (Fig. 157).

But the first design of Michelozzo had straight wide steps, and nothing can be reconstructed with perfect certainty. The end next to the street, however, a kind of portico, and two pavilions at the corners, are no doubt part of the first design. The garden ground rises gently from here to the house. The round form of the stairs is repeated in a lower wall that has a wider arch, wherein the lawn of the parterre intrudes for some distance. Round the house on the other side runs a pleasure-garden, in front of it is the orangery, and there is a park in the style of the sentimentally inclined eighteenth century. Such is the present condition of this garden.

The villa was not completed before the death of Lorenzo, and in the following years, so fateful for the Medici family, it can hardly have been proceeded with, therefore the final condition of the garden may be ascribed to a somewhat later date. This is suggested by two things: an apparent reluctance to make a real terrace construction, to which the site of the house on a hill pointed; and the very sparing use of water for ornament, in spite of the fact that a water conduit in a long arch had been introduced, coming down from the hills near by, for the irrigation of the meadows attached to the villa. Of this conduit Lorenzo and his friends, constantly singing the praises of the place, were never tired of boasting.

About the same time that Poggio a Cajano came into existence, the Crown Prince Alfonso, a friend of Lorenzo's later life, built his famous summer residence at Naples, Poggio Reale. It seems that Lorenzo, an acknowledged authority in matters of taste, sketched the plan for the architect Benedetto da Majano. The relations brought about by Lorenzo's spirited visit, in the cause of peace between Florence and Naples, to Alfonso's father, Ferrante, were the first step towards the change of property. And there is a great similarity between the ground-plans of Poggio a Cajano and Poggio Reale. Both villas have the towers projecting at the corners; at Alfonso's villa all four sides are united by loggias, but at Lorenzo's the loggia is only in front. Instead of the one great central hall at Poggio a Cajano, there is here a court in the middle, which, according to the meagre plan of Serlio—admitted even by himself to be inaccurate—is the very heart of the building. The plan with a section of the court is the only record preserved of Alfonso's country house, which has entirely vanished from the face of the earth.

Serlio's description, as well as his plan, is only concerned with the court. He makes it a rectangular floor, with a hall of two stories adjoining; one goes down by steps to a rather lower room, where the king liked dining with a select company of ladies and gentlemen. When the fun was at its highest, hidden springs were opened at a sign from the king, and in one moment the whole court was under water, and the guests got an involuntary bath. They took pains, however, to have dry walls and beds in the bedrooms close by. Alfonso was a Spaniard, and one may suppose that he brought these water tricks from home, and perhaps he took the Generalife as his model. It is probable that this court, where he liked so much to dine in merry company, was decorated with trees and lawns, or at least with plants in pots. Of this Serlio says nothing, but once, in remembering past joys, he breaks out: "O happy Italy, thy light is lost through our discord."

This cry of pain is apropos of the French rule established at Naples after Charles VIII. made his romantic conquest. Nothing was so charming to the young French noble, when he looked on the splendour of the south, as the fine gardens and airy halls of Poggio Reale. And though Serlio, owing to want of space, gives no account of "lovely gardens with their various parts for fruit-trees, their many fishponds, aviaries, and the rest," we do hear from the French chronicler exclamations of delight. The most intimate picture of the gardens is in a French poem, *Le Vergiez d'Honeur,* though it only enumerates the wonders and gives no sort of arrangement; it mentions summer-houses, loggias, flower-gardens, little lawns, fountains, streams, and antique statues; there is a park for medicinal herbs and a fruit-garden, both of which are larger than the Bois de Vincennes; there are vineyards, a deer park with a well large enough to supply the whole town, and grottoes—the finest the poet ever saw. We have to take great pains to be able to reconstruct this garden. The actual flower-gardens here, as elsewhere, must have been very small; but there is no doubt that Alfonso was a real friend to plants and flowers, for we are told that when he fled on the approach of Charles in 1495 he found time to take with him to his place of exile in Sicily "toutes sortes de grain pour faire jardins."

Jacob Burckhardt shows more botanical than architectural interest in descriptions of gardens of the early Renaissance. No doubt the ground-plan was simple, but we must partly blame a certain want of skill in the writer; for when later on a real artist like Cardinal Bembo gives us a charming garden picture, we are able to trace a scheme that is regular despite its simplicity. In the last years of the fifteenth century a young Venetian, at that time living at the court at Ferrara, published his lovers' dialogue in the garden of Asolo (*Gli Asolani*) at the friendly court of Caterina Cornaro, Queen of Cyprus. Bembo's description is poetry, of course; but it shows character and individuality, and is undoubtedly a faithful account of the garden at Asolo. He says that on the occasion of a wedding feast for one of the court ladies, a little group of the party left the others at their midday sleep, and went out in the garden which was in front of the dining-hall.

> This charming garden was of wondrous beauty. On both sides of a pergola of vines, traversing the garden in the form of a cross, wide and shady, there ran to right and left two similar paths. They were long and wide and strewn with bright gravel. On the garden side, they were shut in, beyond where the pergola began, by hedges of thick yew. This yew reached breast-high, so that one could lean against it and get a wide view of all parts of the place. On the other side tall laurels stood along the wall, pointing upward, but with their tops forming the arch of a half-hoop above the path, and closely clipped so that not a single leaf ventured to push out of its proper place; and outside the walls nobody at all could be seen. On one side of the garden at the end there were two windows framed in white marble: the walls here were very thick, and from either side one could sit and look from above over the plain. On this path strolled fair ladies, accompanied by youths, and protected from the sun. Admiring this and that, and chatting as they walked, they came to a little meadow at the end of the garden, full of freshly-cut grass, and scattered over with flowers; at the far end were two clumps of laurels placed irregularly and in great numbers, looking very quiet and venerable, and full of shade. Between them there was room for a very lovely stream, cleverly hollowed out from the living rock, which was the termination of the garden on this side. From it poured currents of fresh cold water, starting from the hill, not springing out high above the ground, and falling into a marble canal that cut through the middle of the meadow and then ran rippling through the garden.

The middle of the garden was marked by a pergola forming a cross, a constant constituent of a garden for hundreds of years, in the seventeenth century exactly the same as when Boccaccio knew it. But here water suffers completely new treatment: the fountain is no longer the chief feature as in mediaeval times, its place being taken by the grotto. This receives the stream from the rock at the end of the garden, which is cut in two at the back by a canal. Unfortunately Bembo does not describe any further irrigation, nor are we told anything about the planting of the four corners framed by the hedge. Perhaps there would be little fountains and clipped box, perhaps fruit-trees planted round. But there is a careful gradation to be seen in the whole picture, from the painstaking arrangement of the front near the house to the careless disposition of the trees at the back, where the rock ends it all; and even the "selvatici" that are not enclosed are symmetrically and conventionally placed—grottoes and canal, and masses of trees grouped in clumps. In front of the little wood, so "venerable and quiet," we have in strong contrast the flowery meadow with its canal, and we pass on to the front part with its severe straight lines where "not a single leaf ventured to push out of its proper place." We see here, far better than in Rucellai's description of Quaracchi, which Alberti designed, how the most important recommendations of the great humanist of the early Renaissance were actually carried out.

All the gardens named so far have been the seats of princes. But the farm garden of the time, according to Soderini's *Treatise on Agriculture*, has a certain artistic leaning of its own. He lays stress on harmony between garden and house (ground-plan and façade), which must be made to suit one another both in size and in the arrangement of angles and corners. The pleasure-garden ought to lie in front of the windows of the villa, which is to stand rather high; behind is a fruit-garden, and then the real orchard with fishponds and meadows at the sides. The view should reach beyond all this, over the level to the sea or the mountains. Every garden, and even every field, must be divided up into squares of equal size, and intersected by wide convenient paths. Next to the villa must be planted evergreens, but they, as well as fruit-trees, must be so treated with sickle and shears that none exceeds another in height, and that they look "like a green meadow." It was an unquestionable postulate at that time that fruit-trees should be planted in quincunx form, after the fashion of the ancients. Like Alberti, Soderini must have his pergolas walled in, or like a vine arbour, "as one prefers them nowadays." And they must be so made that the *padrone* can walk right out to the field without being bothered by sun or bad weather. In the garden stand vases of flowers and dwarf fruit-trees of every kind. In winter he must have flowers, but at first these are medicinal plants, and in his demand we recall antique ideas. The custom of writing the owner's name in box on the beds seems to him a very agreeable one, and he praises as modern the custom of setting out coats of arms, clocks, animals, and human figures in immortelles, and in thyme or other sweet-smelling plants. In this Soderini displays acquaintance with old authors, especially Pliny, and he also shows an attempt to take Alberti's advice. At this stage of progress the farm garden remained fixed for centuries in Italy as in the North, and in the gardens of the poor the old conditions hold unchanged even to the present day.

In Italy, development seems to have been very unequal in different districts: things moved slowly in districts where the nature of the land meant few estates as round about Venice, and where the need for mere business cultivation was greater, as in Lombardy. Scamozzi, a pupil of Palladio, was a writer of the time when, at the beginning of the seventeenth century, the art of gardening in the rest of Italy had reached its highest point. He had, however, seen but little of the magnificence of Roman gardens, then supreme; at any rate his meagre descriptions do not advance a step beyond those of Soderini. Furthermore, he has no notion of farm gardening, but, like Bembo, has in his mind the gardens at Asolo, and only gives the name of Park to the orchards because of their great size. But since Bembo's time these, like the other gardens, seem to have remained as they were. Scamozzi, who spent a long time on the other side of the Alps, possessed great influence in the North. His writings were widely circulated, and until the eighteenth century he passed for an authority with many makers of gardens, and for the Italian garden this book, which appeared in 1615, meant a pause lasting for a hundred years.

Garden architecture was in a backward condition about 1500, for the ground-plan was of the very simplest nature, and there was nothing more than hedges, leafy walks, and clipped trees. The size of an ornamental garden was still very small, whereas the gardens for fruit and vegetables, so soon to be carefully hidden in the background, were large and important. The princely estate is at its best, however, in the park. In the early years of the Renaissance the love for keeping wild and strange beasts in cages, and the desire to have them to hunt, grew enormously, as it had done among the ancients. For this purpose a wide place was desirable, where large aviaries and fishponds could be kept up. Men of smaller property had such features close to their villas—Soderini, for example, having fishponds on both sides of his own house.

A park of this sort could not of course have any pretensions to artistic unity, but we may suppose the place was mapped out and ornamented with a few things here and there. Thus the French poet describes the park at Poggio Reale as having barricades for wild beasts, and meadows for domestic animals at grass, and also aviaries and vineyards. There is a fountain "big enough to supply the whole town," and there are also grottoes. Concerning other parks, such as the one belonging to the Cardinal Aquileja, and the one set up by Prince Ercole close to the gates of Ferrara, we only know the actual fact that they were there. In Filarete's architectural romance, he has much fanciful description of a park, but the whole tale of buildings and gardens of fabulous size and splendour must have been born of his own imagination. He gave the reins to an architect's dream, so often balked by hard fact, for at the court of Francis I. the Renaissance ideas of Sforza certainly never materialised: the hanging gardens of his Prince Zagaglia are just an attempt to outdo in a fairy-tale the gardens of Semiramis in Diodorus.

The park is more easy to understand, for here the writer had in his mind that in front of the gates of Milan. Filarete's princely park was encompassed by a great wall, and lower ones intersected the gardens to separate the wild beasts from the tame; also there were large ponds for ducks and herons, as these birds were hunted with hawk and falcon. Watch-towers were set up on artificial mounds, from which to view the chase. At the top of a large round hill at the place of the wild animals the prince had a church built in a thick shrubbery of pines and laurels, and inside it a hermitage, where a hermit lived. It had always been considered an advantage to have a holy man living on the estate. No doubt the romantic and picturesque surroundings were mingled with a certain sentimentality, for a hermitage, like ancient ruins, seems to be a thing people like to come upon, and this kind of feeling was natural at a period when men were so much concerned with their own past. In the love dialogue of Bembo one of the speakers says that he found a pretty round wood on the rocky hill at the end of the garden, where a hermit had made his abode among the thick shady walks.

In this matter, as with the passion for ruins, we find only isolated instances among Italians at this time; and yet the mighty remains of their forefathers in Rome excited their imagination, and the longing to possess as their own some real antique objects not only led them to construct gardens round genuine ruins, but also round "faked" ones. Vasari speaks of a sham ruin in a *barchetto* (small park), which the prince put in the old restored castle of Pesaro. Inside it there was a handsome staircase like the one at the Belvedere in Rome. Polifilo also, in his *Hypnerotomachia*, gives a most romantic account of a ruin of this kind (Fig. 158). But these seem to have been exceptional, for in Italy there was such an immense quantity of genuine remains of the past that the disguise seemed too poor and thin.

FIG. 158. PICTURE OF RUINS

The ruins of ancient sculptures had quite a different bearing for the garden. Such objects were continually coming to light out of the rubbish heap, and soon there were so many of them that the houses could no longer contain them all. Presently the idea came that they would make suitable ornaments for a garden. At the beginning of the fifteenth century the first timid attempts were made to put them there. Learned men and artists first created garden museums. The humanist Poggio relates how he set up antique statues in the little garden that he laid out about 1483 at his own villa in Terra Nuova, and how his friends, the other speakers in the dialogue *De Nobilitate*, laughed heartily at him. He had put up these marble statues, they said, instead of his beggarly ancestors to give an appearance of nobility. Proudly did the great humanist take up the jest: he had won his nobility, he said, by finding and collecting antique remains. But what was now criticised as a novelty, and was scarcely to be found in gardens of the middle of the century, was destined to become a universal custom, as the soils of Italy and Greece delivered up ever more and more of their buried treasures. Poggio's things passed into the Medicean collection, and (together with another of Niccolò Niccoli and certain acquisitions of Cosimo's) laid the foundation of that enormous collection of art treasures. Cosimo himself was already putting antiques in his gardens: there was a Marsyas at the door of his palace at Florence. Lorenzo put up a casino, and laid out gardens on the Piazza San Marco, where later his second wife would sit when she was a widow. Here, we are told, "in loggia, in private rooms, and in the garden arbours" he set out his antiques, and established a drawing school, wherein Michael Angelo learned the art of sculpture from the study of these ancient statues.

The works of Mantegna, which Lorenzo thought worthy of a visit in 1487, also belonged to these early collections, and enjoyed a great reputation. The beautiful round court at Mantegna's house in Mantua, now poorly preserved, had niches where the artist liked to set up his well-beloved treasures. Bembo also, rather later, after his separation from Leo X., made a garden at Padua, where the summer-house served as a studio; and round about it among espaliers of lemons and oranges of various unusual kinds he set up his collection of antique statues.

Rome, comparatively late in developing the new form of gardens, was also behind Florence in ornamenting her gardens with statues. It was the man who made Rome's future, and stood on the very threshold of the new city, Pope Julius II., who set the finest example. Already as cardinal he had vied with the Medici in collecting statues, which he placed in the garden at the great Penitentiary at S. Pietro in Vinculi. When he became Pope, he transferred them to the Vatican, and placed them in the court of Innocent VIII., undoubtedly the Belvedere, and thereby laid the foundation of what is probably the most important collection in the world. The court was at that time for practical purposes greater than it is now, for the corridors and corner rooms where the statues are had not then been made. In the vast space there was a garden, which the ambassadors from Venice, who were left there for a while in 1523, described in glowing colours (Fig. 159). "At the end of the loggia" (the present Chiaramonte), they say, "one steps into a very lovely garden, one half of which is full of bright flowers and laurels, mulberries and cypresses, and the other half paved with square terra-cotta tiles. At each corner stands a fine orange-tree, and there are a great many of these, all in excellent order."

Nothing but the Belvedere can be meant, for this alone lies at the end of the loggia, and we can only talk of two halves if the whole thing hangs together as one. All the same, this garden is no more than about thirty metres square, and the treatment of it as half orangery and half pleasure-garden is surprising; but it only goes to show once again

FIG. 159. THE GARDEN COURT OF THE BELVEDERE

how small the space allotted to a garden always was, even when there was a large area of ground to dispose of. It also shows how the fancy of spectators could fashion great things out of little.

An asylum was found here for those statues that for centuries to come were to be the Mecca for the votaries of art, and a starting-point for learned research into the art of the ancients. At that time the statues were well arranged. Of the eight groups and single pieces, which made the whole collection in 1523, six stood in alcoves that were attached to the wall; in the middle of the south wall was the finest piece of all, the Laocoon, between the Apollo and the Venus Felix, and opposite to the Cleopatra and a river god. The two colossal figures of the river gods of Nile and Tiber, now in Braccio Nuovo, stood in the middle of the garden on pedestals, from which fine fountains played. Later there were added others, such as the torso of Hercules. Thus care was taken to consider the nature of each separate work; and in green surroundings, with running fountains and sweet odours, they gave a very different picture of beauty and poetry from that which they give to-day in their cold room indoors.

Naturally the cardinals were soon eager to emulate the Pope. The pictures of early museum gardens have been preserved in certain sketches made by artists from the North, who visited Roman palaces in the sixteenth century. One of the oldest and most important was laid out by Cardinal Andrea della Valle as a hanging garden at his palace, Villa Capranica, which Lorenzetto built (Fig. 160). On the façade he had antique pillars set, with their pediments and capitals. In the sketch, which shows the gardens in an unfinished state, there are statues on the longer side placed opposite one another in two rows, and the short sides are filled in with open colonnades, sarcophagi treated like friezes, and other fragments of reliefs, and below these are pieces of the Ara Pacis, let into the walls. The intermediate walls, which are open to the air, are covered with lattice-work, concealed by green climbing plants set in boxes. A significant novelty originated in these gardens: according to Vasari, the cardinal allowed well-known sculptors to come and restore the statues; and this had such an effect in Rome that it was imitated everywhere.

A great part of the collection of antiques passed later into the possession of Cardinal Ferdinando de' Medici, who had them set up at his villa at Monte Pincio. This happened at the time when Rome was being encompassed by a wonderful girdle of lordly villas, in which the statues formerly in the town gardens, and strictly confined there, were now to display themselves as the finest of all treasures. This great development of art could occur quickly in Rome, because in the earliest decades of the century the city had taken an important step in its history and growth through the revival of terrace schemes in the garden. The different surroundings of the Middle Ages had let terraces disappear and be forgotten; and it had been thought that a garden was at its best when it was quite flat. Even Alberti never asked for terraces, in spite of the increasing demands for villa and garden that were made in his day. He only gets from Pliny the idea of the slightly raised situation of the house, and the gentle easy slope of the paths, and in that Soderini fully concurs. Filarete's fanciful terraces are only the feverish dream of an architect copying the antique, and they lack any sort of foundation in reality. It is most characteristic of him that, as soon as he has anything real to talk about, as in his work *De Architectura*, he does not say a word about terraces apropos of that double garden which he is laying out behind his own villa. In the same way at Bembo's garden in his *Asolani*—which could so easily have been turned into a terrace-garden, situated as it was against a rocky wall —the whole place seems to be quite flat. And this is the same as we have found it in all the Florentine villas.

In spite of this, the terrace was not quite unknown in Italy at a much earlier date, being used always for vineyards. Moreover, poets sometimes mention it, and so early

FIG. 160. THE GARDEN OF STATUES AT THE VILLA FIRST CALLED DELLA VALLE AND LATER CAPRANICA, 1553

as the middle of the fourteenth century we have a very attractive description of terraces in Boccaccio's *Decameron*. On the fourth day the poet represents the ladies as in a valley surrounded by six hills, which form a circle round the place "as though it were made not by nature but by an artist's hand." The sides of the hills, crowned with little castles, slope in steps as at a theatre; and these steps are all in circles, one above another. On their southern side there are vines, olives, almonds, cherries, figs, and other fruit-trees, and on the north there are woodland trees. At the bottom of the valley, which is only accessible from one side, there stands a thicket, as fine as the best artist could make it, of pines and cypresses, giving on a flowery lawn with a cistern in the middle, which is fed by little brooks that rush down from the rocks.

This valley, partly vineyard, partly orchard, and partly wood (whereat the poet was so surprised because it was nearly as good as art), had certainly no direct connection with a garden; but we have a model here which is able to show how, a century and a half later, an important movement might be effected, viz. the introduction of a terrace scheme into gardens that the artist really made. As a fact, this is no more than the revival of old ideas, for the ancients had fine terraces in their own gardens. But, above all others, the Italians not only rediscovered this plan, but developed it for the other gardens of Europe in original and unexpected ways. We ought to bear in mind how far removed from their thought was the idea of the villa of the olden time—that is, of one huge enclosed estate forming a congruous whole—and we ought to realise that there is nothing in literature or in monuments to justify us in assuming that the ancients employed architectural stairways of an imposing kind in the same way as the artists of the Renaissance. The want of unity in the whole plan was increased by having no terraces in their gardens.

The Italian Renaissance artists, taking as one of their chief tasks the creation of the terrace in architecture, could boast that they were the first to grasp the significance of stairways in the garden scheme. It is only by its alliance with the terrace walk that architecture really appears in the garden, linking it with the house. The lengthened lines are easily perceptible to the eye: the axial line is prolonged in the chief garden walk, and the balustrades on the terraces supply the horizontal cross-lines that the eye desires: thus both are united in the scheme of the steps. The walls between give a good opportunity for grottoes on one side, and for a repeat of the façade-plan with pillars or corridors on the other.

By the use of this great novelty, Rome becomes influential in garden history, though comparatively late in experiencing the passion for buildings of a Renaissance type. At a time when in Florence the rich burghers were striving to outdo each other in the grandeur of their palaces and country houses, Rome was still a small unimportant town huddled together on the bends and curves of the Tiber. Mediaeval Rome was desolate, "the very recollection of ancient times was all but vanished. The Capitol was a hill where goats grazed, the Roman Forum was a pasture for cows." Still, as early as the days of Nicholas V., the conception had arisen of making Rome, through its buildings, the greatest city in the world. In his famous speech to the cardinals on his death-bed, Nicholas conjured them, for the honour of God and the Church, to see that Rome was filled with noble edifices.

It was not, however, until Julius II., a Pope of dauntless energy, sat in the papal chair, that such plans could materialise. At the beginning of his rule he summoned the greatest architect of his time, Bramante. Everything was to be done to carry out the projects of Nicholas V., and to build up the Vatican so that in size and area it should be the greatest palace in the world. St. Peter's was to be a wonder on earth. Innocent VIII. had already enlarged the palace with new buildings, and added the Belvedere, the battlemented castle, above it. Julius commissioned Bramante to connect this pleasure-place (which, because of its high situation, was also called Tower of the Winds) with the palace below.

Bramante fulfilled his task in a wonderful way (Fig. 161). What he had to do was to find a means of passing from a garden terrace attached to a villa, without losing a sense of proportion, and always going gradually, to a palace court. He treated the uppermost terrace entirely as garden ground, and yet he could not and would not have it joined on to the Belvedere, for the house was much too small to serve as the end piece after so long a stretch as 306 metres. He therefore decided on another decorative erection—a garden

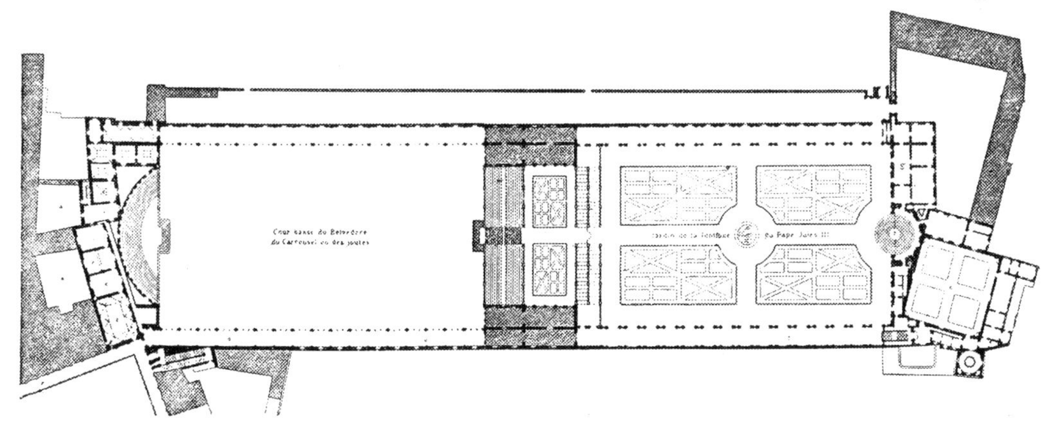

FIG. 161. COURT OF THE BELVEDERE, ROME—SITE-PLAN

structure within the colossal semicircular niche, with a loggia on the top, giving the finest possible view over landscape and town. On both sides of the garden there were colonnades, open on the inside, but walled in outside so as to give that feeling of seclusion which the garden needed. In the middle the Pope had an antique shell fountain set up in the second year. The length of 306 metres was given to Bramante, and he chose to have the comparatively narrow width of 75 metres, because of the situation of the somewhat high terraces; seen from below the width seemed quite the right proportion (Fig. 162). Bramante continued the colonnade of the upper garden terrace at the same height, and the building that was thereby necessitated below (completing in the third or bottom story a three-storied façade) gave all the effect of an inner court on the lowest floor. A semicircular end was here also attached to the chief palace in correspondence with the one above. There were theatre seats in half-circles, by the help of which this court could be used as an arena, and there were straight wide steps as well leading to the second terrace, which provided seats for spectators, and would accommodate a great many—it was said 60,000—if the open loggias on both sides were used as well.

Drawings that were made about 1565 prove that this place was really intended for entertainments. In that year Pius IV., who completed the great building, held a festival in honour of the wedding of his nephew Hannibal of Hohenem with Hortensia Borromeo, and the court which had just been finished was inaugurated with much pomp and pride and with a wonderful tournament. There are two pictures, one showing the place in full festival, spectators crowded shoulder to shoulder, and the tournament in mid-career; the other showing the court full of statues, and also the beautiful stairways, the first leading up from the lowest terrace, quite straight, and the second a set of wide steps with balustrades leading on to the highest terrace. Between them in the dividing wall there was space for alcoves flanked with pillars, in which probably there were fountains playing. The uppermost terrace was laid out with various beds. The whole makes a very effective building, and it loses nothing in beauty or depth of meaning if we seek out the sources

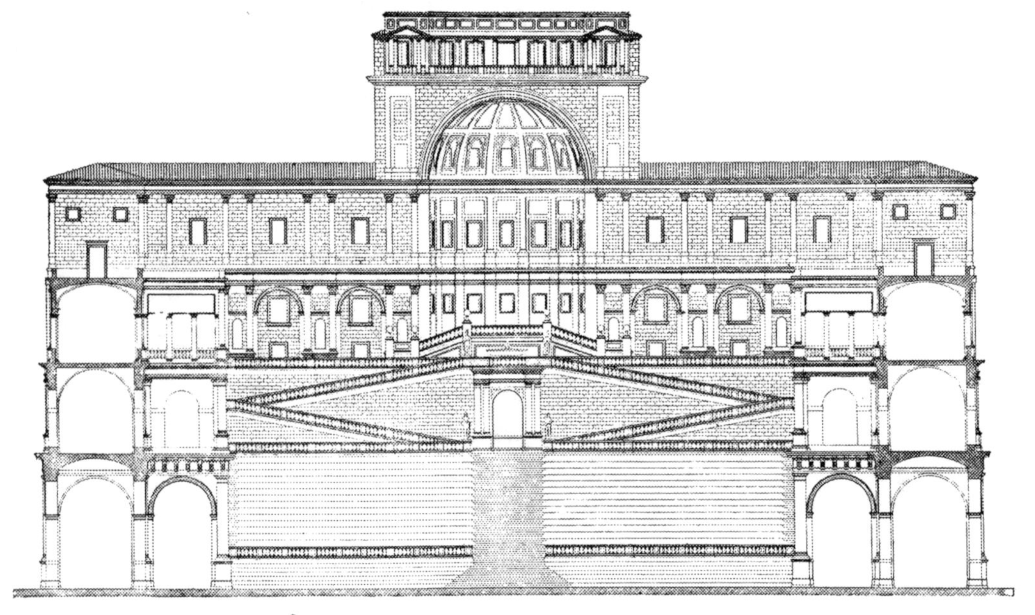

FIG. 162. THE BELVEDERE, ROME—CROSS-SECTION OF THE COURT

of those ideas which the architect borrowed from antiquity, and assimilated with his own original work.

Rome to this day exhibits as a garden scheme the immense alcoves that look down from the imperial palace on the Palatine into the so-called stadium, where Bramante's treatment achieves the desired end of giving a view over the garden and beyond. As has been already pointed out, this was one of the favourite objects among Romans right up to Byzantine times. Similarly, the colonnades on both sides are taken directly from an antique model. How far Bramante was inspired thus in his great scheme for stairways, it is more hard to determine. The worthlessness of Ligorio from the historical point of view has already been remarked upon in reference to his plan at Monte Pincio, but it might have served as a study sketch for Bramante. Then again we have to remember that at Constantinople in the court of the Sigma there was an ornamental staircase leading from court to hall, where stood the emperor's "box," and this served much the same purpose as Bramante's grand straight stairs that led from the court to the first terrace. It is quite possible that some inspiration came to him from these and similar models.

In spite of pressure from the impatient aged Pope, Bramante did not get on fast enough to be able to see more than a small, meagre beginning of the work before his death. A view of the fortuitous condition of things at the time of his death is shown in a picture (Fig. 163). The eastern colonnade only was completed, and that in such haste that it often fell down later on, and very likely some of Raphael's frescoes were buried under it. Again we call in the aid of the Venetian legates to visualise the state of the place nine years after Bramante died (Fig. 163). They hurried through the loggias, and said that these were so long that if one man stood at the top, another at the bottom, and a third in the middle, they could not recognise each other. All the same, they observe three terraces, the lowest

FIG. 163. COURT OF THE BELVEDERE, ROME, AT THE TIME OF BRAMANTE'S DEATH

one turfed, the second a *monticello* (rising ground). On the top one there were clumps of trees. The corridor they walked through is supported on one side by pillars; on the other balconies are built out for the sake of the view over the "prati di Roma."

The report of the ambassadors says nothing about steps, but their silence does not mean that steps were non-existent; for they hurried with all speed through these parts of the Vatican so as to be able to enjoy as long as possible a dearly-bought permission to see the antiques in the Belvedere. The western corridor was finished by Ligorio, architect to Pius IV., nearly half a century after Bramante's death. This Pope had a careless and artistic nature, and loved gaiety and feasting. But the place was only undisturbed for a very short time, and soon ceased to provoke envy and wonder. The short period of the Pope's rule proved fatal, as so often in other cases, to the works of art which he created. A Pope who had enjoyed his own completed work was often followed by a very different person, and the difference came out in their artistic surroundings. To Pius V.

the feasts of his predecessor were both impious and hateful. First the theatre, and then all the rest of the court that contained any heathenish statues, had to be stripped, and wagon-loads of these made their way to Florence and other towns.

Care was taken by Sixtus V. that the old splendour should never return; for on the second terrace, that cuts off the lower court on the garden side, he erected the so-called library crosswise, and so dealt the death-stroke to Bramante's great masterpiece, only twenty-five years after its completion. The uppermost garden did, to be sure, receive its greatest glory in the seventeenth century with the pineapple fountain. Paul V. had this sacred object (from the old basilica where it had stood under a small temple in the atrium)

FIG. 164. VILLA MADAMA, ROME—THE UPPER TERRACE WITH THE POOL GROTTO

moved away from its place to make room for the structure of St. Peter's. The great alcove in Bramante's garden, at that time still empty, was worthily adorned with it, and the garden from then onwards has borne the name of Giardino della Pigna. But through this crossway building the place had become little more than an inside court; and even the Braccio Nuovo (put up in front of the second terrace in the nineteenth century) could scarcely do further injury to the cause of art and beauty.

Though Bramante could not accomplish all the different parts of his work, he exercised a far-reaching and lasting influence on future development in gardens. It was inevitable that rich men in Rome, who were ever more and more responsible for the form that the new city was to take, should see their model in the Vatican. More than half a century before, Pope Nicholas had dreamed of setting up a palace so vast that all the cardinals could live within it. Now they were beginning to build, not only their own town palaces, but also handsome country houses. Busiest of all were naturally the nephews of the reigning Pope.

They staked their all on the duration of their kinsman's life, and the short time had to be used to the best advantage. Their haste enriched Rome with numerous works of art, but many beautiful and lordly places were either never finished or soon left to fall into decay, and nothing could suffer more severely than garden sites.

Leo X.'s nephew, Giulio de' Medici, was in the forefront of it all. He brought with him a passion for villa life that he had inherited from his family. In front of the Ponte Molle stretched the long range of hills, rich in water, of Monte Mario, which half-way up has a moderately large flat terrace with a lovely view; on one side stands the town, far enough off to give the feeling of open country and yet easily reached by the well-kept Via Flaminia. On the other side the river with its bridge winds through the green plain stocked with vines, and behind is the encircling Sabine Range. The cardinal was enchanted, and resolved to build a castle in the grand style. He summoned the best artists of that day, and no less a man than Raphael undertook to plan and construct.

FIG. 165. VILLA MADAMA, ROME—THE ELEPHANT ALCOVE

If fate had allowed the building to proceed, a jewel comparable with anything the Renaissance can offer would be now before our eyes, but changes of fortune have left only the ruins of a fragment. Scarcely was it built than it was destroyed in a savage attack; for when Giulio became Pope as Clement VII., the villa was scarcely half made, and two years later it was the first victim of the revolt headed by Cardinal Pompeo Colonna, who tried to show his revenge in a flaring light by burning the Pope's darling villa on Monte Mario while its master was a prisoner at the Engelsburg. Such part as was burned was afterwards rebuilt; but the whole villa was never completed. It has its present name, Villa Madama, from Margaret of Parma, who when travelling by used to stay there. To-day it is a melancholy picture of decay, yet in it one can trace the features of ineffaceable beauty.

A set of original plans, some by Raphael, some by San Gallo, have been found in quite recent years, and they help one to understand the complicated architecture of the place. Though much must be simply guesswork, one can mentally reconstruct Raphael's part of the building, of which only a bare half was ever completed. Of the gardens two terraces are preserved on the side of Ponte Molle (Fig. 164). The upper one is cut off at the south end of the narrow side by a fine, triple-arched loggia, and on the north there stands a high wall with a door in it, flanked by two gigantic figures. On the side next to the hill there are three imposing niches in the wall, of which the middle one shows portions of an old decoration, viz. a sort of covered-in box, and at the back an elephant's head (Fig. 165), which pours water out of its trunk into a highly ornamented bowl. There are festoons looped across, fastened with masks, and out of these again flows water into basins fixed lower down.

In the other niches stand colossal antique statues: in the first, the one next to the loggia, there is the so-called "Genius," which was taken to Naples by Margaret of Parma with most of the other pieces. In the exedra of the loggia itself is a seated Jupiter, still to be found there in the eighteenth century but now wanting. Perhaps this upper terrace was first of all intended as a garden for statues, because planting unfortunately did not interest the designer. In the middle, a little to the side of the axis, is a very large rectangular tank. From the terrace, stairs, of which only the one on the south is preserved, lead to a very much lower one, almost entirely taken up with a cistern of great size. A well-arranged series of grottoes runs along the wall from north to south, and this is all there is to be seen.

FIG. 166. VILLA MADAMA—RECONSTRUCTION OF THE EAST GARDEN

Of the plans of San Gallo (who to begin with worked with Raphael and then managed the building of the place) we have a large hippodrome outside the door of the giants, planted with chestnuts and fig-trees (Fig. 166). A smooth elongated level indicates that this part was actually carried out. On the east an elegant double stairway leads to the low-lying orange-garden. One of the corresponding stairways leads down to a third terrace, which, if we may rely on slight sketches, should include a wide flower-garden ending in a semicircular fountain. The southern side of the villa and all garden parts appertaining, do not even seem to have been touched. But a drawing of them has been found, perhaps from Raphael's own hand (Fig. 167). In this design the garden is again in three terraces. The uppermost one contains a square flower-garden. The next, a round one (possibly a rosary), has a fountain in the centre. The lowest and largest terrace is an oval having two fountains, and with wide stairs leading from one terrace to the other.

The whole scheme, most intelligently thought out, looks like an illustration of what Alberti asked for—circles and semicircles in the garden—and all the more if one looks at the building plans. According to the last plan put forward, the central feature was to be a round court, and the finished half makes the south façade a semicircle. But on the hillside there ought to be, behind and above, another half-circle of a theatre, overlooking the court and the whole of the villa. Around this middle court there ought to be rooms attached to north and south, and gardens stretching in front. Lastly, belonging to the complete scheme is the nymphæum, which was to be in a fold of the valley between Monte Mario and Collina del Romitorio, extending from east to west on the north side of the hippodrome.

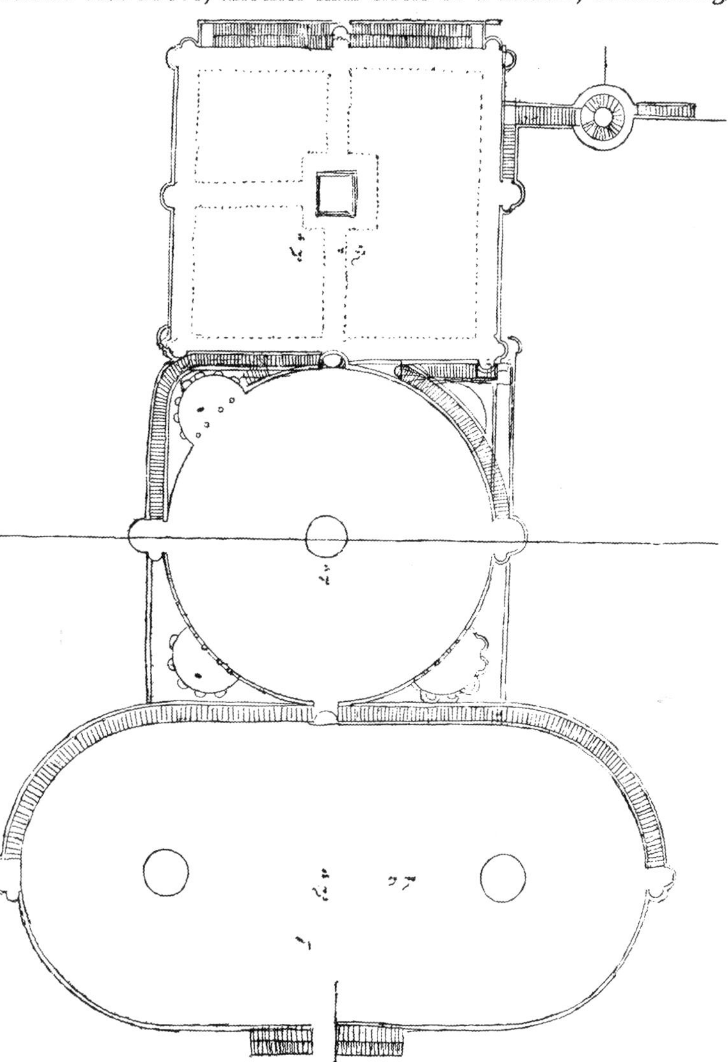

FIG. 167. VILLA MADAMA—RAPHAEL'S DESIGN FOR THE SOUTH GARDEN

There are certain remains which show that this part of the garden was also actually made, just as Vasari describes it in his life of Giovanni di Udine. Raphael had not only consulted with his pupils about the inside decorations of the villa, but he is also the artist of the elephant grotto, wherein he worked in admitted rivalry, according to Vasari, with the decorations, then just discovered, of the so-called Temple of Neptune, thus gaining great honour (and pay as well) from the cardinal. "Then," says Vasari, "he made yet another fountain for a country scene (*selvatico*) in the depths of a woodland cleft; he skilfully contrived that the water should trickle in little streams over porous stone, to look perfectly natural, and on the top of the blocks of stone he set a lion's head with maidenhair fern and various grasses cunningly woven about it. One would hardly believe how charming this looked in the pleasant *selvatico*, which was indescribably lovely in all its details." The cardinal was so enraptured with this piece that he made the artist a Knight of St. Peter.

The impression we get of the place as a whole from these miscellaneous plans is quite overwhelming. Now we can interpret the enthusiastic tales about the villas of these days with their "amenissimi giardini." We now see realised, by the help of old writers, both the fancies of antiquity and the works that have perished, all that Alberti and men of his kind desired to see. A painter, the greatest of them all, was the first who conceived this beauty as a whole, and throughout we feel we have an artist's work. What is not yet here is the seclusion of later Italian gardens, and also the close relationship with the house. Gardens are composed of at least three terraces, and these are connected just as the slope of the mountain requires them to be. For the architect is still far from imagining that he can actually alter the lie of the land; even the terraces with steps between them are not throughout orientated by one main axis. The nymphæum, for instance, is on one side in a little valley by itself. It is the same with the treatment of water, which has to be in separate places for the different garden groups, and however attractive in special parts, does not present the beauty and grandeur of one imposing whole.

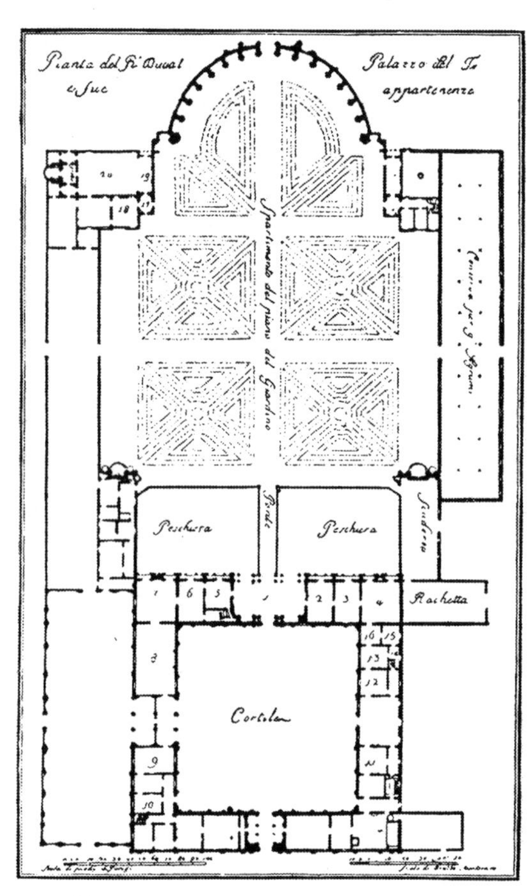

FIG. 168. THE PALAZZO DEL TE, MANTUA—GROUND-PLAN

Unhappily most of the country houses of this golden age shared the fortunes of the Villa Madama, and with gardens it went worse than with buildings. Nothing finished, everything destroyed or altered—that is the story of all the gardens of the first half of the sixteenth century. It needs a very careful scrutiny to make a picture, from our scanty material, of those years when so many great artists created numbers of fine and characteristic works in the way of garden architecture. As a fact, Rome itself after the death of Leo X. was so factious, restless, and actually unsafe, that men's minds began to turn to the peaceful life at a country house. Then the great artists turned their backs on the city for a while, or perhaps for ever, and were received with open arms by other princes in other towns. Raphael's best pupil, Giulio Romano, who after his master's death became the architect of Villa Madama, found plenty of work and also a permanent home in Mantua, the town of the intellectual and artistic family of Gonzaga. Frederick, son of that noble lady, Isabella Gonzaga, received the artist with much show of favour, and one of his first commissions was the erection of a pleasure-house south of the town, the Palazzo del Te (Fig. 168).

Giulio Romano had a very different task here from the one his master left him at

Villa Madama. He had to design a country house on a flat ground, with no view, and it had to depend for all its beauty on itself alone. The amount of bright decoration in painting and sculpture used for chambers and loggias gives us an idea of the *joie de vivre* that characterises this family, as well as of the rather extravagant fancy of the artist, to whom his patron gave *carte blanche*. The splendour of indoor decoration leads us to inferences merely about the gardens, which are unfortunately almost entirely gone. The court and façade, now covered with an excruciating yellow distemper, were once adorned with bright frescoes. We know from a description of the sixteenth century that there were gardens round two sides of the house. In front of the present façade, on the north side looking towards the town, there were fine ornamental gardens. On both sides of the loggia at the entrance there were two little gardens, such as the Italians call "giardini secreti"—small secluded flower-gardens, which must later on have had a special dividing wall, when the ordinary walls between the parts were given up. These walls were ornamented with niches, but we know nothing further of the "large and very beautiful garden" in front. A spacious meadow reached all the way to the town.

The only garden that still shows some signs of its former state is the one in the court east of the house, with either buildings or high walls round it. Stepping out of the middle arch of a large triple-vaulted loggia, whose architecture and ornament, according to Strada, "surpasses anything one could dream of," one stands on a bridge that goes over a cistern between house and garden, and is as wide as the house itself. The canal takes the place of a mediaeval castle moat—a last survival of a style that Italy had finished with long before the more northerly nations. The garden itself had pretty flower-beds on a lawn even at the end of the eighteenth century. At the opposite side it was finished with a half-circle of columns. This end-piece is on the same level as the flat garden, and seems to have had from the start a rather dull, heavy look. But in Romano's mind there was originally a more open, cheerful, not to say frivolous idea for this end, as we know from a little picture that he himself painted on the window wall of one of the rooms overlooking these gardens. This room was one of the first to be painted; even in 1527 we hear of it as nearly finished, and so it would be ready before the garden received its last form. In the picture the semicircular ending is shown with the pillars and arches, exactly like the garden loggia, but round it is a balustrade walk, on which persons are seen enjoying the view of the garden; another balustrade ends below the curve in a broad stairway of four steps leading to a water-cistern with a fountain. On both sides there are standing out of the water some curious little chests like small houses, the significance of which has not been cleared up.

Romano likes to have his architectural work, which he is chiefly concerned with, always treated in an artistic style. The design of the unfinished building, still only at a stage of preparation, of the Villa Madama, holds a proud place in a corner of the Sala di Costantino in the Vatican. In accordance with the artist's original plan, the garden ought to have a pond round it. The feeling of seclusion would be helped by the cheerful pillared halls, which allow one to see out over the green surroundings. Above the corridors of the lower story the garden façade of the house was made beautiful with frescoes by Polidoro Caravaggio, depicting "divers natural forms, trophies, cornices and balustrades."

Even now the gardens are of a mediaeval fashion—really indoor gardens in an enclosure. The attraction towards the terrace form was still wanting, so Romano could do nothing in that direction in spite of his example at Villa Madama; he used all his art and all his Renaissance cunning in the secondary parts of the structure—in ornamenting loggias and alcoves with stucco and mosaic; and for such work as this the town house of Gonzaga supplied the very finest models. Here too there were only indoor gardens, now entirely destroyed; and hanging gardens resting on fine arches, to-day all gone but for the joints of the vaulting and a couple of niches, are the same thing carried farther. Indeed, everywhere that a piece of garden might have opened on the water, some wall or pillared walk shuts off the view. The walls themselves, however, were decorated very gaily with mosaics and stucco. The best ones dated from the happy times of Isabella d'Este; and her rooms, well preserved, express her personal taste. Romano was especially anxious to imitate at the Te Palace the little casino that Isabella had; there were two rooms, and round them a small but good court for flowers. Isabella kept the place sacred to the arts, especially music, and it still has the air of a calm retreat suitable for quiet music. A similar casino, but of larger size, was built by Romano at Te in one of the corners at the back of the garden. Comparing the two, we can see the bolder but somewhat coarser spirit of the later Renaissance. The fine mosaic and the tender outlines of stucco in Isabella's rooms are replaced by stronger lines and colours. Most of all is the tufa conspicuous; it is used to cover the alcoves, though but sparingly, and this is the first appearance of a grotto decoration destined to take a prominent place in the gardens of the future.

Although the Palazzo del Te has often been destroyed, it is not lost to living eyes, whereas Marmirolo, which Romano also built for Frederick I., and which was much admired by Leandro Alberti for its size and splendour, has entirely vanished from the face of the earth. It was not a mere small pleasure-house with dining-halls and bedrooms such as Te, but was an extensive building capable of entertaining the princes and all their retinue. Above all in the gardens the fruit-trees are commended, with great bunches of grapes hanging beneath the lopped branches. "Here anyone who is heated can get cool in the shade, and also in the teasing play of waters which the prince never failed to surprise his guests with—sprays of rain springing from unseen sources." This family of princes owned many another house of entertainment; the plans were supposed to be by Mantegna or some great master employed by the Gonzagas; their gardens were indeed renowned, but except for a mere list of names which can at best only give us a notion of great activity in building, everything has perished.

The traditions of this noble house were conveyed to the court of the Prince of Urbino by Isabella's fair daughter Leonora Gonzaga, when, at the age of sixteen, she married Duke Francesco Maria della Rovere. After many vicissitudes of fortune, which involved husband and wife in exile and much danger, the brave leader thought he was about to enjoy some happiness and honour. In July 1529 the Venetian Republic made peace with Pope and Emperor both, and offered proofs of their esteem and gratitude to their commander. At last the noble lady hoped to keep her beloved husband by her side in days of calm delight, and in her joyfulness she felt she could not mark this blessed prospect better than by building a country place that she would like to have as a museum, such as Castiglione was said to be in the time of her ancestors. And Pietro Bembo caught the true meaning of her idea, when for the motto of this new house (inscribed like a coronet on the frieze) he chose the words; "Fr. Mario Duci Metaurensium a Bellis redeunti Leonora Uxor Animi Ejus Causa Villam exedificavit."

In this way arose the new palace of the Villa Imperiale at Pesaro, that jewel of the High Renaissance, which inspired all contemporaries with enthusiasm, which exceeded the dreams of Bembo, which Titian (friend of Francesco, for whom he painted his best pictures) saw as a guest, which the poet Bernardo Tasso, Torquato's father, made his home for many years (Fig. 169). But now this paradise (as Leandro Alberti called it) is only a shadow of its past beauty, and the "amenissimi giardini," that Vasari admired, are gone, leaving no trace behind. But the pleasure taken by so many noble spirits has borne fruit for a later day. Our imagination is helped by plans, copies, and verbal descriptions, so that we can get an idea of what it was. Leonora, as we know, found to her hand an old castle of the early Renaissance. The tradition of the antique villa was preserved in the inside arrangement of rooms round a central pillared court that enclosed a marble fountain,

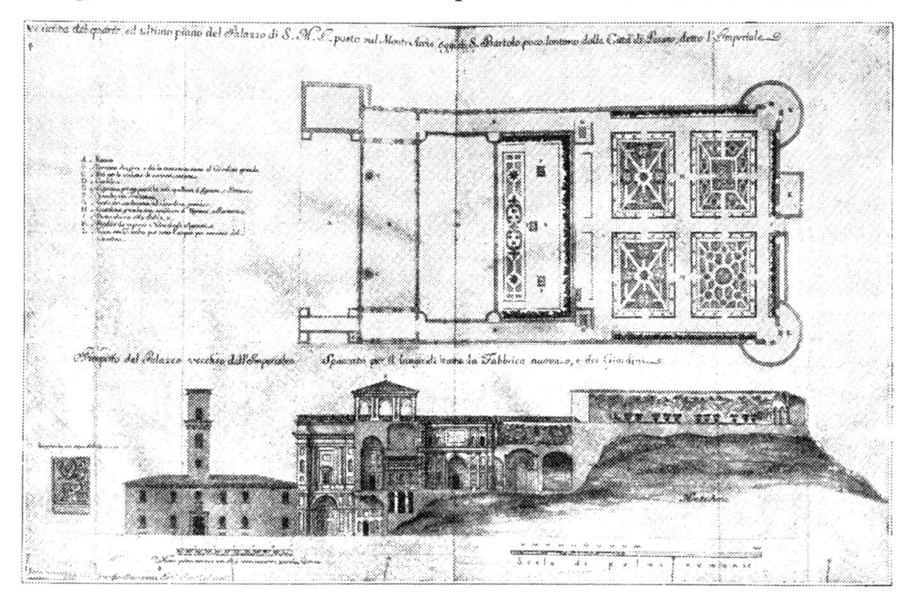

FIG. 169. VILLA IMPERIALE, PESARO—GROUND-PLAN AND CROSS-SECTION

but the outside showed the marks of a mediaeval castle in warlike times. Leonora had the rooms inside painted with the martial deeds of her husband. Most of the frescoes were done by Girolamo Genga, who was also architect for the new house, which was a little to the west of the old one that lay at the foot of the hill, and clung to the slope behind.

It is worth noticing that the architect who designed this house of the golden age was also a painter, and one whose great skill became famous by his scene-painting. Genga's building was narrow from back to front, all the rooms faced outwards, and the best rooms seemed to be almost in one piece with the garden. It even looks as though the place was not meant to be a dwelling-house at all: it gives the impression, as did the Te Palace, of a stage for a theatre, a background for the festivities. We see the villa in this light in the description of a contemporary, Ludovico Agostino, an academician who came from Pesaro. He made an expedition with six of his friends to look at the villas in the vicinity of the town, in order "to cast aside the grievous anxieties that were spoiling his life." This was in 1574, in the last years of the rule of Guidobaldo II., who succeeded after the sudden death of Francesco Maria.

The seven men first visited the Villa Imperiale, which stands in one of the most glorious positions on the Adriatic Coast, upon the lovely Monte San Bartolo. Entering from the east, they first strolled through a wood of lofty oaks, then on to a wide meadow with a wall round it, "which was gay with many sorts of flowers." At the entrance stood a kind of lodge containing two rooms. On the south the ground fell away, and this was useful for a garden of citrons and other fruit planted against the wall. Above this "prato" there was a wood shaped like an amphitheatre, and through the prato one came to the chief entrance of the old palace, which had the mediaeval form of a tall tower. By one way we can pass on directly from the private rooms of the prince into the orange-garden. This garden group, belonging to the old palace, shows mediaeval features, especially in the bordered flower-meadow. To the friends, as they approach, the only things that belong to a later age are the white marble statues, which seem to them to be the sole inhabitants of this villa.

An utterly different spirit apparently reigns here since Francesco Maria died. The garden is higher on the hill and joined to the upper story of the old building by a covered way. It was possible also to walk from the meadow into a court which was approached by a beautiful avenue. The gardens proper rise behind the house (which touches them at the first story level) in three terraces right up the hill. The lowest terrace was devised as a sunk court, planted with laurels, and made brighter with two fountains. To this part one entered from the house by a triple-arched loggia, which at one time had been a favourite room in the house with its ornamental painting and covered-in roof. This was further ornamented with a marble fountain and an ancient statue of a youth, probably the Idolino, which Genga himself put up and restored. Opposite the loggia one passed through the wall used as the palace façade into a grotto lighted from the terrace above. The back of this was a *fontana rustica* of tufa. A great many water-devices made this place lively, for behind there were tiny bathrooms, and stretching along the whole width of the grotto was one large water-cistern; round this seats were put, and the grotto was used as a summer dining-room. Over grotto and basin lay a garden terrace of the same height as the *pian nobile* of the house, accessible from side wings on either side of the court.

The flower-garden (called in the documents "giardino secreto") had a stairway in front, and behind this was a pretty lawn with a great deal of topiary work, of which three ships cut out of myrtle were famous. There were three fountains, and between them stood tubs of laurel and other plants. The dividing walls showed espaliers of lemons, oranges and other fruit. On the third terrace was the chief garden, in a square with a wall round it, also planted with fruit espaliers. According to the old plans it was divided into four fields, the paths shaded with pergolas of vine and clipped laurel, and on the beds myrtle, box, roses, and rosemary. This garden, reaching as high as the house roof, was again accessible from there by ways that led along the side wings. Through a door in the back wall one walked hence by a gently rising path through an oak wood into a spacious meadow enclosed by a wall and planted with olive-trees. Here stood the last and highest object, a little casino, the so-called Vedetta, painted inside with arabesques. Outside there was a walk with a balustrade round this pretty building, with steps leading up to it. From this spot one could enjoy the view over the sea and the landscape. The vedetta was planned in Agostino's time, but unfortunately was so badly built at a later date that it fell down and had to be taken away.

Looking at the picture as a whole, one is struck by the fact that the gardens are far more enclosed and symmetrical than those of Madama and the others. In the centre axial line are the court, the first and second terraces, and the clump of oaks; the meadow is behind, and at the end of all is the vedetta. The view from the roof paths must have shown a very imposing general garden scene. The use of a dividing wall for a fine grotto had never before been carried out so boldly. Against this there is the striking lack of steps, and it almost looks as though Genga has taken pains to avoid them. Neither house nor garden is anywhere accessible or joined together by stairways visible outside, but these are turned off into spacious side wings, with considerable waste of space. In this arrangement no Roman influence can be detected, while the bold terraces lead us unmistakably to the model of the Palazzo Ducale at Pesaro, rising up on its colossal terraces. Water also is handled by Genga with some reserve and timidity. There are only single fountains in the loggia, the court, and the first terrace. Indeed the great reservoir, constructed to work so well as an object of art both at Madama and at Te, here is treated like the stairways, and hidden underground behind the grotto. And the planting of the gardens (the orange, lemons, and myrtle playing so great a part) reminds one of the Early Renaissance, although we also have the ornamental terrace, leading up to the myrtle-garden, and closed in with a clump of trees behind; consequently we get both the harmony and the dissonance of two artistic schemes.

The Villa Imperiale was only one, though doubtless the queen, among a circle of villas that, as Agostino says, "vied with each other in beauty and grace," and were only a few hours' journey apart. He gives a fine picture of festal delights and pleasant, cheerful company. The seven travellers were received everywhere in the kindest way, and met with the most friendly hosts and brilliant society, each host trying to outdo the rest. First of all the duke himself acted as cicerone, showing them the beauties of his villa and then leaving them to do whatever they pleased. At his table the meal was enlivened by clever talk on subjects which they had in common, and afterwards came poetry and music. Towards evening the friends strolled off to another villa, where they would pass the night. The next day was given up to fowling, and when they came back it was to find a company of cheerful people, who bade them welcome, and at once started some interesting conversation. Agostino gives a little sketch of each villa he visited, but not more, unfortunately, than a ground-plan and a section, so that we see nothing of their gardens and are told merely of unimportant details. Each villa has a peculiar charm of its own; and by connecting paths the guests pass through cypresses and shrubberies from one to another. This excursion among villas is simply a little extra bit of Castiglione's great work *Il Cortegiano*, and exhibits a happy side of the Renaissance; the cheerfulness of villa life is like a wreath of flowers round the history of all these men, so clouded by gloomy fate otherwise. Some of the villas on the hills round Pesaro still retain certain traces of their old traditions. One beautiful garden is seen with four terraces belonging to the Villa Caprile (Mosta), which itself has been entirely modernised. Below it is closed in with a cypress wall clipped into the form of battlements.

Charmed by descriptions such as this, the eye seeks for the few relics of the great days of Italian art, but of gardens there are seldom more than fragments left. On a narrow hill-ridge south of Florence stands the fine Villa dei Collazzi with its view over both sides of the Florentine plain, built in 1530 for the Dini family, Raphael's friends. This has led to the mistaken idea that Michael Angelo made the plan and the painter

Santi carried it out. Not more than three-quarters of the building as planned was ever put up, and therefore the real centre part of the facade, a grand court, looking out on an ascending terrace, now appears to be on one side. Through this court one passed into a gigantic hall which covered the whole length from back to front, as it did at Poggio a Cajano and other Tuscan villas of that date. A remarkable seclusion and privacy (suggested by the main entrance beside which was a Florentine fountain) is united with an equally remarkable size in the proportions of the reception hall. Round Florentine steps lead downward, and the upper terrace comes to an end with a wavy wall that has niches at the side. In a very much lower terrace one arrives

FIG. 170. SAN VIGILIO ON THE LAKE OF GARDA

at an ugly stairway at the side, which certainly does not belong to the original plan of laying out.

Far more interesting in what remains of gardens is the Villa San Vigilio on Lake Garda. This was the place where artists of the first half of the sixteenth century could most easily go, where their individuality would be least fettered. And in this work of art we see a special individuality, made as it was by the great architect San Michele about 1540, and built at the order of the Veronese nobleman, Agostino Brenzoni. His task was to build upon a little tongue of land stretched out on the loveliest part of the lake, just where the narrow part widens into coves (Fig. 170). Opposite and similarly placed is the ruined darling villa of Catullus on Sirmio, to the west; the great sweeping line of the other shore gives a feeling of grandeur and even of severity to the beauty of the scene. On the projection there was a little chapel, a Gothic structure dedicated to St. Vigilius, and close to it an inn. Here San Michele solved a difficult problem, and changed the notion of a fortress into a villa, cheerful though less large. The inn and the chapel, which occupied the tip of the little tongue of land, were built round, and joined together with a garden terrace. The inn then had as chief façade a two-storied loggia, and enclosed, opening outward in a semicircle, a little harbour, the chapel being taken into the villa garden. This garden had to accommodate itself to the irregular shape of the promontory, and the house stood within it.

The two-storied loggia was repeated at the villa, showing no ornamentation outside, though it may have been painted at one time—a place with a view so incomparably beautiful that it was to those who took their pleasure there a real "mistress and queen of the peaceful lake." In front of the villa and leading down to the water there was a garden

FIG. 171. SAN VIGILIO, ON THE LAKE OF GARDA—RONDEL IN THE GARDEN

of cypresses, which the clipping-shears converted into a fortress with battlements, walls, towers, and bulwarks. The stone wall round the garden carried out the idea, so universal in Northern Italy, of "swallow-tail" outworks. Unfortunately the main garden adopted in the nineteenth century the very popular style that was known as English. But happily one precious jewel was preserved: there is a hill on the side towards the chapel that was treated as a small castle. The approach is made by paths that circle round it and are themselves hemmed in by castellated walls. At the very top there is a round enclosure (Fig. 171) containing twelve tiny chapels with busts of Roman emperors in their several niches. From the side front of the house a path leads upward, planted with tall old cypresses. It is unlikely that these cypress avenues, which are the chief beauty of San Vigilio to-day, were part of the first plan, for it was not till the seventeenth century that there was any appreciation of the majesty of dark masses of trees, and even in the days of the High Renaissance cypresses were only treated as individual trees (as Alberti advised) or as topiary work. The scanty descriptions which we find speak with great admiration of "broad

FIG. 172. THE GARDEN OF CITRONS, SAN VIGILIO

paths between laurels and myrtles," and "charming gardens of citrons, oranges, and lemons," but nowhere of avenues of cypress.

Besides the main garden there was, in the usual fashion of the day, a series of little separate ones, and one of these at least has kept its old form. Towards the land near the wall and the chapel there is the low-lying citron-garden (Fig. 172). A large pergola built on pillars that support the golden fruits of these "godlike trees," opens out on a simple small lawn with myrtles. The path in front of the loggia ends in an alcove with a Venus, which still stands in a raised place, with two figures of boys, as at a fountain. There is a second statue of Venus with a dolphin between the pillars, and after her the garden, according to an inscription on the outside wall, was named. On the north was a second garden, also enclosed and with two entrance gates. This was once planted with the apple of Paradise, and therefore was known as Adam's Garden.

The most noteworthy of these separate gardens has, however, only preserved its doors and inscriptions. It lay to the side, cut off by the road. The founder of the house himself describes it thus:

> There is a third, the so-called Garden of Apollo, full of lemons and oranges, and in it there is also a strong tall laurel, the most beautiful on the whole shore. At one side of it is set up a head of Petrarch, and from his hollowed eyes there springs a fountain, which waters the foot of the laurel so that the stream trickles through to the roots. On the other side (so that the laurel stands between the two) there is a great figure of Apollo in very fine marble. Petrarch in an inscription thus addresses Apollo:
>
> "O may thy glance inspire the spring
> Of this our love, as now my tears
> On hidden roots such blessings bring
> That Earth all sweet and green appears."

These inscriptions, which speak to us even to-day as we pace the garden, were very dear to the master's heart. Agostino Brenzoni was a scholar, a humanist, and a famous advocate and Doctor of Laws. Neither the inscriptions nor the statues are now in their old place, yet they bear witness, more than they do in any other garden, to the joy that men felt in the revival of the past. The spirits of Virgil and Catullus haunt the place. Tribute is paid to Catullus at the entrance to the terrace between chapel and guest-house; there, where one looks across to his property at Sirmio, in a shrine close to the Christian chapel, his bust is set up, and below it one reads the verses, slightly recalling his own:

> Venus grieving behold, and all the love-gods beside her,
> Grieving for loss of the lyre of Catullus, sweet singer of old.
> All the Muses and Graces and Nymphs are piously weeping,
> Pouring their holy tears at a shrine that is sacred to him.

Few of the statues in this garden are genuine antiques. Like the verses, they are mostly the slight trivial imitations favoured in the Renaissance time. But the humanistic spirit of reverence for the past, and the kindly affection for contemporaries, are perfectly genuine. This feeling is expressed excellently in an inscription wherewith the master greets his guests:

> Whoever thou art, visiting this house,
> Observe these twelve rules:

1. Honour in the sanctuary the best and highest God.
2. Leave your troubles in the town.
3. All girls must be banned.
4. Keep your servants' hands off the garden.
5. Meals must not be luxurious.
6. Drink the cup that will quench your thirst.
7. Turn your mind to joyful things.
8. Lift your heart in honest play.
9. Fill your hands with boughs, flowers, and fruit.
10. Then go back to town and duty.
11. Let not Brenzoni's invitations bring harm to your host.
12. Let the honour of the place be its noblest law.

We get to know Brenzoni by his friendliness and hospitality, when in the year 1552 he received a company of young friends, mostly students from Padua, who came like the party from Pesaro twenty years later, on a holiday tour, and met on the shores of Lake Garda. One of their twelve days (described by Silvano Cattaneo, a member of the party) was spent at San Vigilio. Although the account of house and villa is lamentably short, the true character of Doctor Agostino Brenzoni stands out as that of a man "noble, courteous, worthy, and honourable."

It was in those years when this work of art appeared on the Lake of Garda that the first monument of science arose in the department of gardening; for in 1545 was founded the Botanic Garden at Padua (Fig. 173), the admirable model for all the later ones, both in the north and in the south. The Republic of Venice was

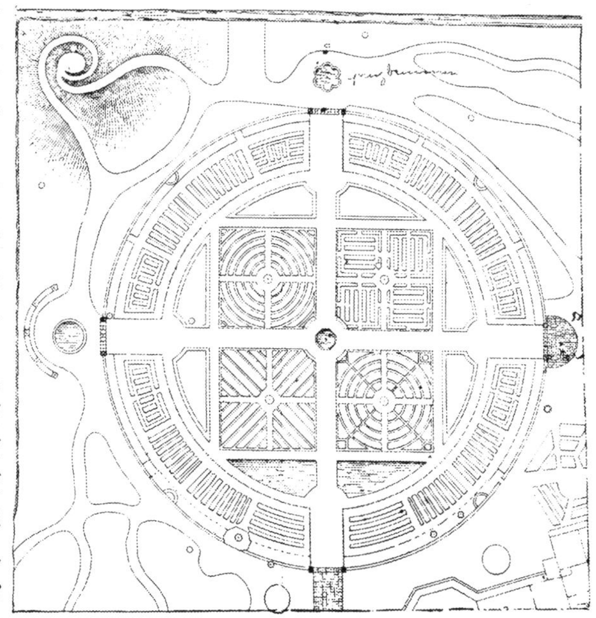

FIG. 173. THE BOTANIC GARDEN AT PADUA—OLD PART

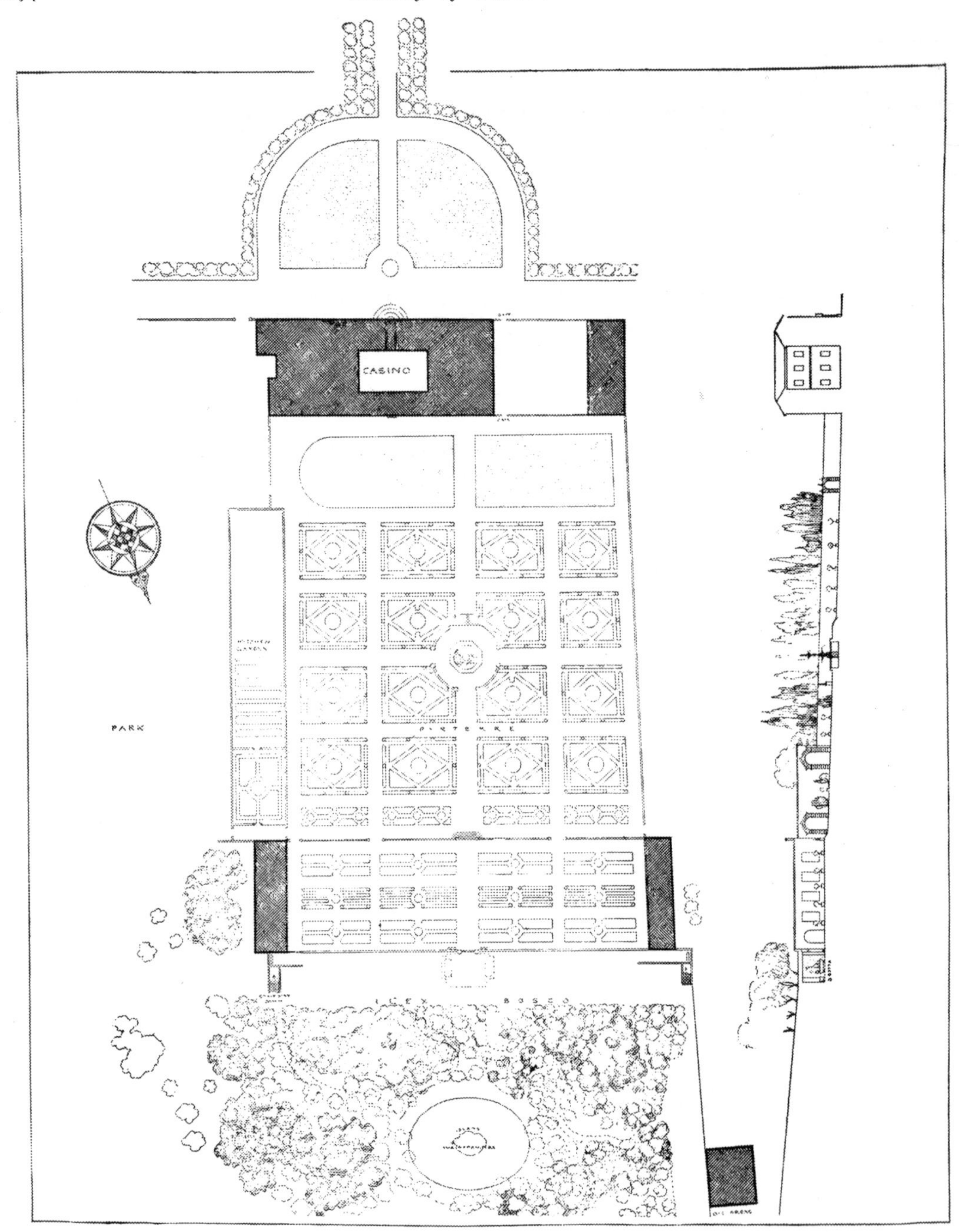

FIG. 174. THE VILLA DI CASTELLO, FLORENCE—GROUND-PLAN

rightly proud of so fine a contribution towards the advancement of learning. Ten years earlier the University of Padua had been the first to establish a Chair of Botany, and hither had come medical students from every part of the world to learn the science that was so essential to the art of healing; and now this garden was laid out on the initiative of Professor Francesco Bonafede. The original scheme of the old part is still to be seen: there are four squares enclosed in a great circle, which has on the outside high barriers and balustrades. Two of the squares are laid out in concentric circular beds —a plan that seems to have been thought convenient for a survey of medicinal herbs, and so has often been imitated. Many of the fountains which are there now are of a later date, but there were certainly some in the middle, and probably some in the side-walks, from the beginning. The statues of Theophrastus and King Solomon behind the fountains must have been erected very early.

Activity broke out once more in the forties at Florence. Scarcely was the rule of the

FIG. 175. THE VILLA DI CASTELLO, FLORENCE—THE LOWER GARDEN

Medici family established, when Cosimo was made duke, before he started on the building of his beloved place, Castello, on the north-west of Florence (Fig. 174). Its garden, in its main lines, is well preserved, though it has lost much of its ornament. A detailed description by Vasari helps us to reconstruct its original form (Fig. 175). The house had already been laid out by Cosimo's father, and the duke entrusted Tribolo, the artist employed at his house, with the surroundings of the place and the gardens. The plans were interrupted by Tribolo's early death, and were only finished in part, but the garden itself, if not finished, was made in all essentials. After so many architects who were painters (at Villa Madama, Palazzo del Te, Villa Imperiale, Villa dei Collazzi) there was now a sculptor to deal with, and his chief interest in art is clearly seen in his work.

The place gained its name, as Vasari says, from a Roman reservoir, which served the

FIG. 176. VILLA DI CASTELLO—THE ANTÆUS FOUNTAIN

water conduit from Valdimara to Florence. Tribolo made use of this when forming two cisterns in front of the garden, at the entrance by the road. They were separated by a bridge, and two fountains on the outer wall discharged their waters into them. An avenue of mulberry-trees, their branches meeting overhead, seems to have passed all the way from the Arno to the house, and over this bridge. On both sides a *prato* made part of the front garden, and in the chief garden behind the house there was also a *prato* on a gently rising hill. It was a favourite notion in Florentine villas that the house should stand in a green carpet of meadow, and this is shown in those that have been preserved, as well as in the drawings made by Zocchi in the eighteenth century. There are a great many ground-plans kept in the Uffizi at Florence, and the *prato* is hardly ever absent.

FIG. 177. FOUNTAIN WITH NYMPH, FORMERLY AT THE VILLA DI CASTELLO, FLORENCE; NOW AT THE VILLA DEI PETRAIA

To the east of the house is a domestic region, and to the left a *giardino secreto,* leading into the orchard, which was shut in by a fir wood, covering the dwellings of the workmen. The chief garden ascended by terraces northward, and was enclosed by a high wall at the side, ending in a cross-path containing niches with statues. From this ornamental green part a few steps lead up to the first terrace, on which stands the wonderful fountain of Antæus (Fig. 176). Farther along is the labyrinth, which is circular, and is surrounded with cypresses and laurels. At one time there was a fine fountain in it, of a nymph squeezing water out of her hair (Fig. 177). A low wall behind this terrace separates the garden from the orangery, which stands a little higher. The next terrace is very much higher, and the wall towards the hill is made use of for a deep grotto in three divisions, kept cool by water continually streaming. There was a remarkable work of art in this place—a King of Rain, in the form of an old man sitting with water flowing from every pore of his skin, beard and hair. Steps at the side led to the upper terrace, where there was a *selvatico* of cypresses, firs and other evergreens, beautifully grouped in the form of a wedge and ending in a loggia with a wide basin in front. Vasari cannot find enough to say about these fine sculptures, which adorned the alcoves in the walls, and in their subjects showed subtle reference to the House of the Medici. But all of them have long since vanished.

To Vasari a gently rising ground—one in which the ascent was scarcely noticed, and

FIG. 178. A SEAT IN AN OAK-TREE AT PRATOLINO

yet which provided a lovely view over landscape and town—seems to have been particularly attractive. Only the highest terrace of all is reached by steep paths, and the real pleasure-garden below is shut in sharply by the wood on the hill. This makes a delightful end to the view as seen from the house. The latter cannot have been finished according to Tribolo's plan, for it is not in the right axial line with the garden: Tribolo would be sure to put a central balcony from which one could look over the whole of the garden. He always had to contend with a scarcity of water, which limited the outlying parts. But when Montaigne saw the garden in 1580, he was delighted with the fountains—especially the full stream that poured out of the mouth of Antæus—the Rain Grotto, and the many playful water-works in the labyrinth. The plantation as he saw it agrees entirely with Vasari's description. He loved the thick arbour-like aisle of evergreen that overshadowed all the walks. He saw the place first in winter, and was almost disappointed to find how little it had changed at the best time of the year.

The chief sign of change here was the disappearance of the variegated flowery pastures of mediaeval and early Renaissance gardens, for, in order to satisfy the love of flowers and plants, the separated gardens, *giardini secreti*, had been invented and laid out, and these are what we find at Castello at the side of the house under the windows of the private rooms. A flower-garden still bore the name of *giardino dei semplici* (garden of simples), and it is obvious that medicinal herbs were grown there, as formerly in the Middle Ages, for there are fountains of Æsculapius set up in the middle.

Vasari and Montaigne both noted another peculiarity at Castello. They describe an old oak bound with ivy, outside the actual precincts, in whose branches Tribolo had fixed a seat reached by steps made of wood. In the ivy there were hidden pipes, which, to the delight of visitors, poured streams of water up into the air. We know of these tree seats from ancient times, and have observed them in the Middle Ages. Montaigne also says that he met with them in his travels in Germany, and often in Switzerland. In Tuscany they seem to have been very common. There is a good example on the side terrace of Villa Petraia, and one of the largest can be seen in a drawing by Della Bella of another Medicean villa, Pratolino (Fig. 178). Thus the water devices at Castello derive from old and distant ancestors. We have seen how, in literature at any rate, the influence of the tree fountain was apparent, having come from the Orient during the Middle Ages. Now, instead of the Eastern magnificence of precious metals and costly wood, we find pipes in the ivy.

At the time Castello was built, the fourth decade of the sixteenth century, terraces had come to be the chief feature of Italian gardens. The plans which Serlio of Bologna expounds in his book on architecture have almost always a highly elaborated scheme of terraces and stairways. The building is generally put half-way up the ascent—in any case it is so placed that it is possible to get a terrace in front to serve as entrance, and the chief garden behind mounting by several terraces.

We are compelled to think of what is done by very different artists in very different places. The middle of the century—a time that is characterised by the beginnings of the

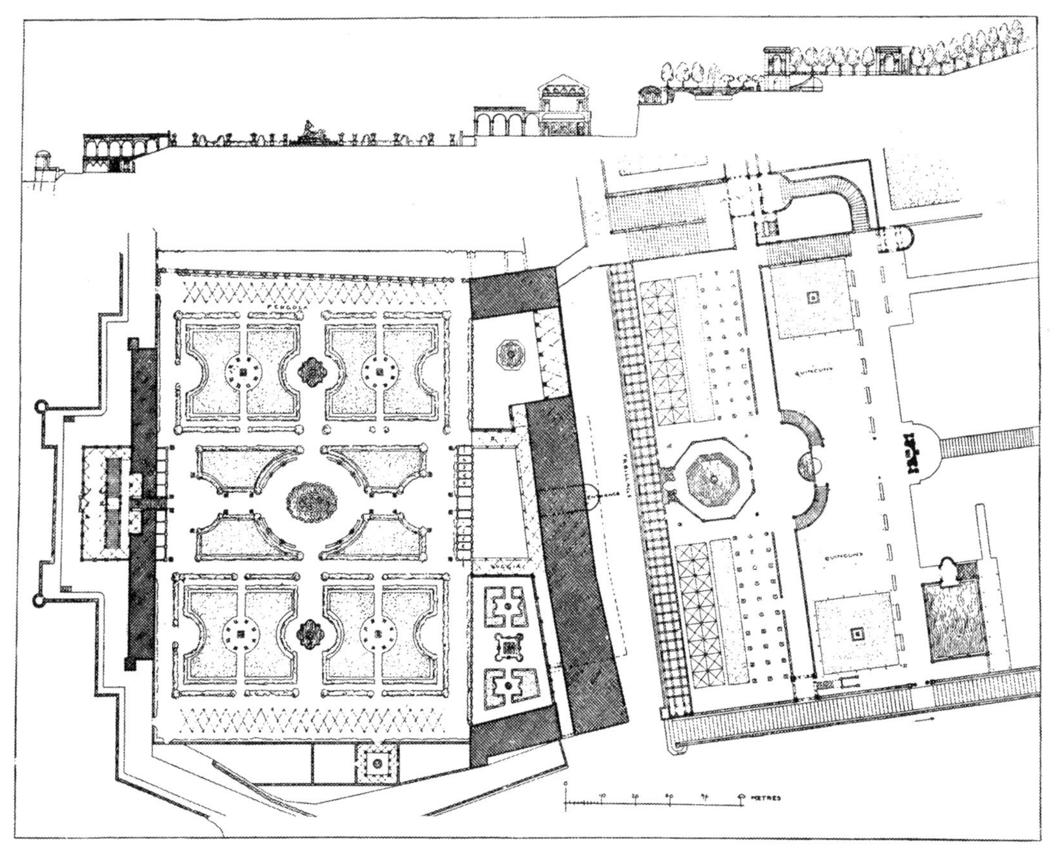

FIG. 179. THE PALAZZO DORIA, GENOA—GROUND-PLAN

Baroque style in architecture—shows at one glance, in any number of works, a complete mastery over all departments of the garden. The leading principle of the new art is expressed in striking words by Ammanati in a letter to Guidi: "Objects that are enclosed by walls must guide and control all things that are planted." That these objects enclosed by walls are first and foremost terraces with their adjuncts of stairways, dividing walls,

FIG. 180. THE PALAZZO DORIA, GENOA—THE CHIEF PARTERRE

alcoves, grottoes, and the connected water-works, will be evident from the examples which have been preserved in great numbers.

Genoa must have been a favourable spot for garden development at this stage, though the hills, rising steeply from the sea, gave no level ground for the gardens, which naturally lay a little outside the circumference of the town; and so for the most part the terraces had to be rather narrow ones. The architect who laid out this ground in the most artistic manner was Galeazzo Alessi. He gave a peculiar style to Genoa about the middle of the century by reason of his architectural achievements.

FIG. 181. VILLA D'ESTE, TIVOLI

After all the destruction suffered in the nineteenth century (in particular by Genoese villas and palaces) there is in what remains one example at least of great beauty which has come down to us from the past, and that is the Palazzo Doria. The great domain that reaches down to the sea from the top of the mountain on which the church of San Rocco stands was presented to Andrea Doria by the Senate. Andrea had the palace rebuilt by Montorsoli, a pupil of Michael Angelo, in the year 1530, and the laying out of the gardens belongs to the following decades. The palace with its airy loggias on both sides, and inner rooms adorned with frescoes by Raphael's pupil Perin di Vaga, lies at the foot of the hill, so that its gardens are in two separate groups (Fig. 179): the fairly level front garden stretching to the sea, and the steep garden on the hill, which is nearly at the same height as the church and separated from the palace by the road.

This cutting into two parts, unfavourable as it was, the architect made use of in the clever way in which he connected the whole garden. He made bridges starting from the first story and leading to the lowest terrace of the garden on the hill. In front this lowest hill-terrace (according to the survey made by Gauthier, a French architect of the earlier half of the nineteenth century) was cut off by a long bower-like avenue, and behind and at the side ended in pavilions. The axial line of the steeply ascending terraces is clearly marked by an octagonal fountain, with stairs, and other fountains in the dividing wall. At the very end is enthroned the Giant, a naked statue of Jupiter in an alcove; and at his feet, according to the inscription, lies the beloved dog of the Doria family, "il gran Roedano," a gift from Charles V. These terraces were chiefly planted with fruit, such as oranges and vines, and Evelyn has special praise for them. A wide gallery, supported on pillars of marble, and balustraded, makes a way by the sea to the flower-garden, and also serves the purpose of making less steep the sharp descent from palace to garden. On both sides there are two *giardini secreti,* now adorned with very pretty fountains.

In one of the gardens Evelyn was still able to see a large aviary, in which, under an iron building, trees two feet in diameter served as an asylum for birds. From the open court steps led into the flower-garden (Fig. 180), which was laid out as a sunk lawn; for it was cut off on the sea side by a high gallery led up to by steps at the side. There were grottoes below. The greatest ornament of this garden was obtained at the end of the century in the fountain of Neptune which is ascribed to Carlone. The god has the features of Andrea, and is represented as standing in a handsome basin, oblong in shape, with eagles, and surrounded by sea deities. The original site can still be seen, as there are some four-cornered beds and two smaller fountains and pergolas. There were more gardens that reached right down to the sea; and it looked as though the Doria family wished to show its power over the city in this huge estate, which stood before Genoa like a barrier.

In this villa Alessi also found his model for the terraces round Genoa and the villas in that neighbourhood, which he and his pupils erected with such knowledge and skill. In most cases the house is half-way up the hill, so that it may command handsome terraces and stairways in front; but unfortunately there is scarcely anything to be seen now of all these villas, and it is only from Gauthier's ground-plans and elevations that we can recover the form of them. At the Villa Pallavicini the front entrance is beautifully worked out: there is a terrace in three divisions with balustraded steps; and two large cisterns on the upper terrace have given the villa the extra name of "della Peschiera."

FIG. 182. VILLA D'ESTE, TIVOLI—THE DRAGON FOUNTAIN

This approach was evidently planted with oranges, and also had arbour-like walks and strips of flower-beds. The garden at the back, where there was a way up, had two fountains in it, and a whole set of plans by Gauthier are very much alike.

There is a variation in the Villa Scassi at Sanpier d'Arena, which has kept its old look very well. The house here is near the road, and the gardens behind it are on very high, broad terraces enlivened by grottoes and cisterns, a casino and a reservoir standing on the top one. The view taken as a whole is most imposing when seen from the palace. In all matters of terrace sites, and especially steps, Alessi had to contend with one great difficulty: the steep, sheer mountain-side about Genoa supplied very few springs, and since his buildings were wanting in this cheerful feature, they appeared rather tame and monotonous, in spite of the wealth of his artistic ideas and devices.

The Villa d'Este at Tivoli shows clearly what a great advantage it is to an artist to have a natural supply of water. This place must always stand out as the finest specimen of Italian gardening in the period of Baroque. A combination of natural and artificial arrangements—nearly always unsuccessful—has here resulted in a work that keeps an ineradicable beauty in spite of every disaster wrought by time and lapses in taste. The spectator is made to feel that knowledge and skill make all things safe. All anxiety, all experimental work, has vanished, for the artist has made use of Nature, and has mastered her. House and garden are the work of a single mind, and woven together into one complex whole.

Cardinal Ippolito d'Este, who went to Tivoli as governor in 1549, was fascinated by the lovely view that opened from the top of the hill towards the Sabine Mountains and the high-lying town of Montorsoli on the north. He wished to build his house on the hill, and the terrace-garden had to be brought into strict line with the chief axis. The ridge of the hill was towards the north-west, but he paid no attention to that, and laid out his garden plan due north, thus necessitating a wall structure of colossal size for nearly half the west side of the garden. Ippolito was not the man to shrink from difficulty or extravagance; and so his garden fell into two divisions, one level, and the other ascending to the house by five steep terraces. The terraces were joined to one another by diagonal paths and side steps. The middle line (Fig. 181), starting from the central gate of the house, is indicated in simpler form by a repetition of the scheme of the great gate in those smaller entrances that occur in the dividing walls between the terraces; they all serve as means of getting into the grottoes. This axial line is emphasised by the large dragon fountain, flanked by immense steps, on the fifth terrace (Fig. 182). There are smaller fountains in the middle of the other terraces. A sixteenth-century drawing by the French artist Dupérac shows the whole garden completed with most of its water-works (Fig. 183).

The architect arranged to have an enormous quantity of rushing water. One part of the Anio, which poured its waters over the hill, was conducted thence into the garden on the east side, and the artist made use of it to mark the cross-paths in an effective way. The flat part and the terraced part are marked off by four large basins (Fig. 184), and the water rises at the east side and ends in an imposing water organ (Figs. 185 and 186) from which a great cascade pours into a cistern below, its roaring sound contrasting with the gentle ripple; to the west there is a projection, but no trace remains of its device, and it may not have been completed. A second crossway line begins with the third terrace and the

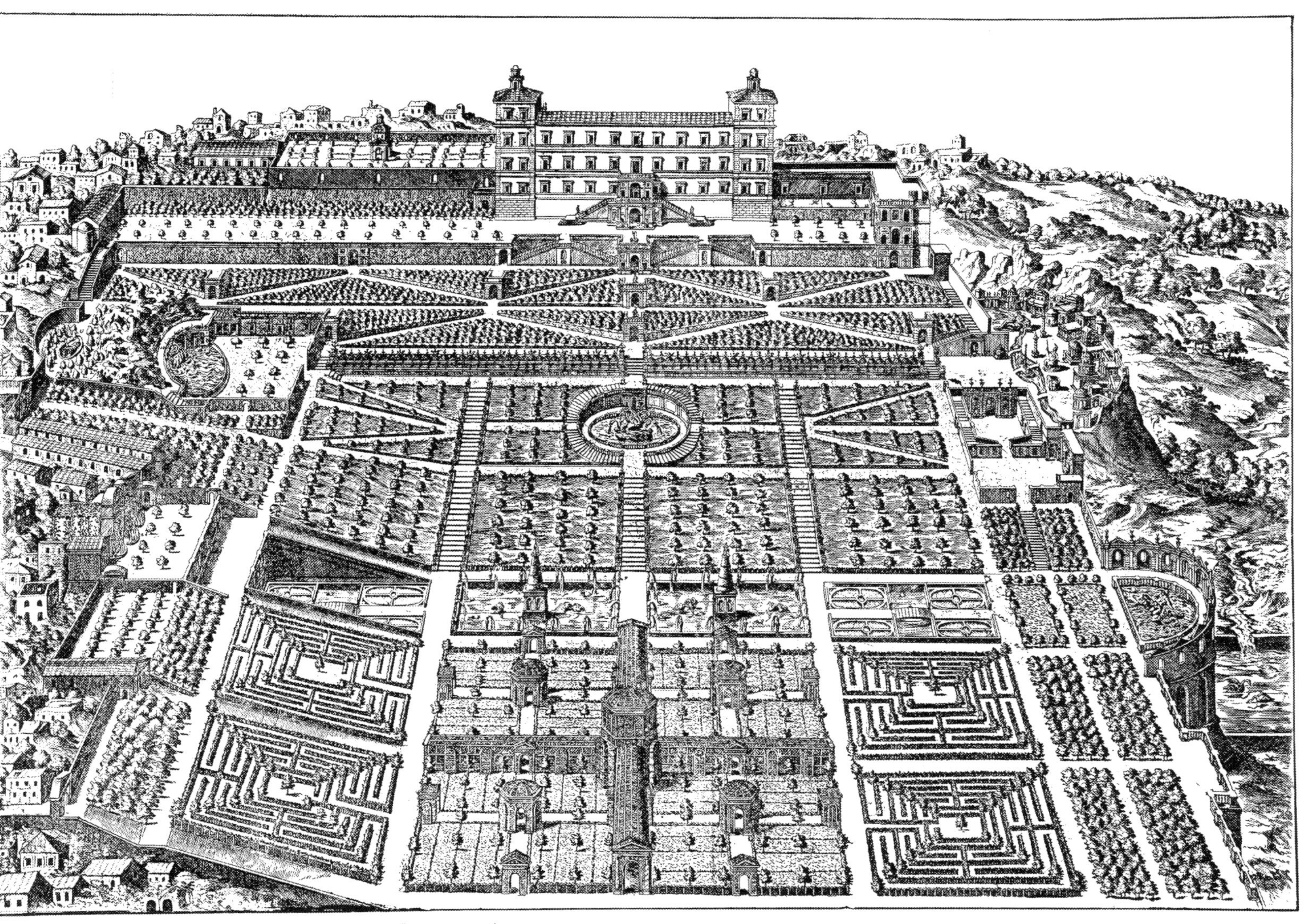

FIG. 183. VILLA D'ESTE, TIVOLI, IN THE SIXTEENTH CENTURY—GROUND-PLAN

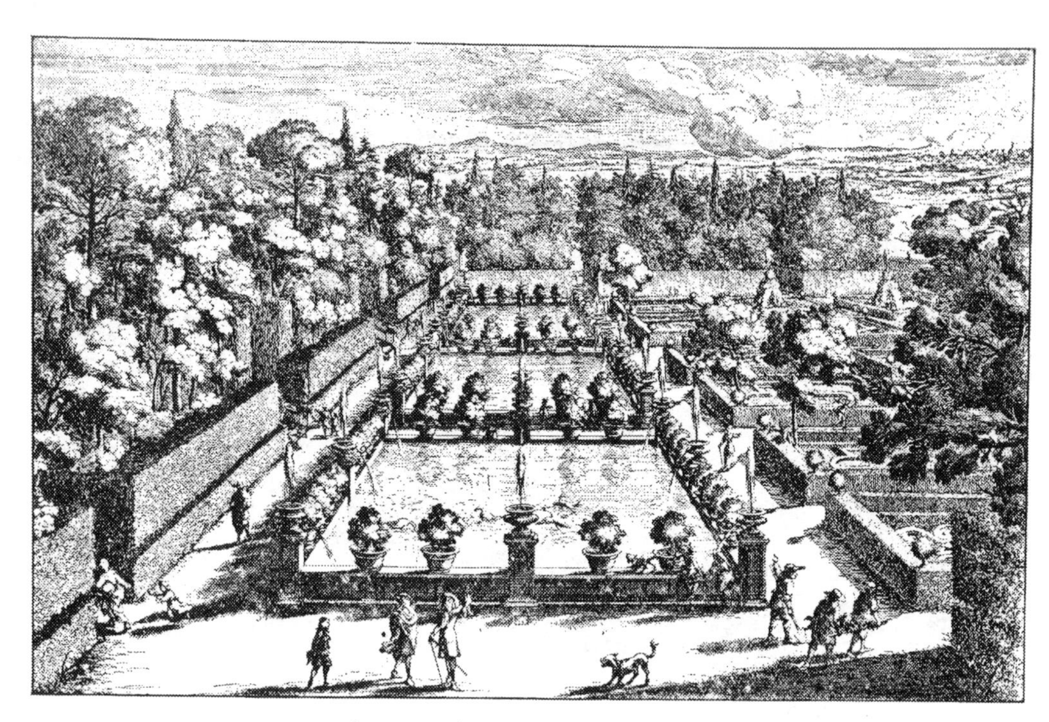

FIG. 184. VILLA D'ESTE, TIVOLI—THE WATER BASINS

FIG. 185. VILLA D'ESTE, TIVOLI—THE WATER ORGAN

eastern water-works. Gigantic blocks of tufa, with a Pegasus on the top, receive the rush of the water. In little irregular niches were water deities, now overgrown (Fig. 187). Under the rock there is a semicircular walk with columns round a huge oval basin, and between them niches with statues. The three remaining sides of this theatre of the waters are walled in, and on the southern side which cuts into the mountain there is another bathing grotto. A narrow passage through leads from this theatre to the main walk (Fig. 188), by the side of which as it passes along the hill there are three canals in steps, with any number of sculptures—eagles, little boats, and grotesques, spurting out their water from one canal into another. There are relief carvings between, depicting the Metamorphoses of Ovid. On the west this path leads to the most peculiar object of the baroque garden, for on a semicircular bay in the wall there stands a puppet-show of a town. In the centre sits enthroned a Minerva, much too big as compared with the buildings. The little town bears the proud name of Roma Triumphans. Under it is another water theatre, including a fountain with artificial singing-birds.

FIG. 186. VILLA D'ESTE, TIVOLI—THE WATER ORGAN AS IT IS TO-DAY

The effect of the level garden, which lies on the other side of the cistern as far as the way out on the north, depended more on the manner of its planting, and consequently has suffered more since it was first laid out. The oldest drawing we have shows the main crossways covered with lattice-work, and in the middle the four paths are united by a temple; also on the four sides there are other smaller temples adorning the beds. These no doubt had flowers planted between the fruit-trees; but Evelyn, who saw the garden in 1644, calls it by its old name of medicinal garden or physic garden (garden of simples). This part, alone and separately bordered, shows the mark of the Middle Ages very distinctly; and the same is true of the four labyrinths lying at the side, except that they are distinctly part of the general plan. Evelyn also mentions four pretty little gardens near the house, but rather vaguely, and perhaps he means at the side, where there may well have been four *giardini secreti*.

The first picture clearly shows how thoroughly we must get rid of the fantastic impression, if we want to know how things were at the beginning, which is made nowadays by the dense overgrown gardens, the monotonous rushing streams, and the tall cypresses. These cypresses especially, which are so important from the architectural point of view, and give dignity to the garden as it is now, most of all in the "Rondello," had as at San Vigilio no place in the original scheme. Instead of avenues we find in the older gardens that the paths are bordered with clipped myrtles and laurels, often with arching branches, or else bowers with lattice-work shading the main walks. After this was added a plantation strictly kept in place, which increased the cheerful festive feeling of it all, and also there was much painting of grottoes, walls, and buildings, that left some traces even in the nineteenth century.

But the chief delight of early visitors was in the lively play of the waters, for which they were apt to overlook everything else. Whether it is Montaigne who sees the garden at the end of the sixteenth century, or Evelyn forty years later, or again in the eighteenth century the Chevalier de Brosses, it is always the water-devices that make the chief part of their description. Montaigne was able to see most of them in action, and he describes the water organ as the most wonderful: below it is a continual splashing rain on the borders of the cisterns. These cisterns seem to be more altered than anything else, and the first drawing shows in each of the middle ones a small building, possibly a breeding-place for water-fowl. The side ones were to be bridged over, but later sketches let this go, and the borders are ornamented with vases and other floral decoration (Fig. 184). Montaigne also takes an interest in the dragon of the centre fountain, and the singing birds in the western theatre. On another side he notices a noise like the thunder of cannon, and on yet another the water makes the sound of fireworks.

FIG. 187. VILLA D'ESTE, TIVOLI—THE WATER THEATRE RUINS

He enjoys, however, most of all—and so does everybody else—the water-tricks which now play an increasing and excessive part in the Italian gardens of the Renaissance. Soon no garden could be found without them, and its charm came to be estimated according to the number of surprises. The most enlightened people, even Leonardo da Vinci, were not ashamed to work their inventions with mechanical figures, thus pandering to the childish fashion of the day. And since it always happened that unwary visitors got wet through, we wonder whether the grand dresses of that time did not suffer, for, according to a handbook, "nobody was spared, however much of a potentate he might be, and many princes and such-like people went there."

If we would understand the garden aright, we must imagine it peopled with a host

FIG. 188. VILLA D'ESTE, TIVOLI—THE MAIN (FOUNTAIN) WALK

FIG. 189. VILLA D'ESTE, TIVOLI—THE LOWER RONDEL OF CYPRESSES

of statues, and not only antiques such as the cardinal had brought from the neighbouring villa of Hadrian, but also copies of Renaissance statues, such as Michael Angelo's Moses, the nude woman by Giacomo della Porta on the monument to Paul III., and many others. Even till the end of the seventeenth century the villa was kept in a good state. It was perhaps in the earlier half that its more serious character was maintained by planting cypresses; and the Rondel of Cypresses (Fig. 189) has only kept the pediments of its wonderful statues and fountains, to be used as flower-stands (Fig. 190).

FIG. 190. VILLA D'ESTE, TIVOLI—THE RONDEL OF CYPRESSES AS IT IS TO-DAY

In the eighteenth century the villa suffered a sudden collapse, for the house, which had never been completed in front, was neglected, the garden ran wild, and the statues were sold. Pope Innocent XIII. bought a great many of the best in 1753 for his collection at the Capitol. At a later date Winckelmann saw the villa in such a state "that one feared that unless speedy help was sent it might soon be like many others which had been robbed for its enrichment." Yet its treasures in statuary seemed inexhaustible; and although Winckelmann procured some marvellous antiques for Cardinal Albani, the value of those that were left was put at 8196 scudi. In the year 1792 a canon of Hamburg could find nothing worthy of praise except "the picturesque beauty in these long-forsaken gardens of cypresses and groups of firs, towering above the thick laurel bushes." In the nineteenth century there was more care bestowed on the garden, but one bit of this secluded spot, the pond on the east, was sacrificed to the spirit of the time—the English taste then dominant, with its twisting, interlacing paths.

In spite of everything an observing eye can detect the original idea by the help of its very clear ground-plan, which makes a complete unity of house and garden for the first time, subordinating every sort of planting to the great skeleton scheme of the whole. The name of Pirro Ligorio has never been expressly mentioned, but one may assume that, as he was architect to the Este family at this date, the design was his, especially as he spent a long time at Tivoli studying Hadrian's Villa. He certainly did get inspiration for his own work from the bold style of the antique, especially in the case of the supporting walls under the Poikile Terrace. The way the little town is placed is also an argument for Ligorio: he was able in a small way to accomplish what in a large way was impossible, the reconstruction of the ruins of ancient Rome. It was a princely castle that Ligorio was to erect in the Villa d'Este; it was to appear enthroned, so to speak, towering above the landscape. Here the cardinal desired to live part of his time with a great company of noble friends, for whom he would hold festival. In his garden a merry, noisy party would wander about and lose their way, seen and admired from below. The many hued picture was

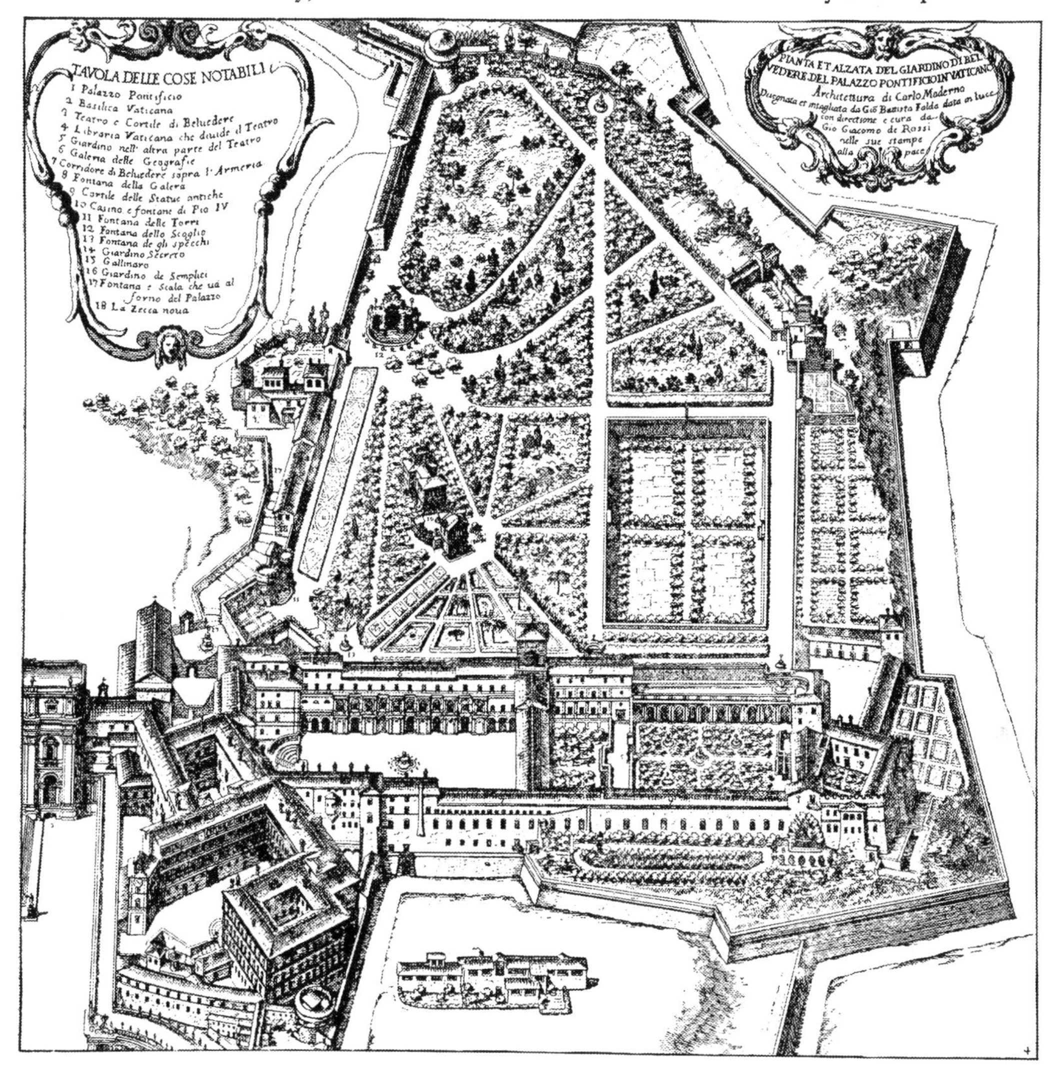

FIG. 191. THE VATICAN GARDENS, ROME—GENERAL PLAN

a wonderful background for feasting and play, and all the sounds of waters chimed in with their gaiety.

The artist made another essay of a very different kind, when after Ippolito went to France he was attached to the Pope's service, and was entrusted by Pius IV. with the building of a little casino behind the Vatican. This commission involved the laying out or perhaps reconstruction of the gardens there. The sunk lawn (*giardino secreto*) may have been made somewhat earlier (Fig. 191, No. 14). This place is very deep in the ground, and has walls that are used as espaliers, with a few grottoes left

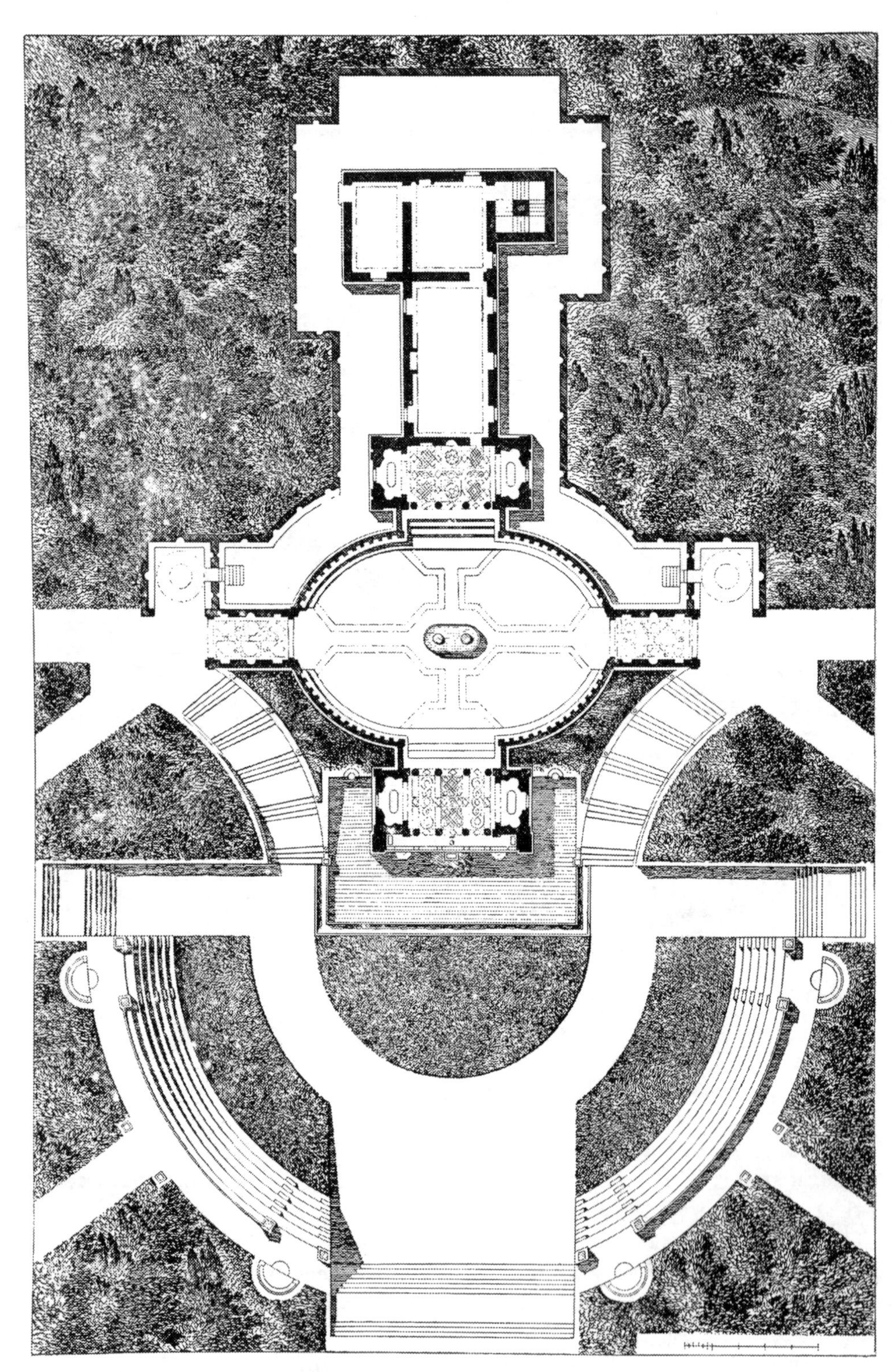

FIG. 192. VILLA PIA, ROME—GROUND-PLAN

between them. The paths, as always in these early parterres, are arched over with foliage; there is a pavilion where they cross, and in the squares left between them there are simple rectangular flower-beds. This is the first example of a sunk parterre, so common later on. They had the same purpose as the *giardini secreti* near the house, that of providing a retreat, slightly sheltered and enclosed, for the master of the house.

FIG. 193. VILLA PIA, ROME—THE APPROACH TO THE COURT FROM THE NYMPHÆUM

The whole plan of the garden is fairly simple, and shows no ambitious architectural notions, and there was no place built where people could sit and rest. Possibly the Pope felt this was a disadvantage, for he bade Ligorio make a pleasure-house farther to the south, called Villa Pia after him, and all the more needed now that the old Belvedere had been turned into a sculpture gallery. It was to be a kind of summer-house for the afternoons, where the Pope might retire either alone or with a select circle of friends (Fig. 191, No. 10, and Fig. 192). There is only one bedroom, which may mean that he was sometimes able to sleep there alone. A real treasure came to light when the architect had fulfilled these demands. The casino itself is preserved: there is a fine front erection with an open loggia, standing like a kind of fountain-shrine at a cistern fed from masks, and decorated with statues and caryatides at the façade (Fig. 193).

FIG. 193*a*. VILLA PIA, ROME—THE INTERIOR OF THE COURT

This place is connected in front with the main building by an oval court. On the side elegant steps lead to round seats, and from them one passes into the court by way of small triumphal gates; in the middle of it is a fountain with a boy's figure, and benches all round to sit on (Fig. 193*a*). There remain to this day most of the carving in relief on the façade, the loggia, and the arched gates; and in spite of the yellow wash that disfigures the building with its ugly coat, there still remain from the beautiful past age delightful grotesques and coloured mosaics on the side of the stairs. "To this severely symmetrical building a tower is added behind on the left as though one final note was needed to extend the impression of grace and charm over the whole place."

The villa was cut out of the woody hill. Its walls were close to the house behind,

FIG. 194. VILLA PIA, VATICAN

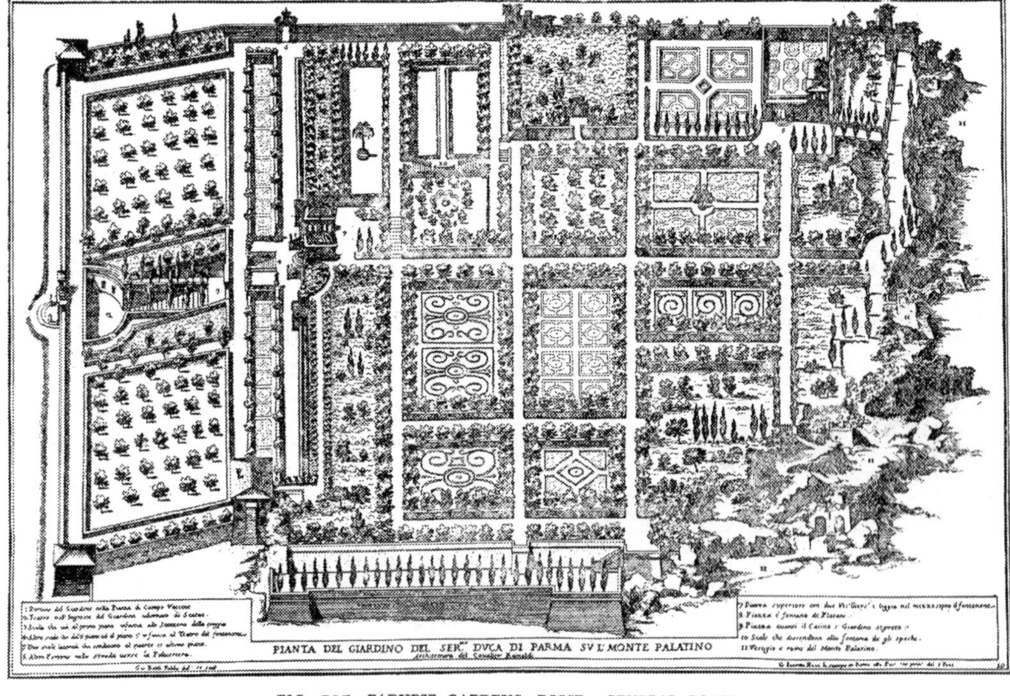

FIG. 195. FARNESE GARDENS, ROME—GENERAL PLAN

hiding the whole of its back (which was never elaborately worked out), and were no doubt concealed in the tops of thick trees, which gave a feeling of peace to the bedrooms where their shade and gentle rustling sound would penetrate. The villa stands by itself, and is really the only thing that constitutes the front view, and according to the architect's plan should open on a sunk lawn with regularly set beds. The semicircular part in front of the fountain house is sunk, with a few steps leading up out of it to the lawn, so that it is suitable for a nymphæum, shady in the afternoons. Garden and loggia both contained choice antiques. Thus the garden was nothing more than an ornamental place in front of the house, which was itself only a garden house. It was left open, and looked very cheerful and bright, but one half was shaded by the dark trees that reached to three sides of the real villa. This was exactly the place for pleasant talk and gossip in the cool of the evening in the hours appointed for leisure, so enjoyed by this happy, lovable Pope. His successor, the stern fanatic Pius V., destroyed this lovely thing. First he purged it, and the Belvedere garden also, of heathenish statues, and then handed it over to his physician who (to suit the Pope's fancy for botany) planted it with palms and foreign plants; and of course the palms were fatal to the original intention of the place, and ruined the view from the villa (Fig. 194).

At the end of the century all the gardens on this hill must have been called Belvedere, unless Duke Frederick of Würtemberg was making a mistake, when in 1599 he saw the Pope's palace. He describes the walk from the Vatican by a long covered path "where there are people walking about and playing ball," and says that one comes to a pleasure-garden "not large but full of fine statues" (the Laocoon among them). "Out of these one walks into another garden, the Belvedere, that has fine pleasure-houses, tanks, and pretty foreign plants, among them a date palm, also a little pleasure-garden for walking in in the summer time, put there for the pleasure of hearing the birds sing." The character of this little wood has been constantly altered according to the taste of the different Popes. Magnificent fountains and grottoes were there in the course of the seventeenth century. The destruction of the woods and complicated walks first began in the nineteenth century, and continued till in the saddest days of the decay of art the very end and crown of bad taste was achieved and set on the mountain top—the erections of the last period of the Papacy, a pleasure-house like a dreary barracks, and beside it the glittering temple which imitates the grotto of Lourdes, a memorial to the victory of Neo-Catholicism. At its feet there lies in sleep, forsaken and lonely, like a wild rose, the artistry of the High Renaissance, wherein the architect knew how to make of garden and house one simple whole in intimate connection and sympathy.

The peculiarity of Italian art that exalts it above any other age and country is its severe and sure development, and that artistic feeling which makes every separate work individual, *sui generis*. Even before Ligorio had carried out his idea of complete unity between house and garden, his great rival Vignola was attempting to create a real garden architecture emancipated from house considerations. He received a commission from Pope Paul III. to make the so-called Farnese Gardens (Fig. 195). Possibly there was a garden in earlier days at the top of the Palatine between the ruins. In the year 1538 Annibale Caro mentions a fountain made of a combination of wooden blocks and *rustica*. This would be one of the fountains, such as we are familiar with from the nymphæum at Villa Madama, that are often used for adorning the more remote parts of gardens. Of

FIG. 196. THE FARNESE GARDENS, ROME—THE APPROACH FROM THE FORUM

this place we hear nothing further; but the present condition of a flat part of the hill and beds enclosed by green hedges may quite well point to the existence of an earlier garden.

The new part, due to the great designer of stairways, is the grand ascent from Campo Vaccino, the Roman Forum, where one walked by five terraces to a level garden region. A great triumphal arch led from the Forum to a semicircular theatre cut into the hill-side, with statues and grottoes (Fig. 196). The central part was interrupted by wide level steps, going up to an immense grotto, which still shows traces of its former stucco decorations, and with a great amount of now useless water bears witness to the number of fountains that were once there. The balustraded terraces get more and more ornamented as they mount higher, with their evergreen hedges and flower-beds.

Especially handsome stairs lead up to the walls on the highest terrace, and reach

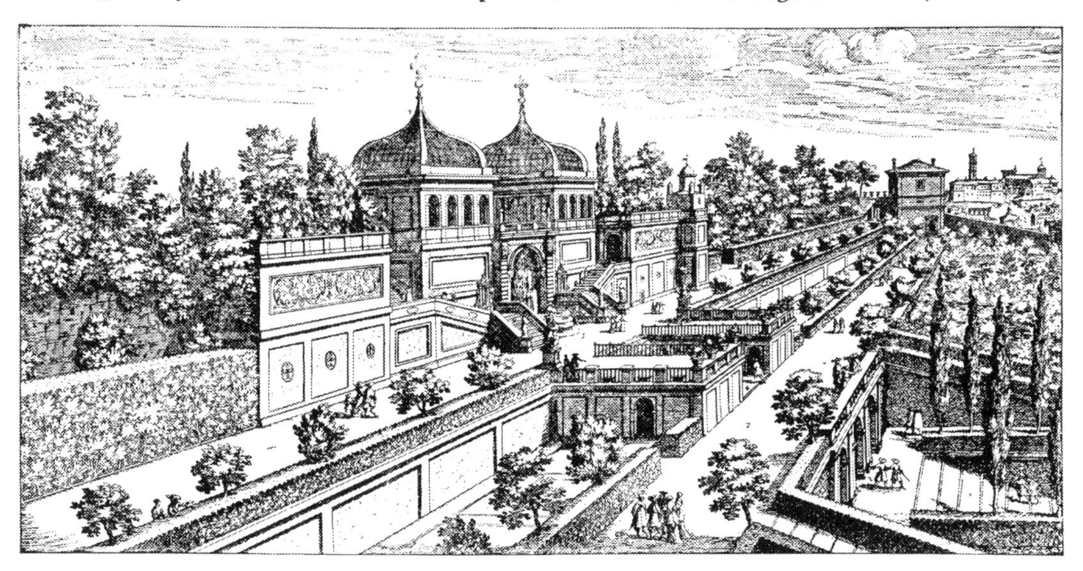

FIG. 197. THE FARNESE GARDENS, ROME—UPPER TERRACES

the top of the garden ground where there are two bird-houses placed obliquely. The reason for their situation is obvious: the architect wanted to present a view from the Forum of his building as a grand finished whole. The terraces come to an end as though forming a palace façade in one piece, with the two aviaries like towers at the sides (Fig. 197). Their oblique position gave an illusion of greater length by a deceiving trick of perspective. This imposing creation of Vignola's fell a victim to archæological zeal, for Napoleon made his first excavations here, and the little that one sees now cannot last long. There are a few remains of decoration, especially pretty *graffiti* on the upper wall, and a few entrances to grottoes, that suggest the beauty of the place as it was. The chief charm of the great level garden lies in the encircling ruins of the Palatine, and the view of the remains of the Forum. From this side the aviaries look nothing but what they really are—little pavilions. The garden is without any sort of dwelling-house, and was simply and solely a place to stroll about in; and only a small casino in the south corner offered any protection from bad weather.

Vignola was also employed by the successor of Paul III., the restless dilettante Julius III., at the Villa Papa Giulio that he built on the Via Flaminia. As at the neighbouring

Villa Madama, a very fine place was designed, and the gardens were to cover the hill almost as far as Ponte Molle. But it is hard to determine now how much was actually carried out, just as one can scarcely find out which part is due to the several artists: Michael Angelo, Vasari, Vignola, and Ammanati. Vasari says that he is responsible for the fine plans of the best court which is still preserved, the so-called nymphæum. This little court is sunk between two larger garden courts, of which the one in front is separated from the house by a semicircular hall, and on its other three sides ends in corridors (Fig. 198). Now it is planted with rose-beds hedged in, which imitate the plan of the originator even if they are not exactly in the old style. Through the pillared hall at the back one passes out to the terrace looking on the nymphæum, which is reached by semicircular steps. On the other side concealed steps lead to the pillared hall which is at the end of the third court. Thus is attained the impression of quiet, cool seclusion that is desired in a nymphæum (Fig. 199).

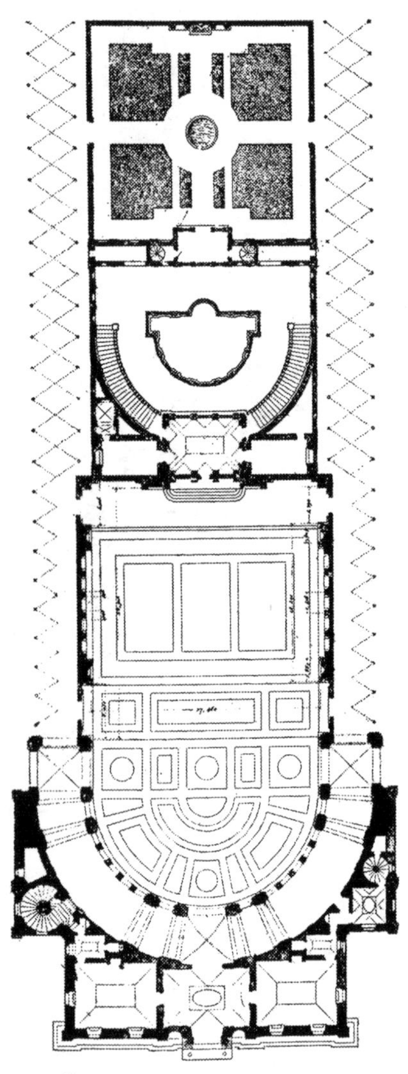

FIG. 198. VILLA PAPA GIULIO, ROME—GROUND-PLAN OF THE MIDDLE GARDENS

As the terrace and the square with the nymphæum and its surrounding canal are both paved, the question of plantation does not arise; but the little fountains always splashing at the grottoes encouraged the greenery, and on the stone floor oranges and laurels were always placed in the summer time. The grotto of caryatides was very attractive, and treated with much delicacy, as also were the borders to the doors that were at the exit of the canal. This cool, secluded place, and a few rooms in grottoes that may be classed with it, are all that have remained of the richly planted gardens of the Pope whose fancies earned so much praise, for the two other courts are quite bare, and one can scarcely see where they were. If more had been preserved of the gardens of Pope Julius, at a villa to which he gave devoted attention for years, we should now possess an example of the suburban villa of the very best period, comparable with what, by the help of our imagination, we are able to see in a villa built at the beginning of the High Renaissance—that is, in the Villa Madama.

A happy accident of fate has preserved one real treasure, though it is far away from Rome and comparatively modest in size. It stands next to the Villa d'Este at Tivoli in expressing the true spirit of this period. This is the Villa Lante at Bagnaja (Fig. 200). We have no information or even tradition concerning the architect, but the fact that Vignola was busy at the time for the Farnese family at Caprarola, which is close by, and still more the well-thought-out scheme that suggests a great architect, cannot fail to lead later inquirers to the opinion that here too we have an example of Vignola's work. At any rate it is not likely that we ought to see, as some do, the combined work of several architects, belonging to a flourishing provincial school, in this villa which is so important in showing the development of garden architecture. Bagnaja and the country round about clearly belonged to the bishopric of Viterbo as early as the twelfth century, and Raniero, one of the bishops of the fourteenth century, had a hunting-lodge

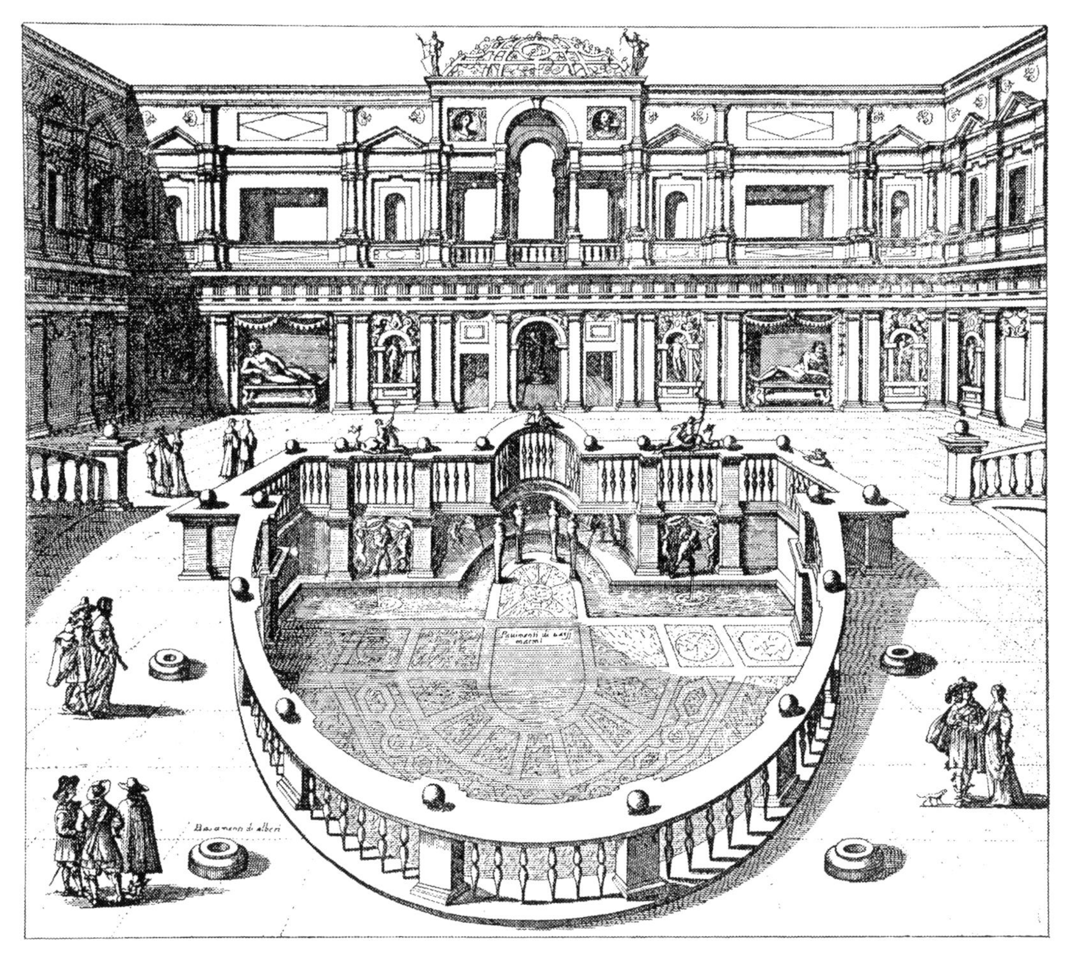

FIG. 199. VILLA PAPA GIULIO—THE NYMPHÆUM

where the Villa Lante now stands, the remains of which can be seen in the stables at the present day. We also hear that a bishop of Viterbo in the fifteenth century, a Cardinal Riario, put up buildings, but these have certainly disappeared entirely. There is no doubt that Julius III. handed them over for a time to one of his nephews, and it is possible that at this time Vignola's plan was made. But the first proper master of the place was Cardinal Gambara, who in the period 1560–80 laid out the ornamental gardens in all their essentials.

The task set before the architect was not unlike that of the Farnese gardens. The villa was to be a place to live in, however, for it was not inside the great city, but close by a little country town, and therefore the plan had to keep the character of a summer residence in the country, even though the building was to be splendid and actually princely in style. So the architect made two houses instead of one, and although they were put up in the most important position they gave to the garden the advantage of keeping an unbroken line the whole way (Fig. 201). It starts from an ornamental parterre and ascends to the highest terraces of all, which are cut out of a thick wood. This was an immeasurable advantage, especially from the point of view of water arrangements, wherein Villa Lante almost stands supreme. The two casinos, both on the first terrace, are backed by the garden, and overlook the parterre, at the bottom of which there is a loggia on pillars. They are in fine, simple form, and give an extraordinarily good and serious effect to the general scheme. If there had been more adornment, such as open loggias, in the top story, they would have looked too much like mere garden pavilions, whereas they keep all the character of dwelling-houses.

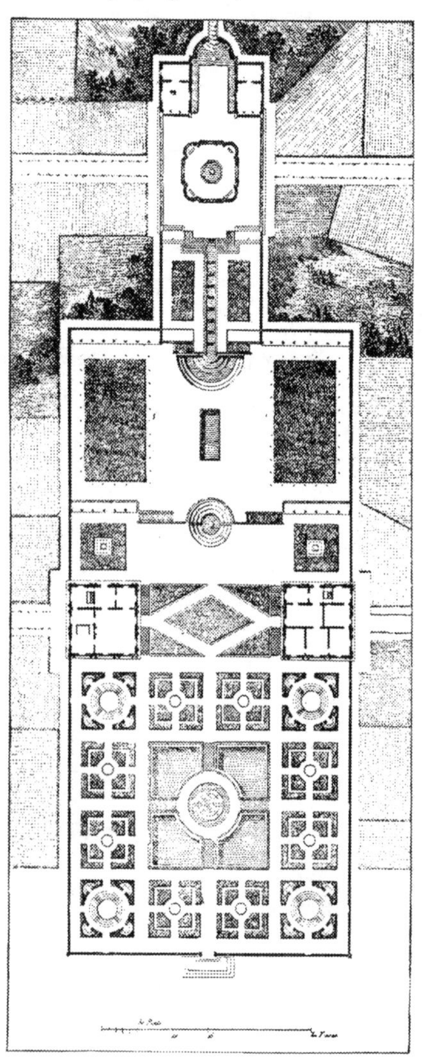

FIG. 200. VILLA LANTE, BAGNAJA—GROUND-PLAN

The first description and earliest mention of the garden comes from Montaigne. He saw the villa in 1580, when it still belonged to Cardinal Gambara. At that time only one casino had been built, although there is no doubt that two had been planned from the beginning, for Montaigne saw the whole water region, which makes the middle line in the garden, already finished except for trifling alterations that were made afterwards. He specially praises the clear, sparkling water in the fountains, and in this respect prefers this villa to the Villa d'Este. He found an authority on fountains at work, "Monsieur Thomas de Sienne," whose inventive genius, according to Montaigne, had already been employed on the Villa d'Este. Soon after Montaigne had seen this villa in its early days, it passed by inheritance to the nephew of Sixtus V., Cardinal Montalto. And in his lifetime appeared the earliest drawing that shows both pavilions and the completed garden (Fig. 202)—in the same surroundings as at present, thanks to the good care taken by the Lante family, who have lived there for hundreds of years.

The parterre on the lowest terrace is very fine indeed. It is cut off from the town by a high wall with a great gate in it, and one can look at it through a wrought-iron grill. The centre of the parterre is occupied by a fountain with water basins round it; these are square, with their corners rounded off, and are balustraded. In the middle of the circular space thus formed stand four figures of nude boys supporting the arms of Montalto, viz.

FIG. 201. VILLA LANTE, BAGNAJA—PRINCIPAL VIEW

three hills with a star on the top, with their hands raised high. The arms show that this group originates with the Montalto family, and Montaigne saw in its place a pyramid of water throwing its foam first high then low. There were four little ships in the four basins, and Montaigne saw with delight four diminutive musketeers shooting water at the pyramid, or blowing through their trumpets with a great noise (Fig. 203). Now the ships bear a freight of flowers, under which are concealed small shapes that are unrecognisable. The upper parterre has formal beds, encircled by a low lattice-work of wood. The enclosing hedges, such as we use nowadays, were not quite unknown at that date, but did not become general till the middle of the seventeenth century.

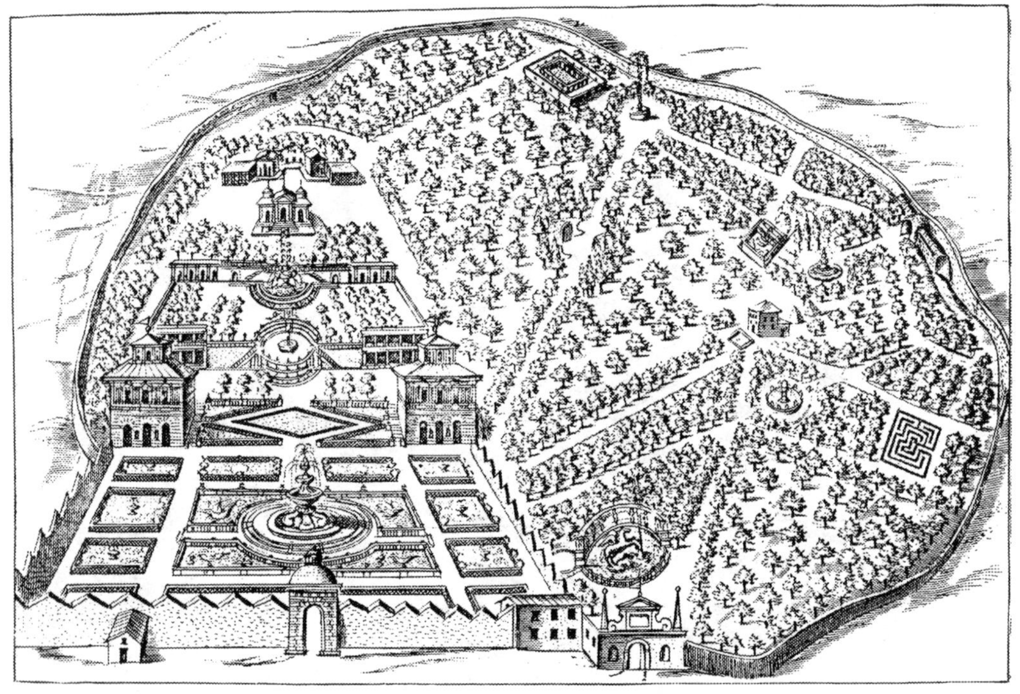

FIG. 202. VILLA LANTE, BAGNAJA, SIXTEENTH CENTURY

From this parterre the garden rises by low terraces. The incline between the two houses probably always had, as it has now, the name and arms of the owner cut out in box. On the terrace behind the houses, which have now become one-storied, we first come to shade, which is entirely absent from the lawn. Two thickets of chestnut-trees with fountains as ornament are on either side. The dividing wall that supports the next step has on it rows of pillars in two stories, the upper ones used as aviaries; but the middle part is kept open for a fountain, which casts its waters in many streams down the steps. The upper terrace is traversed through the middle by a long narrow tank, and at the end there are two mighty river-gods in the wall beside a semicircular watch-tower, at which a crayfish (Gambero) pours out water from his claws, while his body, made like a long canal, carries on the straight line to the terrace above, which now leads us to a beautiful fountain with hedges and seats all round.

The uppermost terrace ends in a lovely nymphæum, a grotto between two open loggias fronting the garden, and this serves as a reservoir for the mountain stream. On either side of the loggias there were again aviaries apparently built as corridors round aplantation of trees with a wire net above: the drawing calls this place "aviarium cum nemore," and it was very likely the same sort of thing that Evelyn saw at Genoa in the Doria Palace. The actual bird-houses are no longer there, nor is the lower part that the drawing calls "cryptoporticus." A few pillars are left standing upright, and in the upper part

FIG. 203. VILLA LANTE, BAGNAJA—PARTERRE WITH BASINS SEEN FROM ABOVE

pigs have succeeded the birds of the air. But the aviary, as we saw more strikingly in the *orti Farnesiani*, has become an important feature of garden architecture. The garden now makes use of all the ideas handed down by tradition, but in such a way that each is individually subordinated to the plan of the whole. In this garden symmetry is for the first time to be found everywhere. Its designer insists that the garden's main axis must coincide with that of the water. Ligorio seems to have taken some pains to avoid this at Villa d'Este, for he made the main axis cut right through the water, so that, in spite of its great abundance, we have to look for it in many different parts and never find one

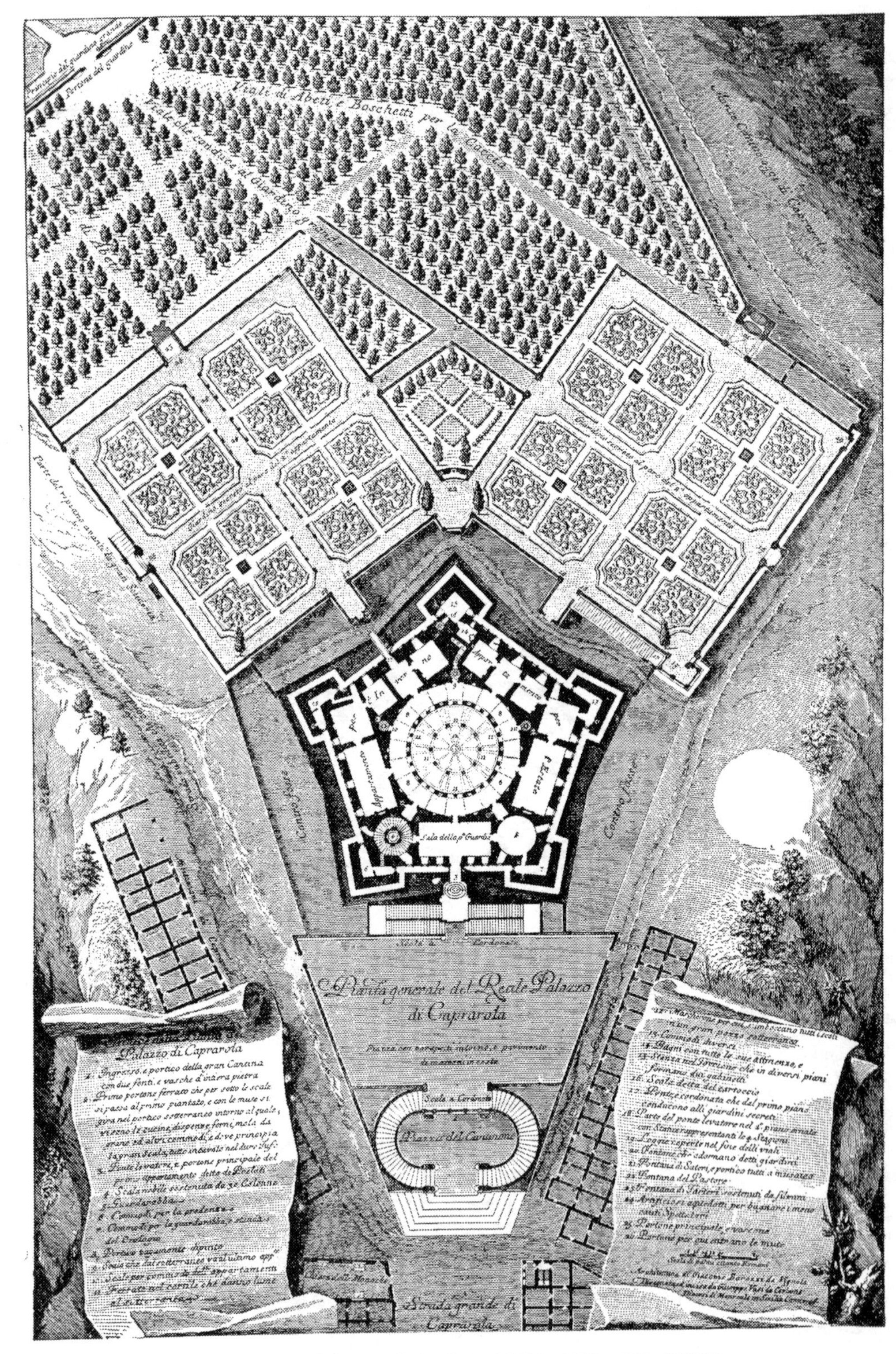

FIG. 204. CAPRAROLA—GROUND-PLAN OF THE CASTLE AND GARDEN

whole scheme. It has already been pointed out how much this particular circumstance (of being able to follow the course of the water the whole way) helps the feeling of symmetry and unity.

It is not, however, only the water; the manner of planting also happily supports the architectural scheme in this case; for the higher we get, away from the low-growing, bright flowers on the lawn (called in the drawing "hortuli con fonticuli"), the nearer we get to shade and the darker kind of plantation, while the uppermost terrace is bounded right and left by pine woods. To the west of the flower-garden lies the park, which is shown in the early drawing, and is still recognisable by the help of the well-preserved fountains, although the woody region is naturally much wilder now. The park is evidently not older than Montalto's time, for Montaigne says nothing about it. Severe symmetry has given way to a laxer plan, and great avenues, some of them converging into a star-like shape, others following the line of the terraces, help to make this impression. The place is enlivened by a number of fountains set up at the end of an avenue, or beside some resting-place. The fine cistern below these fountains makes a striking appearance at the lower entrance, which is cut out of the hill in a picturesque fashion and prettily ornamented.

FIG. 205. CAPRAROLA—CARYATIDES ON THE GARDEN BRIDGE

We have here one of the earliest examples of Italian pleasure-parks, and it is the more remarkable because very few of these are now to be found. More and more the great woodland parks for animals disappeared during the sixteenth century; people began to enclose their property more thriftily, and this we shall find signs of even in the seventeenth century, though in a somewhat different form. One meets but seldom with such a thicket as we have found here, which is a park neither for animals nor for plants. It is too small for either of these; being only about twice as big as the flower-garden, and there are no cages of any kind; so it must be intended merely for a pleasant spot to walk about in. Anyhow the Italian parks, except only a few that are fast disappearing, have either quite run wild or been remodelled after the English fashion. All the more thankfully do we hail the care shown in the preservation of every part of the Villa Lante.

Whatever is thought of Vignola's share in the villas just described, nobody can refuse him the glory of having created the most notable country house of the Renaissance for his old patrons the Farnesi—the Caprarola at Viterbo. Certainly a ground-plan was in existence before, and the elder Cardinal Alessandro Farnese had had a fortress built by Peruzzi; the younger San Gallo no doubt took some part in its design or in its execution. When Vasari says in another place that Caprarola is due to the design and invention of Vignola, he is quite right in so far that Vignola converted a fortress-like ground-plan, a pentagon with protruding corners, into an imposing yet cheerful pleasure-house (Fig. 204). At that time people were very fond of making experiments with ground-plans, and Serlio recommended great variety in the form of country houses, saying once that after much consideration he had come to think that the form of a windmill would not be at all bad. The plan that he appends shows a certain similarity to Caprarola: there is a central building with projecting corners, and a circular court.

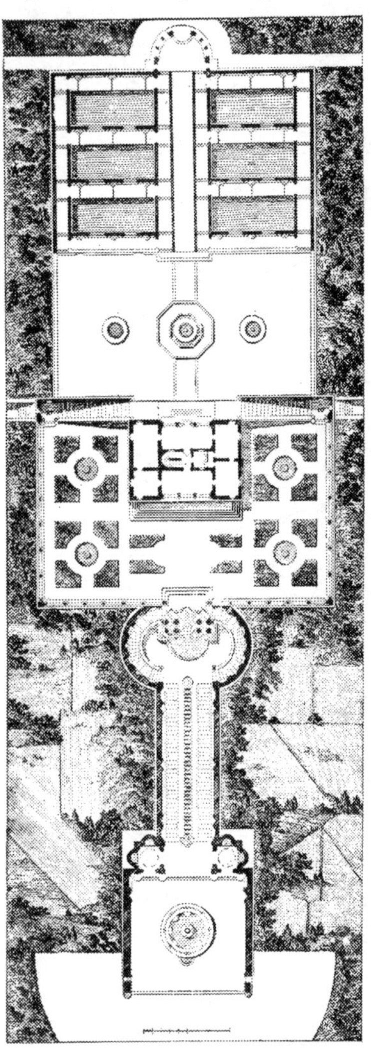

FIG. 206. CAPRAROLA—GROUND-PLAN OF THE CASINO

Serlio had thought of this villa as having gardens all round it, but Vignola put a grand terrace-approach on one of its sides, and added balustrades, grottoes, and fountains in the dividing walls. One has to imagine the wide, sunny and bare spaces of this gigantic approach filled with the many coloured, noisy retinues of the cardinal princes. There are garden parterres, levelled and square, only on the north and west of the façade that comprises the obtuse angle; these are at the same height as the first floor, and connected with it by bridges, which, judging by the drawing of the fortress, evidently work like drawbridges. We have met with garden bridges before, joining the *pian nobile* with a garden at the same height at the Palazzo Doria in Genoa: people liked the plan of going straight out from the main rooms on the same level. The parterres still show their old style and plantation, such as we are familiar with from the sixteenth century, the paths overarched with foliage cutting the garden crosswise, the fountain as centrepiece, and the beds edged with box and planted with fruit-trees: all this shows very little change from the Middle Ages.

Fine grottoes are at the end in the main axis of both bridges. Magnificent planes and cypresses beyond the balustrade with the caryatides have stood there, according to tradition, from the time the first design was made (Fig. 205). The western parterre cuts deeply into the hill, and the triangle left between the two is treated as a high terrace. Unfortunately nothing is left of the western garden except a narrow terrace above the grotto. The park on three sides of this garden has gone absolutely wild, and this is all the more regrettable because the view of the front garden with the house on the hill at the west side making one connected whole, is thereby completely wiped out and lost. Nowadays the charming little casino

FIG. 207. CAPRAROLA—PRINCIPAL VIEW OF THE CASINO

(Fig. 206) appears quite out of touch with the greater house on the hill. Like the Villa Pia, we may perhaps think of it as a retreat where the master could go with a few chosen friends to escape from the noise and pomp of a princely household, and like Pliny's garden house by Laurentinum, "amores mei, re vera amores."

Anyone might apply this title to the beautiful work of the Late Renaissance when he approaches the delightful secluded casino, where the ascent is now made through an avenue of firs. A broad stairway with shallow steps mounts between grottoes framed in *rustica* that are in the side walls (Fig. 207). A narrow bright stream runs down the middle, controlled and confined by a series of stones that form a sort of chain—the same idea as at Villa Lante. The water falls into a fine basin below, which throws up a fountain, as from a spring. Above the steps are two imposing river gods, shooting water from cornucopias into a vase, whence it is received in a bright basin. Above this fountain region there is an airy pillared loggia divided into three parts and reached by ascending a few steps behind the river gods. These are flanked by pretty fountain groups in the flower-garden proper, that lies before the casino, which in front has two stories only. This is enclosed in a low wall, which has on it at regular intervals caryatides of the Hermes kind, both male and female figures carrying vases on their heads (Fig. 208).

FIG. 208. CAPRAROLA—STAIRWAY OF THE RIVER-HORSES NEAR THE CASINO

The front of the garden is ornamented with two river-horses at a fountain in the centre of the box-edged beds. Beside the casino one goes up steps, with dolphins on the balustrade spouting out water (Fig. 209) to the highest of the terraces, behind the casino. From this side the casino looks like a one-storied loggia, and there is only one fountain in the middle still remaining of the water-works. Only very careful examination has shown that there were three water-tanks on the three terraces serving as reservoirs on each side. Their water, as Montaigne says, has come from a distance of eight miles: it is only indicated now by the holes in the masks. Three gateways (Fig. 210) adorned with niches, fountains, and vases lead hence to the wood, which gives a dense background to the picture.

The carrying out of this casino is certainly not the work of Vignola, for Odoardo Farnese did not inherit the upper part of the property till July 1587. But the likeness to Lante is striking. There can be no doubt that special bits, such as the river gods, and the chain canal, have come from there. In the individual parts one perceives the insistent approach of Baroque; the work is coarser and the emphasis stronger. Sculpture, like architecture, is more mixed up with natural surroundings in the vegetable world. In spite of this the place cannot have been made very late, for the treatment of water avoids the massive style, and so would point to the sixteenth century, though the completion of it all may not have been till the seventeenth, for this is certainly suggested by the great quantity of sculpture. The architects—among them Girolamo Rainaldi, of whom we are to hear a great deal more in connection with garden architecture—are strongly influenced in this matter by the spirit that rules in the Villa Lante, even if no plan made by Vignola

FIG. 209. CAPRAROLA—THE APPROACH TO THE CASINO FROM THE SIDE

was ever laid before them. The same influence is apparent in the neighbourhood, especially in the plan of Villa Giustiniani a Bassano near Sutri, which, preserved by Percier and Fontaine, is the only one we have left. The garden, as at Caprarola, is joined to the house by a bridge; the water-line, as at Lante, ends in a loggia; and the parterre in front of the house shows where it has come from by the style of its beds and its copious ornamentation with sculptures.

There is another villa which, being a long way from Tuscany, cannot be shown to depend directly on these, yet has features that are so reminiscent of this group and so unlike all the others in Tuscany that it must be included in the set of villas near or at Viterbo. This is Villa Campi (Fig. 211), which stands on a hill above the picturesque little town of Signa near Florence, and is very little known. The dwelling-house is divided into two small pavilions, just as at Lante, standing above the highest terrace with a sunk parterre as flower-garden, and a gentle slope in terraces between the two (Fig. 212). The pavilions are much smaller than those at Lante, and cannot have entertained a large party. Their small size, with special *giardini secreti* at the side, correspond with the pretty and delicate arrangement of the garden. In the main axis there is a single elegant fountain above the exit-steps of the lowest terrace, from which broad stairs lead to the wood of olives that clothes the hill. The terrace stairs and climbing paths were at one time decorated with statues, and the main line was particularly marked

FIG. 210. CAPRAROLA—CASINO AND GARDEN BOUNDARY AT THE BACK

out with steps. A perfect picture is produced by these simple means, and probably it belongs to the sixteenth century. But the noble avenues of cypress, which in regular lines stand round the cheerful garden on the north side as well as the others, like a dark edging, may perhaps belong to a time that is important later on, when Italy first felt the French influence that brought even great parks into the line of the central axis.

This Villa Campi cannot count as one of that proud series of princely houses which did so much for future development by reason of their actual size and as models of elaborate art. But the small places show even more plainly what knowledge and cunning had been acquired, and what a graceful ease; and scarcely ever has purity of style been better maintained than in this case. Families which have been continually prosperous and have handed down their houses from generation to generation have been apt to alter their gardens according to the varying taste of the times; and it needs a careful scrutiny indeed to detect signs here and there of the original lay-out. The neighbourhood of Lucca is full of villas, and their beauty makes one feel they are gardens of Paradise: they give the visitor wonderful finds of little fifteenth-century waterways, which have been kept for the sake of their architectural charm; but the original plan of it all is irrevocably lost, for there are no drawings, and travellers seem to be so blinded by the supreme beauty of the Medici villas in Tuscany that they pay no attention to this little pearl.

Pratolino (Fig. 213) is one of the villas of a Florentine prince that roused the enthusiasm of travellers even in the late eighteenth century, and was only utterly destroyed in

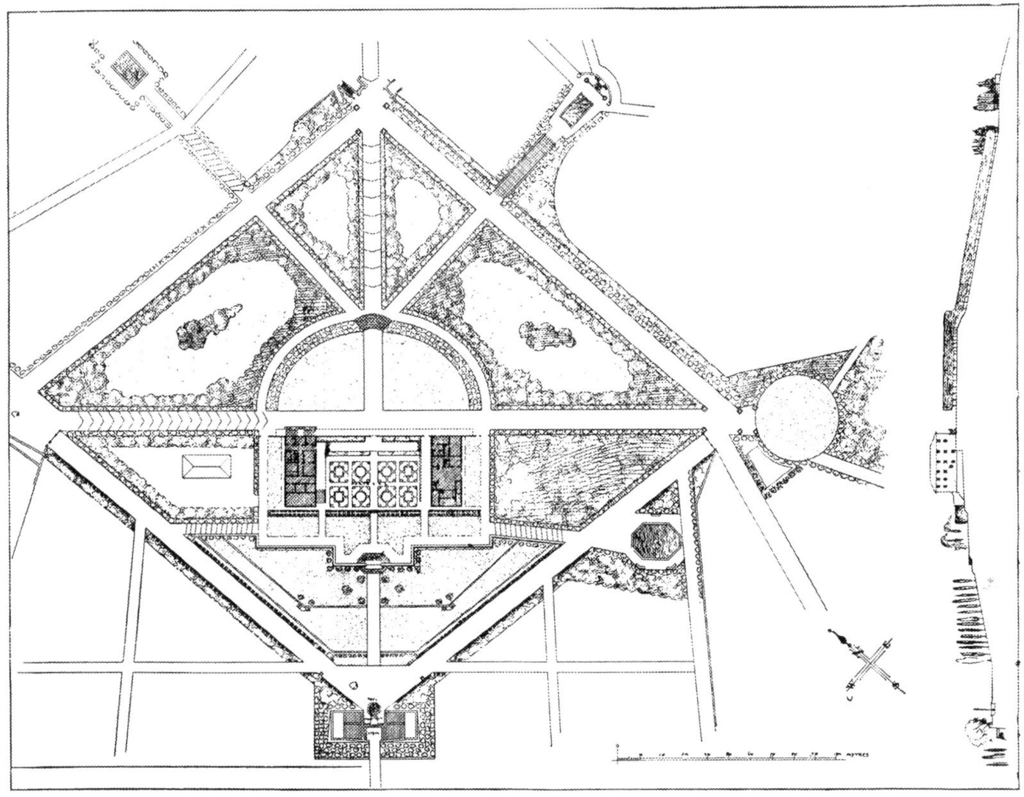

FIG. 211. VILLA CAMPI, FLORENCE—GROUND-PLAN

the nineteenth. It stands on the road from Florence to Prato. There are more descriptions of this than of any other well-known villa, not excepting the town villas in Rome. It was built for Francesco de' Medici, Cosimo's son, by Buontalenti, who (in the words of Montaigne) "emploie tous ses cinq sens de nature pour l'embellir." It is said that this prince, who loved solitude, aimed at making it a paradise for his fair Bianca Cappello, a place suited, according to Evelyn, for all the joys of summer. After many adventures he had won his beloved lady, and very soon after he made her his wife.

When Montaigne saw the villa in 1580, twelve years had passed since it was started, and now it was finished. Half in surprise and half in disapproval he notices that the prince intentionally chose a very inconvenient place, very mountainous and barren, very arid and without springs, merely to have the glory of getting his water from five miles off, and chalk and sand from another five miles. Montaigne says that the villa was made

FIG. 212. VILLA CAMPI, FLORENCE—VIEW OF THE PARTERRE

as a rival to Tivoli, and compares the beauty of the two, giving Pratolino the advantage for the liveliness and brightness of its water—which he had likewise praised at Lante—

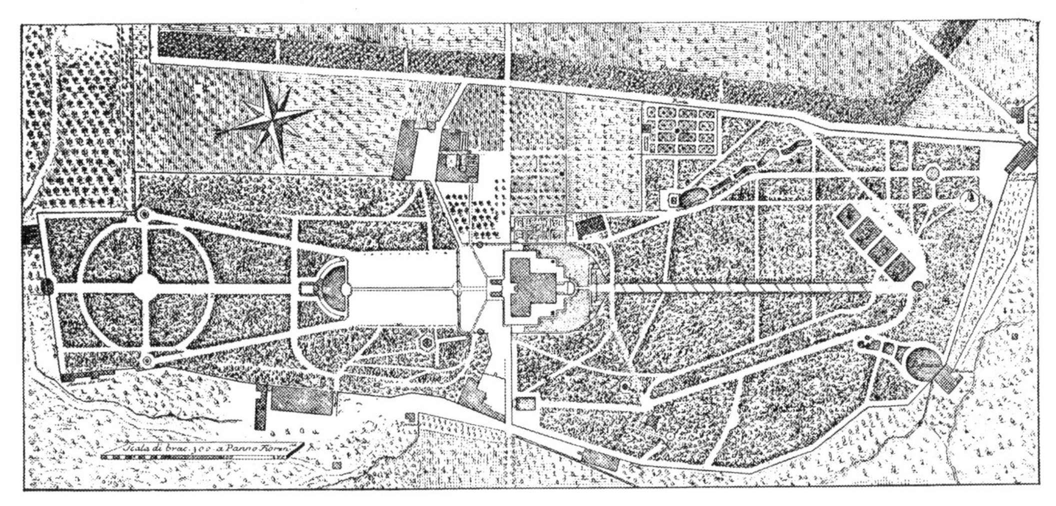

FIG. 213. PRATOLINO, FLORENCE—GROUND-PLAN

for he thought that the yellow Tiber spoiled the full enjoyment of the Villa d'Este. From Montaigne we get another trait of the Early Renaissance feeling, for he cannot quite approve of the grand impressive style, the wild aspect, that distinguishes the waters at Tivoli.

FIG. 214. PRATOLINO, FLORENCE—APENNINO AT THE FOUNTAIN

Although Francesco de' Medici may have desired to possess a place as beautiful as the famous seat of the lord of Tivoli, the task he presented to his architect was a very different one. The ground, though not steep, is intractable, and a terrace series like that of the d'Este villa could not have been made. As a fact Florentine villas had not taken very kindly to this triumph of Roman architecture, but seemed to prefer long, far-stretching avenues; and the terrace and marked crossways are rarely found. But something is missed for the eye as well as the foot, as Montaigne no doubt feels when he says that "there is nothing level in the garden." The house was half-way up the hill, and, like Poggio a Cajano, had round it a broad flat balcony or platform with a balustrade, and curved Tuscan steps leading down to the prato, which lies on b th sides as at Castello. The prato behind the

FIG. 215. VILLA PRATOLINO—STATUE OF APENNINO

house rises in the form of an amphitheatre, and at the end is the colossal statue of the Apennino, a figure seated on a rock, pressing down with his hand the head of a monster, so that water gushes out of its mouth like a cascade, and falls into a semicircular basin (Fig. 214). Near the giant's head (which has a little room inside it) there hovers a winged dragon turning the other way, and shooting out water into a great basin. This is the only group that remains from the early state of the villa.

The statue of Apennino (Fig. 215) is a marvel of execution and design, and deserves to be attributed to Gian da Bologna. Behind there are three shady avenues leading to a round labyrinth, described by Vasari as like the one at Castello. At the extreme end we have a Jupiter with eagle and golden lightning, which marks the crowning point of the whole garden above a fountain that serves as a reservoir. Various groups of fountains enliven the side paths of the garden, which becomes smaller towards the top of the hill. The chief walk of the garden slopes to the south near the house. Streams shoot out from the low balustrade, making overhead a long watercourse, such that a man can ride on horseback under the foliage without getting wet, and "when the sun is shining you can see a rainbow in the air with all its natural colours." At the end was the strange but oft-repeated group of the washerwoman wringing out her clothes so naturally that the soapsuds splash about. The admiration evoked by this effort is shown in the fact that the piece is constantly ascribed to Michael Angelo.

The garden on the sides of this great path appears to have been not too symmetrically arranged: it contained a tournament ground, a great many fountains, fish-tanks, aviaries,

FIG. 216. BOBOLI GARDENS, FLORENCE—ASCENT TO THE AMMANATI COURT

and artificial mounds with hedges and grottoes. The chief attraction at Pratolino was in the grottoes under the wide flat balcony around the house. Many places lay claim to description only for their whims and fancies. One heard music and noise of every sort and kind, and by the help of water-tricks one saw men and beasts moving about—Silenus, Syrinx, and Pan: "If you go near he stands up and pipes, but is seldom seen; many people, they say, have been terrified when he stood up unexpectedly." A Samaritan woman fetching water from a well was very much admired, and also a love-sick shepherd playing on his flute. And the water-tricks are endless. "Never," says the Duke of Würtemberg, "need a man ask for water if he sits on a bench, he will get it soon enough"; and in the eighteenth century Jagemann exclaims, "Woe betide the ladies who sit on the seats." From the detailed descriptions of these jokes we cannot help feeling surprise and wonder at the astonishing cleverness used in the devices.

Pratolino had acquired most of its statues by Montaigne's day, and in this special matter its garden marks a sort of turning-point. Comic statues were soon to be allowed by Alberti; but the number of antiques had so far kept everything else out, and we hear very little of the comic style in the gardens of the High Renaissance. In Pratolino, however, the inferior works are much to the fore, such as the woman wringing out the clothes, or a peasant pouring water out of a pot with a laughing boy looking at him. It is not that antiques or sham-antiques vanish entirely from the garden, but it was thought better to put the most valuable specimens in the newly set up museums, or at any rate in the magnificent public gardens. At the end of the sixteenth century the Idyll

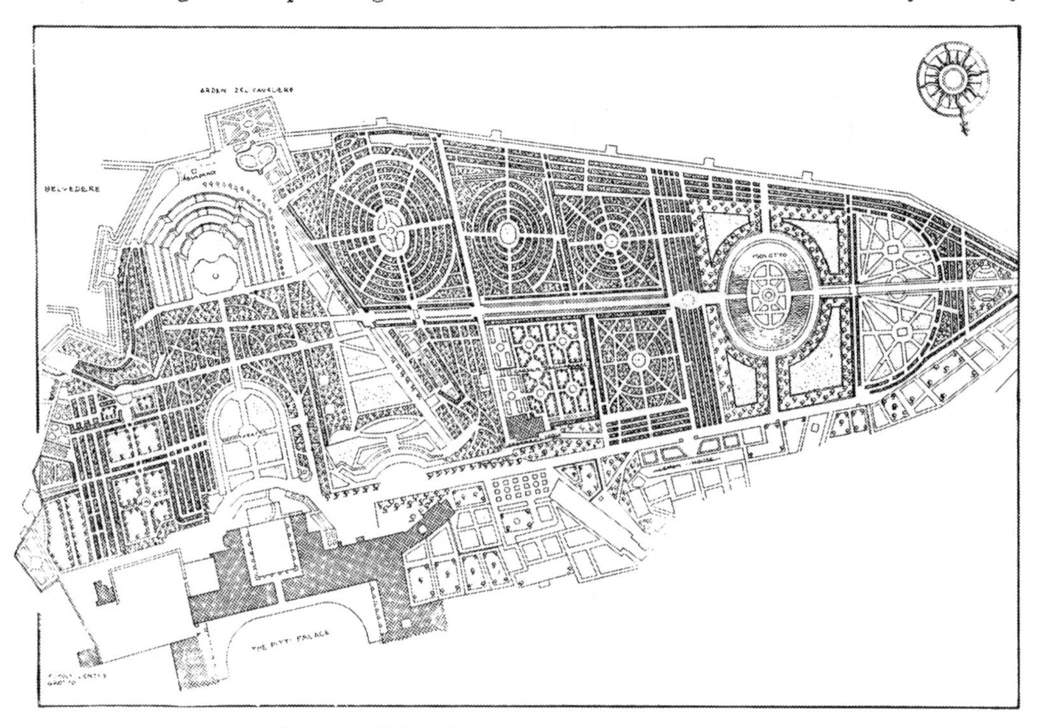

FIG. 217. BOBOLI GARDENS, FLORENCE—GROUND-PLAN

appeared in the garden at much the same time as in literature, and even antique scenes were treated in an idyllic or a comic way; Apollo and the Muses are seen playing on the flute, or Cupid suddenly twists round on his plinth, and spurts water into the face of the spectator.

Yet another novelty is introduced at Pratolino—the extensive use of tufa in the grottoes. Montaigne praises it as something quite new and unknown in his country: in Italy it had already been used for fountain work in country places; but grottoes were still ornamented till the middle of the century almost exclusively with coloured mosaics, which gave quite a different effect to architectural designs. To judge by the scanty remains we have, this sort of ornament was the most conspicuous at the Villa d'Este. At the Palazzo del Te an attempt is made by Giulio Romano, but still in modest fashion, to introduce tufa as an ornament for walls. After this a picturesque covering of tufa seems to have become the ordinary thing, and later on, to have adopted very fantastic shapes.

FIG. 218. BOBOLI GARDENS, FLORENCE—FOUNTAINS AND AMPHITHEATRE

Pratolino is the offspring of a prince's whim: it grew quickly, and was very little altered. Of very much slower development was the garden that the Medici family laid out behind the Palazzo Pitti, their Florentine home. The history of the Pitti Palace is a tangled tale, but the accounts we have of the Boboli Gardens are so meagre that it is very difficult to arrange them even chronologically. In 1549 the proud Spanish Princess Eleonora de Toledo, wife of Cosimo I., bought the palace from a great-grandson of Luca Pitti, its original founder. The middle section stood already with seven windows in the width of it, but the present court and the garden façade did not exist. Eleonora seems to have begun the laying-out of the gardens without delay, for she sent Tribolo (who was busy with the work at Castello) to Elba to fetch away a huge block of granite, twelve braccia (six metres) in diameter, to make the basin of the fountain of the Prato at the Pitti Palace; and according to Vasari, Tribolo also made the whole plan of the hill site, as it still is, arranging everything with excellent judgment in its own particular place, though a few things have been altered since.

After his return from Elba in 1550 Tribolo died, so the work was carried out by others. In consequence of his death not only was the completion of Castello held up, but all the work at the Pitti was delayed, and it was not till 1558 that Ammanati began energetically on the court and the garden façade (Fig. 216). We may suppose that it was at this time that the garden was first laid out (Fig. 217). It is divided into two parts, quite separated. The chief one takes its start in a trough of the valley, and proceeds along the main axis of the Ammanati court as far as the town wall. The court, round the garden façade, is sunk, and its size gives a good idea of the might of the Medici family, for the Strozzi Palace could actually stand in this one court. In the terrace that divides it from the garden there is a grotto with statues and water-works, formerly very lively with mechanical tricks and surprises. Above, on a wide flat balcony, there is a lovely fountain with a little cascade (Fig. 218), and behind, a little higher than the court, is the amphitheatre (Fig. 219). It is a prato like the one at Pratolino, and embraces the whole width of the court. The seats that ascend all round are balustraded above and below, and beside the upper ones there are niches between the pillars at even distances filled with statues which first stood in gaps cut out of the green hedge. In the supporting wall there were corresponding pillars and alcoves, but one cannot know for certain whether there were grottoes behind. This prato was a place for games of the nobler sort, and there are several drawings by Della Bella that show games going on at the wedding festivities in 1628.

Buontalenti has been called the artist of this part, and the likeness to the far simpler prato at Pratolino is undeniable; but it must have been made in intimate association and sympathy with Ammanati, for in 1568 he was busy with plans for its plantation. The ground rises rather steeply from the sunk floor of the amphitheatre in several terraces, but these are not utilised in the Roman fashion. The modern form of the place seems, moreover, to demand a cascade passing down from the large fountain on the second terrace and ending in an oval basin below, whereas what we find is a mere strip of lawn, and the lower basin-shaped hollow is also turfed. There is no information to be had as to whether there was a cascade here or a water stairway. It appears that the fountain up on the hill was carved by Lorenzo Stoldo as a triumphal car of Neptune, to be used in the great festival of 1565, and that it was set up in its place soon after, before the year

FIG. 219. BOBOLI GARDENS, FLORENCE—THE AMPHITHEATRE

1568 (Fig. 220). In the centre of a large basin that now occupies the whole width of the terrace, Neptune stands on a rock, aiming with his trident at a fish, while sea deities hide in the hollows. At the present time there are several grass terraces above the basin in a sort of amphitheatre. But the fountain terrace must have been larger originally, so that it might have round it sweet-scented flowers in beds, and clipped shrubs to make a pleasant setting. On the hill behind, and visible from the palace, was a niche showing the Medici arms. Water flowed from pipes in the dividing wall, between projections, into a canal where twelve stone dogs stood on guard. In the year 1638 the rising ground was topped at the town wall by a statue of Plenty standing in a green alcove.

FIG. 220. BOBOLI GARDENS, FLORENCE—THE NEPTUNE FOUNTAIN

According to Vasari, at the time of Francesco's marriage festivities in 1566, "the very large gardens were adorned with innumerable statues both old and new, as well as many streams." But from this observation we do not learn when it was that the enlargement of the garden, below the west side of the hill towards the Porta Romana, took place. The fortified town wall limited the extension in the main axis, and so it had to be shaped to a right angle alongside the western wall, and the ground showed a wedge shape towards the Porta Romana. In this part of the garden terraces have been left out entirely, and the main axis is marked, in Florentine fashion, by an imposing cypress avenue (Fig. 221), which at first was full of statues. The cypresses were of course not there at the beginning, for we know from a drawing that they were not planted before 1640. In their place were the still existing roofed avenues (Fig. 222). A water-way ran parallel along the wall.

It is clear from examination of the place that on both sides water was poured out from the heads of winged dragons into canals that ascended by steps and flowed to

FIG. 221. BOBOLI GARDENS, FLORENCE—THE CYPRESS AVENUE

FIG. 222. BOBOLI GARDENS, FLORENCE—THE COVERED WALK TO THE PORTA ROMANA

balustrades below. Cambiagi describes a fine labyrinth in one of the thickets to the left of the great avenue, cut off by pretty little walls of tufa that held drinking-troughs for birds. There were different kinds of animals to adorn the inside, and the chief object of interest was the so-called Isolotto, set up at the foot of the incline, where the garden is really level, in the line of the chief axis. This is a large oval basin with a round island in it, whose chief ornament is Giovanni da Bologna's wonderful Oceanus fountain, described by Burckhardt as "more simple and majestic than any other fountain in Italy or the whole western world."

FIG. 223. BOBOLI GARDENS, FLORENCE—VIEW FROM THE ISOLOTTO

We cannot say what was the original appearance of this remarkable basin. Vasari says the design was his own, and a German handbook for travellers describes it in the seventeenth century as follows: "Cut like an oval with an iron railing round it. In the middle there is a small island, and a cupola covered with greenery, all sorts of trees, and a medley of flowers: it is reached by a little bridge." In the year 1627 the Oceanus fountain had been long put up, for Bologna died in 1608, and in 1599 the Duke of Würtemberg had seen it; but the handbooks are apt to copy one another, and it is very probable that we have in this one a picture of Isolotto previous to the erection of the group at the fountain. There is nothing here but the centre-piece, so often repeated, of the pavilion overgrown with green. But long after the Oceanus fountain was set up great changes were effected, as we are told by Bologna's contemporary and biographer Raffaello Borghini in the year 1618. Unfortunately he does not say what these *varie mutazione* were, but probably they were done at the time when a new phase of building began at the palace, when its middle portion was (in the year 1620) enlarged to its present width.

Cambiagi describes the surroundings of Isolotto as much finer than they are now: at the foot of a great avenue (Fig. 223) was a decoy for birds such that when the net was spread the path to Isolotto was completely blocked. From this spot started little walls decorated with tufa, and narrow paths in mosaic, where one was threatened with spurts of fine rain. The entrance and exit of every wall were marked by a pair of dogs or lions carved in stone, and in the middle of the path a very tall fountain came leaping out of a star made of stone. Paths paved with mosaic led to the crossways (Fig. 224).

The large grotto in front of the present entrance belongs to the sixteenth century:

the tufa shows a strong tendency to the Baroque in its idyllic scenes with human and animal forms, and in its pastoral scenes, where a short time before, in this unlikely spot, there stood the unfinished Slave figures of Michael Angelo. The grotto consists of several parts, and gets its light from above. It must have been built between 1579 and 1587, and among the artists who worked at it there is mention of Giovanni da Bologna; to this day the fountain basin in the middle of its first hollow room is adorned with the graceful form of his Venus. Towards the end of the seventeenth century the place had fallen into decay, and it was only re-established in the nineteenth, so far as we can tell (Fig. 225).

FIG. 224. BOBOLI GARDENS, FLORENCE—THE CROSSWAYS

That portion of the garden is quite destroyed which formed the connection on the north-west between the two parts of the house. It is only by careful search that one finds signs in the garden, now run wild and thickly overgrown, that in the seventeenth century there existed a somewhat important terrace site with balustrades, statues, and good dividing walls, which must have contributed another change of feature to the garden. On the top there would have been either a flower-garden or an oblong-shaped prato.

The garden was changed and added to under nearly all its owners, especially in the seventeenth and eighteenth centuries, because of the building of the two casinos at the top of the eastern division of the main garden. Both of these were pleasant little places, but had really nothing to do with the main plan, since they represent an entirely different feeling about art. The garden passed through its most dangerous crisis at the beginning of the nineteenth century, when under French influence it began to change its ideal into that of a picturesque park—indeed it had already partly made this change. Fortunately Duke Ferdinand had the good taste on his return to restore the old garden as far as possible, in spite of his fancy for the picturesque style. A great deal may have come to grief by then, and the statues, so wonderfully extolled for their number and beauty, have almost entirely vanished.

An important matter now is the plantation. It has been already said that the great avenue of cypresses must have been planted not earlier than 1640; anyhow the poor drawings we have depict the divisions by this path as indefinable low hedges, possibly belonging

FIG. 225. BOBOLI GARDENS, FLORENCE—VIEW INSIDE THE GROTTO

to a labyrinth. When Evelyn saw the garden a little later, he spoke well of the topiary work and the pillars by the hedges and also near the fountains, the fish-ponds and the aviaries of the little wood; also every later visitor likes to say that this garden is green the whole winter through, because of its laurels and cypresses. It is probable, therefore, that the thickets at the side are densely planted so as to add a border of dark colouring to the parts that are architecturally laid out. All flowers are expelled from the chief garden, or else confined to the pretty beds round the Neptune fountain, or to the land on the Isolotto. But all the same the Dukes of Florence were wonderful flower-lovers, and it is said of Cosimo I. that he was not ashamed to plant fruit-trees and flowers with his own hand. A narrow strip that runs on the north beside the thicket was treated as a special flower- and fruit-garden, and there to-day we find forcing-houses and kitchen-gardens of every sort and kind. The aviaries, animals' cages and fish-ponds that travellers speak of will sometimes have found cover here, sometimes in the thickets, as we may judge from the much better accounts we get of Pratolino.

At the time when the Medici family was at the height of its glory Florence was undoubtedly the leader in all matters of villa and garden, but when the greatness of their house was on the wane it fell rather into the background, while Rome took the foremost place. From the middle of the sixteenth century the predominance of the Church made conditions in Rome extremely favourable for the development of gardening. Paul III. was the last of the Popes who contrived to found an independent principality for his nephew, and after the fall with Paul IV. of the House of Caraffa, no Pope would have ventured to attempt it. All the more did such "nephews" strive to set foot among the noble families of Rome, and how could they do this more effectively than by showing their power and might in grand country seats and fine buildings? Consequently there arose, besides the palaces in the town and the proper country houses outside—in the same way as the "hortus" of ancient Rome—the town villa, a really good casino or small house with a garden site that constantly became larger and more widely extended. Certainly there has been no other town in the whole world that has so preserved the stamp of beauty by its gardens, all the time from the sixteenth century to the nineteenth. It is only through the development of Rome, in the last decades, into a modern capital that one after another of these gardens has been sacrificed, like flowers plucked out of a wreath, so that nowadays we can only, with very few exceptions, see their vanished glories in our dreams.

In addition to the political cause, there was one other at this time—but not before this time—favourable to the formation of large gardens in the precincts of the city. For a surprisingly long period Roman building had been confined to the valley of the Tiber: on the hills they were short of the best thing of all, water. The huge supply from the old watercourses, from which the present city enjoys only a seventh part, was now drying up. The single conduit made by Sixtus IV., called Acqua Vergine, had for a long time been unable to satisfy the demand. The Popes who followed him had always improved it, reconstructed it, and eventually made it as good as it could be made. But it was Pope Sixtus V. who had the great merit of having made a second conduit, Acqua Felice, and afterwards Pope Paul V. a third, Acqua Paolo; then at last there was an adequate supply, even for the hilly districts. For the first time Romans were able to make use of the hills, with their many advantages in position and especially in charming views, for laying out larger gardens and thus satisfying their great love of fountains.

In this matter the Pope stood first. The Vatican lay so low that it had never been a healthy place to live in, and never safe from fever, so Gregory XIII. gave orders that "a home should be prepared for himself and his successors in which they could enjoy a perfectly pure air." He then summoned Flaminio Ponzio, who in 1574 began to build the palace on Monte Cavallo, the Quirinal. The Popes had before this set up a sort of summer recreation place there, apparently in the Benedictine cloister; at any rate it is said that Paul III. stayed there a few days before his death in 1549. Certain cardinals had gardens there, and Cardinal Carpi brought a fine collection of statues to his, which was also furnished with pergolas and grottoes, and was a model for all to copy. Also Cardinal Ippolito d'Este of the great villa at Tivoli had had gardens here before 1550 that were designed with artistic lattice-work by Girolamo da Carpi. Some of these passed

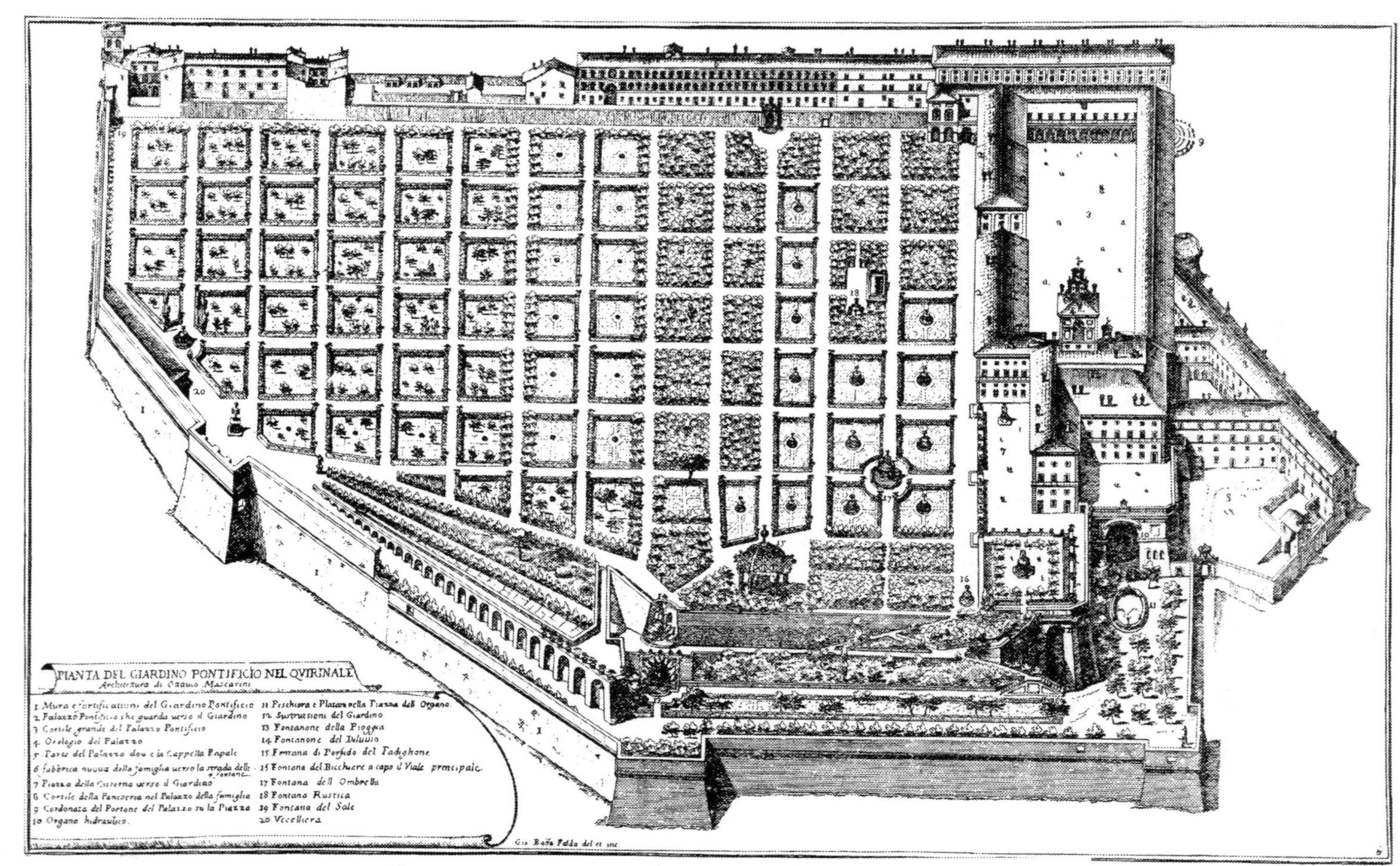

FIG. 226. THE GARDENS OF THE QUIRINAL ON MONTE CAVALLO, ROME—GENERAL PLAN

into the Pope's hands either by gift or sale, and came to form part of the present estate of the royal garden.

As late as 1580 Montaigne reckons among the most beautiful gardens in Rome those of Cardinal d'Este on Monte Cavallo. The palace itself was not finished by Paul V. till the beginning of the seventeenth century, but the garden must have been at the height of its beauty in the last decades of the sixteenth. Duke Frederick saw it in 1599 and describes it in exactly the same way as Evelyn does forty-five years later. He particularly praises the magnificent view over the whole of Rome, the precious trees, herbs, and foreign plants, "the very wonderful and peculiar water devices," of which he can mention scarcely a tenth part. Still, neither the water stairways nor the great tanks interest him so much as (under a semicircular arch) "a grand organ piece with four

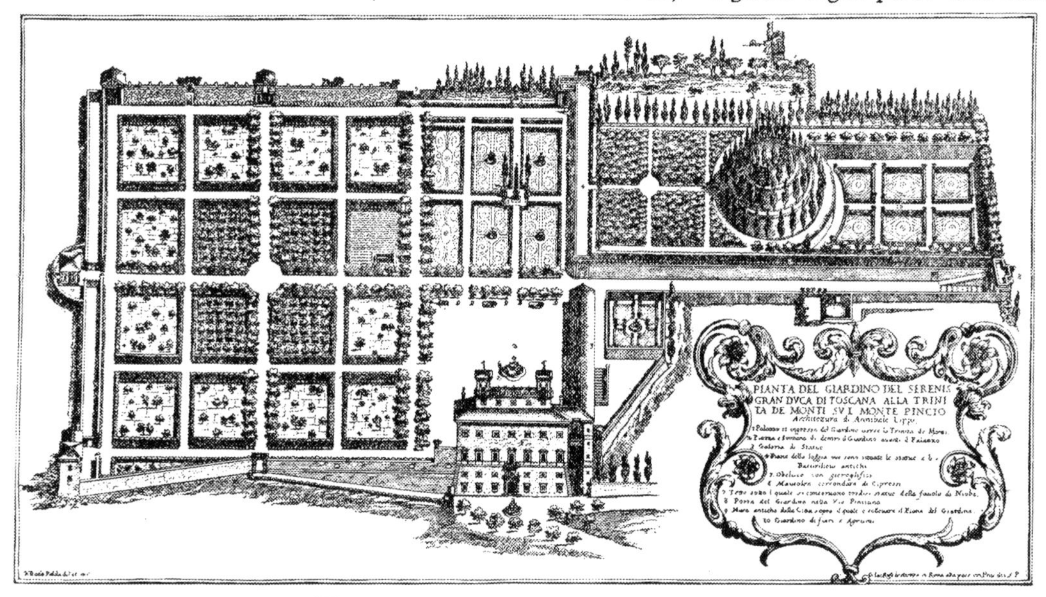

FIG. 227. VILLA MEDICI, ROME—GENERAL PLAN

registers most cleverly made, which if you turn the water on with a tap begins to play by itself just as if a good organist was playing on it." He also praises a wood of laurels, and many walks arched with greenery, as having everywhere beautiful water-works and figures of marble. He ends his story with the melancholy reflection (more often felt about the works of a Pope than about other people's) that anyone who begins to build is sure to die before he has finished.

The sunk grotto with the organ and the large tank in front with plane-trees round it has been ascribed to Maderna, and this part which is not connected with the buildings by any general plan gives to the garden a sort of design of its own. The largest stretch on the level shows dense thicket and paths with hedges (Fig. 226), which were particularly admired by Evelyn. At his time they were somewhat above the height of a man, and made of myrtle, laurel, orange, and even ivy, but in that case they were trained over lattice-work. Evelyn also admires the water stairs and the organs, and says of the former that he saw in one of the underground natural grottoes a basin made out of a gigantic piece of ancient porphyry, wherefrom a cascade flowed down steps to a grotto whose roof was inlaid with mosaic work. Evelyn was told that the gardener was paid 2000 scudi a year for keeping the garden in order; but though he regards this as one of the best and most attractive in Rome, its advantages cannot be found in any consistent artistic plan, but rather in the ornamentation of particular parts.

The garden of the Medici at Monte Pincio came into existence almost at the same time as the Quirinal. It is essentially a great terrace made for the view; a rather narrow strip between the old town wall and the Field of Mars, it lies high on Monte Pincio, the garden mountain of the ancients. But the mighty palace of the Quirinal never succeeded

FIG. 228. VILLA MEDICI, ROME—PRESENT APPEARANCE

in uniting house and garden; that was achieved by the architect in his limited space here by making an unusual alteration. The place was on the site of the ancient gardens of the Acilians, and before Cardinal Francesco bought it had changed owners several times. When, about 1550, Ligorio made his sketch of an antique stairway a certain Cardinal Crescenzio possessed a villa here, and soon afterwards it was bought by Cardinal Ricci da Montepulciano, a favourite of the Medici, and he began building a villa according to the plan of Annibale Lippi. But this villa seems to have advanced slowly, for when Cardinal Francesco bought it, it was in a very unfinished condition. He then enlarged the garden on the north-east side by adding a piece of ground that had belonged to a lady of his own house, Catherine de' Medici. Thus it was he who brought the villa and garden to its present fairly well preserved state, and it must be regarded as his work (Fig. 227).

On the side next to the street the house has an extra story below, which gives it the character of a palace; and there is a sort of sober quiet treatment of each floor that seems to suggest that the façade and its supporting wall would wish to serve as a protection for the precious place on the hill behind. The garden façade is quite different; indeed it gives the gayest, most festive picture that the Renaissance has to show (Fig. 228). We pass from a pillared front hall, divided into three parts, to a semicircular balcony, in the middle of which stands a Mercury fountain by Giovanni da Bologna. There were statues in the hall, and in front of the stairs between the pillars were the two antique lions that have now found a place in the Loggia dei Lanzi at Florence. The whole façade keeps its pretty,

FIG. 229. VILLA MEDICI, ROME—STATUES IN THE PORCH

light character by having its top story set much farther back than the others, together with a turret at each end and some antique relief work.

As was said before, Ferdinand de' Medici bought in 1584 the whole collection of the Cardinal della Valle, and a great part of this relief decoration came from it. Ferdinand, moreover, like della Valle, and like many other men, intended to convert his villa into a great garden museum, and everything—gardens, buildings and all—was designed to serve this end. After early modest beginnings the greatest skill had been attained, and it is important to observe that again it was one of the Medici family who produced the finest example at Rome. Thus the façade, with its fine effective decoration of antique sculpture (Fig. 229), shows the promise that the garden is to fulfil. The only exception to the antique style is a wing in front of the house on the right-hand side, for this wing is the boundary of an open space reserved for knightly games. It has a fountain in the centre, and statues at the side.

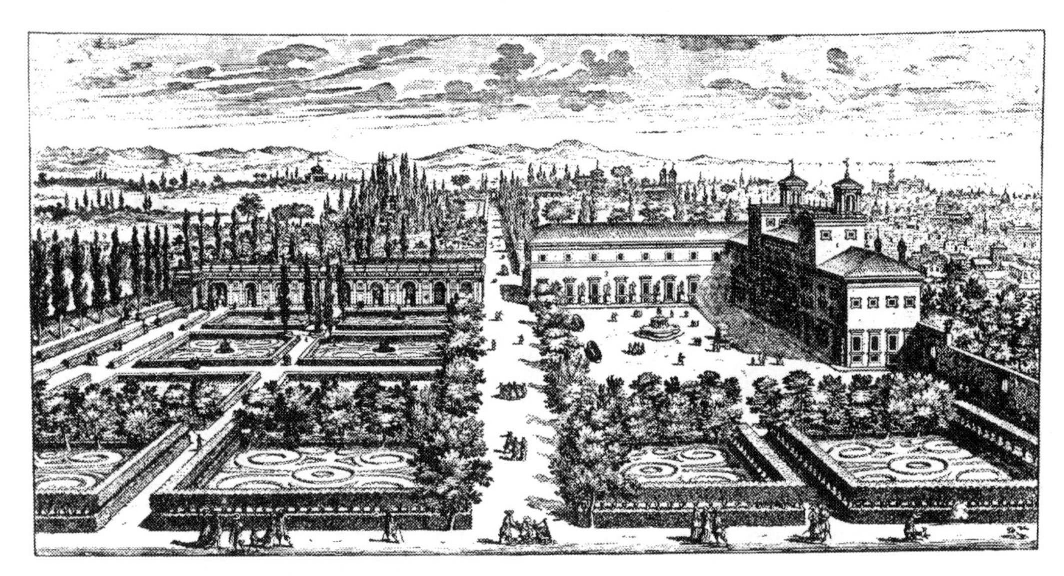

FIG. 230. VILLA MEDICI, ROME—BIRD'S-EYE VIEW OF THE CHIEF GARDEN

The chief garden (Fig. 230) extends from north-east to south-east at right angles to the axis of the house, which has to end at the high town wall. This garden was supplied in Ferdinand's time with the finest antique statues and an obelisk. Some specially good statues were set in niches which were made in the high wall of a terrace on the south, and between them the wall was covered with orange-trees. On the old town wall the sleeping Ariadne reclined under a pretty loggia. Along the Passeggiata del Pincio of modern days there stood under an open roof of a semicircular building the daughters of Niobe grouped round the horse that was with them when they were found; they came into the cardinal's possession in the year 1583.

The garden has preserved its earliest division into formal square beds, but the old pictures give a good view of the ever-changing style of its plantation. The earliest drawing we have belongs to the sixteenth century (Fig. 231), or at any rate to the time of Ferdinand's rule, and shows the change from the dignity and taste of the cardinal to that of the grand duke. The beds are now bordered with low hedges or else with trellis-work of wood. In the axis of the house they have little fountains in the centre, on the outside a circle of dwarf fruit-trees, and the space between is filled up with pretty geometrical beds of flowers. The plan calls the flower-garden by the usual old name of beds for simples. The space on the north, now covered by a dense wood of oak, was at that time mostly planted with fruit-trees. The paths are even now partly covered over with foliage, and have pavilions at the corners where they meet. The close plantation of the northern part appears first in the drawing that dates from the middle of the seventeenth century. The Duke of Würtemberg in 1599 finds the broad walks "enclosed with hedges as with a rope."

The southern end of the garden is peculiar, and has in part remained as it was. It consists of an important high-built terrace enclosed in a fine balustrade above the dividing wall mentioned before, which has statues in niches. Here stood the old round building on the hemicyclium that is now in the garden of the Sacré Cœur. Cardinal Ricci probably found it in ruins, and anyhow it served him as a convenient foundation for a circular hill planted all round with cypresses—seen by Evelyn when they were cut into the form of a fortress, with a pavilion and a fountain on the top of the mound. The earliest drawing shows a straight stairway on the garden side, fountains on either hand, and a spiral path running round.

This plan of an artificial mound we saw in ancient Assyrian times as well as in the Middle Ages. It was not at all common at the period of Italian Renaissance, but was used continuously in northern countries while the formal style was still popular. Its original intention was to provide a fine view from enclosed gardens; but in the south it was no longer needed because of the extension of terraces and the high situation of country houses. The Villa Medici would have had a fine view without the addition of this extra height, and the prospect seen from the high gardens was always praised—on one side the town with its domes and towers, on the other the beautiful garden land, at that time sprinkled with any number of little villas which were gradually absorbed into one large one, the Borghese. Gurlitt describes it as "a garden to look out of rather than to look into," but the splendour of it must have been increased by the noble works of art and the garden full of fruit and flowers in strong contrast with the sombre cypresses covering hills and terraces. And the view must have been magnificent—in its own way absolutely unique.

Later on, when Tuscan grand dukes gradually lost interest in the Roman villa, the precious statues were one by one removed to Florentine collections, though the last Medici of them all, Jean Gaston, did once make a great renovation. In the year 1801

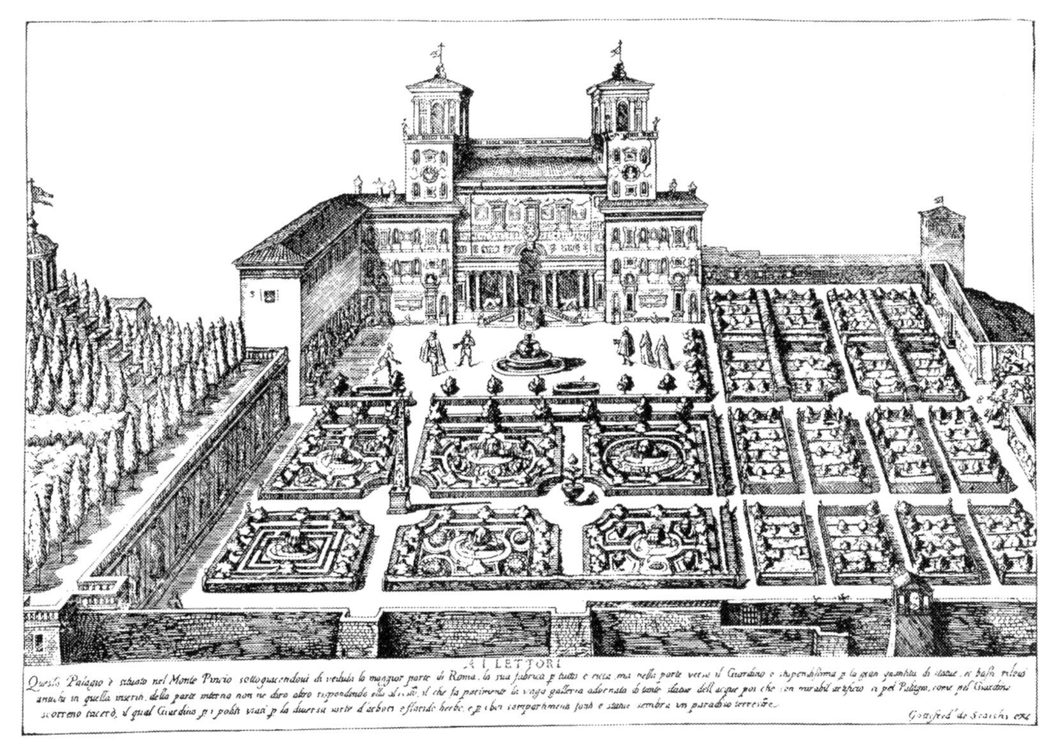

FIG. 231. VILLA MEDICI, ROME—EARLIEST DESIGN

the villa was converted by Napoleon into the home of the French Academy of Art, and has remained so to the present day. At a later date an attempt was made, by setting up a few casts there, to charm back to the place some trace of its former beauty.

Those who founded the Villa Medici enjoyed the great advantage of a fine water supply, for the Acqua Vergine was restored by a long succession of Popes to its former efficiency as a conduit, and watered the Hill of Gardens (*Collis Hortorum*). According to the first drawing, the fountain on the hill of cypresses leapt to a height of twenty-five ells from the pressure of this conduit, and Evelyn mentions with astonishment that the fountain at the entrance to the palace throws its waters to a height of fifteen ells.

On the east side, towards the Esquiline and Viminal, there was a very scanty supply,

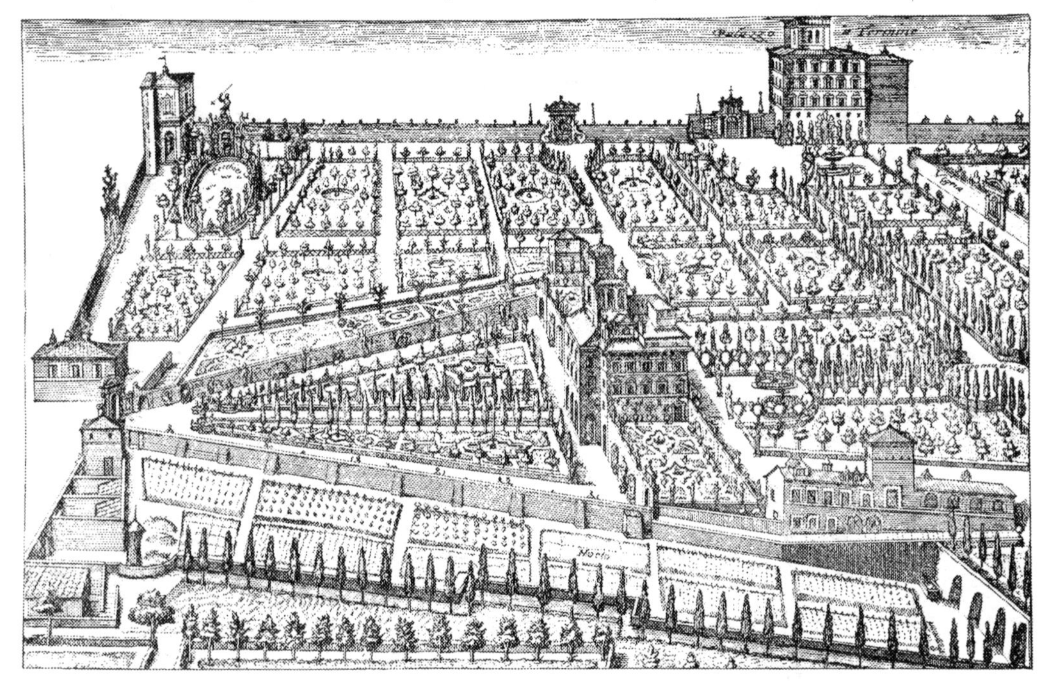

FIG. 232. VILLA MONTALTO, ROME, WITH GARDENS

and it was just here that another cardinal, Montalto, elected to build a villa in the grand style. This cardinal made a victorious exit from the Conclave in 1585 as Pope Sixtus V. To no other Pope have so many legends been attached as to this one, originally Felice Peretti, who worked his way up from the lowly estate of a shepherd boy to the highest position in Christendom, always believing in his luck, which he thought his father had bestowed on him with the name Felice that he liked so much to hear. This Pope is described by Ranke as a man of frugal life, who quickly remedied the ruinous condition of Church finance.

In spite of his economy, however, he indulged his fancy for building. The villa he started as cardinal he brought to completion as Pope, and in the handsomest way. It was adorned with choice works of ancient art, although at one time he had scornfully opposed the cult of antiquity. There is an anecdote that relates how the cardinal suddenly was compelled to give up building his villa, when Gregory XIII., himself a great lover of building, was provoked by hearing that a mere cardinal was making a finer villa than the Pope's. Then Montalto thought it advisable to hold back, feigning illness. But at this point the architect Domenico Fontana, who also believed in his master's good luck, bought the entire place so that he could continue the work, and later on the Pope heaped honours on his head.

This story shows how important to the Romans was the size and splendour of this garden. But one reason why it was interrupted must have been the deplorable want of water in this part of the city, and the first order given by the Pope on the very day he ascended the papal chair, was that Domenico Fontana was to undertake the construction of the water conduit, Acqua Felice. Thus did he inaugurate a work that, as Ranke says,

FIG. 233. VILLA MONTALTO, ROME—THE APPROACH TO THE CASINO

"brought more honour and glory to him in the town than was paid to any other Pope." This gigantic work filled him and the architect with joy and pride when two years later it was completed. All Rome took part in the jubilation, and Torquato Tasso wrote a pompous poem in its praise, following the water on its path below the earth till gleefully it greets the sun which once shone upon the mighty Augustus. It was felt that now something had been made worthy of antique glory, and so it should be celebrated in an antique style. The Pope had one arm of the conduit taken through his garden and thence to the Quirinal, the other was to pour forth its streams at the Moses fountain at the Thermæ.

The Villa Montalto (Figs. 232, 233) which only a few decades ago stood "beautiful and dignified in wild surroundings" is now completely swamped in a sea of houses, stretching from the Esquiline and Viminal between the church of S. Maria Maggiore and the Thermæ. The architect, Domenico Fontana, was the first of a whole family of

FIG. 234. VILLA MONTALTO—PRINCIPAL ENTRANCE AND CYPRESS AVENUE

architects, who gained their chief reputation by the laying out of gardens in the seventeenth century. Like his master, Domenico had risen in the world; from a bricklayer's apprentice in the service of Sixtus V. he had become his architect, and finally he was almost exclusively employed. The casino at the villa was not large, and its façade was only approached by one entrance loggia, and Fontana's particular ability is shown in his spacing, which is remarkable in the massive features of the garden grouped about the Palazzo Felice. One can imagine nothing more magnificent than the approach to the casino from the side of the Esquiline near the church of S. Maria Maggiore (Fig. 233). Through a large gate one steps into a garden where three splendid cypress avenues diverge; and at the point where they end there are two grand lion fountains, which seem to be fastened together by a kind of clasp or chain (Fig. 234).

The middle avenue leads directly to the entrance loggia with its three arches, and we must think of it as described by Evelyn—full of statues, inscriptions, reliefs, and other antique marbles, "such that one could not imagine anything more splendid and gay." The palace has *giardini secreti* half-way to the height of the first story, and there is a higher terrace behind. There are various grottoes in the walls, and behind the casino there is a semicircular level part with a fountain and vases put up between hedges (Fig. 235). A large cypress avenue starting from here crosses another that began at the second gate at the baths, in a round place where a great many water devices are assembled. These two main avenues reached the wrought-iron gates of the flower-garden, and were then continued in the park which surrounds this garden on the north and east; and both of these avenues, which end in hills with statues upon them, give a *point de vue* very clearly, as they are straight as well as long.

If we compare this with all the other gardens we have examined, the continuity is obvious and significant. We have always found hitherto that in the flower-garden at least there has been a formal distribution of beds in squares and bordered with hedges, or else paths arched over with foliage. Then in the gardens of Ligorio, Vignola, and others, the architectural effect has been obtained by either water or terraces. Here for the very first time the artist is working in a larger style with perspective. Long avenues are made with definite endings, architecture or sculpture. The beds must be arranged to suit the form of the avenues; large, open, or semicircular spaces at the beginning of the avenues increase the grandeur of the outlook, as Evelyn particularly notes. Their size is increased once more by the extension of the flower-garden into the park; the importance of this feature will be seen later on. The many beautiful fountains—unhappily all entirely perished—are something more than an adornment of separate parts of the garden or avenues, for they really dominate the whole composition, and are a leading motive. Villa Montalto is in this respect not only the first in a particular style, but it is destined to be for long the only one. More than half a century later Fontana's ideas are united to those of the French garden, and proceed to what is still more important and built on a larger foundation.

Villa Montalto had not only its own *points de vue* to rejoice in, but also the help of buildings outside. This attraction was not the least that was felt by the Pope when he chose the site for his villa, seeing that on one side there were the mighty towers and cupolas of the church of S. Maria Maggiore, on the east, looking through a great avenue, the little church of San Antonio, and on the west the ruins of the baths of Diocletian which extended right up to the park. Ever since Alberti's time the view from the

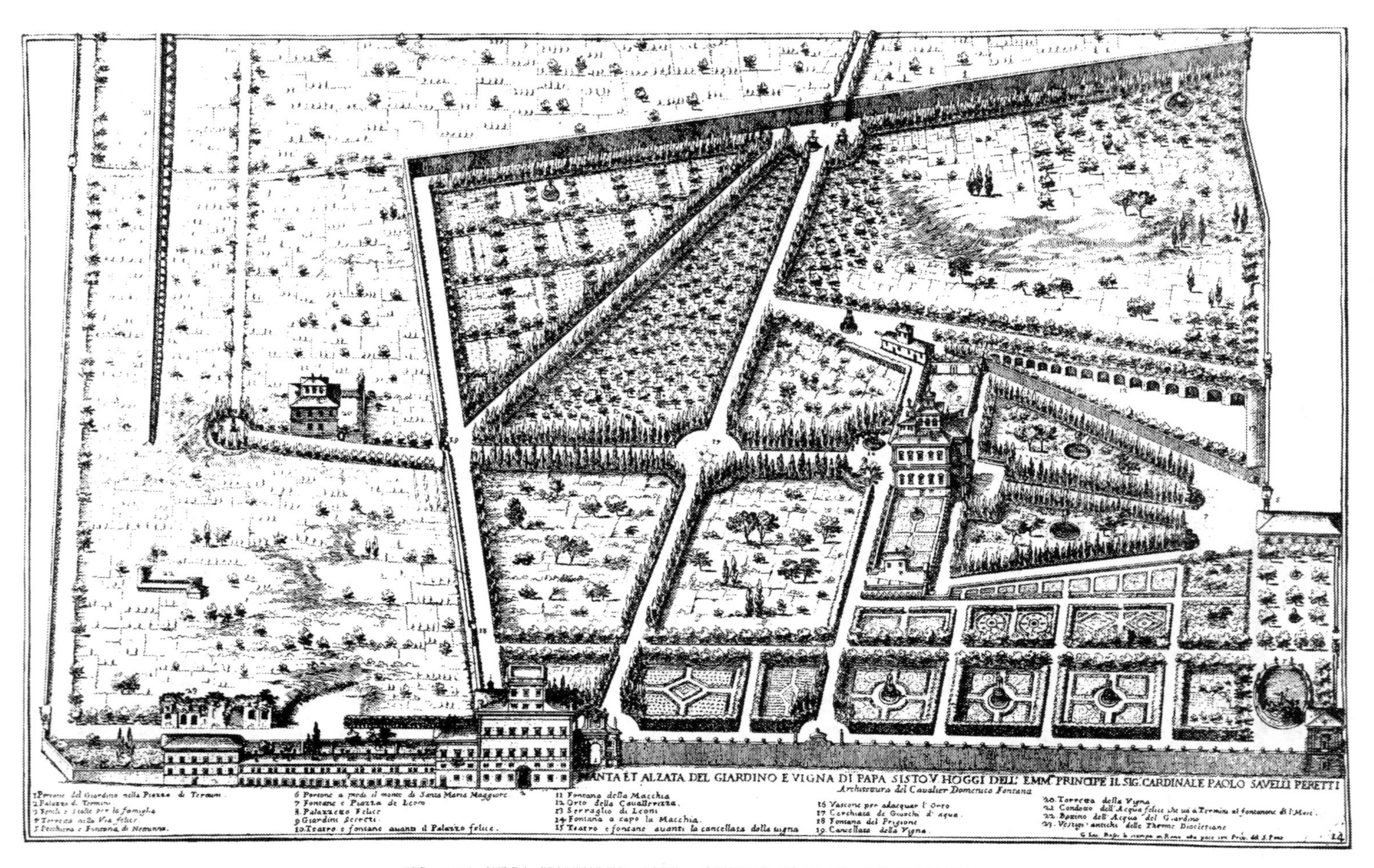

FIG. 235. VILLA MONTALTO, ROME—GENERAL PLAN OF THE GARDENS

garden had been an important consideration, and in a country that was so rich in the beauties of landscape scenery, it was easy enough to find a view. But Rome offered more than this with its abundance of ruins and churches, and it did not take long to see how a villa would be enriched by a view of such architecture.

A church is actually a component part of the Villa Mattei (Fig. 236), traces of which, though much blurred, we can still perceive on the hill of Cœlius. This little church,

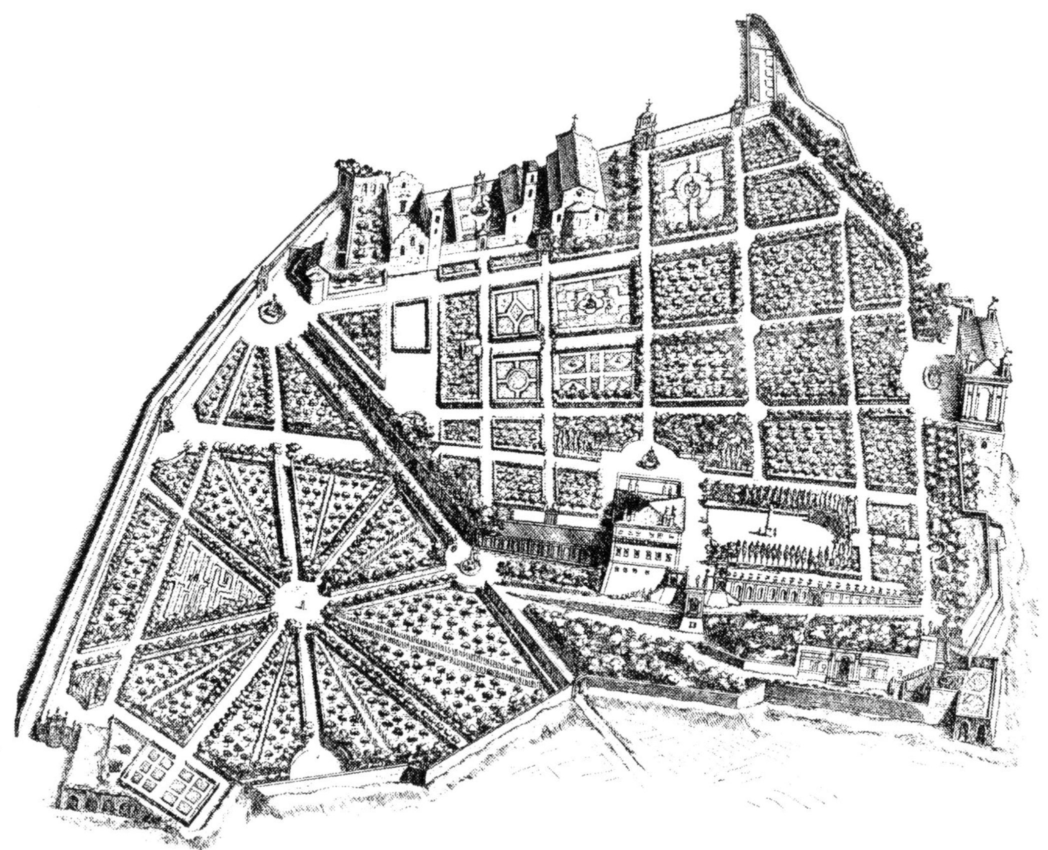

FIG. 236. VILLA MATTEI, ROME, IN THE SEVENTEENTH CENTURY—GENERAL PLAN

S. Maria in Domenica—also called *della Navicella* from a small ship that Leo X. had once put up there, an imitation of an antique find—has the villa garden on three sides of it. Beside its façade is the entrance gate, and this ancient basilica, with its pretty front hall (ascribed to Raphael), represents a casino, or a lodge for the palace which stands in the middle of the garden. Also on the east is the picturesque round building of S. Stefano, also dating from early Christian times and built on ancient foundations. As a fact, it is on such foundations that the whole garden is built; and the old walls, plainly recognisable by their *opus reticulatum*, are made use of as terrace walls, and part of an old conduit in one corner has at some time been entirely absorbed into the garden. Towards the south the eye glides over the Baths of Caracalla to the Campagna backed by hills.

This fine place was selected as a villa site by Cyriaco Mattei, a scion of an old Roman

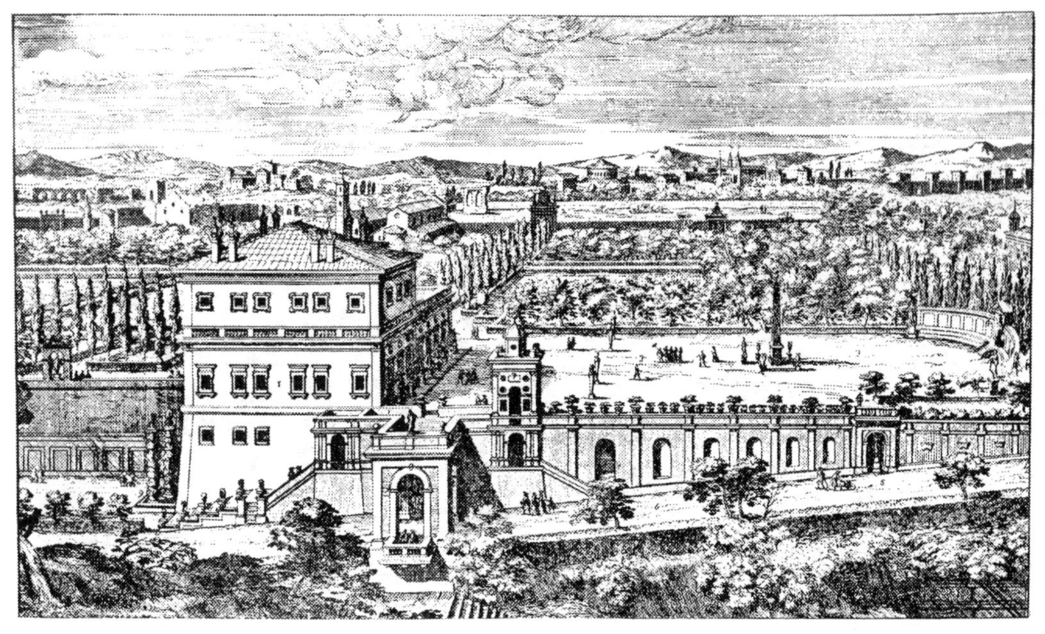

FIG. 237. VILLA MATTEI, ROME—PALACE WITH PRATO

family of nobles. It must have been begun before 1582, and indeed nearly finished by then. At that date Rome presented to this citizen an obelisk which had once stood at Aracœli, but had been removed when the Mattei family erected a chapel there. Cyriaco set up the obelisk in the place of honour in his so-called amphitheatre, a place in circus form, which adjoined the casino at the side, and on one rounded end had a colossal

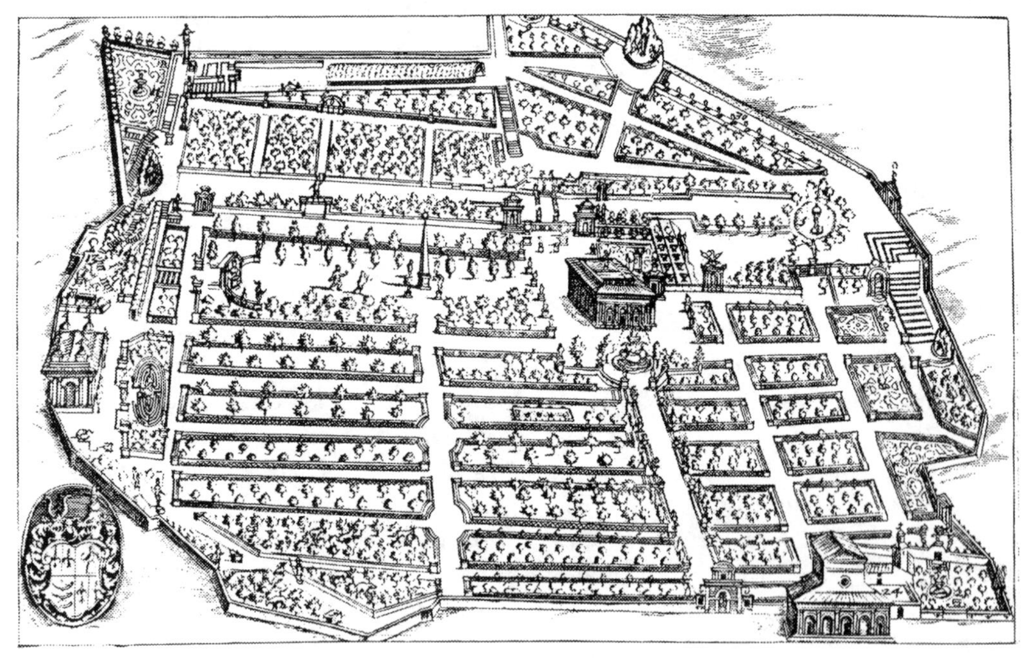

FIG. 238. VILLAMAT TEI, ROME, IN THE SIXTEENTH CENTURY—GENERAL PLAN

bust of Alexander. This circus-shaped prato with cypresses round it serves in a sense to identify the Villa Mattei (Fig. 237), for whereas in Florentine villas we constantly found such places, the idea had not yet penetrated to Rome, and therefore all travellers make remarks about the Villa Mattei.

As the place has now entirely disappeared, one can only guess at its size from the sketches that remain, which are not always to be trusted. Still, there was certainly enough room for an entire Lenten procession, which came round every year after its pilgrimage was over and had its first meal here. On the right side of the circus in the middle of the garden stood the casino. According to the first drawing (Fig. 238) it must have been much

FIG. 239. VILLA MATTEI, ROME—THE EAGLE FOUNTAIN

smaller than it is to-day or than it appears in the engravings of the later seventeenth century (Fig. 236), and the original garden is certainly not large. It was not till the seventeenth century, or at any rate not till after Cyriaco's death, that the park-like region on the north-west was thrown into it, and laid out in the peculiar form of a star, which is fairly well preserved; at that time the ruins had not been included in the garden grounds. But it was later still, in the nineteenth century, after the place had passed out of the hands of the Mattei family, that the great extension took place and embraced the park on the west and south which was now laid out in the English fashion.

The garden of Cyriaco was, like the Villa Medici, first and foremost a garden museum, and the great characteristic of Montalto was wanting. The beds are of the usual four-cornered shape and planted like those at Villa Medici. Corresponding to the circus at the side there is in front of the house a round space with a fountain in the centre, from which leads the wide walk to the entrance of the church. The ancient foundations are utilised for unsymmetrical terraces on south and east. Though the space is small, the drawing gives us countless objects that are worth seeing—seventy-seven antique statues, fifty-three busts, the Eagle (Fig. 239) and other fountains, two of which, at either end of the garden, were joined by a bridge over a stream which ran behind the house. On the south stood a little casino with a labyrinth in front; aviaries, even a thicket containing animals made of stone and a shepherd painted in natural colours, all delighted such travellers as the Duke of Würtemberg. Then at the entrance to the prato were two life-sized dogs, also painted in natural colours.

Such fancies are a sign of the love for the *genre* style, so important, as we have said before, for this period. The Duke of Würtemberg and Evelyn too relate that Cyriaco in his will left a sum of 6000 crowns for keeping these darling ornaments, adding that if this sum was not used for its appointed purpose, all the rest of his great possessions were to pass over to a side branch of the family. "So much," Evelyn concludes, "are they concerned with their villas and pleasure-houses, even to excess."

Some of the villas of which travellers, especially Montaigne, tell us, cannot be observed otherwise, for many have vanished even to their very names, and some have been completely built round, house and garden both. To the latter class belongs the villa of Cardinal Riario in Trastevere, which Montaigne calls one of the loveliest villas of Rome. In the year 1729 the Riarii sold palace and garden to the Corsini, nephews of Clement IX. This family had the house rebuilt in grand style by Fuga, and the old parts were utilised only as side wings. But in the garden a very pretty part of the old villa was preserved—a water stairway with a fountain at the end of the garden where it rises to the Janiculum. This stairway, originally made exactly in the middle axis of the old casino, is now sideways to the house, but Fuga has very cleverly improved it by an oblique line in the façade.

The whole architecture of this end of the garden is specially good and dignified; high steps lead up to the third terrace, and here the water stairs are first interrupted, for a fountain stands at the top, making the final feature of the grand view from the house. From minute and careful search it appears that traces of terrace-sites remain even on the side of this stairway. Whether Montaigne saw the whole place at that date—he describes it no farther—has to remain uncertain, but the architectural style and also the remains of the sculpture seem to point to a later one, for it is only in the last decade of the sixteenth century that we find similar places. The Villa Riario would seem to date back to Julius II., if not to Sixtus IV., so that the first plans at least are quite gone.

The town villas of Rome in the sixteenth century always suffered from the disadvantage of not having room to expand in the way the grand style needs. Churches, town walls, old streets, even antique foundations, had to be considered, and it required a man with the impulsive determination of Sixtus V. to subordinate all these elements to his great garden plan. "If you want to see anything really wonderful, go out of Rome to Tusculum (now Frascati)": so says the handbook of Sprenger, of Frankfort. The modern Roman had rediscovered the much-praised Tusculum, where Cicero loved to stay, with its mountain slopes full of lofty vegetation and unsurpassed water, with its beautiful views of Rome and the sea. He wished to escape thither from the dust and turmoil of the great city.

It was in the second half of the seventeenth century that Sprenger visited the villas and wrote about them. Frascati must at that time have given a lovely picture, even

FIG. 240. VILLA FALCONIERI, FRASCATI—THE APPROACH TO THE LAKE

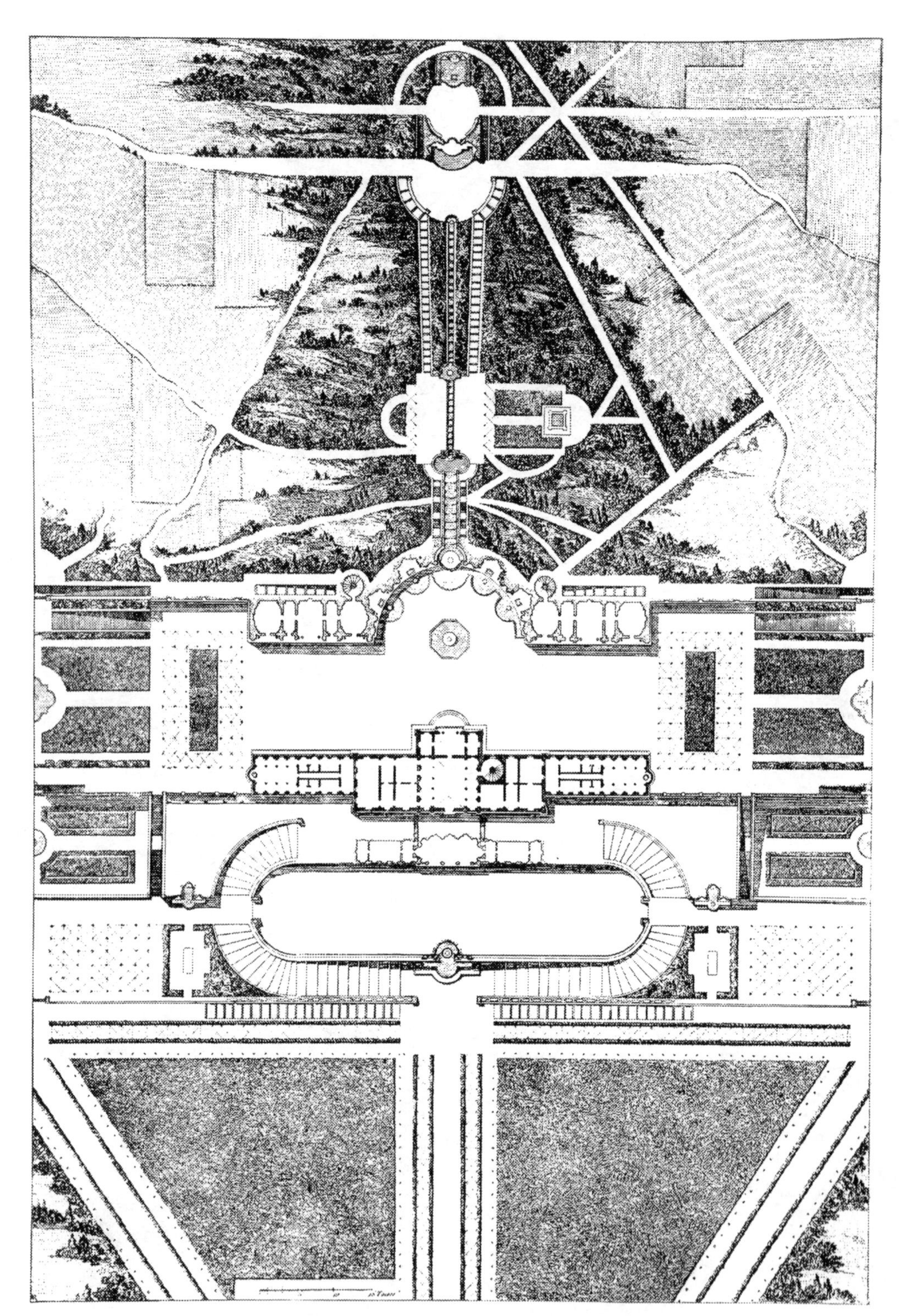

FIG. 241. VILLA ALDOBRANDINI, FRASCATI—GROUND-PLAN

surpassing antiquity; for from one drawing we get a comprehensive bird's-eye view of the place. All the villas that exist to-day, though for the most part spoiled and even ruined so far as gardens are concerned, were then there in all their beauty, but mature enough to display their utmost charm both in nature and art. The oldest sites belong to the sixteenth century; but some of them, such as Villa Falconieri (called La Rufina) (Fig. 240) and the present Villa Lancelotti, were so rebuilt in the seventeenth and eighteenth centuries that their present condition really belongs to the latest period, both house and garden. Thus in all essentials the villas at Frascati belong to the first half of the seventeenth century Here we see, even more markedly than in Roman palaces and gardens, the pride and magnificence of the Popes' nephews at the time of the Counter-Reformation. Most of the villas were built during the reign of the Popes, and we ought to realise that at this period the garden stood next to church architecture as leader of the arts.

The town palaces in Rome were imposing for their size from the very first decades of the seventeenth century, but they became ever more and more uniform, alike in the whole and in details. Indeed there is an inexhaustible abundance of detail of the most various kinds both in the garden itself and in the decoration of the buildings, and the sculpture of the Baroque period and style grew to what it was in the garden. There is doubtless in indoor life a certain exaggeration, and also a certain oppressiveness and weariness. But when we get into the open air, and find ourselves allied with that world of nature which is treated more and more as the beautiful green background of the whole, the art of picturesque expression no longer offers us what is wearisome or discordant, but brings to us a feeling of wide spaces and the glowing brilliant colours of an Italian sky.

Among the villas of Frascati, Aldobrandini (Fig. 241) is unquestionably the first, showing as it does the finest skill of the time in full flower. Taken as a whole, it is the most beautiful, and has served as a model for many others. It may be regarded as *villa urbana* in the ancient meaning of the term. Cardinal Pietro Aldobrandini, the powerful nephew of Clement VIII., had a garden or perhaps a villa in the town on the slope of Monte Cavallo, which is still in existence, though hard pressed by an ever-growing region. This garden, of no intrinsic importance, became notorious because a copy of a very famous Greek painting was discovered there, the so-called Aldobrandini Wedding. It was found in a cellar, and the cardinal had it taken up and placed in the dining-room with a wooden porch to protect it. There it was seen in 1625 by the Spanish ambassador and connoisseur, the Duke of Alcalá, who had a copy made for his own house, Casa de Pilato, in Seville.

When, in 1598, the cardinal decided to build a villa at Frascati, he was at the height of his power. The dukedom of Ferrara had just come into the hands of the Church as an inheritance from Lucrezia, the sister of the last legitimate duke; and the cardinal looked on this event as a triumph for the papal chair, as a pledge of peace that he gave to Christendom, and felt that he could not celebrate it better than by founding the superb villa at Frascati with part of the wealth that had come to his share. This is explained in a Latin inscription over the semicircle of the theatre. The likeness of the cardinal in bronze still stands on the chimney-piece of the banqueting-hall, which runs the whole width of the building and opens on both façades. His face is short, bearded, with an expression of great power and energy, but with a benevolent look also.

The cardinal employed Giacomo della Porta, a pupil of Michael Angelo—an architect

FIG. 242. VILLA ALDOBRANDINI, FRASCATI—THE APPROACH TO THE FRONT

who here began his last work and continued it until his death in 1604. The actual house (Fig. 242) is not one of the largest at Frascati; its width is imposing, but there is very little depth. These country villas have quite a different importance from that of the casini in the town; for weeks and even months at a time they were continually putting up a host of people. But not wishing to make the place look too heavy, the owner decided to use the grand terraces which extended in front of the house towards the city, and at the sides also, as domestic rooms underneath the building. The kitchen chimneys, which of course had to be close to the house, were relegated to the end of the side terraces, and there treated as ornamental turrets.

FIG. 243. VILLA ALDOBRANDINI, FRASCATI—THE FOUNTAIN OF THE LITTLE SHIPS

The *sine qua non* for a country villa at that time seems to have been the covering and concealing of all that was not really necessary, for fear anything should mar the picture of the whole place, which must be perfect. The large tract of land given up to farm produce and the kitchen-garden was planted round with hedges cut to above the height of a man, so that anyone walking or even driving with carriage and horses would fancy he was in a pleasure-park and never suspect that there were olives, vines, vegetables, nay, even cornfields, behind. These avenues were well stocked with statues, fountains, and other ornaments, consequently the eye need never weary of long stretches.

Goethe himself is surprised at the householder's wisdom that he finds in Southerners, who do not, as Northerners do, he says, "waste a huge expanse of good land in a park which flatters the eye with bushes and trees that are quite unproductive." Thus we find in front of Villa Aldobrandini and between the entrance and the city, an important tract of country that was certainly at one time used for cultivated plants. It is shut in by a wall in front, made both strong and pretty by a border of stone and wrought iron. At the central door, according to Specchi's plan, there ascends a wide path with thick straight-cut laurels, protecting the pedestrian from the sun and even from the rain. On the right and left there are avenues of firs with hedges that cross the farm region. The drawing gives an entirely festive appearance to the villa from the town side, and the stairway at the end of the middle path reminds one of Palazzo Farnese at Caprarola, though this site is better and more cheerful. The stairs are gay with oranges in pots and with fountains

FIG. 244. VILLA ALDOBRANDINI, FRASCATI—POOL IN THE FLOWER-GARDEN

(Fig. 243), and these are also found on the terrace balustrades at the side of the house. In the middle axis as well as at the sides there are grottoes in the dividing walls with pretty water-works and fountains. The house on this side has only a sober, rather plain façade; and though the interrupted gable ought to make it more interesting, the effect is ugly.

The whole front relies on the approach, and the value of the picture is enhanced by two dark clumps of oak severely cut back on both sides of the upper terrace; here the oaks and firs do much to help out the architecture. A sunk parterre for flowers is at the side of one clump, but it is hidden away, perhaps as the only irregular feature; and a pretty ship fountain—a favourite fancy since the copy of the antique ship was set up before the Navicella—is in the central place between avenues of arching foliage (Fig. 244). The house loses one story on the hillside, so that the great reception-room with its balcony on the first floor opens on a most lovely distant view from over the little town towards

Rome, covering the Tiber Valley as far as the sea. If one walks through the room an unexpected fairy-like view appears on the other side of the door. There is a level, semicircular space, to which one goes down by a few steps cut into the mountain-side.

The division wall which seems to support the hill shows a row of pillars and niches, and between them grotto-rooms which are full of the familiar water devices. In the middle stands Atlas, holding the world on his back, and on each side of this fountain-piece is a ribbon with the usual inscription. A star—the family device—pours water at the top of the balustrade from its tips, and from above there plunges foaming down over wide stairs

FIG. 245. VILLA ALDOBRANDINI, FRASCATI—THE THEATRE

a flood from a still higher terrace, where at the side two columns stand, from each of which a single stream rises, falling again in spiral fashion round the columns and thence to the balustrade that borders the water stairway. Then, glittering in many greater and smaller cascades, the water at last seems lost in the thick growth. If one mounts from the semicircular theatre (Fig. 245) to the terrace where it is interrupted by the wide stream of the water stairway (Fig. 246), one stands opposite to the garden façade with the gay loggia in the uppermost story. The path leads to the next terrace by the side of the stairs, and the water is connected by a fountain of tufa, and plunges in a wide current over a wall which is ornamented with niches at both sides and with figures of peasants (Fig. 247). Finally, on the highest terrace of all, a natural grotto made of tufa receives the water in cascades, after it has been conveyed for six miles underground.

In this water system has been brought to pass what has been tentatively aimed at

FIG. 246. VILLA ALDOBRANDINI, FRASCATI—THE WATER STAIRWAY

FIG. 247. VILLA ALDOBRANDINI, FRASCATI—RUSTIC FOUNTAIN

for long enough. From the first timid beginnings, as Bembo describes them in his *Asolani*, by way of the pleasant essays in Villa Lante we have come to this proud conclusion, which men of that age (who saw the streams and fountains in their undiminished force) gazed at with astonishment and rapture. This water is bordered by lofty hedges, and it seems as though from the very beginning these were made out of the dense *selvaggio*, which was traversed by straight formal paths, even when the hedges were not actually clipped. Moreover, in certain places there do appear to be signs of box clipped into various shapes. The idea of rising above fancies and follies to a simpler form of fountain has now been actually expressed. Following the two figures of peasants on the last terrace but one, we arrive at a *fontana rustica* (Fig. 248) at the top, and yet we find, until we come to the obviously "natural" forms and the close thicket in the woods, that there is a very severe style of symmetry for the buildings, the water arrangements, and the laying out of paths.

FIG. 248. VILLA ALDOBRANDINI, FRASCATI—NATURAL FOUNTAIN

We must once again emphasise the fact that, in spite of the use of such expressions as *selvaggio, fontana rustica*, or *fontana di natura*, everything in the Italian garden is subjugated to the firm feeling for architectural style, and that it is only much later that influences quite foreign to the Italian spirit caused the great change and even revolution to the style of the picturesque. Among the architects of this garden Domenico Fontana is named first, and one would like (after his leadership at Villa Montalto) to ascribe this place to him; but as a fact in these villas (whose speedy completion was a necessity for a nephew) there was generally a whole staff of architects employed, whose chief would probably be the architect at the palace, with several others looking after the irrigation, or the water devices and games—the grottoes, and the like.

In addition to the Villa Aldobrandini, the Villa Ludovisi, later called Conti, and now Torlonia, was particularly noticed by the copperplate engravers of the seventeenth century. This villa came into existence after the year 1621, but as it is in its chief features

FIG. 249. VILLA LUDOVISI (TORLONIA)—THE ARTIFICIAL WATER ARRANGEMENTS

FIG. 250. VILLA LUDOVISI (TORLONIA)—POOL ABOVE THE WATER STAIRWAY

an imitation of Villa Aldobrandini, it may very well be put along with it. The construction of the water-stairs (Fig. 249) above the wall ornamented with niches (which is not semicircular in this case but straight) reminds one of the Villa Aldobrandini, though the details are simpler in the working out. The ground above the stairs is level, and has a large oval basin and fountain, with well-made balustrades, and other water playing on them (Fig. 250 and Fig. 251). None of this water arrangement is in the axial line of the house, but at the side of it in the park, and goes down by steps that are impressive and good to look at, but perhaps too large to suit well with the whole picture. The house stands separately in a

FIG. 251. VILLA LUDOVISI (TORLONIA)—THE UPPER POOL

garden terrace with fountains (Fig. 252), scarcely at all connected with the park: seclusion like that of Villa Aldobrandini is here attained only in this one way. The villa of to-day, Lancelotti, has taken the idea of the semicircular theatre without being able to imitate the water scheme. The whole place was evidently of a late erection, and only began with the rebuilding of the palace in the eighteenth century, for the architectural detail shows an unmistakable classical leaning (Fig. 253).

Many charming spots are to be found in the small villas, each of which has its individual attractions. The garden of the Villa Muti seems to have undergone many changes, unless the general sketch we have is very incorrect. There is a plan in the house, unfortunately undated, of the garden in existence when it was made; its chief beauty lies in the pretty arrangement of terraces and steps, but there is no union of the parts to make one comprehensive whole. There is a charm almost fairylike about

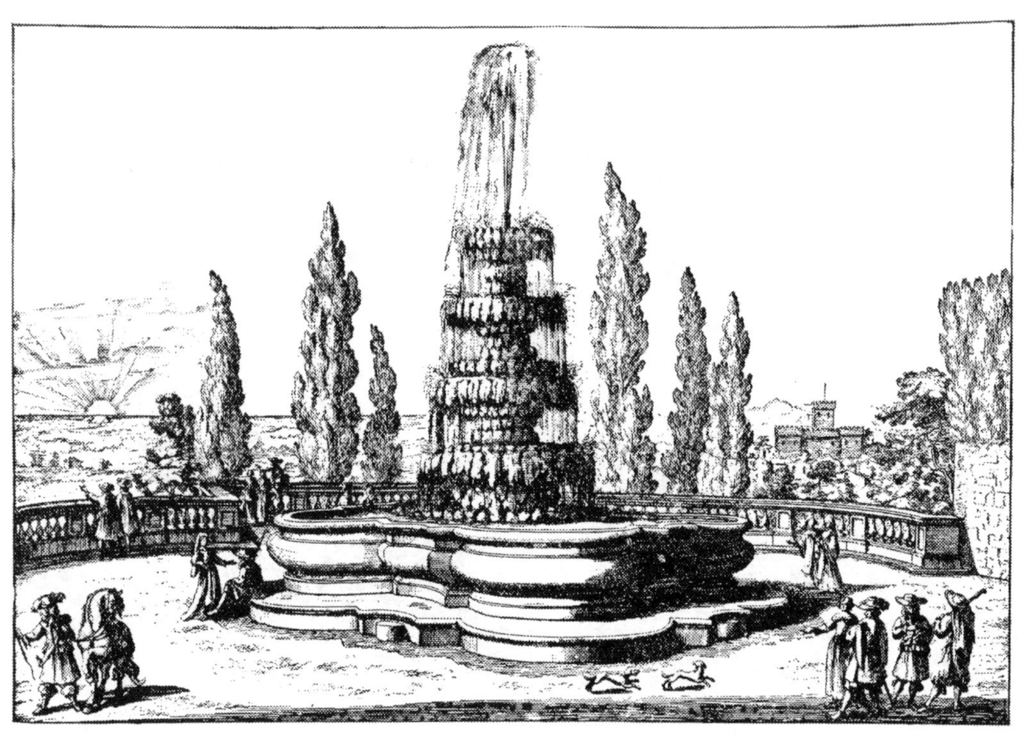

FIG. 252. VILLA LUDOVISI—THE FOUNTAIN ON THE TERRACE

FIG. 253. VILLA LANCELOTTI, FRASCATI—THE PARTERRE

the look of the plants which have outgrown all training (Fig. 254). The lovely avenues of oak that cross the farm-land, with an olive plantation behind, show most happily the success of these rows of fine trees with their border of hedge in concealing what lies within.

Yet another of the small type of villa may be mentioned, Belpoggio, now called Pallavicini, for the place, still very attractive, is quite recognisable from the Rossi drawing. The house stands on a wide terrace (Fig. 255) with grottoes round, and is approached by

FIG. 254. VILLA MUTI, FRASCATI—THE MIDDLE TERRACE

a fine avenue. There are four fountains, and the symmetry is helped by two avenues of oak-trees, one on either side. Severe formality is shown in the continuation to other garden terraces, fountains in the axial line, overarching paths round and through the lowest terrace, and pavilions in the centre and at the entrance.

Mondragone is by far the largest of the villas at Frascati. As early as 1567 Cardinal Marco Sitico Altemps (Marx Sitich von Hohenems) had a villa built by Martino Lunghi, but it seems to have been a very modest affair, and it first reached its present size and importance under Scipio Borghese, the powerful cardinal nephew of Paul V. The Altemps family exchanged this villa for the Palazzo Rospigliosi on the Quirinal, where even a short time before there had been very important ruins, acquired by Scipio Borghese, of the Baths of Constantine. To a man of vigorous nature, in an age of vigorous activities, such ruins were only tiresome obstacles, and he had them all cleared away, and on the top built palace, casino and garden. But scarcely was this finished when he was attracted by other schemes, and he exchanged his villa on the Quirinal for the Villa Mondragone.

A whole army of architects (always at the service of the Borghesi), Ponzio, Vasancio, Girolamo Rainaldi, Giovanni Fontana, were now summoned to make out of a little summer-house the gigantic structure with its 366 windows. In front of the house lies a terrace of enormous dimensions (3 in Fig. 256), which is really only (as at Aldobrandini) a roof for the kitchen department below, and is similarly furnished with chimneys to suit the style. The three-shelled dragon fountain stood on the central semicircular projection, where the double coat of arms of the family was held in place by four eagles above and four dragons below. From the terrace the farm is visible, traversed by avenues (1 and 2) which were once decked with very fine statues and fountains. The cypress avenue, now majestic in its age, leads to the chief entrance, and semicircular stairs enclose the end of a space immediately in front of the terrace. On the notion of the semicircle the whole of the Villa Mondragone is conceived. If one walks through the great court (6) behind the palace, which on the right hand has as a sort of winter gallery a small low building, and on the left is separated only by a wall from the flower-garden, one proceeds by way of arcades into an amphitheatre (12) cut deeply into the hillside, which rises imposingly to the view. But east of the great court, and quite shut in, there is a charming flower-garden (9 and 10), and this also ends in a semicircular theatre (Fig. 257), raised on a terrace.

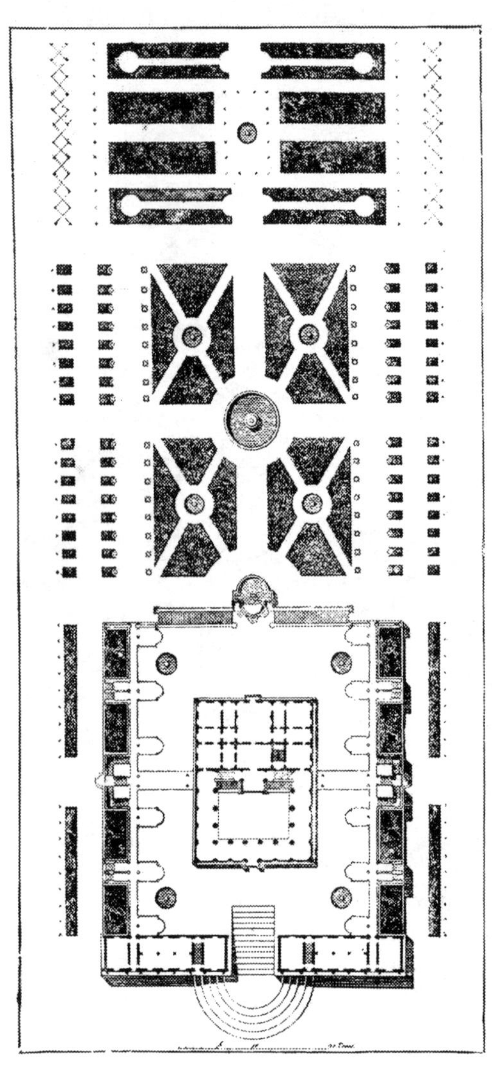

FIG. 255. VILLA BELPOGGIO, FRASCATI—GROUND-PLAN

This feature shows, with niches cut deep for the sake of the perspective, the very clear connection that obtains between the garden scheme so often used at Frascati and a permanent theatre decoration. It was not long since Palladio's Olympic Theatre was finished, a work much admired for its perspective; and all over Italy the larger towns were beginning to build great theatres. To be sure, these garden sites can hardly have been used for festival plays. The want of seats for spectators is against it, and there was generally a fountain in the middle, which would interfere with any performance. The garden theatre proper we shall soon be able to distinguish by its side-scenes cut in the greenery. In Mondragone there is also a fountain in the middle of the terrace, and water-tricks concealed all about. Steps lead down to

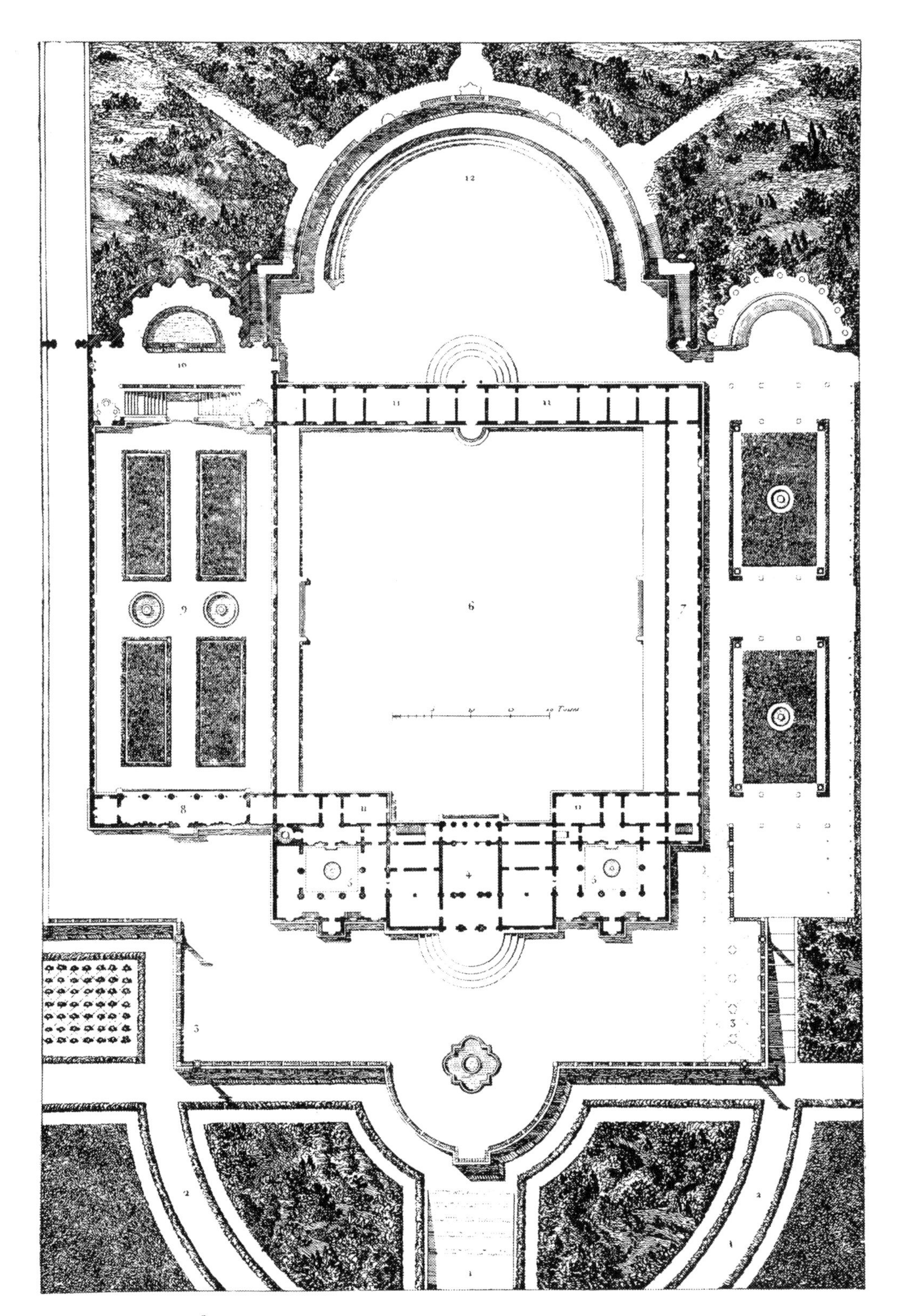

FIG. 256. VILLA MONDRAGONE, FRASCATI—GROUND-PLAN OF THE FLOWER-GARDENS

the garden by the dividing wall, which is adorned with mosaic and *graffiti,* the middle marked out with niches. The garden is divided into rectangular beds with box hedges and fountains, and shut in at the side by high walls with round niches and busts on the top, the lower part hidden by hedges. The north end is made by a grand loggia (8) which is certainly Vignola's work: nothing is left of it but the empty wreck of the garden. In the bare unplanted court pupils of the Jesuits in their ugly clothes now play about, utterly careless of the beauty that was there before them.

But great as was this house and garden, the love of building and the pride of the Borghesi could not be satisfied with Mondragone. Cardinal Scipio bought another place almost as large close by, and built a palace for his sister, no doubt simpler, but still

FIG. 257. VILLA MONDRAGONE, FRASCATI—FLOWER-GARDEN WITH THEATRE BEHIND

so richly furnished that Lalande in the eighteenth century thought it was the best of them all. Both these villas are poor in the matter of water. The Borghesi, strangely enough, had always confined themselves to making beautiful special fountains, and paid no attention to the general water scheme. Villa Borghese is in strong contrast to Aldobrandini and Mondragone, being laid out in a peculiar, personal sort of way, and possibly the lady whose home it was to be preferred it so. The place, with three wings, enclosed at the back a sunk court with a remarkable nest of grottoes and stairways. At the back was a semicircular bit, originally, as Rossi's drawing shows, a theatre like those in the other villas at Frascati. In the middle a pretty avenue leads up to the hill. Everything here is so blurred and perished that a general impression, with no picture to support it, cannot be obtained. On the other hand, the terraces on both sides of the wing in front are preserved. These are at the height of the first story as at Montalto, and lie above the porticoes like a hanging garden, the inner courts being enclosed by them.

There were always flower-beds here, as there are to-day, and also orange-trees. As

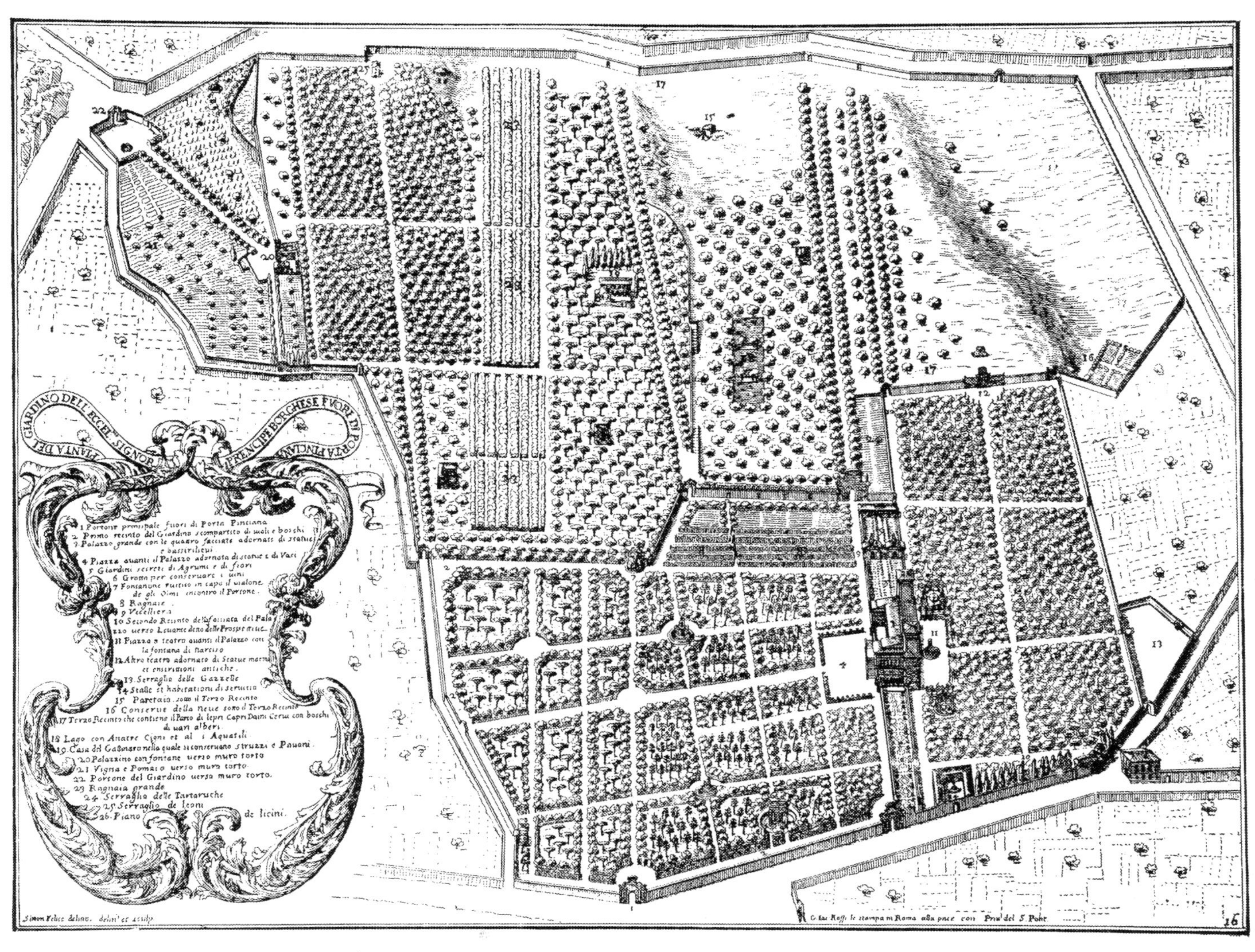

FIG. 258. VILLA BORGHESE, ROME—GENERAL PLAN IN THE SEVENTEENTH CENTURY

flower-garden there is a well-kept *giardino secreto* now extant in bordered beds at the side of the villa; an octagonal basin with hedges cut to the same height, and seats, and a pergola shading the main walk and now overrun with glycine (wistaria); all these make the picture attractive and homelike. At this villa also avenues traverse the whole of the farm domain, mostly planted with oaks, which have in parts kept their connecting hedges, and accompany the visitor from the entrance gates up to the palace. The two estates were united by the Borghesi with avenues and thickets contiguous to each other, so that both combined to form one great place, the greatest of its time. This was entirely to the taste of the family, who above all things desired to make an impression of greatness and power. Their activity in Frascati was only an offshoot of what they did in Rome.

Under the rule of Pope Paul V. Rome had really taken on that appearance which

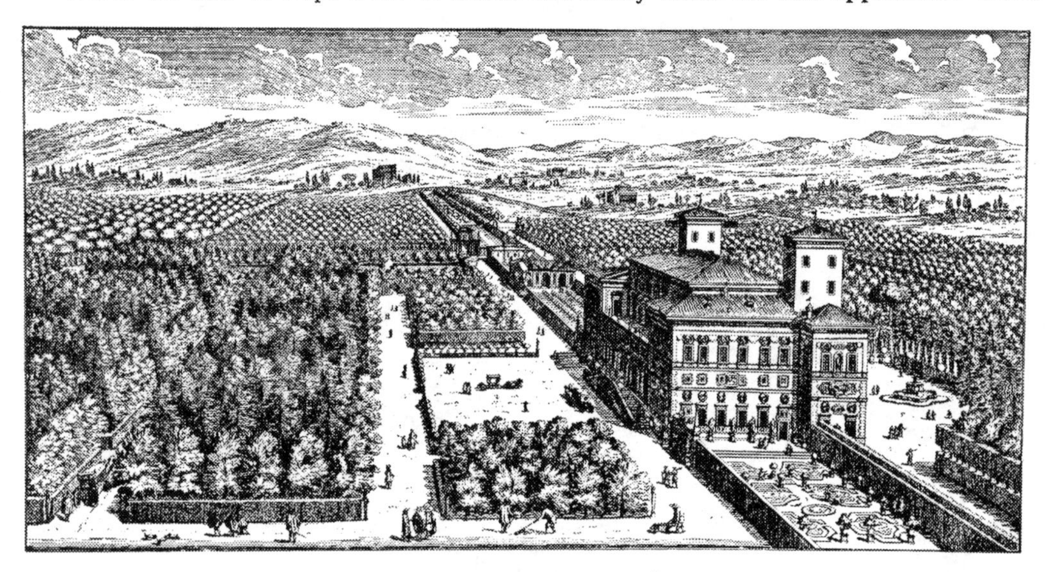

FIG. 259. VILLA BORGHESE, ROME—A BIRD'S-EYE VIEW

was maintained till 1870. It was by Paul V. that the last and greatest of the conduits was brought to Rome, and the Acqua Paola really is what it is called by a contemporary historian of the Pope, "no longer the play of pipes, for it breaks out in a stream." On the height of the Janiculum the four arms of water rise from the ground, five times as strong as Acqua Felice. Rome was henceforth the richest capital in the world in water. It was quite clear that the villa built by Paul V.'s nephew had to be the grandest of all. On the north Scipio bought what was for those days a very large estate, with a circumference of three miles (Fig. 258). Once again the family of Altemps were his predecessors in possessing the greater part of this villa, but of the first lay-out there are only a few entrance-doors to be seen, which are attributed to Martino Lungo. The casino at any rate, and with it the garden, was certainly made by Scipio. The villa remained till a short time ago in the possession of the family, for the Borghesi were among the very few in Rome who kept their property for centuries, and from time to time restored it. Naturally this was a good thing for the villas, and in the eighteenth century and the first half of the nineteenth, when most of the others were in a very sad state of neglect and decay, the villas of the Borghesi, especially this Pincian villa, roused the greatest enthusiasm among travellers. But this undeniable advantage meant the almost complete ruin of the original gardens of the beginning of the seventeenth century.

It is only by the help of old drawings and certain descriptions by travellers in the seventeenth century that we can recover anything of the past scene: the territory that Scipio acquired embraced only about three-fifths of the present circumference. The great drop to the south of the so-called *giardino del lago* towards the town wall was at that time only occupied by small houses and villas, and the part between the old gate and the

FIG. 260. VILLA BORGHESE, ROME—THE FOUNTAIN OF THE RIVER-HORSES

so-called *muro torto* (22), as far as the present entrance at Porta del Popolo, was the garden of Villa Giustiniana, at that time much praised for its beauty. To-day we need only say of it that a colossal bust of the Emperor Justinian stood there, since the family liked to remember he was their ancestor. These new parts were first acquired by Camillo Borghese at the beginning of the nineteenth century. Things went worse in the eighteenth century, when the old park of Marco Antonio Borghese became an English park with a pronounced classical tendency. At that time were made the lake garden, the hippodrome, and the temple; the smaller building was converted into a mediaeval castle, and the casino rebuilt. Only the two gardens before and behind the casino remained as they were, so far as ground-plan went. But neglect has more to answer for in this state of growing decay than have any deliberate changes.

The old garden as laid out by the cardinal was surrounded by two walls, so adorned

FIG. 261. VILLA BORGHESE, ROME—THE FRONT FAÇADE

with small houses, pyramids and turrets that they looked to people coming from a distance "like a little town complete in itself." The garden was in three separate parts, with walls round each and wrought-iron gates. The main entrance (1), "a rifle-shot from Porta Pinciana," led into the first garden in front of the casino (Fig. 259), but not right up to it, only to a cross-path that ended in a grotto in the rock with a fountain. It was only later that the beautiful river-horse fountain was put here (Fig. 260). The chief avenue leading to the casino crosses this entrance walk in the middle. We can see in this garden how the effect of a complete architectural whole has quite disappeared, and the casino is placed where it is in order that it shall not be seen till we reach the garden. This is quite level in front and divided into formal squares of plantation, which at an early date were certainly full of different kinds of trees. Also firs, cypresses, myrtles and laurels have been noticed with approval, making up the different bordered squares. On both sides of the main walk there were round places with simple but pretty fountains, each place hedged in and ornamented with statues. At the side there is a little circular temple above an artificial ice grotto, that serves as a wine cooler. This grotto still exists, but at that time it was a kind of dining-room in the open air, and it kept cool in the hottest summer weather. The casino was built by a Flemish architect, Vasancio, in a style that shows a marked likeness to the garden façade of Villa Medici. Probably this casino was not a dwelling-house at the beginning, but only meant for the reception of large parties.

The great square in front also corresponds with what we have at Villa Medici, designed as a fine ascent and surrounded by vases and statues. The façade with turrets (Fig. 261) has here also antique relief work and busts instead of any architectural decoration. On both sides of the building there are *giardini secreti* for oranges, rare flowers, vegetables and fruits. In the middle of the more northerly of these gardens there is a bird-house in two sections (9 in Fig. 258), like the one in the Farnese Gardens. Behind the casino is the second flower-garden, and there is a place corresponding to the one in front of the house, with a fountain in the middle (11 in Fig. 258) topped by a bronze figure of Narcissus (Fig. 262). Round this are sarcophagi, statues, and sphinxes. In front of the exits of the paths, which run through the square thickets into which the whole space is divided up, stand Hermes statues of colossal size carrying fruit. The northern wall shows the theatre decoration, with more antique inscriptions and statues in niches (12 in Fig. 258). From two windows there is a view over the open hunting-fields adjoining. This is wonderfully effective as a contrast to the highly decorated part, for even the semicircular place is not without seats all round, and carvings. An obelisk stood in the middle with an eagle on the top. The east wall was covered in the same manner.

All the rest of the estate was a great park with many cages for wild beasts—unusual foreign creatures—birds, fish in a large pond, bird-houses and hunting-grounds for rabbits, hares and deer. But everything, with the one exception of the hunting-park, which was altogether uncultivated, was planted in an entirely formal way, and traversed by wide straight avenues. The aviaries also (23 in Fig. 258) are formal covered walks with wooden stakes and nets thrown over. Each home appointed for a pair of animals has a special surround with its own entrance gates. At the *muro torto* on the site where the *giardino del lago* now is, there was another part planted with formal thickets of oak, by the side of which was a small casino, adapted to various ends, a garden-house or at times a little dwelling-house for the family (20 in Fig. 258). Here also was fine ornamental sculpture, and as there was a second entrance to the villa, no doubt entertainments were provided for visitors.

FIG. 262. VILLA BORGHESE, ROME—FOUNTAIN OF NARCISSUS

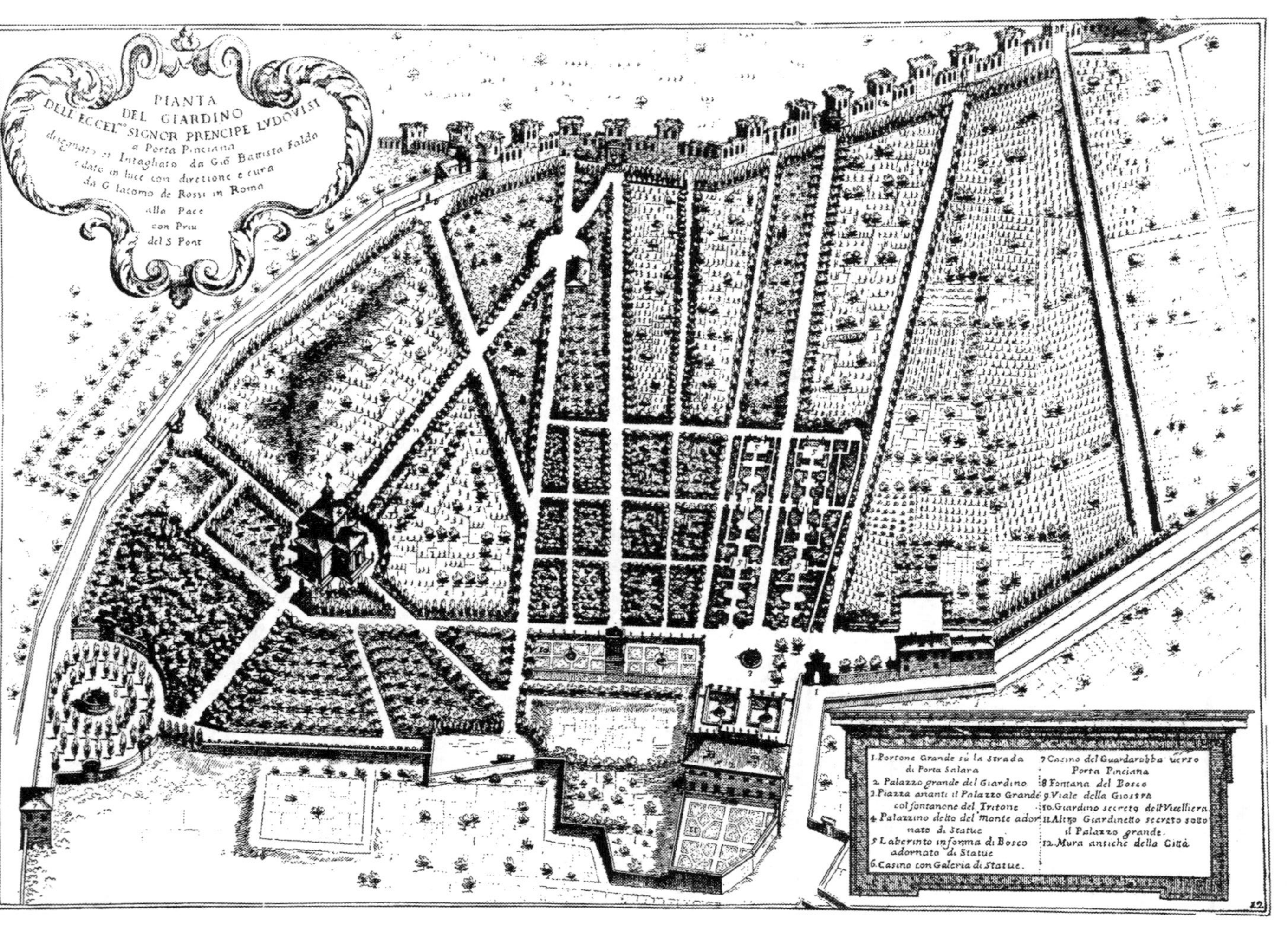

FIG. 263. VILLA LUDOVISI—GENERAL PLAN

One entire novelty fitting well with his feudal instincts was created by the cardinal: precisely in front of the town gates he laid out a great park that was exclusively devoted to the nobler forms of sport. This was a harking back to mediaeval times when princes were wont to make huge deer-parks just outside the gates; in the sixteenth century the custom had died out. But Scipio was clever enough to unite these feudal precincts with casino and flower-garden, set out with every refinement of luxury and the choicest works of art. The gardens had to accommodate themselves to the very severe character of the park. There was no more to be seen of the gaiety of a Renaissance garden; here

FIG. 264. A TREE IN THE OLD GARDEN AT VILLA LUDOVISI, ROME

were the massive effects of oak, laurel and cypress, densely set about the casino; for these were the suitable background, in their solemn colouring and ceremonious form, for the great festivals of that period. Even though the outlining of separate bits of the park had a practical intention, it must have had an archaic effect: the monumental entrance doors gave to the whole a fine architectural sense of structure. Mediaeval, too, must have been the aspect of the heavy double walls equipped with turrets and pavilions, which surrounded the whole of the precincts and gave them grandeur.

When the work was done, the cardinal opened the place freely to the Romans, setting up an inscription:

Whoever thou art, now be a free man, and fear not the fetters of the law. Go where thou wilt, pluck what thou wilt, depart when thou wilt. Here all is for the stranger more than for the one who owns. In this golden age that promises security to all men, the master of the house will have no iron laws. Then let him who evilly and of set purpose shall betray the golden rule of hospitality beware, lest the angry steward burn the tokens of his friendship.

This inscription is the conscious exhibition of a sentiment beloved by Scipio, *noblesse oblige*. It is not that the cardinal was doing anything wonderful in throwing open his garden—the inscription made by Brenzoni for the villa at San Vigilio shows the same spirit—but Scipio did this, like everything else, in the grand style. From the earliest days of the Renaissance it was customary in Italy for people who owned works of art and gardens to give the public free access to them. Montaigne, after enumerating the best gardens in Rome, breaks out into an exclamation of wonder: "And these are beauties that are open to anyone who likes to have them, who may sleep in them and even

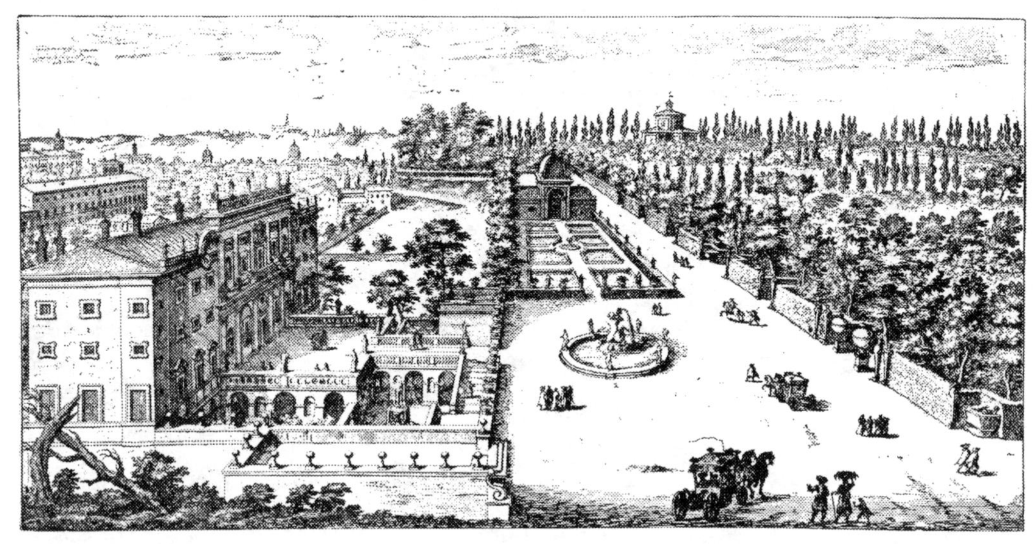

FIG. 265. VILLA LUDOVISI, ROME—A VIEW OF THE GARDEN

take his friends, if the masters are not actually there." Moreover, meetings of all sorts were held, and in clerical circles people combined pleasure with edification, and these gardens became a favourite theatre for sermons and theological disputations. But apparently learned doctors were not so liberally disposed as other noble owners, for Montaigne complains that famous men made people pay highly for admission to their debates. His description, an extract from Roman history long since past, ends with the words, "I have no use here for grief or melancholy." These words awake an echo in the hearts of those who fare to Rome in any age. Winckelmann expresses the same thought in words as happy: "I feel freer than I have ever been in my life, I am in a measure lord of my lord and his castles of delight, I take my way whither and to whom I will. In this place people understand better than we do the value of life and its secret: they love to enjoy and let others enjoy."

The seventeenth century shows no loss of creative power in this gardening art, but with time there is some dearth of new ideas; on the other hand there are traces, although very faint ones, of a foreign influence, which comes from France. The villa (Fig. 263) built by the Ludovisi, that nephew-family which turned the Borghesi out of the Papal

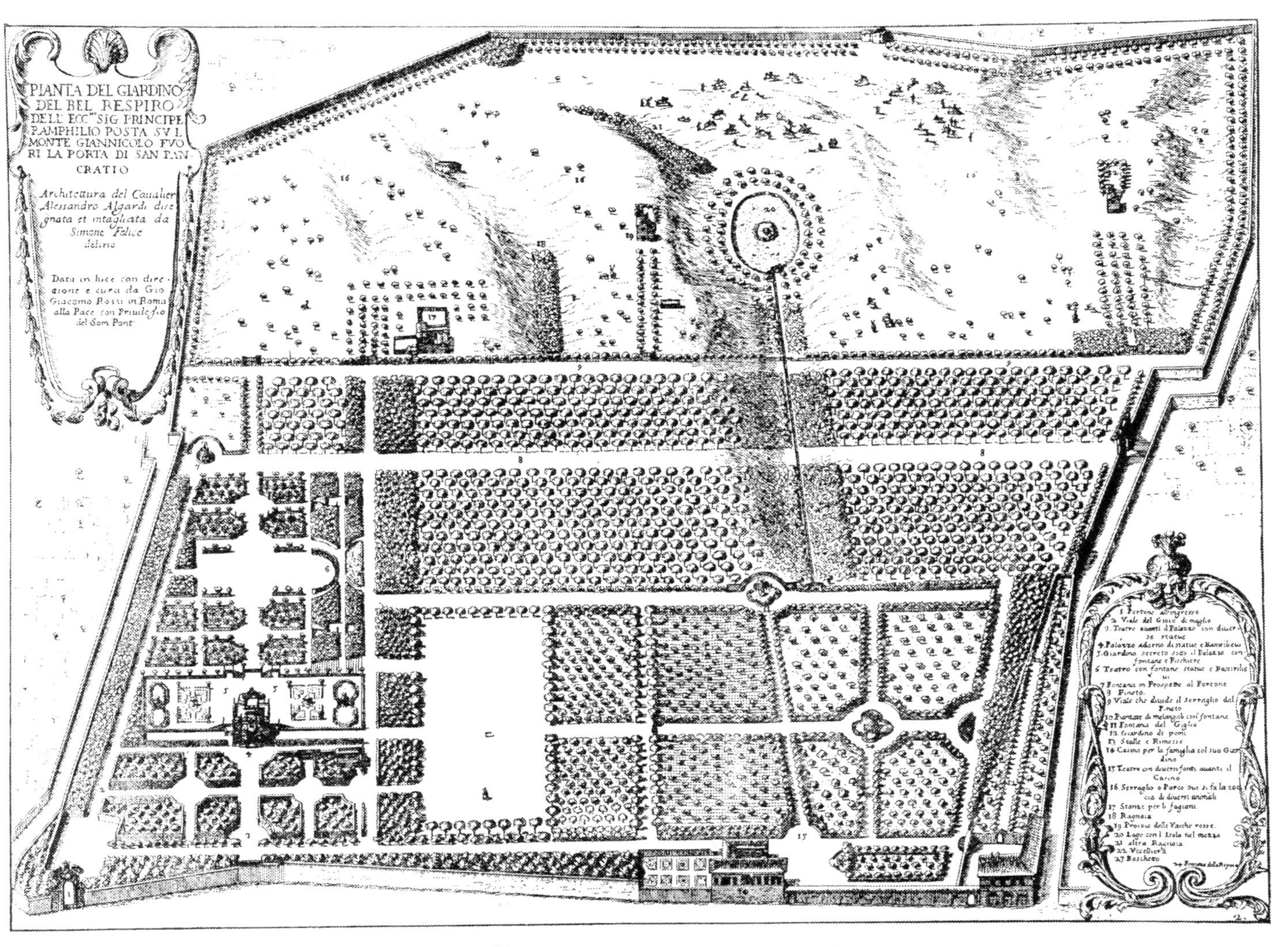

FIG. 266. VILLA PAMFILI, ROME—GENERAL PLAN

chair, has certainly not vanished from the memory of the living representatives of that race—indeed, many of them live under its shadow—but the place itself has completely disappeared under the sea of elegant houses of the modern capital. Except for small traces in the old garden (Fig. 264) all that remains is one casino, a little central building which, set on the highest point, was once surrounded by park-like ground belonging to the villa. It was in the middle of a piece with statues, to which broad paths led, making the shape of a star. At the points of this star there were particularly fine specimens of sculpture, an ancient tomb, a colossal statue of Alexander Severus, and others: the arrangement of these gives the desired limitation to the view of the beholder who is looking down from above.

The woodland (*selvaggio*) dominated by the little rotunda made one side of the villa, and the rest of it was round the rather small dwelling-house, which was to the

FIG. 267. VILLA PAMFILI, ROME—THE SUNK PARTERRE

south near the entrance at Porta Salaria. The garden front (Fig. 265) cuts through a sunk court with two fountains for ornament, by a terrace on the height of the first story. Below the terrace there were grottoes. Above these one arrived at the great open-air parade, demanded by Roman ceremonial at that date, and in the centre stood a round fountain of the Triton. At the side, but unattached to the house, was a *giardino secreto* divided as in Villa Borghese by a house for birds and dominated by it: the house looks at the back upon a wonderful parterre of flowers. On the other side of the parade is a *laberinto in forma di bosco ornato di statue,* which really means that spaces and paths have been cut in the thick of the greenery. The rest of the garden is laid out in dense clumps as at Villa Borghese. The main entrance, at the side of which stood a casino containing a picture-gallery, opened on a grand avenue with a statue at the end. The villa was cut off on the north by the picturesque town wall. All these familiar traits are handled with much skill in detail, but the picture as a whole needs the creative touch, which we could always find before: the parts here are put beside one another, but do not spring from one common centre.

Not much later than the Villa Ludovisi, before the middle of the eighteenth century, the Pamfili, nephews of Innocent X., built, in obvious rivalry with the Villa Borghese, the second large Roman villa in front of the town (Fig. 266). It lies on the Janiculum, in front of the Porta Pancrazio, and surpasses the Villa Borghese not only in circumference but also in the wealth of special sites. Unfortunately it met the fate of its predecessor, for by reason of the greater extent of its parks it suffered a fundamental change to a style that was not really natural to it, and so it came about that only a very few of its fountains and water arrangements reveal the features of the old design. For a Roman villa of that period it is unusually good in the treatment of the purely ornamental

FIG. 268. VILLA PAMFILI, ROME—THE SITE OF THE THEATRE

garden, which though altered in detail has kept in great measure the same form as of old. The casino and its own garden were at the north-east corner of the old villa, and everything on the east of it is territory that has been annexed later. The chief entrance gate leads to a long avenue, whence one gets a lovely view of the dome of St. Peter's and the casino garden farther to the north. On the south the garden façade of the house—at first covered with antique reliefs—opens on a parterre, which serves as a *giardino secreto*, being on two of its sides cut out of the hill (Fig. 267). In the dividing wall there are niches and statues and also espalier fruit-trees.

The whole picture is made fine and festive with fountains, flower-beds, and pots standing on the balustrades. The beds have no doubt changed very much since they were first planted, but they must have been the sort that showed arabesque patterns in box, filled in with flowers. (It will be shown later that this plan was an Italian invention, which at this period found its way into France.) From here one passes by steps to a flower-garden which is one terrace below, laid out with beds and tree clumps, with hedges round. In the crossways axis by the side of the park there is a very good and pretty theatre place (Fig. 268). The fountain belonging to this site is on the park terrace, and makes a good connecting-link with the various parts behind. The park was originally laid out in two sections divided by a railing with doors in it.

The northern section, nearest to the real flower-garden, has much of the character of a pleasure-ground. It is an immense open place by the side of the casino, and was probably used for games; at the back of it there are shrubs and an orangery, with central

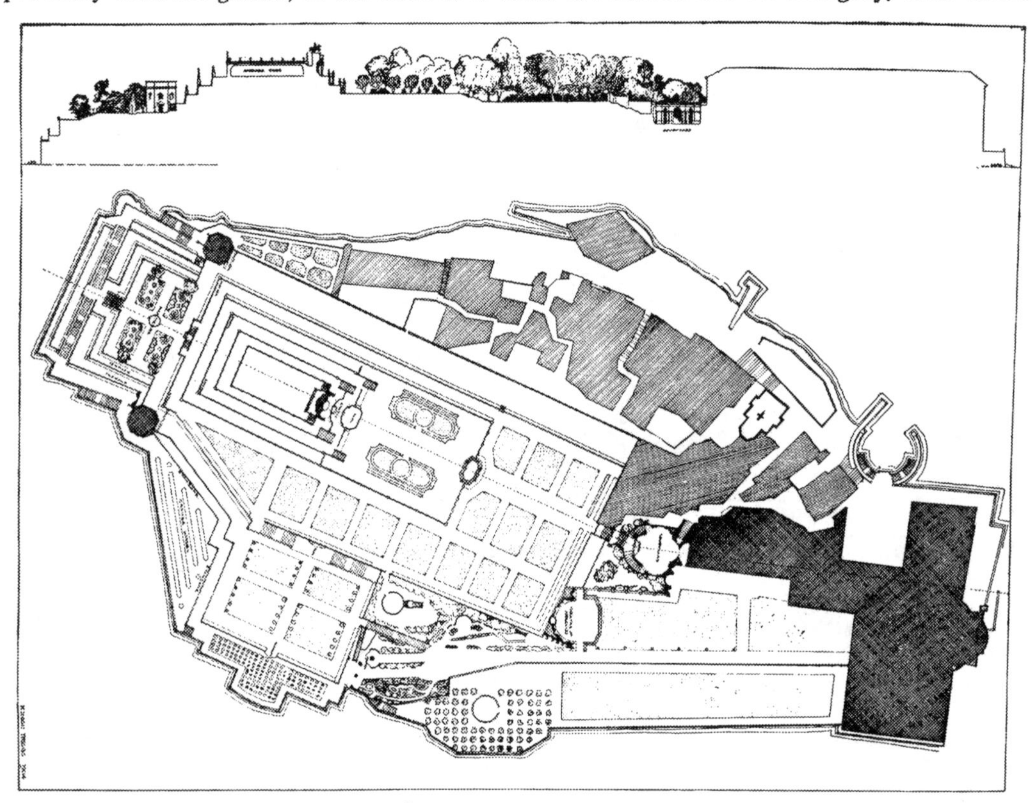

FIG. 269. ISOLA BELLA—GROUND-PLAN

fountain and statues. In this part there was also a special little *casino della famiglia*, as there was at the Villa Borghese, close to the wall of the terrace that has the fine view, and having a *giardino secreto* of its own and a small theatre. Of the very attractive gate-casino, or lodge, only part has been preserved: it lies entirely in the angle of this terrace, and shows some remains of statues and fountains. Stretching across the whole width of the park in front on the south was the much-admired pinetum, which must have created the impression of an enormous pillared hall with a green roof, so regularly were the trees planted. The second part (separated from this) was intended for a large animal-park with many kinds of enclosures and cages to serve as domestic quarters for the different beasts. But this hunting-ground also shows that it was intended to have a connection with the other part of the park just described: there is a large lake, oval in shape, with an island in the middle of it, surrounded by several rings of trees planted at regular intervals. From this pond a canal makes its way straight as a die through the pinetum in a dip of the valley; it has several bridges over it, and ends in a water theatre, the terrace above being topped by the fountain of lilies, which remains a beautiful thing to this day.

The place is unique, certainly for an Italian garden, and so one may perhaps suspect French influence here. We shall see later how the canal idea developed in the French garden, though with very different results, and it is not impossible that the French notion has slipped in without any deliberate planning. There is no doubt that certain characteristics, such as the star shape of the park at the Villa Ludovisi, gave occasion for the fanciful

FIG. 270. ISOLA BELLA

theory that the great French garden artist, Le Nôtre, himself designed both these gardens; but Le Nôtre did not come to Italy with a made reputation till 1678. He first stayed there some time in the forties, when he was entirely unknown and had no experience in garden matters. Here the cart has been put before the horse, for Le Nôtre came not as a teacher but as a learner, and in these villas he found the incitement and material that led him to his great activities.

The creative Italian spirit is powerful enough to master new tasks with its traditional skill, though a light-hearted way of ignoring difficulties added to the danger that stood in the way of all art everywhere. There was in some quarters an exaggeration of bad taste which even penetrated into Italy in the late seventeenth century. The Counts Borromeo possessed a summer villa on the Isola Madre in Lago Maggiore, and a few remains of the old garden in this heavenly island point to the sixteenth century. One

FIG. 271. ISOLA BELLA—THE TEN TERRACES OF THE GARDEN

FIG. 272. ISOLA BELLA—GROTTO AT THE END OF THE TERRACES, LAND SIDE

FIG. 273. ISOLA BELLA—THE SIDE TERRACES

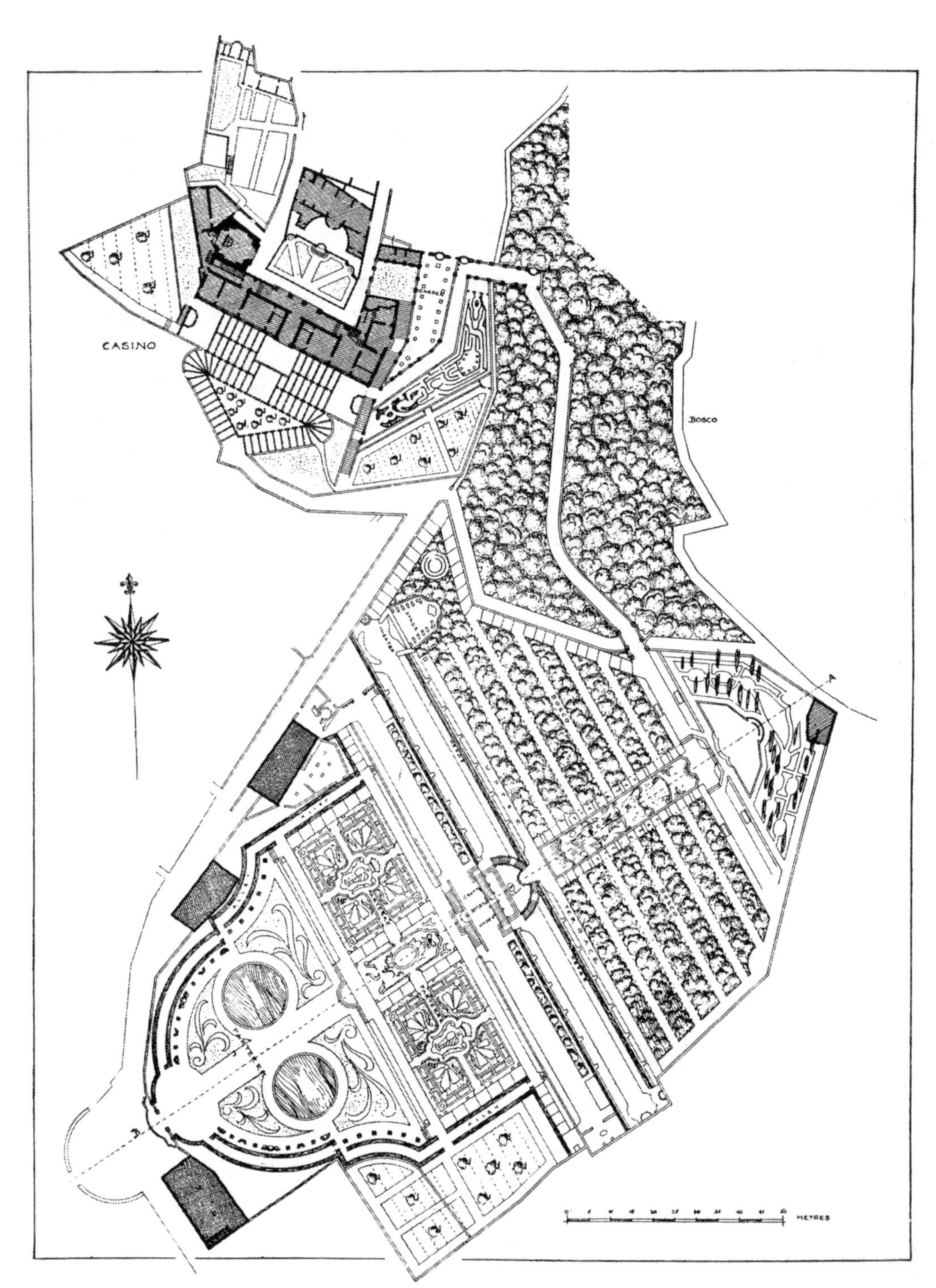

FIG. 274. VILLA COLLODI NEAR PESCIA, LUCCA—GENERAL PLAN

front of the house with its protruding side loggia, and certain ideas that recall Alessi, the wide basin in front, and terraces simply articulated (which are almost unrecognisable now), a fine dolphin fountain up against the wall, all these meagre traces give way before the complete change the island has suffered into the style called "picturesque" of a botanical garden. At the present day the tall tropical plants form a delightful contrast to the sister island, Isola Bella (Fig. 269). This was a mere flat rock of shining schist until Count Vitalione, partly led by caprice and partly attracted by almost overwhelming

FIG. 275. VILLA COLLODI NEAR PESCIA, LUCCA—THE TERRACE GARDEN

difficulties, was impelled to move his chief residence here from Isola Madre, to build a mighty castle with a garden behind, and in this way with a real Asiatic lordliness to defy Nature and erect upon a flat surface terraces and hills with high substructures (Fig. 270).

Neither the house nor the remarkable harbour in front was ever finished; but the garden still stands, scarcely altered, as a memorial of its creator's ambitious and defiant spirit. On the side next to the lake the garden is in ten terraces (Fig. 271). In the ornamentation of the five upper ones, mounting by steps that get smaller as they go up, it is possible that the architect had in mind some account of Babylonian hanging gardens. The narrow terraces were on three sides, the dividing walls were planted with lemons, and in front there were low clipped bushes. On the balustrades were a number of statues. Beside the house these five terraces are like a single wall of fantastic pattern, decked with grottoes and shell-work, and topped with a unicorn, the device of the Borromeo

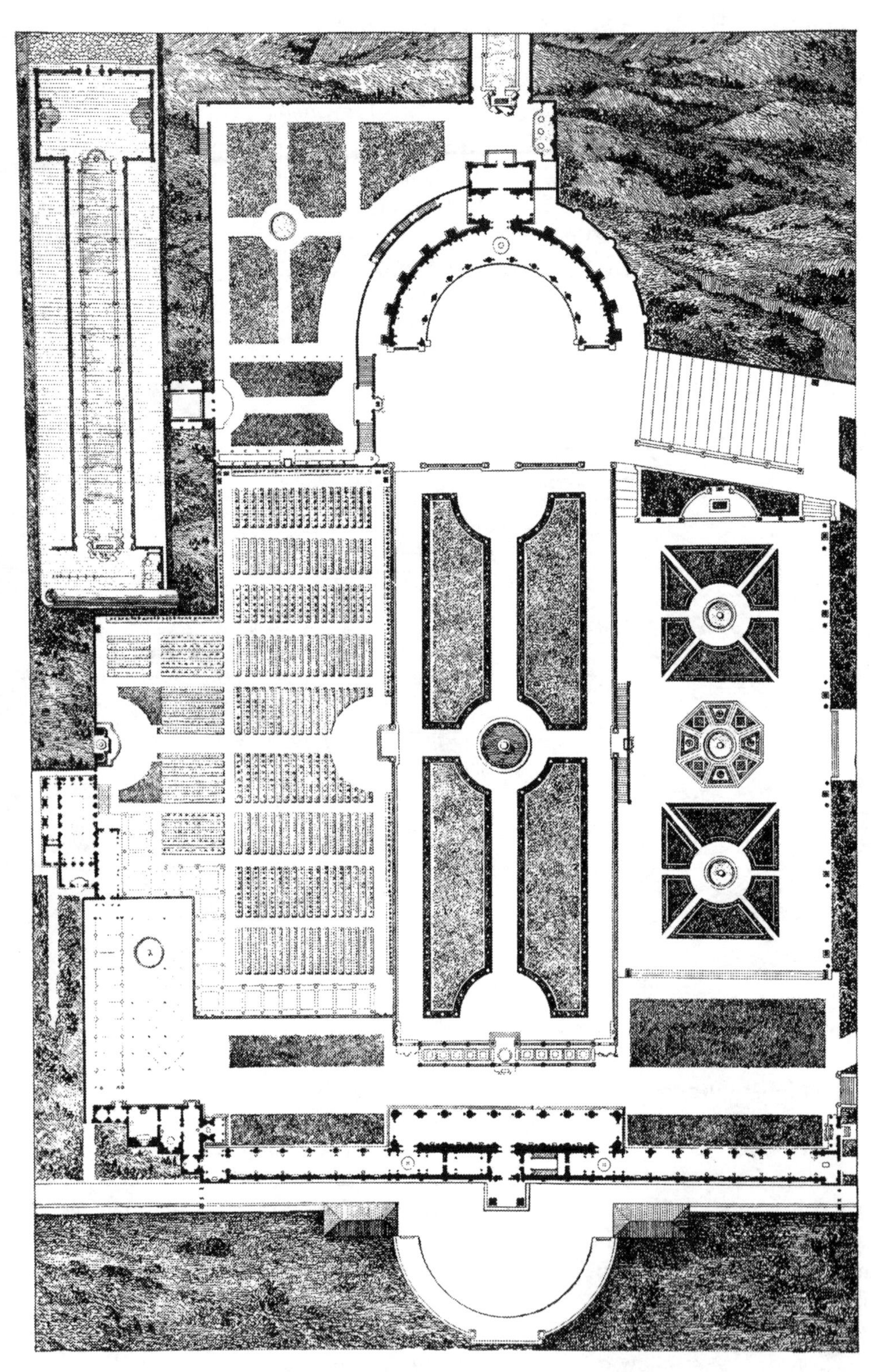

FIG. 276. VILLA ALBANI, ROME—GROUND-PLAN OF THE GARDEN BETWEEN THE CASINO AND THE COFFEE-HOUSE

family (Fig. 272). This wall forms the end of a planted court, whose entrance from the other side skilfully conceals the break in the axis of house and garden. On the side by the lake the steep frontage is flanked by two strong round pavilions of two stories each; a wide balcony projects below, and is laid out as a parterre.

The lower set of terraces, which have to accommodate themselves to the shape of the island, are clad in a stouter, more sombre vegetation, to support the structure, so to speak, by what looks like a basement (Fig. 273). These have, growing by the side, both boskets of laurel and orange gardens, and they also have natural grottoes. Many of them have peculiar borders to give a more personal and intimate character, as a contrast to the grand spectacle of the garden at the front. They lead up to the side of the house, which finds its chief attraction in the magnificent view from the two terraces. The upper one is closed in on one side by a semicircle treated with extreme delicacy, and a Hercules fountain, which serves as *point de vue* for the row of principal rooms. The garden is wanting in water, which plays a very poor part in the grottoes at the end of the terraces. Still, if any garden can afford to dispense with so enlivening a feature, it is certainly this one, for it is intimately connected with the lake all round it, and is helped by the contrast of colours and the unceasing change of the water's surface, shut in by towering hills. The charm of the island's quiet beauty has always made a special appeal to a Northern fancy. In Jean Paul's *Titan* it is revealed to the eyes of the hero, who is drunken with the joys of youth: from the opening scene onward, it takes every hue and tone that can be lent to it by sunrise, sunset, and the glistening moonbeams of a night in spring.

FIG. 277. VILLA ALBANI, ROME—A GARDEN WALK IN THE PARK

In these late Baroque gardens of Italy we see the final subjugation of that spirit of seclusion and private enjoyment which prevailed in the mediaeval garden. It now resigns —at any rate in its ornamental parts—that feeling of private life which it had before, and has become like a picture hung up outside and inviting inspection. This feeling is expressed in another garden just as at Isola Bella—in Villa Collodi, which has kept its form scarcely

FIG. 278. VILLA ALBANI, ROME—VIEW OF THE COFFEE-HOUSE FROM THE CASINO TERRACE

altered since its foundation in the second half of the seventeenth century. Anyone, as he first approached the Villa Collodi near Lucca by its great entrance gate, and at one glance beheld the curious picture of this garden, full of rich colour, must have exclaimed in astonishment. Before him lies a parterre of level ground with a large circular basin in the middle (Fig. 275); high hedges disguised as battlemented walls with pots and vases shut in the parterre at the side, and round it there are gay flower-beds in pretty arabesques, and box variously clipped into many different shapes. On a second parterre gradually ascending

FIG. 279. VILLA ALBANI, ROME—THE PRINCIPAL PARTERRE SEEN FROM THE SIDE

he sees the name and arms of the owner set out in box and various coloured stones. The effect is made perfect by hedges, pediments, and flower vases, and in addition to the figures made of box there are numerous white statues in baroque surroundings (Fig. 274).

The garden is above in five narrow, steep terraces. Just as in the garden at Este, so here, the middle axis is marked out by niches and grottoes. The dividing walls and formal stairs are bordered with balustrades, and show an exuberance of figure decoration. As climax to the scene there is a foaming, glittering water stairway, with two female forms, Lucca and Florence, standing above it, while below two swans spout water into a basin. Above the steps there is a large cistern, and at its upper end a colossal hovering figure of Fame, who appears to be flying hastily out of a dense wood, while from her resounding horn she pours streams of water into the basin below. No tall tree is left uncut. The narrow terraces are uniformly enclosed by hedges, and adorned with niches and statues. At the end of the third terrace there is a theatre, with roof and side-scenes cut out of the greenery, and statues in the walls. Close to the water stairway (which is also enclosed with a border) there is on both sides a thick plantation of oak, overtopped by the lordly cypresses on the upper terrace. Behind the figure of Fame, which stands above a grotto, are the baths, retired and almost hidden: though at one time they were fitted out with every kind of refined luxury, music-rooms and the like, they are now the only part of the whole garden that has completely fallen into decay.

FIG. 280. VILLA ALBANI, ROME—WALL OF CYPRESSES, WITH STATUES

One may easily believe that in a garden such as this there would be no lack of merry water-pranks. Visitors who entered the labyrinth lying in a concealed flower-garden between the house and the garden proper had to be wary, for a person wandering there might get suddenly wet through from spurts of water springing from a hidden side-path. The house itself lies on the top terrace at the side of the garden, with its whole width fronting towards the town, but it is very narrow. Its founder, Garzoni, who died in 1663, seems to have completed the greater part of the inner decoration (which is remarkably well preserved) and also of the garden. This is indeed a picture full of pride and beauty, which makes us able to interpret many a garden now run wild or entirely perished, that flourished at the classical period of Italian taste. For here we find Nature under complete control, and grown to be a wonderful architectural creation, with all the world of plant life subdued to ornament. And though in particular details we may detect bad art and poor taste, the Villa Collodi taken as a whole finely represents, like the Isola Bella, the splendid garden-craft of the North of Italy in the second half of the seventeenth century.

Roman gardens, however, were simpler and showed more distinction, and their creative power was destined to survive not only the days when French taste ruled the world, but those when there came about in England an actual revolution overthrowing all the most fundamental canons of architectural style in the garden. The Villa Albani (Fig. 276), though built as late as the middle of the eighteenth century, has such a thoroughly Roman character that in spite of its late origin it may fitly be taken as the

FIG. 281. VILLA ALBANI, ROME—A COVERED WALK OR PERGOLA

last link in our Italian history. For Germans the very name of it has an intimate sound, because it was Winckelmann who advised Cardinal Albani, the founder of the villa. Though Winckelmann may have had little to do with the form of the garden, yet the classical feeling lived in him and because of him had its effect in all the many parts of this whole. The cardinal followed Roman tradition in that he considered house and garden first and foremost as frames for his antique art-treasures. So Villa Albani stands in the series of villas, Medici, Mattei, and Casino Borghese, as the last of all to flower.

The garden has the double axis form, suiting the two entrance gates. Of these the *Entrata nobile* on the Via Salaria leads first to the thicket at the side of the casino, which perhaps has felt the influence of the star formation of Villa Ludovisi and other predecessors, and the points of the star are marked by statues (Fig. 277). The main avenue leads first, from a granite pillar with the family arms above, to a pair of steps that mount to a higher terrace. This terrace is simply laid out with stretches of grass and fountains, whence one proceeds on a stairway to the chief parterre lower down, and this in turn is shut off by a hedge on the far side from an orangery, where the view comes to an end with a beautiful fountain. In former days one saw a grand prospect over the Campagna to the blue boundary of the Sabine hills, but now the roving glance sees merely what it would avoid—a mass of lodging-houses close at hand. The second main axis goes through the casino, which with spreading pillared halls stands high above the great ornamental parterre (Fig. 278). One goes down by steps, finding the beds, with arabesque designs in box, arranged in the style of the day, connected together by the eagle fountains in the central axis, and round them oranges in coloured pots, just as they were to be seen in the earliest gardens (Fig. 279). A semicircular portico, reminiscent of the theatre at Frascati, stands at the end, with a couple of rooms in it, and this corresponds with the pure classical style of the house in simplicity and distinction. The third and lowest of the terraces lies behind the portico called the coffee-house.

In this part, which leads along a canal to a second entrance gate, we clearly see the suggestion of the rococo style of garden that is coming hither from the North. Later on this place has to be considered again, as it falls out of line with the ornamental garden proper. The cardinal ornamented it with the most exquisite works of art, which had for the most part been collected for him by Winckelmann. Even to this day we find here and there a Hermes or a bust set up on its tall plinth; and there are statues also standing among the green hedges, which are still clipped even now (Fig. 280). The most conspicuous of the antiques were kept in the palace itself, in the coffee-house, or in the so-called *bigliardo*—really a small casino near the side of the house, built upon a terrace on the same level behind a group of oaks, where at the present time a bust of Winckelmann is set up. Like the pergola with its water basin (Fig. 281) all these places are of a dignified plainness and simplicity.

The villa is a work of art which has remained faithful to the ideas that belong to its own nationality, and has preserved the Roman character at a period when the new French spirit had already wrought destruction on many noble works—and that even on Italian soil. It seems as though the art of the past, which took into its service two such men as Cardinal Albani and Winckelmann, has been able to encase itself in some protective shell, garment, or frame, whereby it rests immune from every infection of unworthy trifling or baroque excesses, even in the most insignificant details.

CHAPTER VIII

SPAIN AND PORTUGAL IN THE TIME OF THE RENAISSANCE

CHAPTER VIII

SPAIN AND PORTUGAL IN THE TIME OF THE RENAISSANCE

"WHAT of Spanish style? Where, in what century, is there a school that can be called national?" These questions occur in Carl Justi's *Introduction to the History of Spanish Art,* and he thus answers them: "One might put forward certain prelates, grandees, magistrates, guilds, as the friends of art in the past—of art that proved their taste to be uncritical and unprogressive—but they showed their enthusiasm rather as spectators than as artists, just as the Arabs were wont to take pleasure in the dance."

These words are more true of the world of gardening than of any other. During the whole period when Spain was under the yoke of the Arab, she was so controlled by the manners and customs of his nation that the Oriental way of life was accepted everywhere as natural and inevitable: therefore gardens were open living-rooms, and in arrangement and ornament were extremely like rooms or halls of houses, which contained fountains and flowers as decoration.

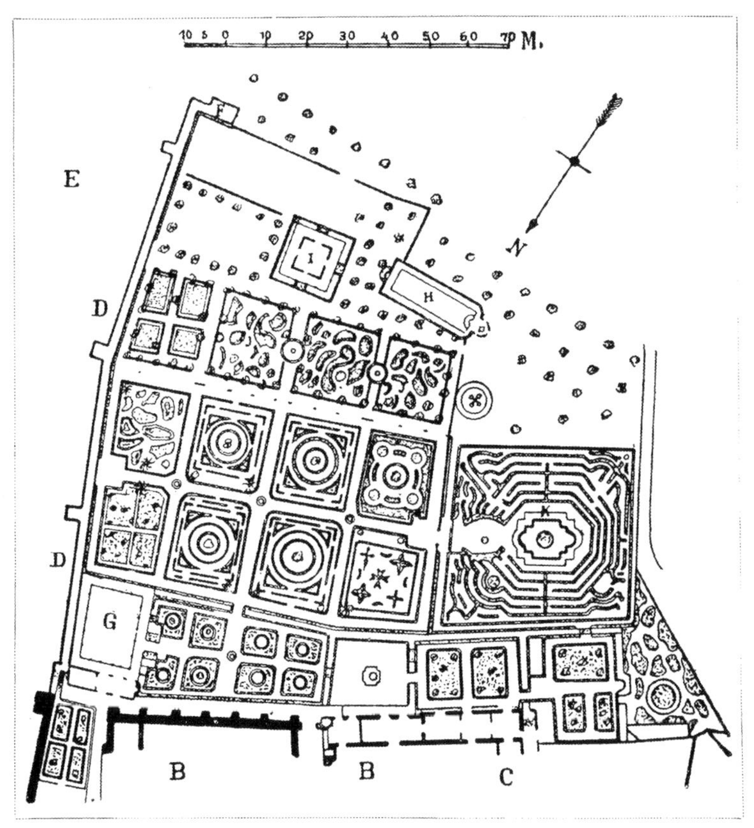

FIG. 282. THE ALCAZAR, SEVILLE—GROUND-PLAN OF THE GARDENS

And Moorish customs had struck such deep roots that for a long time after the struggle for power in Spain had ended in a victory for Christendom, the castles and pleasure-houses of Catholic kings and their nobles were with very few exceptions much the same as the Moorish ones. It is true that historians endeavour to distinguish separately a style called Mudejar, made use of by Christianised Spain until the sixteenth century; but what we really find is that Arabian architects have discovered a way to combine their usual style of ornament with something from the Italian Renaissance, just as they had previously done in the case of Gothic structures. For the ground-plan of the house is Oriental-Arabian as it was before. The living-rooms are grouped around a central court to which a few side courts are attached by great towers, and on one side—occasionally on two—there is a garden adjoining. It is as much as possible shut in from the outside by portions of other buildings, by galleries, and by high walls—and this moreover at a time when in Italy the outdoor country house has succeeded in being entirely in the open.

FIG. 283. THE ALCAZAR, SEVILLE—PATIO DE LAS DONCELLAS

The Alcazar in Seville is an excellent example (Fig. 282). When Peter the Cruel (1353–64) built his royal palace here, it was over the ruins of the old sultan's castle, which covered a much greater area, and with its garden reached as far as the Guadalquivir, though now only one of the old fortresses remains, the so-called Golden Tower. The king employed Moorish workmen in rebuilding, and very probably he was able to utilise some of the Arabic remains, in any case as models. During the next five hundred years there was constant building and restoration going on about this old kernel in the centre; but never did the Moorish feeling for complete seclusion, and dislike of the open, disappear. The beautiful chief façade was regarded as a kind of façade to a court; and this led to smaller spaces being grouped around the central one, which was called the Patio de las Doncellas (Fig. 283). This is a paved court; but there would be pots with plants in it just as there were in the courts of certain private houses, which are to be seen at the present day. There is no fountain in the Patio, but it may have been lost when Charles V.

FIG. 284. THE ALCAZAR, SEVILLE—VIEW OVER THE GARDENS

FIG. 285. THE ALCAZAR, SEVILLE—ANOTHER VIEW OVER THE GARDENS

rebuilt the place. Other courts, such as the Jardins de la Maria Padilla (the beloved lady of Peter the Cruel) and the Patio de las Banderas, whose surroundings are certainly modern, were perhaps always planted with oranges and palms as they are now.

The flower-garden proper lies south-east of the buildings (Figs. 284 and 285). Don Pedro (Peter the Cruel) first made it, but its present style dates from Charles V. in the sixteenth century. Still the same thing is true of the garden as of the buildings, that in spite of changes and improvements made by later generations, the fundamental

FIG. 286. THE ALCAZAR, SEVILLE—THE GALLERY OF PEDRO I.

idea has never been obliterated which shows it to be a near relation of the gardens of the Alhambra and the Generalife. For these also complete seclusion was a *sine qua non*, and hence the high walls, which in Pedro's time had galleries all round (Fig. 286). Now they are adorned with grotto-work in the baroque style, beneath which is concealed old work of many kinds. The entrance was at the side, without any architectural connection with the house (A, Fig. 282), and into a sunk garden. From a large pond at the side (G) there was a way down to the baths of Maria Padilla, before which lay a small parterre, the great garden sinking to the west in several terraces. When Charles V. made a second entrance at the back, at the end of the middle walk, it was really a pavilion in Mudejar style (I), ornamented inside with *Azulejos*, which means blue tiles (Fig. 287). At the end of the eastern gallery is a garden-house (F, Fig. 282) leading to the fruit-garden (E). Beside the pavilion of Charles V. was another large reservoir (H) which was named after his mother Joanna, called the Crazy.

At the intersections of the tiled walks stand fountains with round seats about them, and in the same way as at the Generalife little spurts of water spring up from unseen pipes to sprinkle the unwary. The separate beds are laid out in geometrical figures and always bordered. A novelty which was probably unfamiliar to the Arabs, the labyrinth (K), was introduced by Charles. It was a very large one, and put on the lowest terrace so that it could be seen from above. In the second half of the sixteenth century the taste for the baroque had introduced all sorts of tiresome novelties, and dripping grottoes, and garden gates, but also had added many beautiful fountains, such as the Mercury fountain in the Jardin del Estanque. But each of these things was really a foreign element in the picture.

FIG. 287. THE ALCAZAR, SEVILLE—THE PAVILION OF CARLOS V.

Charles V., who in this garden abandoned himself entirely to Moorish influence, in another quarter yielded his Spanish territory as entirely to the influence of the Italian Renaissance. The first attempt at this art on Spanish soil was the palace that the emperor built on the Alhambra hill; but it is in ruins, and we cannot glean anything to speak of from such garden sites as we now see. Charles had spent the first months of his happy married life at the Alhambra, and always hoped to see this fine place again. We know the condition of the town in those days; and of the lovely country houses with their cool gardens and rushing waters; for Navigero, who was the follower and friend of the prince, has given a very charming description of them. In the next year Charles began the erection of a new palace which was altogether Italian in type. The villas of the period, especially the Villa Madama, must have directly influenced the ground-plan of this palace: the central building was in a square with a great circular pillared court in the middle, and there is no doubt that such plans played an important part in the minds and memories of theorists and artists alike. They would think of Leon Battista Alberti's counsels, Raphael's ground-plans, Serlio's villa designs. Whether the immense court of the imperial palace was left quite empty and only used for tilting and shows, or was partly planted and adorned with flowers, must remain an open question. It calls to mind the central court for games at Poggio Reale, which a Spanish prince built on Italian soil. Charles never succeeded in holding festivals here, and he never saw the building, which remained a ruin. Evidently, however, some work was done on the gardens, which apparently were planned to ascend in terraces.

In the early forties there seem to have been preparations made for a visit from the emperor. At that time Luis de Mendoza had a sort of fountain arrangement set up at the ascent to the castle (Fig. 288), and it has been preserved as a memorial of these parts of the place. Close under the Moorish Torre de Justicia stands an elaborate wall, divided into sections by six Doric pillars, having between the second and fifth a highly decorated fountain attached to the wall, and a small basin in front into which water is spurted through the masks of three river gods. In the reign of Charles this approach to the Alhambra was the only one: we know its beautiful decoration, but not the places it went through. It was Philip II. who made the convenient carriage road which was planted, no doubt by his orders, with dwarf elms, for he was the first to introduce the tree into Spain and liked to have it widely spread. Laborde in the nineteenth century speaks in praise of

FIG. 288. THE ALHAMBRA, GRANADA—FOUNTAIN OF CARLOS V.

this approach, planted with elm-trees. Philip also planted the fine park of the Alhambra with elms, though the later plantation with the trees we see to-day dates only from the time of Wellington. The pretty garden sites on the terrace at Torre de la Vela perhaps belong to the earlier time (Fig. 289).

In every one of his kingdoms Charles V. was a great lover and patron of the garden. With all men of his generation he shared an intense interest in botany, and when he retired, sick and weary of rule, into the monastery of San Jeronimo at San Yuste, he devoted the time he had left in this world either to mechanical amusements or to gardening. The weary monarch had built for his own use a modest though not undignified dwelling on the southern slopes of the hill below the cloister. His bedroom was touching the church, so that from his bed he could participate in the Mass through a glass door. But his work-room looked towards the south, and was completely surrounded by the unfinished garden which he had made himself, while over it were the lovely hills and slopes of Verga de Plasencia, covered with chestnut-trees, mulberries, nuts, and almonds.

Close to these rooms, which were reached by a corridor dividing the north from the south, there was a large garden terrace: half of this made a sort of hanging garden over the ground floor of the house which abutted on to the side of the hill, and Charles converted the terrace into garden ground, and planted it with oranges, citrons, and sweet-scented herbs. He had obtained from the most distinguished botanists of the time the plants that grew in every part of his mighty dominions, and these he had had cultivated in all the

FIG. 289. THE ALHAMBRA, GRANADA—JARDIN DE LOS ARDIVES

main centres: Vienna, the Netherlands, and Spain. In the middle of the terrace there was a fine vivarium, and trout were caught and kept in the mountain brooks, and used for the royal table on fast-days. At the other end of the corridor on the west there was a similar terrace, and hence one reached the cloister courts on the same level. A gently sloping path led from the east terrace to the lower garden. It was said that the emperor, being so ill, did not want to go up any more steps, but the disinclination for terrace steps in the early Renaissance gardens may have had something to do with it.

The lower garden had been the orchard of the monastery, which the monks ceded to Charles, making a new one for themselves at the north-east of the cloister. The emperor had a wall to protect him from curious eyes and undesirable visitors, and a door that let him through to the monks' new garden whenever he pleased. The lower orchard showed exuberant vegetation, with ornamental trees and fruit, and the trunks of citrons and oranges which reached right up to the windows of the emperor's room, and delighted him with their colour and perfume. In his last illness he enjoyed the pure pleasures of this world, and on good days he would walk through the gate of the high-walled cloister garden, and step out on a level path into a park full of ancient oaks and chestnuts, beneath which, on the green lawn, pastured the cattle, kept for milk in the royal household. There were little oratories along the path, and a hundred paces off was the hermitage of Belem, where the emperor rested awhile the day before he entered the cloister.

The emperor's life as a monk only lasted two years. When towards the end of 1558 his son Philip received his father's will, he found a request that he would set up a worthy tomb. A year previously, after the victory of St. Quentin, when one of his generals had destroyed a monastery of St. Laurence, the king had vowed to build a handsome new monastery to that saint. These two ideas were now combined with a third one, which

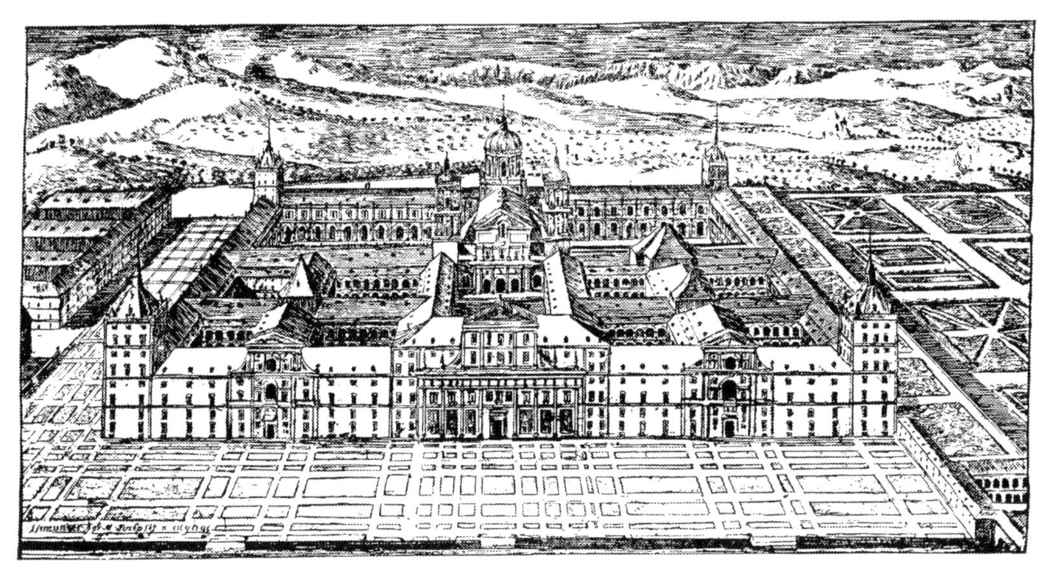

FIG. 290. THE ESCORIAL—GENERAL VIEW

had long ago taken root in Philip's heart, and now through the noble example of his father had gained new importance: it was to weld together a monastery and church built for a royal tomb with a princely palace, and so to create a place of hitherto unheard-of magnificence. Thus arose the plan of the Escorial (Fig. 290).

The final act of reparation made by Charles impressed the whole world with a feeling of reverence and surprise; for it was thought to be the most noble step a man could take. It looks as though the son had wished, by building a royal cloister residence, to produce the impression that he also was minded to combine the monkish life with the royal power, and unite the two in his own person. The actual idea of connecting a cloister with a royal dwelling was not foreign to the Spanish court, for long ago the kings had possessed a retreat at the monastery of San Jeronimo that lay above the prado before the gate of Alcalá at Madrid. The royal rooms were close to the sanctuary. The combination of religious exercises with the pleasures of a festival was congenial to the Spanish character. "The chapel is the first room that Spanish kings set up in their houses." When Philip II. ordered the restoration of Aranjuez, he began with the chapel, and the old Aranjuez was originally a summer-house for the Brothers of Santiago. "In this place ceremonies, audiences, and other formal occasions alternate with devout exercises, just like the times of sleep and waking, and the one seems to summon the other." Such is the description of the court of Philip V. by an Italian, and it holds good in a different way, but more intensely, of his grandfather's.

This state of things was quite foreign to the Italian temperament, though a certain pride in having a hermit at one's country place, with a view to getting some peaceful pious support when a stay was made there and therefore more happiness, was certainly to be found here and there even in Italy in Early Renaissance days, and a landowner liked to build such an asylum in his park. We shall find this disposition more marked in France; but there was at a very early date a kind of playful theatrical feeling about it, which extended over the whole of Northern Europe, till in the end to have a hermitage in the park was a sort of unwritten law. But on one occasion there was a great desire genuinely felt by Spaniards of the fifteenth and sixteenth centuries to unite the refreshment of the body with the most profound devotion of the soul. The gigantic expression of this desire is seen in the Escorial.

The first architect who made the plans for the royal cloister, Juan Bautista de Toledo, died soon after the foundation stone was laid in 1563. His successor, who really built it, was Juan de Herrera. It must have been more than chance that both he and the gardener Marcos de Cordona, whom Philip then called to begin on the garden as soon as the foundation walls were up, were men who had stood faithfully by his father's side to the last. Juan de Herrera, after his apprenticeship to architecture, had become an officer and gone to Yuste in the emperor's bodyguard, but after his master's death he went for his education to Toledo. When his teacher there unexpectedly died he succeeded to the place and was worthy of it. The grouping of the parts of the building had to satisfy the very far-reaching demands of the king. The whole forms an immense square, and through the chief entrance in the middle of the west front we come to the court which faces the entrance to the church, having on the left the college and on the right the cloister. The chief chapel with the high altar stands above the east front, and grouped round it are the private dwellings of the king's family. The king's own rooms, in contrast to the other buildings, are very simply arranged, for his wish was to stay here as monk and not as king. In any case the whitewashed walls which strike visitors to-day can tell us nothing, for Charles himself had tapestries from Flanders to hang on his walls, and among the private properties that he ordered Philip by his will to have sold, these possessions were valued at seven million gold thalers. The wall-coverings of his rooms in the Escorial were included.

With the thought of his father always in his mind, Philip had his bedroom so close to the church that he could take part in the Mass from his own bed, and there he died with a crucifix in his hands. Like his father, he took care to have his private dwelling isolated from both cloister and town; and because it projected over the east front he had a garden round it on three sides. A large wide garden terrace lies in the front of the whole east façade, and extends, equally wide, along the south of the house. It was a large parterre laid out in box and flower-beds, compared by visitors of old to hanging gardens: the best feature was the lovely view of the mountains. The walls beside this terrace seem to have had at their base originally only kitchen gardens; at any rate we hear of nothing else in older accounts. The lower park, with the Casa de Abajo, a pretty pleasure-house, is not earlier than the eighteenth century.

Originally the gardens were full of statues and fountains, which also served to ornament the numerous inner courts. There were no fewer than seventy-six fountains and seventy-three statues which the king set up here; for he was the first whose love of art induced him to import the precious treasures of the Italian Renaissance to Spain. Among the smaller courts there is the very largest court, not only of the Escorial, but of the

FIG. 290*a*. ESCORIAL—PATIO DE LOS EVANGELISTAS

whole world, the Patio de los Evangelistas (Fig. 290*a*). In the middle there is a little octagonal temple with four square basins at the corners, above which stand statues of the four evangelists, their attendant symbols supplying the basins with water. Round about is a bordered parterre now laid out with box in arabesque patterns of the eighteenth century, but originally a brighter picture of geometrical beds and flowers. In this sacred spot, which answered to the innermost soul of the king, he was fond of lingering, and ever more so as he grew older and more gloomy.

It was he, however, who quite at the beginning of his reign began to rebuild the mediaeval fortress of the Alcazar. He had the great square on the south front laid open and an imposing façade built in the Italian style, but left the palace on the river side, where it stands on the steep hill, to preserve its mediaeval appearance. In front of the west side, at the foot of the ascent, Philip found an extensive park, and did not want to spoil it for the chase. It has been said that Charles V. had rooks brought over from the Netherlands and that they nested in the tops of these trees. Little is known about the gardens that Philip laid out at the north-east of his palace, the King's Garden, the Queen's Garden, and the Garden of the Prioress. All perished, whatever gardens there were, in the great fire of 1747, and for long decades everything lay waste. At the north side there must have been a space large enough for an amphitheatre where bull-fights could take place.

There is more information about one favourite spot that the king made. On the south side, near the Torre d'Oro and only approachable from that direction, he made a little garden for statues. There was a terrace with fountains laid out as a parterre, and all round Philip set up half-length figures of Roman emperors, which had been sent to him as a present in 1561 from Cardinal Ricci da Montepulciano, who had then begun to build Villa Medici in its later form. These and a copy of the Boy with the Thorn were brought from the cardinal by the artists themselves, who may also have helped the king to set them up. Philip's wish was to make this garden of statues an imitation of Italian ones, but he found he had none too many pieces even when the whole series of emperors arrived from the Pope, and put them all up in the same garden. Later on a bronze bust of himself was added, also one of his stepbrother, Don Juan of Austria. In the garden-rooms which looked out on the terrace of statues the king hung his favourite pictures, which Titian painted for him.

In this quiet remote corner of the great sombre palace he spent hours of untroubled peace and enjoyment. For this gloomy ascetic, whose stern figure has appeared in the world's history painted in the darkest colours, had a tender human side. This part of his character is not only shown in the quiet, humble joys of life: in his children, in nature, in his garden, or towards his faithful servants. It comes out very clearly in some letters that have lately been discovered, written by him during the occupation of Portugal in the years 1581 to 1583, to his two daughters, girls of fourteen and fifteen. Nothing could be more surprising, because so unexpected, than the tenderness and fatherliness which we find here; there is almost a mother's insight into the nature and needs of his young daughters. With affectionate words that are most unusual in the sixteenth century he writes to them of his delight in nature and his longing for his home and his children.

It is indeed a different picture that appears in these letters from the one we know of a tyrant despising mankind, for here he is seen longing to hear the nightingales in the gardens at home, and reminding his daughters of the evenings when they sat chatting together on a garden seat in the moonlight. He had taken with him his own servants and an old woman who was his cook, and he enjoys their gossip, and has a bunch of flowers brought in every day by his gardener. He greatly admires the gardens at Lisbon, and has plans drawn of them. The first sweet orange, presented to him in Portugal, he sends to the girls and is eager to hear what they think of its taste. In these letters a great part is played by his gardens at home and the care to be taken of them.

In the year 1558 the king had bought a great tract of woodland on the other side of the Manzanares, opposite his residence, and had it made into a park with fish-ponds, pheasantries, aviaries, and cages for wild animals; he also had a hunting-box made, called Casa del Campo (Fig. 291), with a flower-garden. But in 1582 he found this garden had been much neglected. He writes to tell his children how sorry he is about it, but says he will take care to get a new gardener who will restore it to its former beauty. Here also Philip put statues in the four corners of the parterre, marking the centre with a very fine shell-fountain and many sculptures. Later on the standing figure of his son Philip was erected there. The unusually large shady trees, admired in the seventeenth century, were also planted by him.

At the east of Madrid the king made another garden, where on a hill stood the royal retreat already mentioned at the monastery of San Jeronimo. Philip had it enlarged by his architect Juan Bautista de Toledo, and the king and queen had thirty rooms each on the east of the choir of the Gothic church of the old cloister. One report says that Philip gave orders to the architect to build the whole place with galleries, towers, moats, and flower-gardens, after the pattern of the country house in England where he had lived with Queen

FIG. 291. CASA DEL CAMPO, MADRID—VIEW OF THE GARDEN

Mary; and we think of Winchester, Hampton Court and Richmond. But unfortunately these gardens at the monastery have disappeared, leaving no trace, and there is no drawing to confirm the report which is so interesting, giving as it does a slight touch of piety and human feeling to the relationship of Philip with England's "bloody Mary."

The ancient retreat of San Jeronimo looks down on the prado, formerly a field-property belonging to the cloister. It had long ago been turned into a public walk, very likely a garden for the people, or a promenade for the townsmen, like those we found in Paris, Florence, and other towns. In Philip's time it was a rendezvous for the fashionable world, especially when in Lent or at any time of mourning the court was living at San Jeronimo, and also when on coronation occasions kings made their entrance into the town. The threefold avenue—now the Salon del Prado—was already in existence, and in Philip's time there were fountains in it. The place then had a world-wide reputation, and in the words of an old chronicler, *Veneri sacer est et amoribus aptus.*

Philip clung to Aranjuez with a peculiar affection, and made it the real summer residence of Spanish royalties. The Order of Santiago had formerly owned at this place a four-square castle that could only be approached by two wooden drawbridges. The kings of Spain had long ago annexed it, attracted partly by the abundance of water, and partly by the tall vegetation, hitherto unknown in Castile, and Isabella had been fond of staying there. At that time the place was known as Isla (island), for it stands on an island made by the Tagus and the Jarama. Charles V. was often here, and by one account it was he who made the gardens. Saint-Simon says he had them laid out according to the old Flemish taste, and commanded that they should always be kept up by Flemish gardeners, adding that his orders were faithfully carried out. The revered tradition of the emperor held sway till his race came to an end; but as a fact these gardens, when Saint-Simon saw them in the beginning of the eighteenth century, had suffered so many alterations that it is extremely difficult to recover the original form. These alterations were made

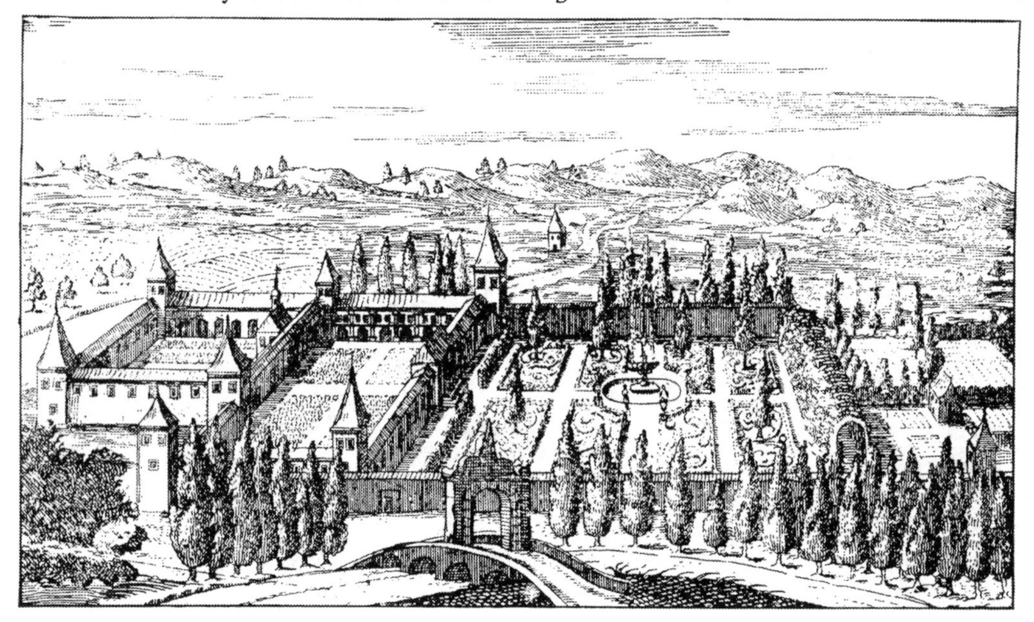

FIG. 292. ARANJUEZ, MADRID—FROM THE NUREMBERG PLAN

by Philip II., who converted the little shooting-box into a permanent summer residence, and greatly enlarged it. And the gardens round the castle may have been quite small still (Fig. 292).

It was very late before the king thought of joining on the Tagus Island, now the Jardin de la Isla, to his gardens. It was only in the eighties that Juan de Herrera built the bridge over the Tagus, and it was then that the garden was laid out. When Philip's daughters send word to him in Portugal, how pretty the garden is now, he cannot remember just at first which island they mean, but it then occurs to him that the new bridge has just been built across the Tagus. All these gardens were drastically treated in the seventeenth and eighteenth centuries, and especially is the hand of Philip shown in the south parterre, the Jardin de las Estatuas, which he decked with statues, in the same way as in Madrid, and with the busts of the emperors; Philip IV. added the whole-length standing figure of his father. Philip II., as we said before, introduced the elm into Spain, and this tree, which stands in many rows of avenues, more often than not, was a great success. Apparently little canals were cut at this time alongside the trees, and the water they supplied has been rightly held to account for their enormous height. In the island garden one of the avenues was composed of plane-trees, and it bore the name of Salon de los Reyes Católicos (Fig. 293), and might perhaps hark back to an older time. The best specimen avenue, known as Calle de la Reina, was certainly planted by Philip: it is so long that standing in the middle one cannot see either end of it. This avenue had reached a considerable height when Velasquez painted it.

The interest taken by the king in building and in art awoke in the grandees at court and in the country a desire to collect. They quickly learned from Italy that houses and gardens owed their chief glory to antique or modern statuary, and Italy must have given up many of her treasures to Spanish art-lovers and patrons. The first Duke of Alcalá gained a name of importance among them; when he was viceroy at Naples he had a fine opportunity of getting antiquities, for which there was not the same enthusiasm in Italy (especially in the days of Pius V.) as in the period of the Early Renaissance.

FIG. 293. ARANJUEZ, MADRID—JARDIN DE LA ISLA, KNOWN AS SALON DE LOS REYES CATÓLICOS

The Riberas, dependants of the duke, owned a house which a generation before had been built quite in the old Mudejar style, and is now called the Pilatus house. Four mighty antique goddesses are in the double-pillared court, put there by Jusepe Ribera, the Spanish Mœcenas of the seventeenth century. Above there are twenty-four busts of emperors, and in the middle of the pavement, adorned with pots of every kind, there is the Janus fountain. Behind on one side of the house was the garden, on two of its sides closed in by buildings, on the other two by high walls covered with tablets, inscriptions and old carvings. The bordered beds were adorned with baths, fountains, a grotto of Susannah—all reminding us of the taste of the Renaissance.

In rivalry with the Medici the duke instituted a sort of academy at his house, where artists and learned men were always welcome. But the dukes of Alcalá were not the only persons who made their houses a centre for art; the Duke of Alva also had his antiques that came from Italy to make beautiful his house and garden at Alva de Tormes. In another place of his, called Lagunilla, not far from Plasencia, much of the garden has been preserved. The castle is in a charming situation on the slope of the Sierra de Bajar, with a flower-garden and two terraces in front, niches in the dividing walls, and steps at the side leading down from the balustraded upper terrace. The park, densely planted with trees, is encircled by the little stream of the Ambro. In the middle is an octagonal summer-house, once lined with mirrors, with a fountain on each of its eight sides. In the year 1555 the Duke of Alva engaged a sculptor, Francesco Camilani, to fill his garden and park with statues.

Another castle of Alva's has been immortalised in the verse of Lope de Vega. The castle and garden of Abadia on the borders of Estremadura were likewise encircled by a little mountain stream, the Serracinos: on one side the castle was mirrored in the water. The garden is not large, but to the poet's thinking is a paradise. It is in squares, and planted with oranges, which cannot tolerate the severe winters of Castile, but were brought to Spain about this time. First of all the poet speaks of a parterre planted with myrtles round it and flowers between: it is chiefly the fountains that he admires among the trees. In one of the squares Alva has a Parnassus, on which animals are seen climbing up to the top, where there is a Pegasus, put up as a memorial to Garcilaso de la Vega, a poet and hero who died early and had been a friend of Alva's youth. Opposite to this, in a

FIG. 294. ARANJUEZ, MADRID—ENTRANCE TO THE JARDIN DE LA ISLA, WITH HERCULES FOUNTAIN

cleverly arranged circle of myrtles, antique marble busts stand; this was once a very favourite garden scheme. In the centre of a basin Lope sees a ship borne by sea deities, and guided by Neptune; on the deck stands a Venus upon a rock supported by four sea gods. In the middle of the garden between the bushes is the chief fountain which Alva had made as a memorial to himself, ornamented with many statues and carvings, partly depicting the deeds of Alva, and partly allegorical. The river cuts through the garden; on the other side four gates enclose it, also adorned with fountains, statues and little summer-houses, some of these grotesque; in the walls there are carvings and inscriptions, and there are also covered avenues and shrubberies with clipped figures.

This description of a garden is the most complete one we have of that time, and it fits well with the many scattered bits of information about other gardens, though it may be a little poetical. But in spite of this, it is hard for us to get any living picture of a

Spanish garden. And this is not only because (out of what really did exist in the seventeenth and eighteenth centuries) far too much has changed or been quite destroyed by fire or earthquake or else through neglect—for indeed this is true and lamentable in our unstable art, whatever the country may be—but in Spain there is another reason, which we can find by looking into the history of her development. There are indeed many beautiful and interesting gardens in Spain, but there is no garden style that can be called Spanish. Spaniards never learned from Italy, as we shall see that the French did, to adopt a new scheme of their own as a complete, consistent whole, nor did they allow the influence from abroad to work uncontrolled upon what they had at home, as did many Northern lands. But, since the great merits of their Moorish gardens made an early Renaissance unnecessary, they allowed their gardens to be laid out by Italian artists; or at any rate they made them themselves so entirely in accord with foreign tradition that they could not be distinguished separately.

FIG. 295. ARANJUEZ, MADRID—THE FOUNTAIN OF THE TRITONS

And so it came about that the seventeenth century brought no change, though in painting there was such an immense upward soaring. And when the court and the whole country awoke from that sleep of the spirit in which it was plunged during the dark reign of Philip III., and the young Philip IV., who was chivalrous and loved beauty and art, came to the throne, it was clear that Spain ought to play an important part in this century of great rejoicings. But for this she needed mighty castles and great gardens, and at home there were no architects who had the experience or originality to satisfy the crying need quickly enough. But from Italy ready help was forthcoming to cement the friendship with the Spanish court. The Medici more than any others sent the king architects, engineers, and gardeners.

The artist to whom the flowering of this art in Spain is above all others due is the Florentine Cosimo Lotti. He was a pupil of Barlotti, and a companion of Buontalenti, but younger. All these artists were employed together at Palazzo Pitti and the Boboli Gardens, where about 1520 a new style of architecture arose. Cosimo Lotti had distinguished himself especially as architect for theatres, decorator for festivals, and architect for gardens; so he was just the man who was needed in Madrid when he appeared there in 1628 with two gardeners from the Boboli Garden. The first thing given him to do by Philip and his powerful minister Olivarez was the reconstruction of the gardens of Aranjuez. To the island garden he gave the character of a grove richly supplied with fountain and ornament. At the entrance we are met with certain signs of the influence of the

Boboli Garden. The Hercules Fountain (Fig. 294) rises from a huge basin, in the middle of which there is an island with a fountain shell, and above it a Hercules swinging his club: from this centre piece there are two bridges leading across the lake. The whole is a simplified version of the Isolotto in the Boboli Garden.

There are fountains everywhere at the crossways of the avenues, standing on round or octagonal spaces; the Triton Fountain (Fig. 295) was painted by Velasquez, but this one has been transferred to the palace garden at Madrid. All visitors in the seventeenth century agree that the chief charm of these gardens lies in the number of fountains, to

FIG. 296. A NOBLE FOUNTAIN AT ARANJUEZ, MADRID

which new ones have constantly been added (Fig. 296). The Countess d'Aulnoy, who saw the place in 1679, found so many that it was impossible to cross an avenue, to step into a hut, to walk on a lawn or terrace, without passing five or six basins with statues. She describes several of them, all tossing the Tagus water on high by some mechanical means. One of them is placed on an artificial mound on a lawn otherwise level; at another the goddess Diana stands with dogs climbing about her, while concealed love-gods squirt water over the animals from the clipped myrtle bushes. Also there is another Parnassus here with a waterfall known as Helicon. Another fountain with streams rushing violently down from three great trees is reminiscent of Arabian water-tricks, and may very likely have come directly from the Moors to Aranjuez.

An engraving by Meunier gives a charming group of trees, with a fountain and many figures in the centre of a circular space, from which radiate four pairs of avenues held together by lattice-work (Fig. 296*a*). The grouping of the trees seemed to the lady's French eyes, already somewhat pampered, to be narrow, but at least there was an abundance of water. At the beginning of the eighteenth century it was said: "It appears under the feet of inquiring visitors, it falls from artificial birds, that are seated in the trees, in a heavy rain which wets you through; and other streams jump out of the mouths of animals and statues and wet you to the skin in a moment, so that you can find no way of escape." Even now, as one walks along the palace front at the side of the Tagus by the path called Los Burladores (the jokers), one can see some of these water-tricks. This garden was enclosed by the broad ribbon of the Tagus (Fig. 298), which ran all round it. It was not made on any particular design of its own. The river gets a very bright effect

FIG. 296*a*. ARANJUEZ—FOUNTAIN AND LATTICE-WORK AVENUES

from a natural weir which rushes down like a cascade to the chief parterre on the east of the house. This bit belongs in its present arrangement to the eighteenth century, and the great Hercules fountain at the end of the lawn was only put there in 1827. The Jardin del Principe and the pleasant garden of the Casa del Labrador belong to a still later date.

After Madrid had actually been declared the capital, a new importance was given to a set of shooting-boxes that stood round this summer residence. The princes of the House of Hapsburg were always great huntsmen, and the chase was famous at Madrid, especially at the extensive landed properties on the north and west. Charles V. rebuilt an old shooting-box on the Manzanares, called El Pardo, with a central block that had turrets at the corners—a favourite idea with huntsmen. Philip II. also was very fond of the place, filled it with pictures, and had transferred there his famous portrait gallery of contemporaries painted by Antonio Moro. There does not appear to have been a garden of any importance at that time. When his daughters described a walk which they took to the place, he wrote that it was perhaps better in winter than in summer, for to him it seemed to be wanting in the greenery of the other castles. It was Philip IV. who first let Cosimo Lotti try what he could do there, and the artist did not take the Florentine garden as a model, for just at the time when he went to Spain the fame of Roman gardens, especially of the villas at Frascati, exceeded that of every other. The theatre wall (Fig. 299), which he put up for the terrace wall of the Pardo garden, with niches, and water playing in them; the water stairway, which cut through to the middle of the castle opposite the wall; the triumphal arch at the end of the upper terrace; the way he embedded the whole ascending structure in the woodland of the park—all these ideas were to be found deeply rooted in Roman gardens, and especially at Frascati.

From El Pardo one could go out into an immense wild park. Half a mile away Philip IV. had an old halting station, Torre de la Parada, made into a little place to put up at, and fitted inside with valuable pictures; a little farther off was the hunting-box of the king's brother, the Cardinal Infante Ferdinand, a place called Zarzuela. Here too there was a terrace-garden in front of the house, and one of the terraces was supported on a great number of arches, "which, seen from afar, gave a very strange effect." From the upper terrace you descended by flights of steps with pierced balustrades. Fountains and streams watered this garden, which most likely belonged to the same age and was made by the same architect as El Pardo.

FIG. 297. TRELLIS FROM THE ROYAL GARDENS, MADRID

Madrid itself was quite neglected during the dull reign of Philip III. He did not like the town, and had tried to pass it over and make the Residence at Valladolid. But Philip IV. spent nearly the whole of his forty years' reign at Madrid. For internal decoration of the rooms in the castle he did everything he could, but we hear little of the enlargement or embellishment of the garden. His interest and attention were soon entirely absorbed in another place, Buen Retiro. He also, like his grandfather, frequented with pleasure the summer-house at San Jeronimo. At the beginning of his reign he had been present at a curious incident. In the year 1624 Charles Stuart had come there on his adventurous honeymoon. The two princes first met in the garden, intending to maintain an incognito until the next morning, when the Prince of Wales was to be received into the town with royal pomp.

Quite near by, the king's powerful favourite, Olivarez, had a little park called Galineria where he bred peculiar kinds of fowl. This minister guided the machine of state with a free hand very much to the detriment of the nation, and insisted on leading the king (who was restless and always in need of distractions) into ever fresh festivities and novel proceedings, to keep his attention from the deplorable foreign policy of his country. In the gloomy Alcazar at Madrid there was not enough room for a holiday crowd, and the love of festivals was taking possession of European courts, one after another, in the sixteenth century. Alcazar could not make a satisfactory background, therefore Olivarez determined to build, in a place that the king liked, a royal house which should be superior to those villas that Roman nobles had lately set up on the borders of their own city.

The territory that he acquired, partly through presents and partly by purchase—and the cloister-ground of San Jeronimo was brought into it—was a mile in circumference. With amazing haste and secrecy Olivarez made in little more than two years the Buen Retiro of Madrid (Fig. 300), which from this time was the main theatre for the history

FIG. 298. ARANJUEZ, MADRID—A VIEW OF THE PARTERRE AND THE TAGUS

of the Spanish court. The architect of the palace which adjoined the old cloister-house on the north, the Italian Crescenzi, could not win himself much glory, for Olivarez was not able to attach great importance to the outside architecture, both because of his great haste and because of the enormous expense. Cosimo Lotti had a much better chance for showing his powers. It was his task to build the theatre, to lay out the gardens, and to conduct the festivities which were soon to be held with a show of unexampled luxury.

The palace, and with it the flower-gardens proper, have now completely come to grief; but if we look at old engravings it is evident to the eye that there is an entire want of unity in composition as compared with what we find in their Italian predecessors—not to speak of French garden schemes that had by now arrived at the threshold of their perfection. It all hangs together with the usual Spanish architecture, not quite vanished in the seventeenth century: the palace rooms were grouped round courts, and the

FIG. 299. EL PARDO—VIEW OF THE TERRACES AND THEATRE WALL

plan was very highly esteemed. But there is more than this: the gardens in front of the façades are treated as separate court gardens each by itself, with walls all round. At Buen Retiro the planted courts of the cloister of old San Jeronimo and the beautiful Jardin del Principe in front of the windows of Philip II.'s old house have precisely similar treatment. It is just like an Italian parterre of the sixteenth century, adorned in the main lines of direction with two shell fountains. North lies the chief parterre of the palace, the Jardin de la Reina (the Nuremberg design (Fig. 302) gives a wrong arrangement of the park), which clearly shows the influence of certain Italian villas—Medici, Ludovisi, Borghese, and others—in that by the side of the parterre a large place is left open for games.

Since 1642 the bronze equestrian statue of Philip IV., cast by Pietro Tacca, has stood in the centre of this square (Fig. 301). Once more the parterre answers to Spanish taste, not Italian, in that a wall encloses it on the park side, and apparently it has no gate leading into the park (Fig. 302). In the walls there are the customary niches for busts. Towards the north are other small parterres beside the rooms of the palace and the theatre house, but all of them completely shut off on the side of the park. The Countess d'Aulnoy thus describes the garden belonging to the palace court: "Four large *corps de logis,* with four larger pavilions, form a complete square, in the middle of which one finds a parterre filled with flowers, and a fountain whose statues pour out streams to water the flowers and avenues, which cross one another, and lead from one *corps de logis* to another."

The world art wonder of that time—for which the rough drawing (Fig. 300) offers scarcely any help—was the park, which in spite of its magnificence had no connected plan. In this work the artist took his model from Italian suburban parks, which this one sought to rival. But the Spanish place lacked the fine views straight through, that in Italy connected the main buildings and the park. The park as such had (according to the accounts we find) to contain every single thing that the long ages of garden development had brought into being. The central feature on the eastern side was the great star, which had eight covered walks leading from an octagonal piece of ground into the different parts of the park. The water constructions were a source of great pride, and the huge pond still exists, 500 by 270 feet in size, with pavilions all round partly used for draw-wells. It was meant as a place for a naumachia, with an island in the middle; and a canal led out of it to other ponds, so that they were all connected, and navigable by gondolas. At the eastern circumference of the garden the canal turned at a right angle by the side of a tennis court where there were also several rows of trees. All over the place there were little groups of trees with a great many fountains, in case anyone wanted a place in which to be quiet. Wide views were to be got from raised mounds on the walls.

By the west wall were the flower parterres. A certain Jesuit priest who saw these in 1638 describes the beauty of the beds in his own idiom:

> Here one beholds beds glowing with colour, where letters are cut in rosemary, revealing the secrets of the interwoven flowers: in vessels of painted Talavera ware, which shame the finest silver, the heads of pinks show bright, with basil all round about, just as if the earth were clad in red and blue gauze, in Oriental fashion. Here too are streams clear as mirrors, paths sown with rose and jasmine, leafless pinks of shining purple hues, meadows where Arabia must have bestowed all her lilies. There is no imaginable beauty that this garden lacks.

Flowers were brought here from every part of the world. In the year 1633 thirteen

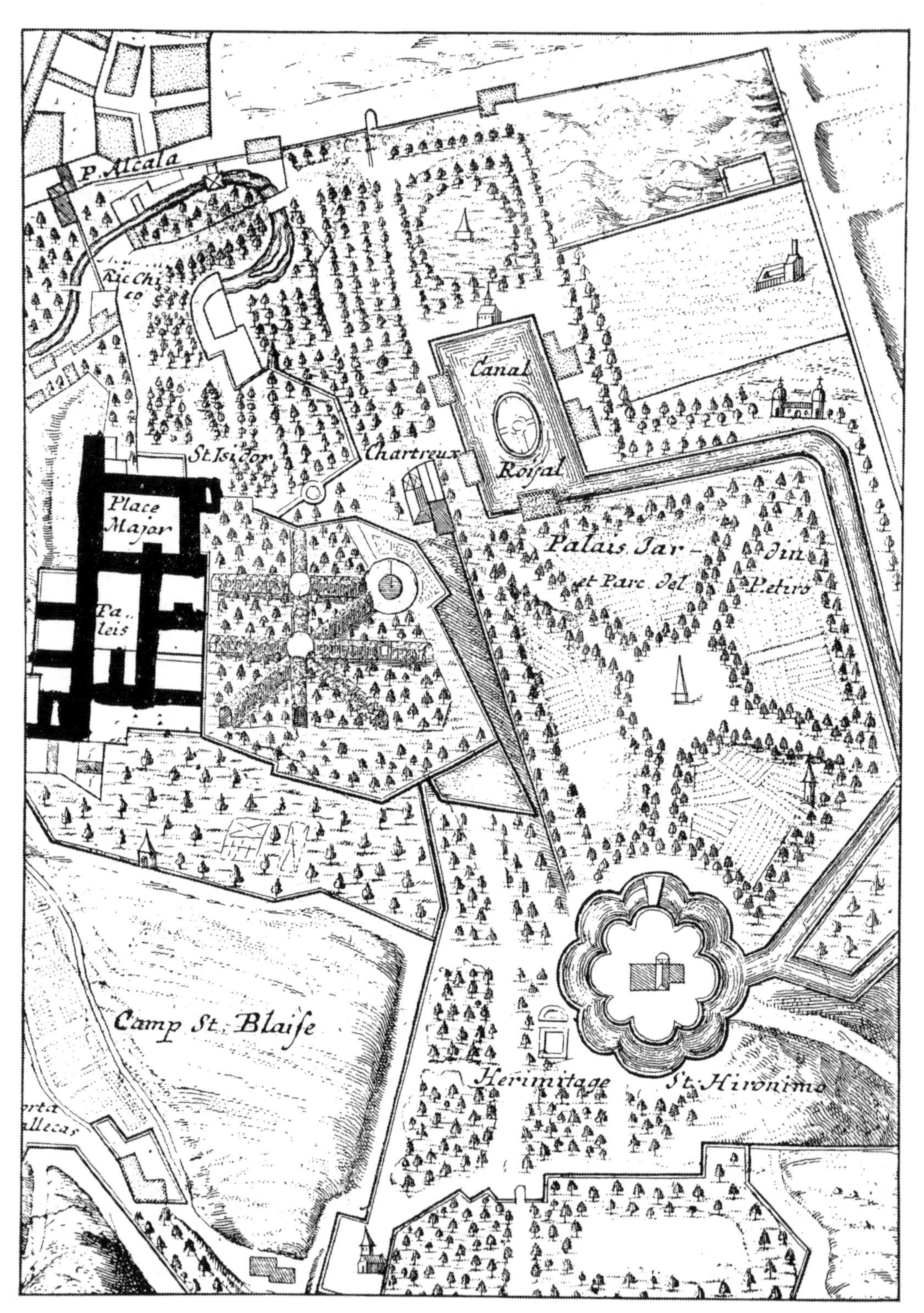

FIG. 300. BUEN RETIRO, MADRID—PLAN OF THE PARK

wagon-loads arrived from Valencia, and gifts of this sort from princes were very sure of a welcome. Cardinal Pio of Savoy sent his own gardener Fabrizio from Rome with bulbs worth 10,000 ducats. There were certain bits on the south that were more like open country, with meadows and trees, "keeping the simplicity of country life and remarkably pleasing."

In two matters destined to prove very important in a future development of the garden, Buen Retiro was the precursor: the custom of making hermitages in the park, and holding festivities in the actual garden. We have already mentioned the intimate connection of pleasure with piety in the Spanish nature. Though the severe fanaticism of the sixteenth century had by this time vanished, perhaps its decay led to more rapid and immediate changes in festivals and church practices at a court where there was always a burning desire for something new. From the start Buen Retiro had the monastery close

FIG. 301. BUEN RETIRO, MADRID—JARDIN DE LA REINA

at hand. And as Philip II. had no objection to the monks looking on from the windows of the Escorial at any shows that he was pleased to give, so it was now the custom for the Fathers at San Jeronimo to be invited to the parties at Buen Retiro. They knew very well that their services would be in request at the confessional after these doings. But the jaded palate did not remain for long satisfied with the older monastic habits, and little hermitages were now scattered about the park, mostly near the outside. These were small garden villas, each with its chapel, a turret for the view, a small parterre, a labyrinth, a grotto, and other *invenzioni boscherecci*. There was one hermitage to Saint Isidoro, next to the palace on the north-west, and others to Saint Bruno, Saint Inez, Saint Magdalena and Saint John the Baptist. The largest was dedicated to Saint Paul (Fig. 303); and in its garden, adorned with statues, stood a triple Narcissus fountain. The house stands in a parterre enclosed by trellises. The hermitage farthest towards the south-east was Saint Anthony's, which was surrounded with moats like a little castle, the outside one bending in a curious way and connecting with the great canal in the park.

All these were joined to one another and to the main park by avenues; inside they were supplied with pictures and all kinds of luxury, and in them lived important members of

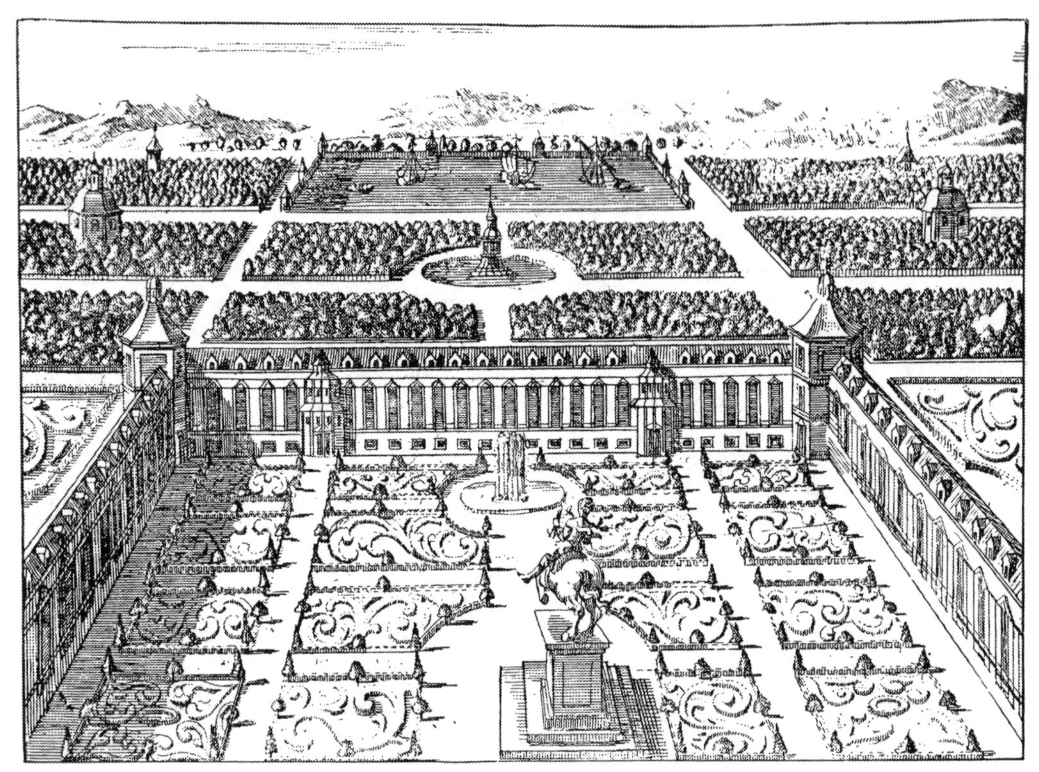

FIG. 302. BUEN RETIRO, MADRID—GENERAL VIEW

the court. In John the Baptist's hermitage Olivarez took up his abode, and tried to make gold with the help of the alchemist Vincenzo Massimi—gold which he very well knew how to spend. There is no doubt that here, where these religious dwellings appeared so conspicuously, the step was taken which gradually made the hermitage a mere fashion. One would like to trace the necessity of this both to the nature of the Spanish people and to the busy work of this particular court; but while these hermitages were springing up

FIG. 303. BUEN RETIRO, MADRID—HERMITAGE OF SAINT PAUL

at Buen Retiro in the seventeenth century, Cardinal Bourbon at his own park at Gaillon had already built in 1560 the *érémitage abondant en tout plaisir*, which exactly answers to the description of the Spanish ones. Perhaps it is only documents that are wanting to show that Spain also had earlier examples of the same.

But there is no question that Philip has the glory and credit of making great and amazing garden festivities at his court before anyone else did it. In the year 1632 Olivarez first received the king at his own villa, but only to present it to him, handing him the key of office on a silver plate.

Until now it had been necessary to drive out to Aranjuez, if there were any festivities going forward. But in 1623, on Philip's second birthday after he became king, there was a great theatrical performance for the occasion, produced by an Italian, one Cesare Fontana, perhaps belonging to the distinguished family of Roman architects. The queen and the court ladies took part. At this time, in elaborate affairs of the kind, there would be enchanted castles, earthquakes, marvellous gardens, fire-breathing dragons, genii hovering in the air, all playing a great part. Everything must be new, and ever again new Engineers indulged in orgies of art. It is a curious reflection that the technique of machinery, now exact and progressive, since in our own time it has been harnessed and subdued to useful ends, came into existence in those childish years of merriment, and devoted its original audacious inventions to the ephemeral delights of the passing hour.

Philip IV. did not find all his engineers and musicians in foreign lands; he had artists in his own country: painters who were ready to do decorations and backgrounds; and above all the greatest dramatic poets that his nation had ever known. The aged Lope de Vega was prepared with a dedication ode for the new villa, and Calderon wrote dramas and prologues and divers pieces that pointed to the creation of Olivarez as the holy city, the blessed Jerusalem. The king had but to utter a wish, and every entertainment became some new link in the chain forged by the pen of the great masters, not to speak of those of second and third rank. In the park the great pond with its island was a favourite spot for these dramas. On the Feast of St. John in 1640 a stage was fixed up above the island, supported on ships, and the machinery and preparations must have been immense, for we are told that in the representation of *Circe* (an odd mixture of Homer, Ariosto, and Tasso) there appeared first the ship of Odysseus, then Circe's wood with the wild beasts and talking trees, then the enchanted feast and the metamorphoses, next all the charms and spells that Circe wrought to ensnare Odysseus: sea monsters, Tritons, sirens, love-gods—everything emerging from the water; and finally the palace of enchantment, which sinks to the bottom accompanied by lightning, thunder and earthquake, as Odysseus falls into the arms of Virtue.

For those spectacles that needed less room the theatre at Buen Retiro had been built in 1637, and this was of great importance for the development of dramatic art in Spain: it was the first Spanish stage except for the people's theatres, which had long been quite common. Only members of the court had the right of entry, but as the first requirement was pomp and glory, Cosimo Lotti hit upon the ingenious idea of building the stage so that by removing a wall at the back the view was thrown open into the park. As a garden scene was wanted for almost every play, this plan gave the advantage of an unlimited space and complete naturalness, and the possibility of artificial lighting as well. It also had the merit of allowing people to sit in an enclosed place and at the same time to see the picture before them in a sort of frame. Lotti appears to have done his very best; and yet his successor, who was summoned to Madrid after his death as garden artist and decorator, Baccio del Bianco, surpassed him in the judgment of contemporaries in one important respect. Baccio had been for some time in Wallenstein's service at Prague, and had made his beautiful loggia and perhaps his garden. When he came to the preparations for Calderon's *Perseus*, he went to the king quite overcome, and told him he would have to be given bed and board, for the performance would take eight days with such elaboration of detail. All the same, in the few hours allowed the enormous amount of work was accomplished without a minute's interruption or delay.

These were the most brilliant days of Spain, when Buen Retiro served as background for the magnificent assemblies of people who were mad for show and festivity. And yet this society, which seemed so highly gifted and so lofty-minded, was stumbling to the edge of a precipice. To contemporaries, Buen Retiro was the Colossus with clay feet. As soon as it was built there arose a general alarm at the way it swallowed up enormous sums, and at the Nero-like unconcern of Olivarez in setting up a barrier on the east for a town, which already on the west was debarred from traffic with the country outside both by the castle and by the Casa del Campo. The rich murmured also because of the plundering of their objects of art, which furnished the spacious rooms of the palace and the gardens. Now, after their complete destruction, the park has been restored to Madrid, just as once in Rome the grounds of Nero's Golden House were restored to the city. It is metamorphosed into a great town park, and bears the official name of Parque de Madrid.

With Buen Retiro the Italian influence in Spanish gardens came to an end. When under the new royal family of the Bourbons the art flourished for a short time, it was the powerful influence of the culture which they brought with them from the North that allowed any new growth to arise.

Portugal

When in the year 1581 Philip II. took possession of Portugal, he wrote to his daughters about the beauty of the gardens in his new province with something like envy. All that Nature so seldom granted, especially to Castile, and then only after much diligent care had been spent, was poured out only too generously in Portugal. The narrow strip of coast-land is watered by the great rivers which are not even navigable till they have left Spain, and the mountains lie open to the invigorating western sea breezes. The wide river valleys, the level table-land, and the mountain slopes, are all alike attractive for villa building and the making of gardens, and the fortunate land offers the most favourable conditions for many sorts of vegetation. Portugal, like other countries, has had to make use of foreign ideas and foreign art to supplement the gifts of Nature. Unfortunately what we learn from literature is too poorly supported by old pictures to help us to a clear notion of how the art developed. All we get from the remains of the past that bears upon our subject is the evidence that certain plans were worked out and then consistently adhered to—plans which are not found so firmly imprinted in the history of any other country.

The cross-roads of the monasteries are very extensively used for gardens at an early date; the royal court at Batalha is mediaeval, and the garden site still preserved with its five beautiful fountains may have been just the same in the sixteenth or even the fifteenth century. It differs in beauty, but otherwise is not very unlike other garden courts of the time in essentials of style. The courts in actual Renaissance days seem to have had more sides to them. In the court of San Francisco at Evora there are boxes placed in the figure of a star and making a pretty pattern with the paths, and a fountain in the centre. The small walls are covered with tiles. Nowadays places of the kind are still called *alegretes*, which may be translated pleasure-gardens. The expression was used as early as Philip II.'s time, for he writes to his daughters in July 1581: "There are little gardens here in different parts, which are not bad, and are called *alegretes*. We will get the plans of them." One may suppose there was a likeness to the *giardini secreti* of the Italians; but perhaps the only idea was to show some fixed design in the laying out, like that found at Evora.

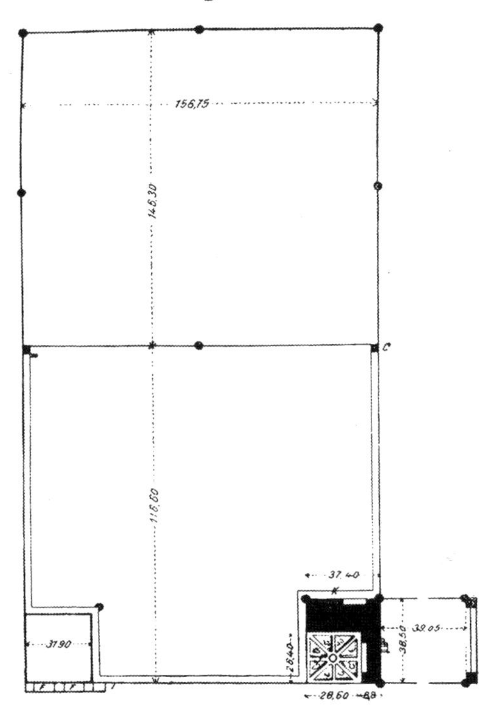

FIG. 304. QUINTA DI BACALHÃO—PLAN OF THE SITE

In one of his letters soon after, Philip particularly admires the monastery garden at Cintra, the *Penha longa*, which took its name from the long rock on which it was built. "They are pretty, and there are numbers of them; there are lovely fountains which I should like to bring away with me." Much is still told of these gardens enclosed by walls beside the monastery, and of a grotto near a fountain that was paved with flags, and other things of the kind. One of the finest cloister courts of Portugal, perhaps of the whole world, is the great court of the cloister at Belem on the outskirts of Lisbon, the chief creation and the best beloved one of the fortunate King Manuel. This court, with an indifferent garden at present, had an unusual one until 1830. The fountain now on the north-west of the court made the centre piece of an island, which was connected by bridges with four others. All of these were in the same pond, which had the shape of a star. The high perpendicular banks were covered with tiles, and the islands were laid out as gardens. Similar to this was the court of Santa Cruz at Coimbra; here there were little chapels on the islands which were round a rotunda of open pillars; these also were united by bridges. Small canals emerged from the pond, and in the extra corners there were hedges and beds. These cloister courts have a certain relationship with the court of the Escorial. But they are not only prior in time; they are also superior in the luxuriance and delicate treatment of the whole garden arrangements.

Of the non-clerical buildings, the villas and their gardens, one treasure has been left from the first half of the seventeenth century, in the property that Dom João de Castro, the great Viceroy of the Indies, made and called *Penha verde*. The simple house stands close to the slope of the Serra, not far from Cintra. The garden, in front of the castle, is a geometrical parterre. What is best worth noticing is the park, which opens behind the house, and has several little chapels in it. At the highest point, half concealed in rock, is the beautiful rotunda of the Mary Chapel, with a ruined arcade round it. The entrance to this small place, lying among lovely trees, is by a door with Sanscrit inscriptions on it. In the middle of this space stands a tombstone which states that here rests the heart of Dom João after the work that in life he devoted to his fatherland. The park seemed wonderful to contemporaries because of the number of foreign plants it contained, brought home by de Castro from his travels, and falling in thus with the botanical interests of the period. His biographer relates that he had the fruit-bearing trees removed from this place (which the king had given him as a reward for his services) in order that he might gain nothing useful from it.

A marked feature in the decoration of Portuguese architecture and gardens is the

FIG. 305. QUINTA DI BACALHÃO—POOL IN THE GARDEN AND PILLARED PAVILION

use of tiles, originally blue but afterwards of different-coloured materials. Here, as in Spain, they were a legacy of Moorish art, but in Portugal the use of them was more widespread and endured for a longer time. These tiles they particularly liked (as we saw in the cloister courts) for the basins, where the reflections gave a variegated effect. So the peculiar treatment of such basins gave a national character to the Portuguese gardens of the sixteenth century. In the first villa that clearly shows Italian influence in Portugal, the country place of Bacalhão, a charming specimen has been preserved. The founder was the natural son and sole heir of the great Admiral Albuquerque. As quite a young man this Alfonso travelled in Italy in 1521 in the train of Donna Brites, when she came to Savoy as the duke's bride. From Villafranca, near Nice, where they landed, the Portuguese gentlemen made a journey to Italy, and there Alfonso had his eyes opened and his taste formed when he saw the pictures of Italian art. Thus when later he bought this property from the Infante and built his castle he chose Italian styles, but his garden showed his own individual taste as well as a combination of Spanish and Italian.

The house is approached through a court flanked by pavilions and includes a little *giardino secreto*, which we ought to call an *alegrete*. This part is at one corner of a very large terrace (Fig. 304), 160 by 156 metres. At the opposite corner there is a large sunk basin about thirty metres long, shut in by walls, against one of which stands a hall constructed with three pavilions (Fig. 305), the rest of it covered with tiles and terra-cotta busts At the other corners are two little pavilions, and below there is a second terrace of the same size, both of them at the present time used for fruit and vineyard gardens. Whether the upper terrace, which shows a rather narrow band running along the wall and somewhat raised, was formerly intended as a flower parterre, must remain uncertain. The ground was very large for that time, and one cannot suppose there were no fountains in connection with it, but they must have entirely perished. Anyhow the whole place, cistern and house alike, has no predecessor similar to it either in Spain or in Italy.

FIG. 306. BEMFICA: QUINTA OF THE MARQUEZ DE FRONTEIRA—THE CHIEF PARTERRE

The basin with walls and tiles that lies the whole length of the little castle of Bibafria, near Cintra, reminds us more of Italy—of Madama, Palazzo del Te, and Petraja. The windows of the best rooms open on the water, and on the narrow side of the basin there is a pretty summer-house, and below this is the garden.

A real treasure is the Villa Bemfica. It lies in the lovely valley of Alcantara, which has ever been a spot beloved by the Portuguese nobility. Here the Marquez de Fronteira, friend and contemporary of the poet Luis de Sousa, built himself a summer-house which was very much enlarged in the eighteenth century, though it still kept the style of the seventeenth when it was first put up. The garden is laid out in geometrical beds, mostly square, confined by box, with flowers and dwarf palms intermixed (Fig. 306). Among these one finds some fine clipped box, in the form of cones and spheres. Five fountains, one of them in the crossways, indicate the design of the parterre, which is adorned with statues in every corner. Here too, however, the eye is arrested by something new and strange, a basin lying at the side, which occupies nearly the whole length of the garden (Fig. 307). In the water stand two statues, and there are two little flowery islands in it, while the high wall that supports a narrow terrace is articulated with three doors, between which are twelve panels with the figures of knights made in faience.

The wall of the narrow upper terrace has no plants, but is decorated with plaques and also has five niches containing the portrait busts of Portuguese kings. The niches are coated with red enamel, and between them are terra-cotta ornaments of many colours in the style of Robbia. On the balustrades there are statues, and the terraces are flanked right and left by small turret-like summer-houses. The basin itself is separated from the garden by a balustrade, and we reach the upper terrace by steps at the side. The place reminds one of Villa Madama, but the whole ornamentation of the basin is of Portuguese

FIG. 307. BEMFICA: QUINTA OF THE MARQUEZ DE FRONTEIRA

origin. A second wall close to the house is similarly adorned, for we find again the plaques of faience, with statues between them, and above are hanging fruits and medallions: in the ground in front a shell is placed, from which we should perhaps infer that there used to be a fountain.

Above all other things these large basins, generally by the side of high walls, are characteristic of the Portuguese garden of this period. The surroundings of a large basin that is set above imposing terraces at the Quinta da Ramalhâo, a place belonging to the eighteenth century, is very individual. When the Infante Dom Pedro, who in the year 1750 became Pedro III., built a castle and added a park to it, it was naturally dubbed the Portuguese Versailles by his contemporaries. But in spite of the fact that a French architect built Queluz, the influence of the French style was so slight that this royal castle was entirely of the Renaissance type (Fig. 308). Queluz is situated two miles to the north-west of

Lisbon, and the river by its side is admitted into the garden. The river bed and the bridges are laid with tiles, and this has a brightening effect on all the vegetation south of them. In the main axis of the house there is a garden court, which is flanked at the entrance to the grand parterre by statues of knights. A water tower stands at the end with a cascade, and the cross-axis of the parterre leads to the bridge over the stream. At the side by the river where the valley takes a downward slope, there is a terrace like a hanging garden at an obtuse angle formed by the castle wall. An important set of stairs leads to the low-lying parts of the garden. At this juncture, when all eyes were directed to

FIG. 308. QUELUZ, LISBON—THE PRINCIPAL PARTERRE

France, Portugal held firmly to her own traditions. The place has now gone wild, and the statues are partly overthrown.

Another scheme for the garden that the Baroque period perfected in Italy made its own peculiar deviation in Portugal: in a long series of lofty monasteries the "stations" were developed into wonderful water stairways. The cloister of Bussaco, in its fine park full of tall cedars, is the first example of a place of this kind (Fig. 310). Straight steps led from platform to platform on both sides of a watercourse which was enclosed in tufa and arranged in a zigzag pattern, with a bridge also edged with tufa put across each platform. Between the steps the water fell in little streamlets, ending in a basin, the uppermost one being filled from the well-house above. In former days this had been the entrance to the old cloister. The rule of the foundation enjoined that "hereby this retreat shall for all time be made pleasant and in accordance with the law, so that the prior each year plant trees whereof he may remove none and cut down none without the consent given of the chapter." Pope Urban added a bull excommunicating any persons who committed a crime against the trees in the "forbidden preserve." Bussaco did not, however, remain sanctified, for modern hotels have pushed a way into the place of the ancient monastery. Still, the beautiful garden picture of a stairway once devoted to religious worship is there to-day in the wood that towers high even now in its grandeur, spared through many centuries (Fig. 309).

In a still nobler way such stairs adorn two cloisters, Braga and Lamego, which

FIG. 309. STAIRWAYS IN THE PARK AT THE MONASTERY OF BUSSACO

were founded in the eighteenth century. The first, Bom Jesus do Monte, has suffered the same fate as Bussaco. It has become a place for open-air cure, and the quiet of the cloister has been lost in hotels and new buildings of every kind. But the stairway of the "stations" (Fig. 311) which lies in the old park still full of wonderful trees, shows the pomp and pride in which the Church loved to be arrayed at that date. The stairs rise in a zigzag on both sides of the water, with two chapels flanking them below. To right and left the stairway, which goes straight up to the first stage, has two pillars ringed round like snakes, and possibly at one time water trickled down these as at the Villa Aldobrandini at Frascati. The platforms themselves are decorated with ornaments, statues, niches, and fountains, and on both sides there are narrow ribbons of flowers. On the highest terrace, which has the cathedral with its two towers at the top, the stairs are of a semicircular

FIG. 310. AVENUE OF CEDARS IN THE PARK AT THE MONASTERY OF BUSSACO

shape and fit round a niche at the summit. This charming picture, which makes the path for those who aspire to grace *abondant en tout plaisir*, has served as an example to other less ambitious places.

The Monastery of Nossa Senhora dos Remedios at Lamego was founded in the last third of the eighteenth century. The stairway is very similar to the one at Braga, in that

FIG. 311. BRAGA, BOM JESUS DO MONTE—STEPS OF THE "STATIONS"

it is cut out of the hillside. The nine stopping-places on the steps were all furnished with chapels, and the kind of ornamentation for each individual stopping-place was copied from Braga. The lower end was formed by a wall with a high rounded niche. Thus the little country had no style of its own that would work out into further development, but it had a number of important ideas both firmly expressed and persistent. France, towards which we now turn our eyes, is the country whose destiny it was to free herself, with an art that was born at home and was truly vigorous, from subjection to the vogue of Italy, and soon to become Italy's most formidable rival.

CHAPTER IX

FRANCE IN THE TIME OF THE RENAISSANCE

CHAPTER IX

FRANCE IN THE TIME OF THE RENAISSANCE

IT is wonderful to see how cheerfully France accepted the leadership of Italy and recognised her aims, while she was from the very beginning conscious of her own style, and guarded it strictly.

In the year 1495 the young and romantic King Charles VIII. began his triumphal march into Italy. He passed with little opposition into the South. His spirit, intoxicated with sheer beauty, received into itself that glamour which the new Italian art diffused over all things in those joyful days of the Early Renaissance. Both he and the young knights, his companions, were loud in their praises, which reached the highest point when Charles arrived at Naples. Here they thought they had truly found the Earthly Paradise. "The king in his gracious favour," writes Cardinal Briçonnet to Queen Anne of Bretagne, "has been pleased to show me everything, both inside and outside the city. And I assure you that the beauty of the places is incredible, with the furnishings of this world's pleasures in every possible kind." And Charles himself writes to Pierre de Bourbon: "You cannot believe what lovely gardens I have seen in this town; for, on my word, it seems as though only Adam and Eve were wanting to make an Earthly Paradise, so full are they of rare and beautiful things."

There was one night, spent by the king at Póggio Reale, that seemed to him and his chroniclers, the poets at court, to be the crowning point of all this glory. "To describe the beauty of this place, one would need have the *beau parler de Maistre Alexis Chartier, la subtilité de Maistre Jean de Meung, et la main de Fouquet,*" says one of his companions. Charles did not bring home much in the way of political gains; the stakes which he won he lost again. But for the education of France this journey of his had far-reaching consequences; it was the birthday of the French Renaissance. He came back with twenty-two Italian artists of the most diverse types, and he gave them a home at the castle of Amboise. With them there arrived, under the care of Nicolas Fagot, the king's tent-maker, a consignment of various tapestries, libraries, pictures, sculptures in marble, and porphyry, weighing in all 87,000 pounds. The French writer who makes this statement goes on to say: "To tell of these would take, not a single sentence, but twenty sheets; for what the upholsterer Nicolas Fagot brought into the heart of France from the depths of Italy was nothing more nor less than the whole of Italian art—that art which was destined to bring forth at Amboise, at Gaillon, and in our entire fatherland, countless marvels, perhaps the choicest that France has ever seen."

Charles felt a burning desire to make Amboise beautiful. Briçonnet had repeated the king's words to the Queen of Naples about the Earthly Paradise, and added that "he now no longer values Amboise as he did." But when the king was home again it was his earnest wish to make another paradise there. The rebuilding of the castle—already put in hand before he took his journey—was continued in feverish haste, so that, when he died suddenly two years later, the historian Commines was able to tell of magnificent buildings begun and carried out at Amboise. Among these was the colossal tower, where one could

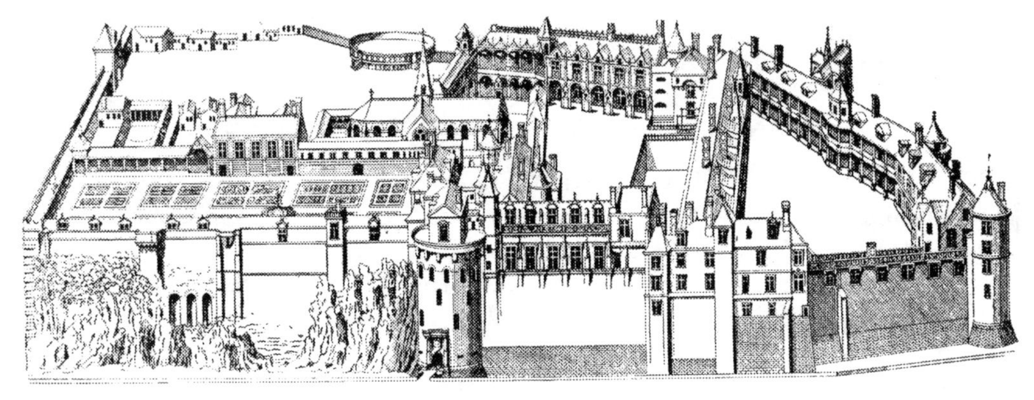

FIG. 312. AMBOISE—GENERAL PLAN

ride up to the top, for instead of stairs it had a slope ascending gently to the upper terrace of the castle.

The death of Charles did not interrupt these activities for a moment; his successor, Louis XII., continued them very zealously in the seventeen years of his reign; while under Francis I. and those who immediately followed him the passion for the work exceeded all bounds. The achievements of eighty years of architectural effort on the castle estate, specially concerned with the study of gardens, are given in a collection of fine engravings by the architect Androuet Du Cerceau. By their help we get a view of such a development of this art as can hardly be said to have occurred a second time in history, and certainly never in a period of abundant work and fresh enthusiasm.

The desire felt by the French to learn the lessons of Italy is shown to some extent by the immigration of Italian artists, but much more by the continual stream of young French architects into Italy, who went in all eagerness to familiarise themselves with modern and antique works. Nevertheless, from the very start France exhibited an astonishing independence and conservatism. First of all, the villa idea never was welcomed on her side of the Alps. The Italian country house had kept the character of a mediaeval fortress much longer than the town house. Early Florentine villas: Careggio, Quaracchi, Caffagiola, Villa Imperiale at Pesaro, the Belvedere at the Vatican, all showed so far the outside features of battlemented towers and walls, and projecting structures below. But this type was soon abandoned, and at the end of the fifteenth and beginning of the sixteenth centuries, the open-air country house, the Italian villa, has appeared with its towers, formerly outstanding projections of the building, now incorporated into the façade; its ground-plan is formal, most of the parts are grouped round a single court, and the main entrance façade is generally made with a pillared hall, as open and inviting as possible.

A comparison with the French style of castle will show what an effect this had on the garden. France, as we have said, had no part in the Italian development, and in spite of great admiration for the flourishing art on the other side of the Alps was for the most part content with imitating decorations. The castles, however, all but kept their mediaeval character, with more or less pleasing irregularities of ground-plan, and with towers at the corners, which no doubt were ever changing into pavilions. They still, however, preserved their independent native style in the chief buildings. The high roofs also, and most of all the moats, kept with a tenacity shown by no other country, were a marked characteristic of the Middle Ages. Moreover, all these features of a fortress were not only found in the castles, rebuilt on old foundations, but at nearly all the new buildings, though here one did begin to see a gradual tendency towards the formal ground-plan. Thus it comes about that the gardens at the earliest castles, especially Amboise, differ very little from the mediaeval type in site and plan.

Charles brought with the artists a Neapolitan gardener, Pasello da Mercogliano, by profession a priest, who combined the care of the garden of the soul with that of the earthly garden. At Amboise he found a little plot to work in that was in its proper mediaeval place close to the burial-ground. Charles had the high terrace widened, so that he could make a larger garden (Fig. 312), and this to begin with was enclosed with pretty lattice and pavilions, round which Louis XII. later on put a gallery, still to be seen in the engraving by Du Cerceau. The fine pattern of the parterre also belongs to the middle of the sixteenth century. In Charles's time the garden was still partly planted with fruit-trees beside the geometrical beds and all round them. We hear of a purchase of fruit-trees made for the garden by Pasello. But on the south, quite away from the real castle garden, which could not be enlarged any more, the king put an orangery, the first that France had to show; and in memory of this—of course it was only the bitter orange—the tenant handed the king a branch of orange every year. The gallery which Louis put round the garden represents an idea that was found in Italy, but there it was soon given up, whereas in France it remained for a long time as a special kind of ornament. One charming pavilion of trellis, not set in the centre here but at the side, was still standing in the middle of the seventeenth century until it came to grief in a storm.

The gardens at Blois were finely set out, but quite differently (Fig. 313); they

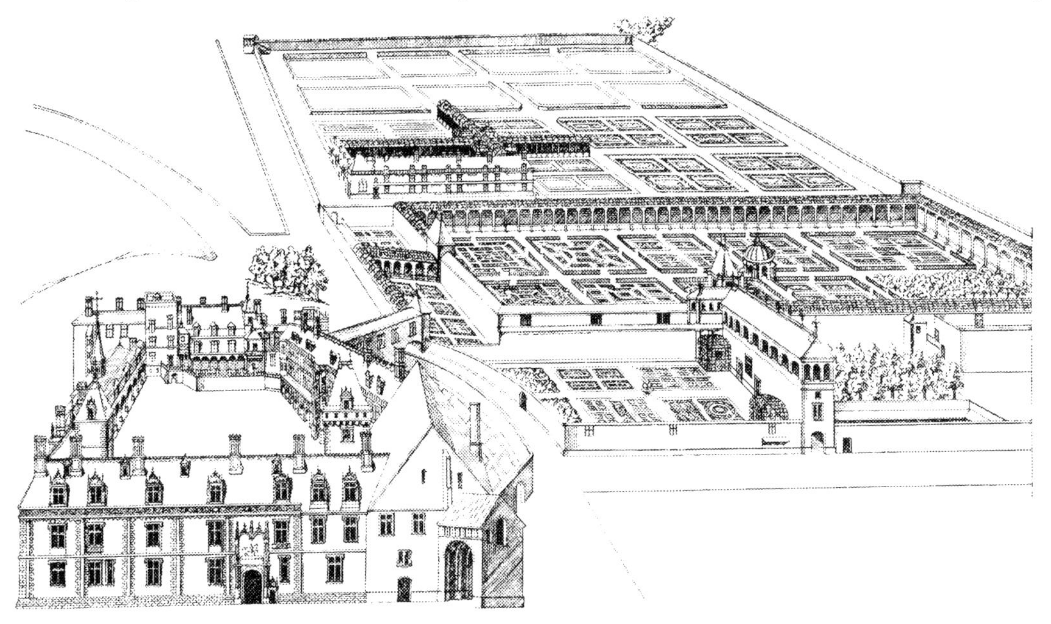

FIG. 313. BLOIS—GENERAL PLAN OF THE CASTLE AND GARDENS

began right back in Roman times. Louis XII. was born there, and he rebuilt the castle on the old foundations. An unusual character was given to the place as a whole by the much-admired wing on the town side which was added by Francis I. Louis moved his residence from Amboise to this favourite place of his, and took with him Pasello, who found a growing field for the exercise of his activities. There was not much garden. The little plot which in mediaeval fashion was under the dungeon-keep, but was in this instance deeply sunk, was continued in narrow strips, and had already been widened into a large piece. This old ground was converted into the lowest parterre of a garden that rose in three mighty terraces with powerful supporting walls. People were not at all afraid of the difficulties of large subterranean works, and for these the immense substructures below the high-built castles of the Middle Ages served as patterns. But the garden on each individual terrace was still a garden all to itself, having no connection whatsoever with the other gardens on terraces above or below.

FIG. 314. BLOIS—CHAPEL OF ANNE OF BRETAGNE

The great Roman invention of connecting steps had hardly established itself in Italy; and France, like all the Northern countries, adopted it late, and then but sparingly. And the different parts of a garden are as little related to the castle as they are to each other. The moat dug by the wall, which goes round three sides of the isolated building, seems to have lost its water very soon and then to have been turned into a fruit-garden. Over the trench Louis made a gallery from the corner of the burial-ground to the middle terrace, the queen's garden, which was laid out as the chief flower-garden. All round it were pleasant wooden galleries covered with greenery, and to make these the king summoned the best joiners in the kingdom. The separate beds—called in the accounts *parquet*—were surrounded with trellis borders: *accoudoir*.

In 1503 a marble fountain, made by the Italian Pacchiarotti at Tours for the sum of 662 livres, was set up in the middle pavilion, which had a St. Michael on the top. The real engineer of the waters was Fra Giocondo, who had come over with Charles, and now received an extra commission for his skill. The fountain was an octagonal basin with two shells above made of white marble: the remains of these are still kept at the castle. The *pièce de résistance* is the little chapel (Fig. 314), a great favourite of Anne of Bretagne; and the king also liked to say his prayers in it. It is the one thing in the gardens that has been preserved. The highest of the terraces, called the King's Garden, had its chief ornament, the pergola, put there only in Henry II.'s day. This was also the time of greatest success and beauty, when the king's table was enriched every day with fruit, and when Blois could boast of wonderful mulberry-trees. Then came the period when this castle and others on the Loire were more and more forsaken for the palaces of Paris because they were too near. For Paris was the centre of political life, and more and more

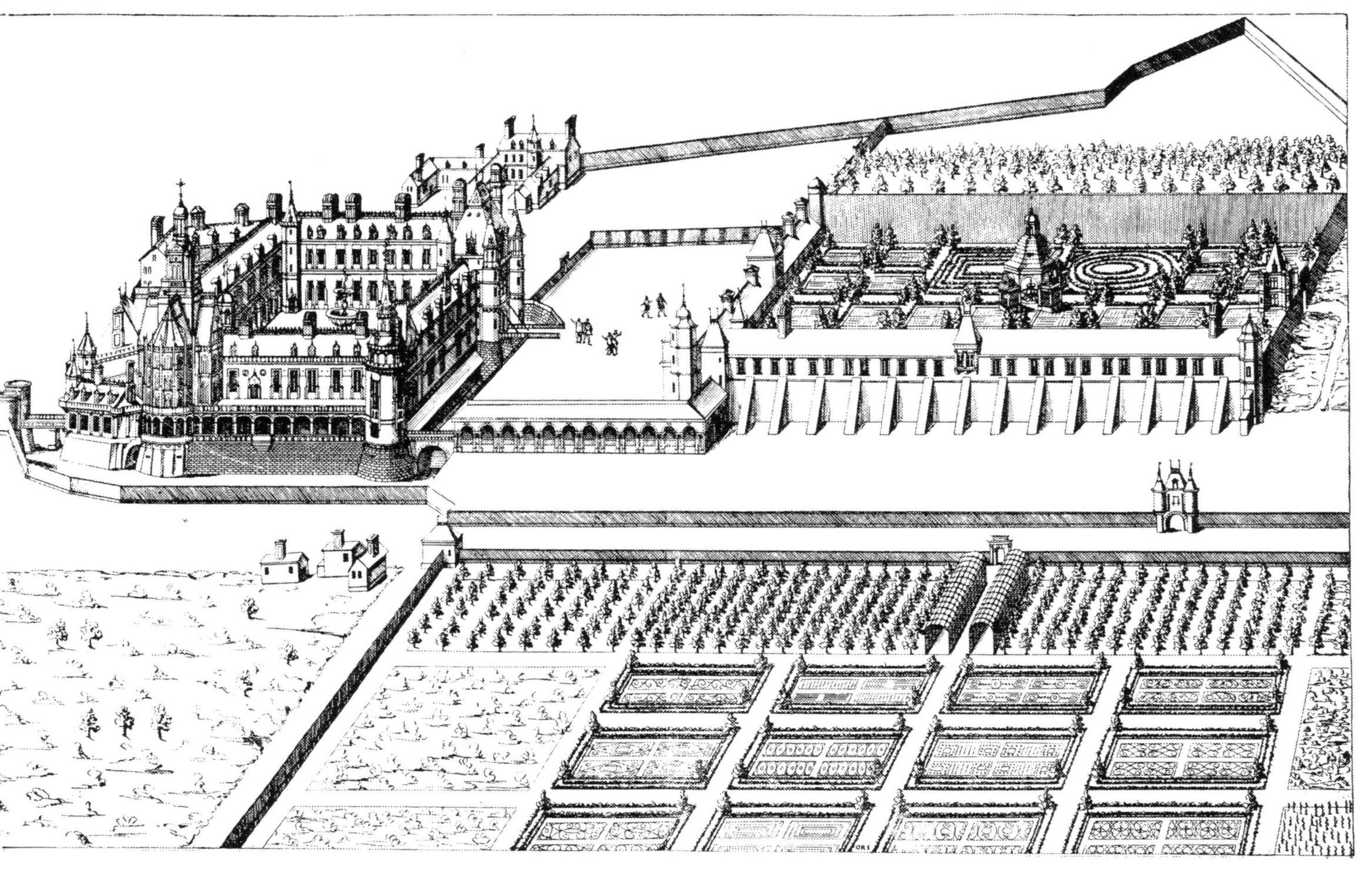

FIG. 315, GAILLON—GENERAL VIEW

attracted the court and nobility. Henry IV., however, cultivated the gardens of the ancient place, and cared for them. Once again they rejoiced in the delights of a royal visit, when great festivities were held in honour of the presence of Louis XIV. in 1668. That, however, was a last farewell, and the beautiful gardens began to decay. It is true that Gaston d'Orléans intended to revive their old-fashioned charm when he introduced fine showy features in a new style, but he only succeeded in destroying much of the beauty of the old castle with his grand but depressing new buildings. After his death the gardens soon degenerated, and now we can see nothing but the buttress walls, and these only where new roads have not made an end of them.

While Louis XII. was making his garden at Blois, his minister, Cardinal Amboise, was building Gaillon near the Bishopric of Rouen in competition with his master (Fig. 315). The mediaeval castle that used to be there had been demolished by the English. The cardinal, as Du Cerceau remarks, carried out his work *sans tenir de l'antique.* In spite of this, however, the ground-plan was quite informal, and there was a moat all round crossed by many bridges. The chief buildings, with turrets at corners and doors, stand about an almost square court with a wonderful fountain in it—a present from the Venetian Republic to the first French Minister, who was also a cardinal (Fig. 316). As a sign of homage this present was brought by sea, *stupendo fonte marmoreo ex Venetorum munere illustrato,* as the inscription says. On the plinth the cardinal had his own and his royal master's arms carved. The wonderful fountain of three shells, of which Du Cerceau's engraving is our only record, was adorned with eight lions' heads and eight masks, with fine figure groups on the middle shaft and a statue of the Baptist on the top. It was the work of a group of Genoese sculptors, with Agostino Solari, an artist who had many commissions at the time, not only in Italy but also in Spain and France, at their head. The fountain was afterwards removed to the archbishop's garden, and was destroyed in the eighteenth century. When one regards the wealth of ideas and the wonderful vigour in the execution of a work like this, it seems that the poetical descriptions of fountains, from Greek antiquity to the Middle Ages in the West, need no more be thought of as the inventions of uncontrolled imaginations.

These gardens of Gaillon are not in real connection with the buildings. One walks through the great court into the main garden, which, as at Blois, forms the middle terrace. Through a turreted gate in a gallery that runs along the garden on two sides, one passes into the parterre: a tree-garden stands on a higher terrace, and the thick trellis makes an effective contrast with the cheerful *parquet* of the garden. From above one looks down upon two labyrinths. "Here," says Sir Henry Wotton in the seventeenth century, "we can follow our friend with a smile as he wanders about picking berries, till at last he will never find his way out unless somebody helps him." The other beds are shown by Du Cerceau to be bordered by box hedges, and planted at the corners with low trees, but originally they had a wooden bordering with little doors in it, just as at Blois. In the box borders there were either flowers or different-coloured clays, slate or terra-cotta.

Though Du Cerceau seldom mentions unworthy trifles, we learn from the accounts that the beds were laid out to depict coats of arms, and all sorts of animals cut out of wood were used to decorate the garden. A masterpiece of joiner's work, a pavilion finer than the one at Blois, because at the four corners were little aviaries, stood in the centre. The lovely two-shelled fountains were protected by the dome, a

FIG. 316. GAILLON—THE GREAT FOUNTAIN

companion work to the one at Blois made in Tours by the same artist: at the top was a statue of John the Baptist. The friendly Louis did not take amiss the rivalry of his cardinal-minister, as later princes would have done, but at times lent him his own gardener. The laying-out of this place was also entrusted to Pasello; and one of his relatives, Piero da Mercogliano, became permanent gardener here. According to Italian custom, Pasello had his kinsmen come out to him from home, and they were mostly, like himself, half priests and half gardeners.

In the axis of length, opposite the entrance tower, there is a small two-storied casino, that may perhaps have had a living-room in it. On the side of the valley the garden is shut off by a gallery, which is a masterpiece of its kind, with its five entrance turrets and windows with gables. It probably had entertainment rooms inside. From the windows there was a famous view of the great garden that lay below, and beyond that of the pleasant valley. But people could not and would not walk straight down into the lower garden. Each terrace had to be an independent whole, with no relation to the castle or to the other divisions of the garden; it must have its own architecture and its own axial direction. So to get to the much larger garden on the third terrace the only thing to be done was to return to the court and climb up by concealed steps from there. It was not till the time of Le Nôtre that the various terraces were connected with stairs. The lowest garden, not in the same line as the one above, was mainly kitchen-garden and orchard. Two beautiful *berceaux*, as they called the barrel-shaped walks made of trellis and covered with greenery, were by the side of the entrance path. Second only to Blois, the mulberry-trees were famous, two hundred of them, their cultivation being one of the most important duties of a sixteenth-century gardener. An attempt was made to grow peaches at Gaillon, but it must be admitted that they proved very expensive. Close to this garden was the vineyard. The park that rose in a sort of terrace-slope led to an unusual and peculiar place, the hermitage; but this belongs to another period and another type of sentiment.

Gaillon must be reckoned as one of the finest products of Louis XII.'s time; but if we want to study the natural bent of French style and its relation to its Italian teachers and guides, we must remember that Villa Madama was built at almost the same time as Gaillon. While in France the gardens held firmly by the mediaeval castle scheme, being separate and self-contained, having no contact with the buildings and no links with one another, in Italy we find the cheerful open loggia, the house and garden all in one. Though the Italian garden is not condemned by the stern decree of an undeviating axial line, the various groups are interconnected with a number of different stairways, and each group part of a composite whole; whereas at Gaillon the actual approach is so poor that Du Cerceau proposed to improve it by adding a proper stairway.

The peculiarity of the gardens of France does not depend upon their markedly secluded character, which is after all negative, nor in the tendency shown later to a sort of licence that it was first to assume and then to change in a remarkable way, but rather on the successful adaptation of another mediaeval feature of the castle scheme, the trench for water. The French adhered longer, and with more conscious intention, than any other country to that type of castle which is encompassed by wide moats. And when at last these had lost their old use as a protection to the house, they remained merely as ornaments. Indeed we shall see that new buildings, without any compulsion from an

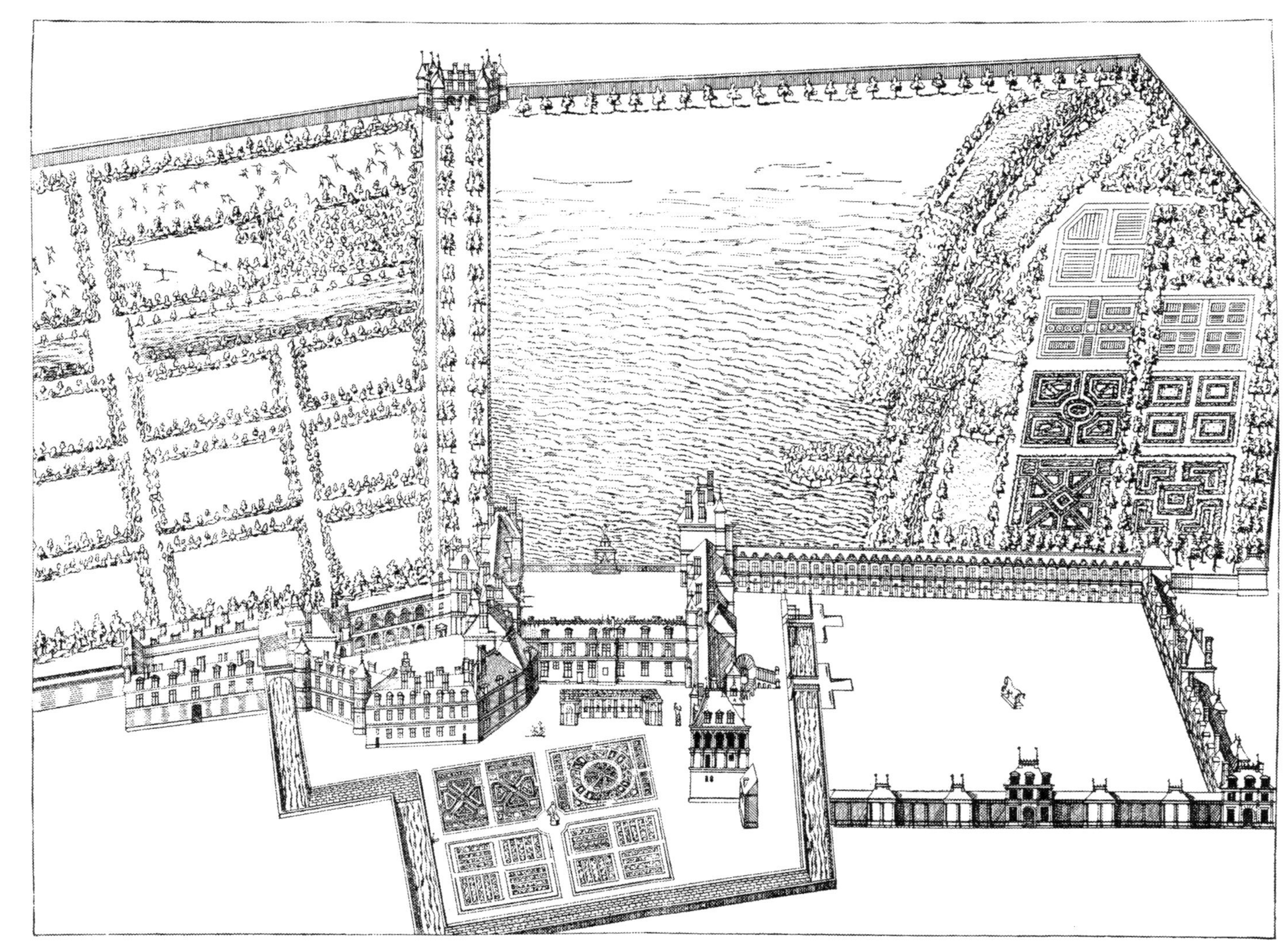

FIG. 317. FONTAINEBLEAU AT THE TIME OF FRANCIS I

old ground-plan, made use of such trenches as decoration, and that they had a great influence on gardens as time went on.

Two castles whose Renaissance style dates from the reign of Francis I give a striking picture of the first phase in the development of this kind of castle in relation to its gardens: these are Fontainebleau and Chantilly. The first was an ancient hunting-seat, in the middle of a great wood (Fig. 317). Here Francis I. built his fine castle, which embraced a number of courts grouped in no sort of order. A wide canal ran round the greater part of the buildings; and as there was a watery marsh close by, it was easy for the king to convert it into the gigantic pond which bordered the castle on one side. Near this pond, which was always famous, and still is, for its supply of carp, there is an avenue of four rows of trees extending from the old house to a pavilion at the entrance gate. On the other side of this beautiful promenade, which is raised like a bank, there was then a garden of fruit and meadow land where the parterre now is. Not only does the wide canal run through the middle, but every *préau* with its great trees is surrounded by a little canal as well.

By permission of Mr. P. D. Hepworth

FIG. 318. FONTAINEBLEAU—GROTTE DES PINS

On the other side of the pond Francis made the so-called *Jardin des Pins*, probably a kind of winter garden, planted for the most part with fir-trees, between which Du Cerceau's engraving shows formal beds evidently laid out in box, and perhaps containing useful vegetables. Here too there was a wide canal along the façade of the castle, the Gallery of Ulysses. At the end of the gallery a grotto under the corner pavilion completed this beautiful garden picture: great arches of *rustica*, flanked by giants (Fig. 318), reminding one of Giulio Romano's work at Palazzo del Te, lead to the inner part, which is adorned with water-fowl and sea-beasts, and has running fountains. There is a little nymphæum in front forming a court with pleasant walls and fountains, overarched with green.

This place is now the only part of the garden of Francis I., which (pretty well preserved) shows what it was before it was changed into an English park. But the real flower-garden was on the north of the castle, inside the canal. This little garden, cut up into four squares with wooden pillars marking them off, and laid out with box, flowers, and earth of various colours, contained as its finest ornament and central point the so-called Diana of Versailles. Apparently Francis put up other statues as well. It is an odd thing that a pergola which ought to give a sort of connection between the garden and the informal façade is, according to the drawing, not in the same line with the garden.

The second castle, perhaps equally remarkable, Chantilly (Fig. 319), stands on a rock in a wide marshy valley full of trees. Throughout the Middle Ages it was a strongly fortified spot. In 1495 it came into the possession of the Montmorency family, and in the years 1524–32 the famous Constable of France, Anne de Montmorency, built the Renaissance castle on the old foundations, retaining the three towers at the corners. Unfortunately the engravings of Du Cerceau do not give a good picture of the gardens, but they are laid out in the same style as at Fontainebleau. The wide tanks round the castle leave far too little room for much garden. On the little island beside the main castle a casino was built later, and in front of this was a small parterre with an aviary at

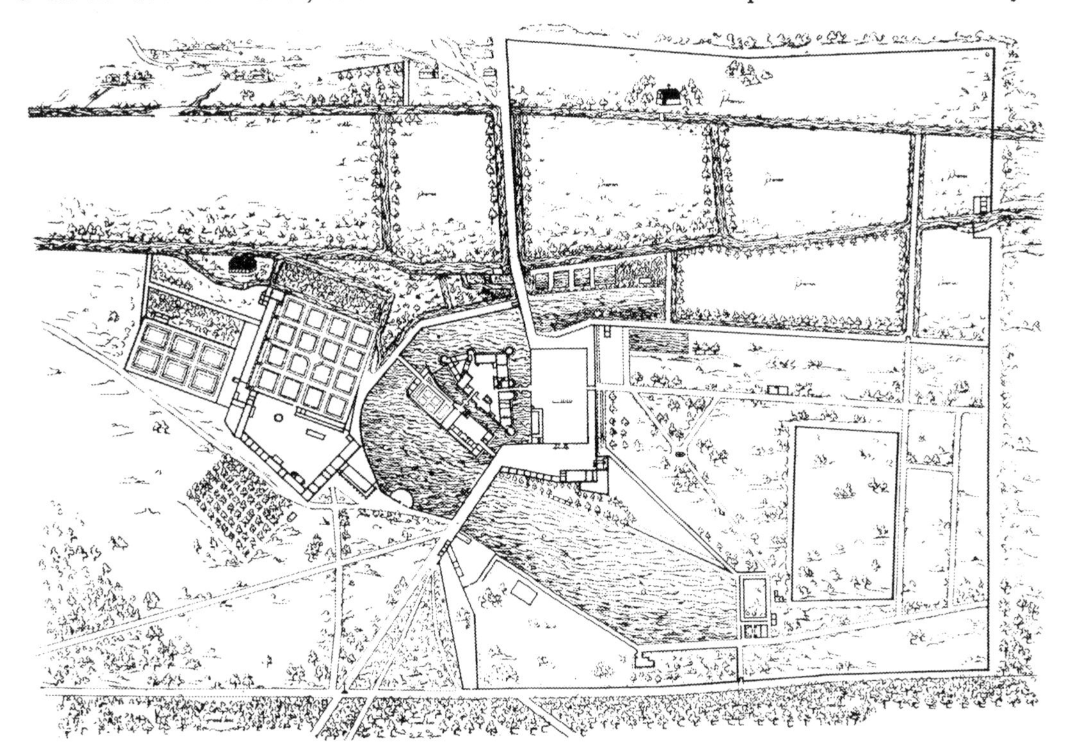

FIG. 319. CHANTILLY IN THE SIXTEENTH CENTURY

the end, which still gives its name to the little garden. The constable made other gardens to the east of the tanks near the kitchen parts. In front of the tennis court—an indispensable adjunct to a castle both in France and England in the sixteenth century—there were several formal beds, and a grotto which has left certain traces behind, while on the other side a high gallery, called *des cerfs*, made the end of this parterre.

The whole place and its arrangement recall vividly the *Jardin des Pins* at Fontainebleau. They also have in common the whole meadow-garden with trees, a four-cornered ground, with small canals round it—a plan that helps the irrigation very much. Certain of the parts inside the canal system were covered with cornfields. The ground-plan is quite irregular, all the garden features are separate as though they were scattered about by chance here and there, but everywhere there is water either in wide tanks or narrow canals, pleasing but scarcely ever artificially wrought, and serving rather as a frame for house and gardens. At a later period the land will be laid under the control of a systematic design, in which water will play a leading part.

Gradually a feeling grew in France for the nearer connection of house and garden, and similarly for uniting the two symmetrically with the ground-plan. A very interesting find of engravings of a villa design—after the researches of Geymülle they may be attributed to Leonardo da Vinci (Fig. 320)—shows how a great artist near the end of his last weary days in France submitted himself to the artistic feeling of that country, and helped forward the scheme of a water-garden: the inscription says that the castle is to follow the style of Amboise. King Francis had prepared a house for the aged master, when he secured him for his country's service in 1516, a place in the artists' colony in the little castle of Cloux, and there Leonardo closed his eyes three years later. The church at Amboise had the honour of receiving his body.

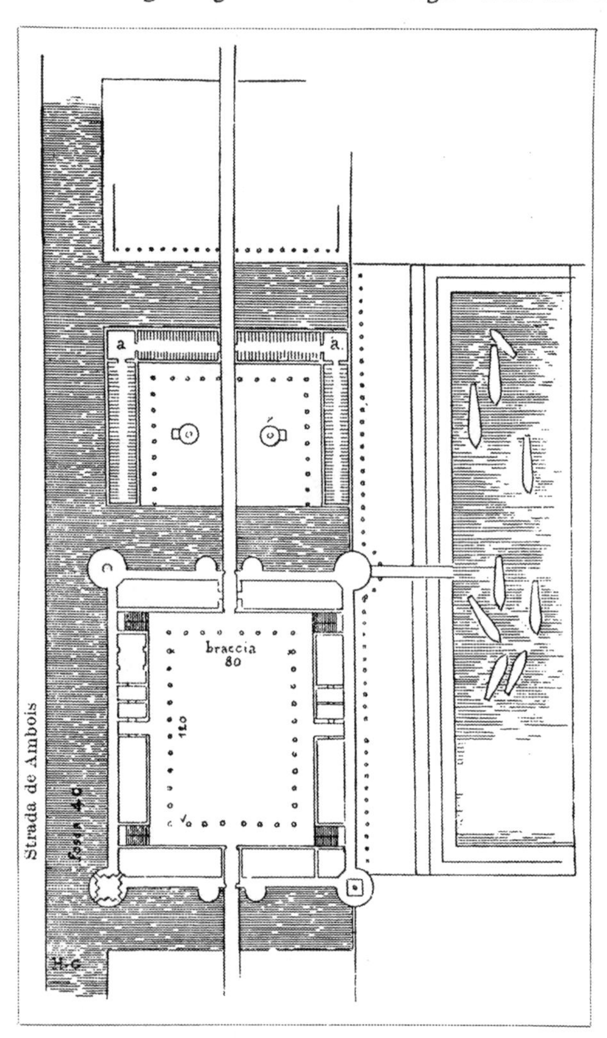

FIG. 320. A PLAN FOR A CASTLE

The ground-plan shows a formality not yet natural to France, but the castle built round a pillared court is flanked by four corner towers and surrounded on three sides by wide moats. A bridge leads to kitchen regions with colonnades round, or perhaps it may be a garden with two fountains, ending with ground strictly orientated in the right axis, and with a canal round it, though unfortunately this part of the plan is not very clearly shown. Evidently there is a corresponding garden on the other side of the castle. By the side of this place there is a large sunk pond, intended by the artist to serve in antique fashion as a show-place for water tournaments. This is an indication that the great ponds at Fontainebleau, Chantilly, and other places were used for this purpose, and explains their immense size.

An attractive little castle that first attempts to carry out the ideas of unity between house and garden is Bury (Fig. 321). It is only a few miles from Blois, and is now a ruin. It was built by Florimont Robertet, the accomplished diplomatist attached to Francis I. He vied with the king in his love of building, and built the place with the money that Francis had given him as a reward for the successful negotiations which he had carried out with the Venetian Republic. Robertet brought his feeling for art back with him from Italy, and perhaps his architect also. The whole place makes an almost perfect square; but there was a wide moat round it, and the mediaeval aspect was maintained by its having seven corner towers and portal turrets on each side of the drawbridge. The gate led into the *cour d'honneur,* which seems to have been grassed over; in the middle stood a bronze David by Michael Angelo, since destroyed, which the owner had received as a present from the Republic of Florence.

A second court for kitchens lies behind this, and these two courts correspond

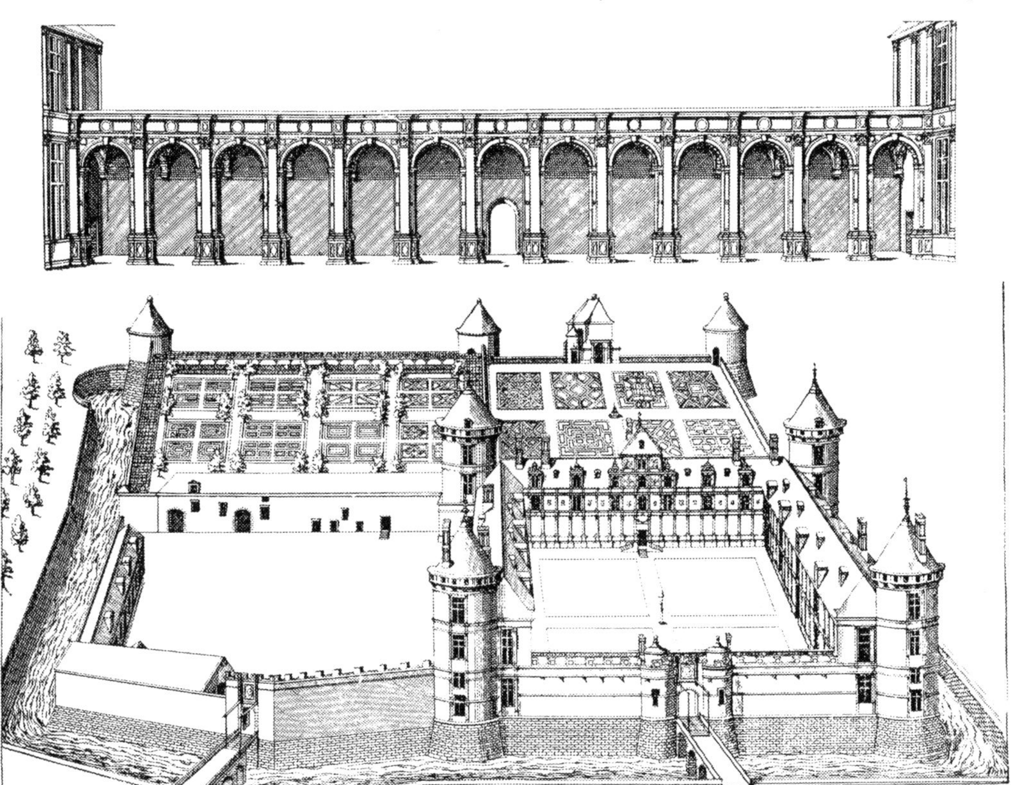

FIG. 321. BURY—A GENERAL VIEW OF THE CASTLE

to a pair of gardens connected by a walk shut in with trellis-work. From the castle façade rounded steps lead to the parterre, which shows eight beds designed by Du Cerceau in particularly fine geometrical patterns. In the middle stands a two-shelled fountain in an octagonal basin, and at the end opposite the steps is a little chapel built out over the walls as at Amboise. It is worked into the plan of the garden in the same way as the one hard by at Blois. The corner towers seem to have been also used as garden houses. The second garden, answering to the kitchen court, is more of an orchard and vegetable-garden, with a cheerful pergola running round it, crowning the terrace walls.

The unity and small size of the whole place at Bury leaves on one side certain more complicated problems of garden development. A series of small castles, all built in the sixteenth century and not completely finished at the date of Du Cerceau's engravings, now lead up to the perfection of what we may best call the type of canal garden that belongs to the French Renaissance. The earliest of these, Dampierre (Fig. 322) near Boissy on the Seine, belonged to Cardinal Lorraine, who bought it from a banker. It lies in a woody valley, and has no view, which Du Cerceau considers a defect. The parterre is in front of the master's house, which is built round a court, forms an obtuse-angled triangle, and is enclosed with galleries that have turrets at the corners. Both castle and parterre have a wide canal round them, and walls. On the other side, taking the whole breadth of house and parterre, there lies a second large square garden, separately enclosed in a narrow canal and divided into formal beds with regular patterns. There are four bridges over it, answering to the chief walks, while, strange to say, the main canal has no bridge at all. A third and a fourth canal, bordered by avenues, separate this whole garden scheme from the park crossed by its formal rows of trees. On the other side of the castle (arranged as in Leonardo's plan) a very large pond is found, no doubt meant not only for fish but also as a place for a naumachia.

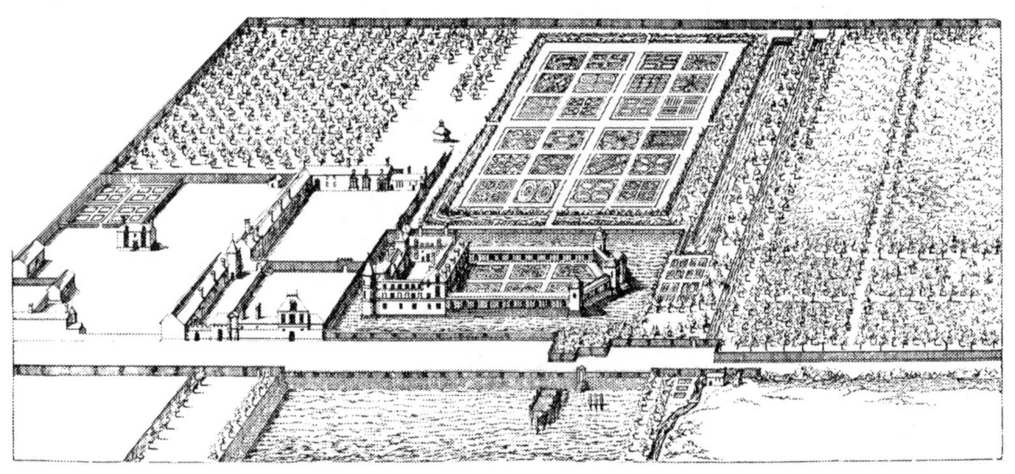

FIG. 322. DAMPIERRE—GENERAL VIEW

A similar preference for a great water place, that gives breadth and a certain importance to the garden, is found in the castle of Valleri (Fig. 323). The building itself kept its old mediaeval fortified appearance on the top of a high hill. The garden lies to the south at the foot of the hill, and is so lost to view that it can only be seen from its own buildings. Round three sides of it runs a pillared hall, flanked on one of the smaller sides by two pavilions at the corners, which might serve as places to live in, as they have every kind of convenience (Fig. 324). The beds are partitioned in the way we know so well, bordered in sets of four. There is a raised terrace all round which gives a pretty view over the garden. A wide canal, rounded at the end, occupies the whole of the middle part in the axis of length, following along the four sides of the garden in an underground conduit: this finally debouches into the immense pond, which far exceeds the whole area of garden ground. If we stand on the terrace, we overlook the pond on the one hand, and the whole flower-garden on the other, while above the roof of the gallery the tops of the trees appear from the orchard behind, which is traversed longwise by three wide canals. The plan of Du Cerceau shows a third bit of garden, with narrow canals round it. Each of these gardens is quite separated from the others by walls, and they have no immediate connection with each other, although they are set in the same line, and have a certain sort of relationship due to their raised walks.

In the last part of the reign of Francis I., there was a striking development of fancy among architects, corresponding with the love of building among the great and powerful, and they shrank from no effort of skill. Francis built as a hunting-box his strong turreted Chambord, and put there trenches and bastions; he went so far as to propose to make a mighty water-castle, with the idea that he might divert one arm of the Loire. According to Du Cerceau the gardens themselves never matched the buildings in magnificence, but the park, which the king enclosed with a wall, was of unexampled size at that time. The estate, which was marked out in 1523, measured 5500 hectares.

A counterpart to this fine hunting-place was another much smaller *rendezvous de*

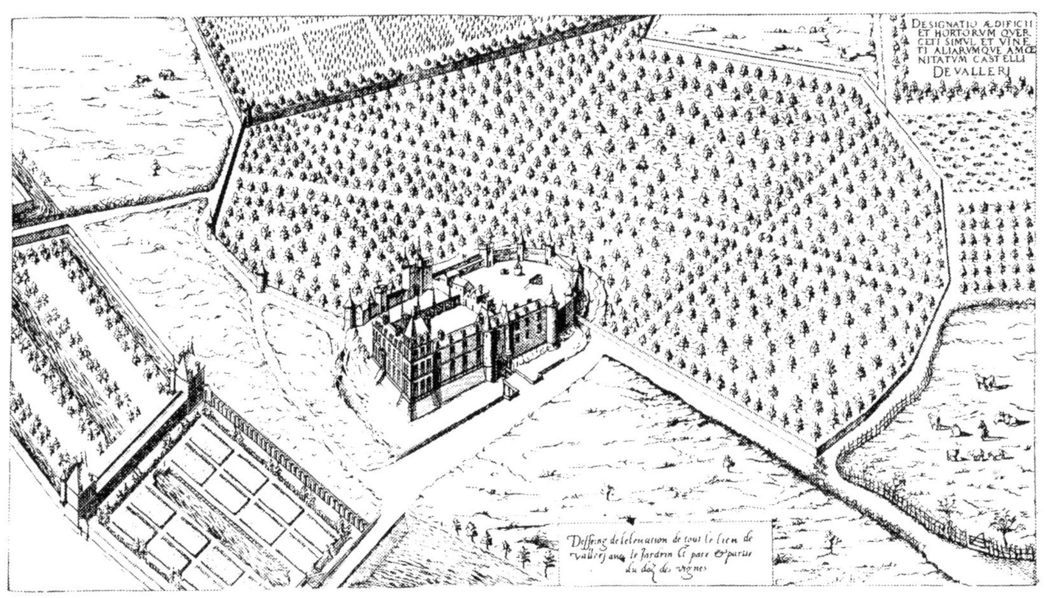

FIG. 323. VALLERI—GENERAL VIEW OF THE CASTLE AND GARDENS

chasse built by Francis in the Bois de Boulogne. The name "Madrid" which it bore till it was destroyed in the course of the Revolution, gave rise to all sorts of tales, and it was said to be put there in remembrance of Spain, and even of the king's imprisonment. No doubt he did bring from Spain the idea of its decoration, which filled the French with surprise. The façade of the castle was adorned with faience and enamel, which would remind the king of the tiles that Spaniards used for their villas and gardens. He had terra-cottas of Girolamo della Robbia put up, but of course by this time an international exchange of art treasures was quite common. When Evelyn visited "Madrid" in 1650, he said the material was mostly a coloured clay, and very brittle, though the colours were very bright, like porcelain or china ware. Whole statues and reliefs, chimneys and pillars, both out of doors and indoors, were made of this terra-cotta. The house had a moat round it, and there was a wonderful view of the river and the Bois de Boulogne. Chinese porcelain, which was destined to influence French taste so strongly for more than a hundred years and the style of French villas, had probably had no effect so far on the ornamentation of "Madrid." Unfortunately Evelyn says nothing of the garden, and Du

Cerceau says very little that is satisfactory. But the fact that Francis intended the place to be used not only for the chase but also for his love affairs, makes one suppose that there were gardens carefully tended in the charming parts outside the castle. After many changes and chances, the little castle did experience a sort of renaissance before its fall, for it was here that Madame Necker held her intellectual and much-frequented salon.

A castle designed by Thomas Bohier, one of the generals of Francis, was indeed a whimsical fancy. The great castle of Chenonceaux was built on the piers of an old mill in the River Cher; and this shows that the preference of the French for water castles was never satisfied without novelty and change. It was not till the castle had become a royal possession, and Henry II. had presented it to his mistress, Diane de Poitiers, that a long bridge was made connecting it with the opposite bank. The completion of the whole thing was due to Catherine de' Medici, who after her husband's death compelled her hated rival to give up the castle. Unfortunately Du Cerceau's plans show little of

FIG. 324. VALLERI—THE PRINCIPAL PARTERRE

the garden, though his descriptions are most eloquent. But since Chenonceaux is to-day one of the few French castles that show, at least in part, what their Renaissance gardens were like when restored, we are able to see that there was no ground-plan that blends the castle and the garden into one. The parterres, situated on the river bank, at the side of the kitchen-court, of trapezium form, which was never quite finished, have canals separating them on one side from the castle, on the other from the park.

Du Cerceau mentions another large garden beyond the river, which has entirely disappeared. The great parterre that exists to-day has a raised terrace round it, which may have had pergolas or galleries, though after its restoration in the nineteenth century it shows no trace of the architectural surround usual at the period. Du Cerceau speaks of an artifice in the middle that is meant for a surprise: there is a small flint with a hole in it that is closed with a wooden tap, and this conceals a waterspout, which dances up six feet into the air if the tap is taken out: this Du Cerceau calls "*une belle et plaisante invention.*" On the other side where there is now an equally pretty sunk parterre, very charming with its flower-beds edged by box, Du Cerceau describes a fountain issuing from a rock (as the place is quite level, we must suppose that it is worked artificially) running into a large basin in several streams. A flower parterre is in front, with a terrace round it, enclosed by a trellis, also with a dividing wall which is ornamented with niches, columns, and statues.

After the second half of the sixteenth century this separation of garden from house is an exception unless there is a special necessity for it, and it only happens at Chenonceaux because of the eccentric character of its ground. France arrived much later than Italy at the deliberate treatment of the garden scheme, but her peculiarity lies in the fact that she came to this in her own way, and was less and less dependent on the ways of Italy.

Du Cerceau says the Castle of Anet (Fig. 325) was one of the finest of that day; it was built by a very famous architect, Philibert de l'Orme, at the command of Henry II., for Diane de Poitiers. No doubt de l'Orme had to use the old foundations of a mediaeval castle, but in his combination of house and garden he produced an architectural whole which ranks with the finest Italian masterpieces of his time. He knows how to make use of all traditional ideas, and also how to adapt them, both to the whims of a pleasure-

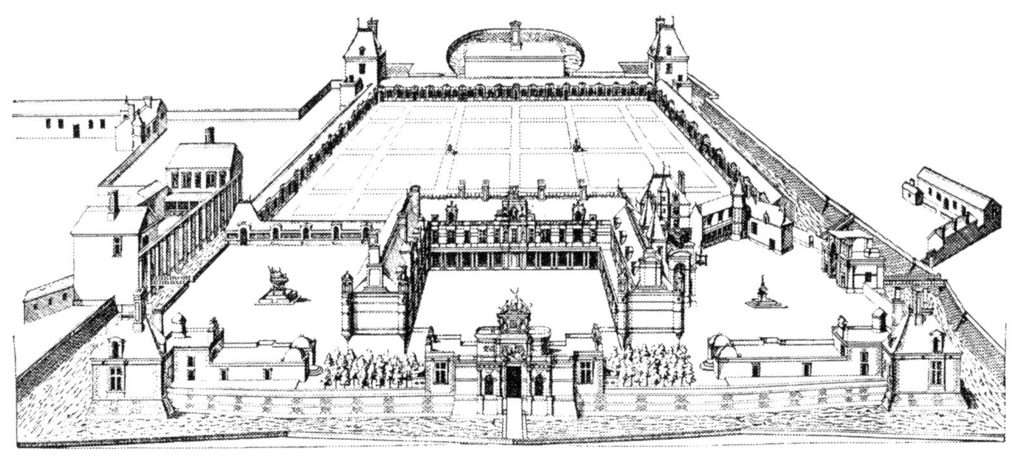

FIG. 325. ANET

loving day and to the caprices of the lady. The wide strip of water round the place, though it is furnished with walls, bastions, and towers at the corners, is only employed as an enlivening ornament. We step over a drawbridge to the entrance lodge, itself a cheerful building on pillars, with Benvenuto Cellini's nymph of Fontainebleau let into its lunette. On either side of the portal there are thickets, whose tree-tops cast shadow as we enter, and suggest to us—as does the stag with dogs on the top of the gate—that we are now in the domain of the namesake of the goddess of the chase, with whose trappings the king's friend loves to be adorned. At the sides of a front court there are two others, each with a fountain in the centre, one of them being Goujon's famous Diana. Through the middle gate, opposite the entrance lodge, the only part of this beautiful building which has been preserved for us, we pass into the garden, which extends the whole width of the three courts.

In front of the castle a broad terrace gives a fine view, and from it we step down to the garden. This has on three sides a *rustica* gallery, "*qui donne au jardin un merveilleux éclat à la vue,*" as Du Cerceau says. There are two fountains orientated to the wings of the castle, and these adorn the fine beds, and also give to the garden, which is a good deal wider than its depth, the balance of a (so to speak) double centre. At the back the canal

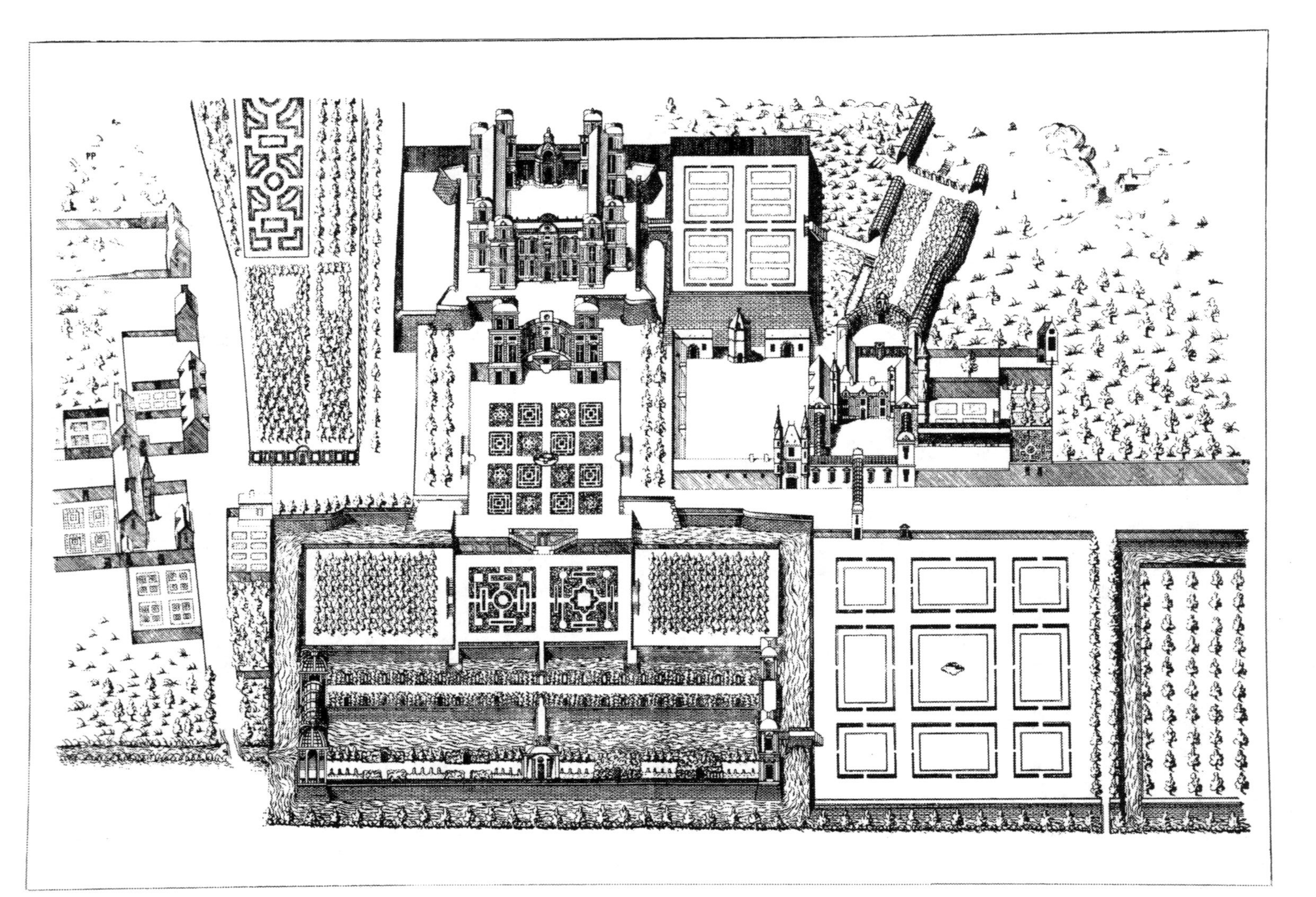

FIG. 326. VERNEUIL—GENERAL PLAN OF THE CASTLE AND GARDENS

widens out into a semicircular basin with a garden house in front that may perhaps have been used as a bath.

Unfortunately in the matter of these further gardens Du Cerceau's engravings leave us uninformed; but he says there were two parks behind the flower-garden, quite separate from one another and fenced in. In one of them there was not only a semicircular grotto, but a falconry, fowl-houses, and above all an orangery, small and pretty, of which Du Cerceau also made a drawing; and this park with its *rustica* façade, corner pavilions, and fountains among the beds is a kind of picture in miniature of the whole garden. The separate parts of the park are called "parquet" by Du Cerceau, as are also the square beds, and probably the word may be taken to mean any piece of garden that is square. These separate parts included "some of them meadows, others bushes, others again shrubberies, or warrens, orchards, or fish-ponds. They were also set apart by avenues with little canals between them. In fact there was everything that could make a place perfect." This verbal description of Du Cerceau's gives us an exhaustive picture of the park of that time, and enlivens the formal sketches of parks that we get in the engravings.

Two castles reached the highest point of development in this style, but they were never quite completed and have since vanished from the face of the earth. These were Verneuil and Charleval. Our information about them comes from Du Cerceau's pictures, and though they may never have been carried through, they rightly hold their place, because in all probability Du Cerceau was himself the architect who designed them both. At Verneuil (Fig. 326) there was a nobleman, a connoisseur—Philippe de Boulinvilliers—who owned a large castle in the charming valley of the Oise. In 1558 he decided to build another on the hills near by, and this lofty erection was equipped with bastions and trenches. The former served as a castle terrace, and in the fashion of the day were paved with mosaic, like the other terraces of the same sort at Anet and at Gaillon.

Towards the garden on the terrace, which widened at the court, there was another gate-erection, an exedra flanked by two pavilions. Stairs led down to a sunk parterre, which had a two-shelled fountain in the centre of its sixteen finely ornamented beds. Narrow and higher side-terraces made a shade for this flower-garden with their avenues of trees. More steps led down to a decorated wall, and encircled a path of grottoes leading to a second parterre, which was perhaps another shrubbery, or possibly a kind of labyrinth. On the side there were also two raised terraces planted with woodland trees, and these, being of the same breadth as the middle parterre, gave considerable width to the garden. A wide canal runs round the whole plot, and bridges cross the canal on either side, leading from the middle parterre to a narrow terrace, which is enclosed right and left by a trellis covered in greenery. A second arm of the canal separates this narrow terrace also from the last one, which brings the garden to an end. In the middle there stands a pavilion approached by the connecting bridge over the canal. On both sides the last two terraces are united, on the right by a gallery, on the left by a pergola with trellis. The view must have been very imposing from these lowest terraces over the different parterres with all their waters reaching to the high exedra; and from the height of the exedra (looking in the other direction) the eye took in the beautiful expanse of garden, while on the left hand there was the old castle at the foot of the hill, which was itself connected with the new castle by a series of not insignificant garden plots coming up close to the hill.

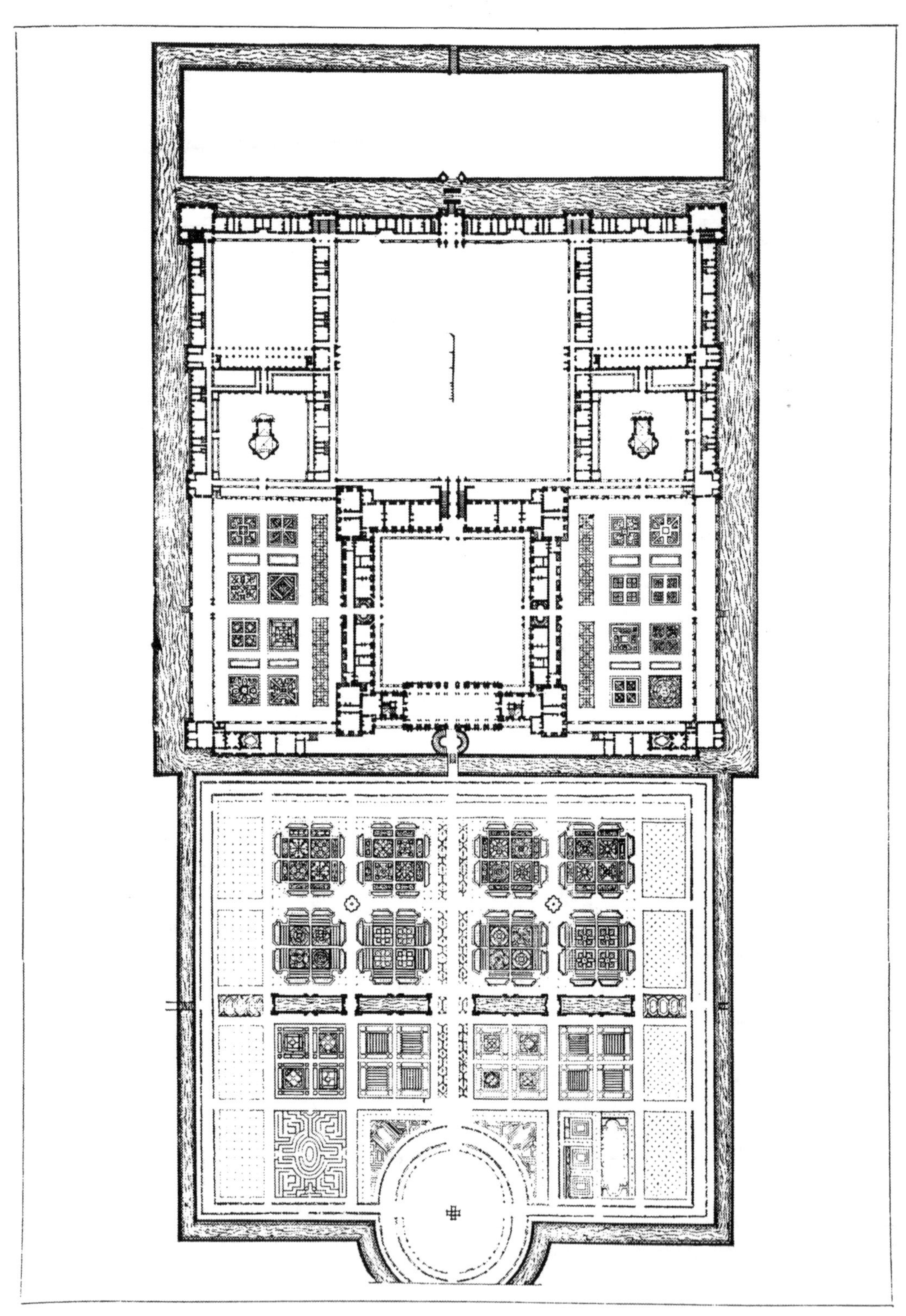

FIG. 327. CHARLEVAL—GENERAL PLAN OF THE GARDENS

This abode of a private person was excelled by the royal castle built by Charles IX. in "Normandy," between Paris and Rouen, if not in its variety of parts, certainly in the splendid execution of one coherent scheme. It was in a valley, called by the king "Charleval," after himself (Fig. 327). We know nothing of this castle except from the drawings made by its founder, and very likely only a small part of his plan ever materialised; but the laying out of the garden began at once, so that when the building was completed the full impression might be given. In this instance also the ground-plan of the whole is indicated by the lines of the canals which passed round and through the castle and garden; through a forecourt surrounded by water we pass over a drawbridge to the great kitchen

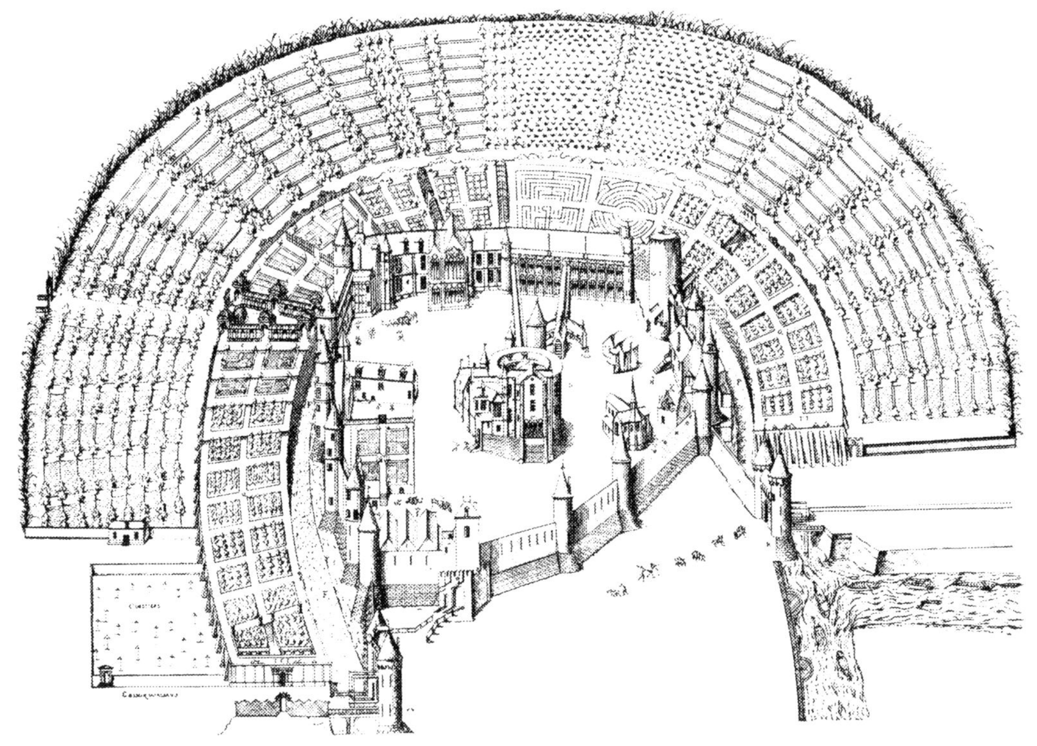

FIG. 328. MONTARGIS—GENERAL PLAN OF THE CASTLE AND GARDENS

part with four smaller courts round it. The square of the castle that comes next has a garden parterre on each side, and these are treated as *giardini secreti*. They have galleries on three sides, and at the castle side are bordered with two fine pergolas.

In the middle of the castle front is a very large garden-room from which there are horseshoe steps up to the bridge over the third arm of the canal, and leading to the extensive garden parterre. This has not only canals round it, but is also cut by four basins rather like canals in the middle. The central path is marked out by covered walks. The ornamental beds are arranged four on each side with a fountain, while clumps of trees by the arcades that accompany the edges of the canals supply a note of solemn shade to this garden of brilliant colours. The garden behind the four basins in the cross-path is treated in the same way, but there are no fountains, and one of the divisions is a labyrinth. The lower end of the garden is a rather long, large piece with a bow at the farther end where the canal is rounded off. This region is noticeable for several rows of arcades, and very likely there was a small pavilion in the middle.

All this fine garden, Du Cerceau says, was only half of what the king had planned, and on the other side of the oblong there ought to have been an equally large flower-

FIG. 329. MONTARGIS—TRELLIS IN THE CASTLE GARDEN

garden. He was right in calling the place "worthy of a monarch," adding, "If this building had been finished, it would have ranked with the first in France." With the plan of Charleval the French art of gardening, joined with the castle scheme, reached the first stage of its growth. Here was the tradition of the Middle Ages at work—not at all in the same way as in Italy—gradually and uninterruptedly effecting the change, which was to bring into being, from the starting-point of the strong mediaeval water-castle, this magnificent offspring of the Renaissance, through the blending and union of house and garden.

Apart from direct progress there are various gardens belonging to this time that bear witness to the great skill of their architect. In the garden of Montargis (Fig. 328) the difficulties in carrying out a formal plan seemed all but insuperable. When Renée, Louis XII.'s clever daughter, after the death of her husband Hercules II. d'Este, left Ferrara in 1560, Montargis was assigned to her for her home as a widow. On the steep mountain-top she found a neglected, ill-arranged castle within the turreted walls, with two little gardens all telling of the forgotten Middle Ages. But Renée, after a time of bitter trouble and subjection, which she had suffered in Italy at the hands of a church she hated, felt disposed to make herself the centre of a Protestant community in her own beloved home. This friend and protectress of Calvin, so highly praised by Goethe, was not only, in the words of her daughter Leonora, "in wisdom and right thinking such that none of her daughters equalled her," and one of the ablest women of her time, but she was also a friend and patroness of art. To the special art of the garden she was already friendly from her long stay in Italy. She summoned Du Cerceau, that he might rebuild the castle and lay out the gardens; and in a curious fantastic way (suited to the nature of the place) he acquitted himself of the task.

He set out the gardens in two concentric terraces round the semicircular walls of the castle, which in spite of the steep slope had moats with plenty of water. On the first terrace was the flower-garden with two rows of beds, box ornaments, and flowers in the middle and clumps of trees at the side, which every here and there were interrupted by leafy paths in trellis (*berceaux*) (Fig. 329). The great difference in level was counteracted by very high supporting walls, which held up the terraces with small crossway walks between. In two places sloping paths led to the second terrace, and these were bordered by fine wooden pavilions and arched ways, in which the master produced his best work. This second terrace was a vegetable and fruit garden, which covered the foot of the hill in regular rows of avenues and formal beds of trapezium shape.

The symmetry of the large garden which Catherine de' Medici made at the royal seat in Paris, the Tuileries, was almost too stiff. Catherine had the castle built by Philibert de l'Orme, with a wall round its garden, so that not only was this separated from the castle by a street outside, but had practically no relation with it, though both were in the same axial line (Fig. 330). The entire evenness of the ground and the little use of water rendered the garden very uninteresting in its first state. Visitors made for the place of echoes at the end of the garden, a semicircular region where voices at certain spots seemed to come from the clouds or perhaps from under the earth. They also found a labyrinth and a sundial in the shrubbery, and best of all a grotto that was made by Palissy—a masterpiece to be spoken of later. In Henry IV.'s time the parterres received particular notice, for the king planted the great mulberry avenues at the side of the garden. The chief important changes really became apparent in the great period of French gardens, at a far later date, under Louis XIV.

Although after the middle of the sixteenth century the complete regularity of the ground-plan became a matter of course everywhere, only to be yielded through sheer practical necessity, it was quite permissible to indulge any sentimental fancy within these limits. Every geometrical form was allowed, such as a triangle at Azay-le-Rideau, and a pentagon at the charming little castle of Maune, which is rather like an attempt at Caprarola in miniature. The plan seems really to have been made inside a circle, so regular is it with its perfect rounds and lines of connection. There is a pretty little garden, but only on one side: it leads out from the front of the house, and contains a sunk basin. Central buildings in any kind of geometrical form are to be found in the large "Ideal-plans" which architects set out on the drawing-board, indulging their fancy untrammelled by

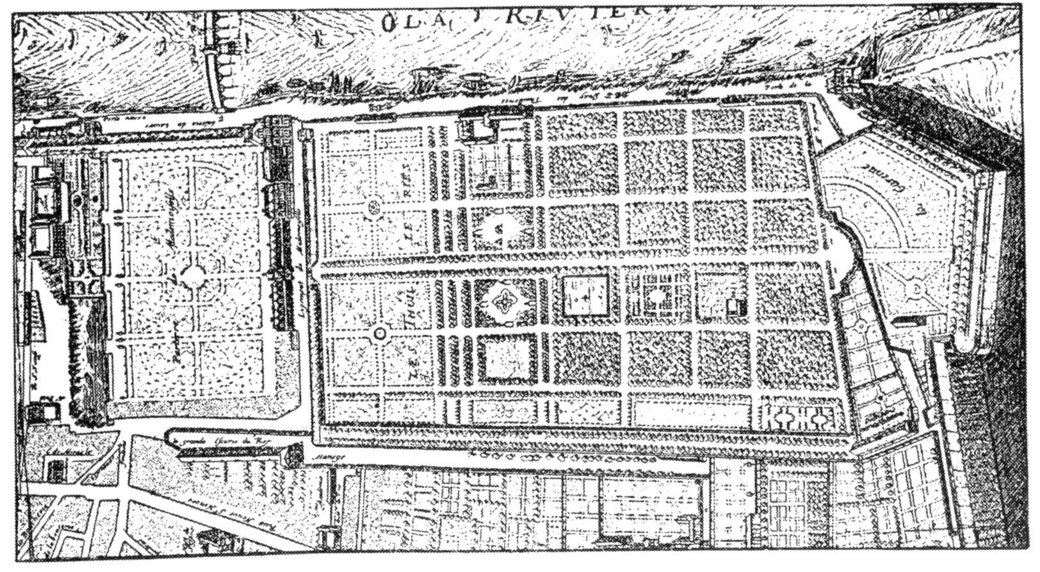

FIG. 330. THE TUILERIES—PLAN OF THE GARDENS

fact, and here de l'Orme and Du Cerceau are especially prolific, the latter allowing himself a number of *toute sorte de jardins* on every side, and then treating them separately as *giardini secreti.*

The descriptions given by poets came out to meet, so to speak, the visions of the architects. Rabelais had already told of a perfectly formal hexagonal place, with towers at each corner and water all round, in his Abbey of Thelemites. In these complete connecting links between different parts of the whole, in the blending of number and rhythm, Rabelais was trying to get a model and a sort of foundation for his idea of a common life on the one hand and the educational training of mankind on the other. He based the desire for harmony, not on any compulsion, but on free will following out the highest principle of *Fay ce que vouldras.* In the middle of his court he has a fountain of the Three Graces—three young female figures, who pour forth waters from various parts such as eyes, breasts, mouth and ears. He would have gardens in all the six sections of the castle. On the side where the Loire flows by there is the flower-garden, with a labyrinth in the middle. On the other side the fruit-trees in the orchard were planted in a quincunx. In the park farther on and near the castle, there was no lack of baths, animal cages, hippodrome, theatre, open and enclosed places for tennis, targets for archery—in fact, all the things that could be found in the parks of that day, or that a learned student of past history could supply. Rabelais does not say much about the real laying-out of the individual gardens; and any special feature is only useful to him in so far as it serves the ends and suits the style of his general scheme of ideal education.

Very different is another Utopian thinker of the French Renaissance, as he lingers awhile over the garden and its style—Bernard Palissy, the renowned potter and handicraftsman. In his little book, with the strange title, *A True Recipe, by which all Frenchmen may learn to add to their Treasures,* he describes "a precious garden, so lovely that it is second to none in the whole world except the Earthly Paradise." Thus does he speak in the dedication to the Duc de Montmorency. Palissy had found in the duke's father, the great Constable who laid siege to his insubordinate native town, a very staunch friend, in spite of the fact that Palissy had been one of the most energetic supporters of the revolt. It was Montmorency who first made it possible for him to employ his manifold talents. The Constable had just finished his castle at Ecouen, which he built in the days of his exile. He commissioned Palissy to set up a grotto in this garden, the passion for grottoes having quickly spread from Italy, where it flourished at the beginning of the sixteenth century, to France.

The common characteristic of grottoes is ornament of the "grotesque" sort, sometimes in structures standing separately, sometimes in ground-floor rooms, and sometimes underground, in this case generally under terraces. In France they started rather late, but one early example is the *Grotte des Pins* at Fontainebleau. By the middle of the century every garden that had any pretension to importance was bound to have at least one grotto of the kind. It was the business of the artist to produce new ideas for them. The grotto at Meudon was especially famous, because it was sung by Ronsard on the occasion of the espousal feast of Claude de France, a daughter of Henry II., with the Duke of Guise. The castle was started under Francis I., but Philibert de l'Orme was commissioned to finish it by Jean de Guise, the first cardinal of Lorraine who belonged to this family. The cardinal's lavish ways and love of pomp were proverbial, so he built this princely seat and laid out lovely gardens, of which, however, we know no particulars till we come to the grotto. This was close to the castle, at right angles with the pretty parterre, raised above the orangery of a later date (Fig. 331).

In many cases the grottoes, even till the time of Louis XIV., served as theatres for performances in gardens, until later on the regular theatres in the open air were set up. Ronsard himself comes with his friend, both in the guise of shepherds, to the enamelled cave of the grotto, where there is a dwelling for the Muses. They admire the lovely

FIG. 331. MEUDON—THE GROTTO AND THE PARTERRE IN THE TIME OF LOUIS XIV.

structure, the plan, the pillars of *rustica,* and the shells decorating the entrance. They look at the separate rooms, and the terraces (perhaps raised paths), the ornamental festoons and arabesques, and the enamel of many colours looking like a meadow strewn with summer flowers.

Palissy did not build the grotto at Meudon; he was working at that time at Ecouen; but he soon appears as master of this style of decoration, and his grottoes (worked out in detail in his writings) actually show the realisation of all that Ronsard describes. Soon after Palissy completed his work at Ecouen, Catherine de' Medici summoned him to Paris, where he began his *chef-d'œuvre* in this field of work—the great grotto in the Tuileries Gardens. This was on an island, approached by large enamelled bridges. No trace is left of either, and perhaps they were never completed, but in Henry IV.'s time they were destroyed. Nor have we authentic information about the grotto at Ecouen. On account of the grotto and its site, Palissy is very particular about the garden, constructed according to his own fancy, "so that one can retire there and find repose in times of turmoil, plague, epidemic, and other visitations." The garden has to be on the level, at the foot of a hill and a rock, to protect it from the bad winds from north and west; it must also have the water it needs, and (as we shall hear later) facilities for all sorts of grottoes. Palissy published his little book in 1563, but he wrote it at a time when everyone in France still demanded a completely level ground at the foot of a hill as the only possible place and where every rise and slope was removed, often at great expense.

Palissy insists that there are more than four thousand fine houses in France where it would be easy to get so desirable a site as that described; but in his garden the dwelling-house is never even mentioned. This garden is a square, quite shut in by itself, and not differing from the mediaeval plan in its main lines. Two avenues cut it across so that it makes four square plots. There are grottoes at the ends of the main walks, in the corners of the garden and in the centre, nine in all; also there is a set of rooms, two stories in height, that are cut into the boundary rock. The lower is a fruit granary, but the upper, provided with balcony and balustrade, can be used at the pleasure of the owner as a cool airy retreat. The direction of Palissy's fancy, at that time so concerned with varieties and changes in natural forms, is shown in his proposal to make human figures appear on this balcony, leaning over the edge and able to move, so that a visitor would bow to them. Palissy was one who experimented in every field; and this idea of his reminds us of the pictures that Mantegna painted on the ceiling of the bridal chamber of the Gonzagas at Mantua.

Palissy's description is important, because it first introduces into the garden the realistic figures that played so great a part in the seventeenth century. The eight grottoes that enclosed the garden were to have amphitheatres round them, probably semicircular; the four corner ones were to give the effect of caves in the rock, and in front were supported by columns or grimacing Herms. There were inscriptions on the architrave, giving praises to wisdom, which alone is pleasing to God: this was in accord with the religious feeling of Huguenots at the Renaissance. The inside of the grottoes is richly adorned with enamel in every conceivable colour, which is reduced to one lustrous mass; for by burning a great fire inside the place the artist has made all the colours run together to produce a fine iridescent effect. This enamel, moreover, by the help of the technique he handled so perfectly, was ornamented with animals of all kinds, treated naturalistically. But here, too, Palissy had a pattern to his hand in the favourite book that he was never without, the *Dream of Polifilo*. In this book there is a grotto of the same kind, with lizards and fishes so natural in colour and movement that you could fancy they were real. Palissy wanted to introduce into another grotto which he planned for the queen-mother—possibly at the Tuileries—these natural animal decorations both outside and inside the canal enclosing the place.

The four grottoes at the corners were so made that on the outside (with so much plantation, shrubs, water run artificially, mossy stones, etc., and if possible a natural rock) "one would never guess there was a human habitation below"; but if one were walking on the prominence of rock above, it was quite a different thing with the grottoes at the ends of the avenues. These he desired to make out of elms, their trunks to serve as columns and their leaves and branches arranged as pavilions with windows, frieze, and roof. This method of clipping trees he rightly thought to be nothing unusual, for in many gardens he had seen at the very least clipped low-growing shrubs in the shape of "cranes, hens, geese, and other creatures; and elsewhere men on horseback or on foot with various armorial bearings, also letters of the alphabet and mottoes."

The inside of these grottoes is enamelled. In the centre there is a pavilion made of poplars, and at the point of the roof a kind of wind-organ; in the middle of the hut stands a stone table with green hedges and seats all round it, and beyond this an aviary made of a trellis of copper wire, with four gates leading out to four corresponding avenues. All this central part is upon a small island made by a little stream, which comes from the hanging rock and flows round the garden. The slight interlacing of this water is very pretty in the garden, designed by the ingenious whimsical fancy of the potter. Palissy only

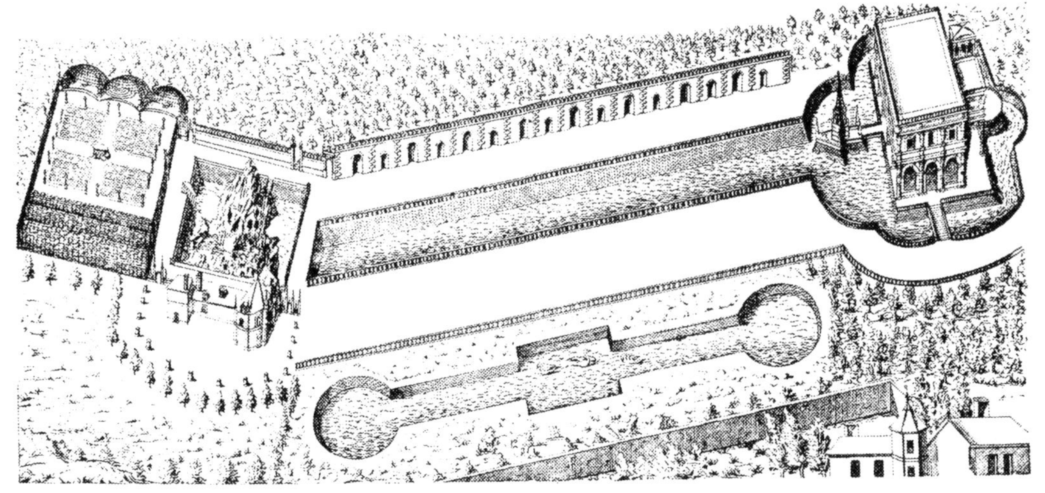

FIG. 332. GAILLON—THE HERMITAGE AND THE WHITE HOUSE

thinks of water as flowing brook or canal, but we remember what excessive use was made of water devices in the Italian gardens of that day. Little is said about any other planting, but all round there is a covered way, and that is all we know. Palissy firmly rejects the notion that this garden of his was only a dream, like that of Polifilo. But what he does admit is that he dreams of a garden "where the saints live as shepherds." Still this particular garden is to be actually made, and is to serve as a model for many others.

The literary feeling which is evident in this work of Palissy's, a mixture of religious sentiment and one that has its source and aim in pleasure, finds material expression in much that was made at this time. Du Cerceau has preserved a masterpiece in the hermitage at Gaillon (Fig. 332). At the end of the terraced park, which stretches westward from the garden, Cardinal Bourbon built in the sixties his *Chartreuse abondant en tout plaisir*. "If," says Du Cerceau, "you walk from the upper garden through the park partly by terraces, partly through avenues of trees, always keeping your eyes on the lovely valley, you come to a little chapel and house and hermit's rock in the middle of water enclosed in a square basin, and all round it are narrow paths where you can stroll for pleasure. To get there you cross a swing-bridge. Close by you find a little garden, and in it are statues three to four feet in height, standing on a great plinth, depicting all sorts of allegorical subjects, and also several *berceaux* covered with greenery; the hermitage here is very pretty and attractive, and as full of charm as any you can find anywhere."

There has already been some anticipation in Spain of the hermitage in a garden, avowedly put there for the sake of enjoyment. Although the whole arrangement of this spot seems so frivolous, its religious character was probably maintained. In the little house lived a real holy man, and the owner of the place kept certain seignorial rights; such as "one dish of meat, two vessels of wine, two loaves, and the first dance after a wedding." There was something of the same kind in the chapels at Blois and elsewhere, but they were oratories with a more serious purpose. In the park at Chantilly there were seven chapels, and the Pope allowed the hermit to grant indulgences. But at Gaillon the hermitage was not the only *lieu de plaisance* in the park. From it a wide-walled canal with balustrades led to a small summer-house round which the canal also ran. A little marble palace, excessively decorated, and known as the White House, contained a large garden-room on the ground floor, which had arcades round it and was ornamented with fountains in niches. Outside this are the trees of the park. The canal is on one side of a wall with balustrades and alcoves, and on the other side there is a second canal with circular ends to it. It often happens that the next generation, and now and then the original owner, allows places of this kind to fall into decay or to be altered. So it comes about that we cannot get further light on Du Cerceau's plans and sketches except from occasional descriptions or comparisons made with the ornamental grounds so common in Italy in the days of the Renaissance.

The theorists, however, contribute more to our knowledge of the way these architectural plans were worked out; for they describe the fine results of this development at the conclusion of the period, and more or less point the way to the future. The activities of all these persons, Olivier de Serres, Mollet, and Boyceau, belong to the time of Henry IV., in whose reign an important movement in garden art took place. They were all proud to have served Henry the Great of happy memory, who (to quote one of the dedications) "himself planted and grafted, and so even now [1652] the great lords and princes of France take the utmost pleasure in doing the same." The earliest of these was Olivier de Serres, whose great work, *Le Théâtre d'Agriculture*, appeared in 1600, and went through many editions. The sixth section of this book, which had great influence on agriculture in France, was concerned with the garden. He desired to have a garden divided into departments for kitchen products, flowers, medicinal herbs, and fruit-trees, with which he included vineyards. From this one sees at once that his chief interest is in what the garden can produce, and that there is little left for the architectural and artistic side of things. Kitchen-garden and orchard—not to speak of vineyards—must take far more space than anything else. When in Chapters X.–XII. he discusses the flower-garden (*bouquetier*) he abandons the architectural ideas of his time, and fills the parterres with flowers and various plants. In his parterre he demands perfect symmetry in lay-out and plantation: the very name "parterre" (incorrectly stated to derive from *partiri*, to divide) is used from now onwards to mean the patterning of the flowers as opposed to the high-growing shrubberies. Here there has to be a great variety in the blending of plants in patterns. For the edges of the parterre they must use sweet-scented low bushes like lavender, thyme, mint, marjoram, and many others. At the end of the century, however, box came more and more to the fore, taking the place of other things because it was evergreen. The inner part of the pattern was made by low-growing plants, such as violets, wallflowers, pinks, heartsease, and lilies of the valley: many of the kinds, at least twenty, with which we fill our beds nowadays, are mentioned by de Serres. A small part is also played by imported foreign plants, even bulbs, in particular parts of the parterre.

To enhance the effect of this variegated picture, clays of different colours were strewn about, and these had to be selected carefully, to preserve the harmony of it all. The higher-growing perennials were used at the edges of the footpaths, but de Serres will have no mixtures, and wherever there are several kinds planted near together, each kind must be separated from the others. In the corners of the patterns there may be cypresses, but he will have nothing to do with other trees with spreading tops, because they spoil the designs. Box is the best for clipping of every kind, and can be made into seats, benches, buildings, pyramids, pillars, men, and beasts, always according to the skill of the gardener. In the parterre the main ideas are harmony and a fine view above and beyond it. There may be a cypress in the middle to act as a sundial, and instead of a tree-pyramid statues

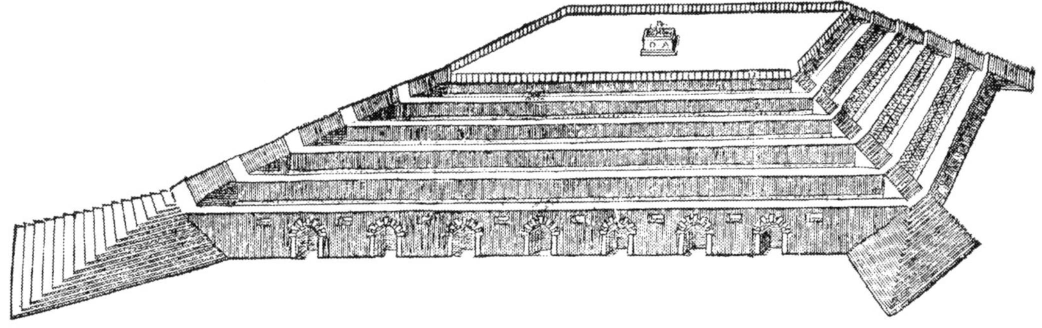

FIG. 333. AN OLD PLAN FOR A BOTANIC GARDEN

of every sort: obelisks, columns, pyramids of marble, jasper and porphyry, can be erected as the parterre. A parterre of this kind is before everything intended to be looked at from above, so it is well if it can be seen not only from the house but from surrounding terraces, as was so well arranged by King Henry at his Tuileries Gardens, where he planted round his parterre a terrace with mulberry walks.

The most remarkable chapters of de Serres' book are the fourteenth and fifteenth, which are concerned with the medicinal garden. In consequence of a scientific interest in botany that was felt in Italy as early as the middle of the century, botanic gardens had been founded at the Universities of Padua and Pisa, and now at the beginning of the new century Northern Europe was to follow their lead. Even in the Middle Ages the care of plants for the sake of their medicinal virtues had gone hand in hand with delight in their fragrance and beauty.

It happened otherwise with the development of the flower-garden, which was both parallel with the medicinal garden and derived therefrom, keeping the name *giardino dei semplici* in Italy till late into Renaissance days. Botanical study gave an unprecedented impetus to these herb-gardens—chiefly by the introduction of foreign plants—and thus they were converted into botanical gardens, though their original intention was never quite lost sight of.

In the northern countries it was the reverse of what happened in Italy, where public botanic gardens were established with purely scientific aims, for it was the private gardens that went ahead. Whereas the *Jardin des Plantes* in Paris was not founded till 1632—and this was nearly a hundred years behind Italy—we hear of a great many private botanical gardens in Germany, in France, in the Netherlands, and in England, all of which enjoyed a great reputation. It was understood that the owners introduced foreign plants into their own flower-gardens and there cultivated them. But Olivier de Serres, in his strong feeling for a sharp distinction between gardens of different kinds, advises the owners to make an extra place for medicinal herbs. "For in this way," he says, "many foreign healing plants have been naturalised with us, rare and remarkable in their useful properties, and hitherto unknown to those who own land." For these private botanical gardens he recommends making artificial mounds (Fig. 333), because he thinks they would save space and money, and also give variety of aspect as each kind needed it, seeing that all four sides of the hill would be at their disposal.

No doubt de Serres was familiar with the spiral hill; he refers to the Tower of Babel and the Pharos lighthouse at Alexandria. In his ground-plans the artificial mound has sometimes a round and sometimes a square form, and its lowest walls give room for grottoes. He does not mention Diodorus's account of the hanging gardens of Semiramis, but his drawing might stand for an illustration in miniature of those mighty buildings, with the stairs at the sides, platform on the top, and grottoes in the walls. We have no first-hand information as to how far he carried out his plans; but the influence of his book was so great that we may suppose many made attempts to create such places as he writes about. Of course his influence was more felt in his main effort to improve farming and husbandry than in the garden as such.

In the discussion of flower-gardens de Serres refers to the king's own gardener, Claude Mollet, who in 1613 towards the end of his life wrote as a theorist after his practical experience was finished. Mollet belonged to one of the most famous families of gardeners of that time. His father before him was in the king's service, and Claude says, "God has granted me His grace, so that for the blessed King Harry the Great I have made many beautiful things." In his book Claude was associated with his sons, who made the plates for him: it appeared a good deal later, in 1652, under the title of *Le Théâtre des Plans et Jardinages*, but it throws a clear light on his own activities. His sons appear to have gone into foreign service, and one of them, André, we shall meet again as a practical and theoretical gardener in England and in Sweden.

Claude Mollet directs his attention first and foremost to the laying-out of the parterre, which, as the ornamental garden, naturally was most interesting to a French gardener. For the most part we find in the planting of it the same ideas as those of Olivier de Serres, but Mollet demands that, among the borders of tall shrubs which go round the squares in a wide band, there shall be many kinds of flowers, so that there shall never be gaps in the bright ribbon, and new flowers shall always succeed the old ones. This principle is pursued by the most modern gardeners. Mollet is proud to think that he used box very extensively, because it was so strong and good to look at. He first made high hedges of box instead of using cypresses, which would not endure the French climate. He supported very high espaliers with dead wood, and had plants on both sides, so that the desired shapes, such as windows, battlements, arches, and pillars, could be more easily obtained. The most important thing that Mollet effected was the great development of the parterre. If we look through Du Cerceau's engravings, we find that all his parterres are divided into somewhat small squares, most of them in a geometrical pattern. It was desired that these should show variety and alternation; it was possible to look at them conveniently from some raised surround, which might be walls, as at Henry IV.'s Tuileries Gardens, or might be terraces; thus all the patterns could be seen separately. But towards the end of the sixteenth century there was a new departure, and Mollet lets us know the exact moment when these small parterres, divided up into geometrical patterns, were united to form a considered, connected plan.

In 1582 the French architect Du Pérac came back after a long pilgrimage to Italy, where he had eagerly studied Italian art, ancient and modern; at the present day his pictures of old ruins are among the most valuable we have. The Villa d'Este seems to have attracted him more than any others that he saw in Italy, and he dedicated to Maria de' Medici, a princess of his own country, a set of drawings which he made of it. On his return he was appointed architect to Henry IV., and carried on the building at Fontainebleau, which the king had so much at heart at that time. Du Pérac was also employed by the Duke of Aumale, who made him (as Mollet says) superintendent of all his castles. He made designs for Anet, which seemed to Mollet the most beautiful of all the French castles, "such that a garden makes one single parterre, divided by wide paths." This new kind of parterre was called *compartiments de broderie.*

FIG. 334. LUXEMBOURG GARDENS—PLAN FOR A PARTERRE

Boyceau gives a perfectly clear example of this style among his drawings of the parterres in the Luxembourg garden (Fig. 334). Here we get an entire sequence of separated plots, and the one complex split up into its component parts: the whole parterre is enclosed so as to form a united design about a middle line which preserves its symmetry. For a design of this nature the wavy lines of an arabesque pattern are employed, and the sharp geometrical lines have entirely disappeared. According to what Mollet says, we must assume that Du Pérac got his inspiration in this matter entirely from Italy. If this is so, very different results came about in France, for in Italy the parterre played a very subordinate part, especially in the sixteenth century. In the Villa d'Este, for example, the movement was towards the development of terraces and artificial water-schemes, while the art of France was directed towards the gardens that were on level ground with canals round them, and concerned with the laying-out of parterres, in which matter at least France took the lead among all the nations.

Another peculiar kind of art that was growing up now influenced the culture of the parterre in the first decades of the seventeenth century. In the reigns of Henry IV. and Louis XIII. expensive flower-embroidery was fashionable, and was used in arabesque patterns for dresses, and later for all sorts of furniture, in silk of many colours, and thread of gold, silver, and linen, with a close imitation of natural flowers. So great was the growth of skill and the love of luxury, that soon the ordinary flowers did not suffice; and people looked round for foreign plants, which were eagerly copied by embroiderers with astonish-

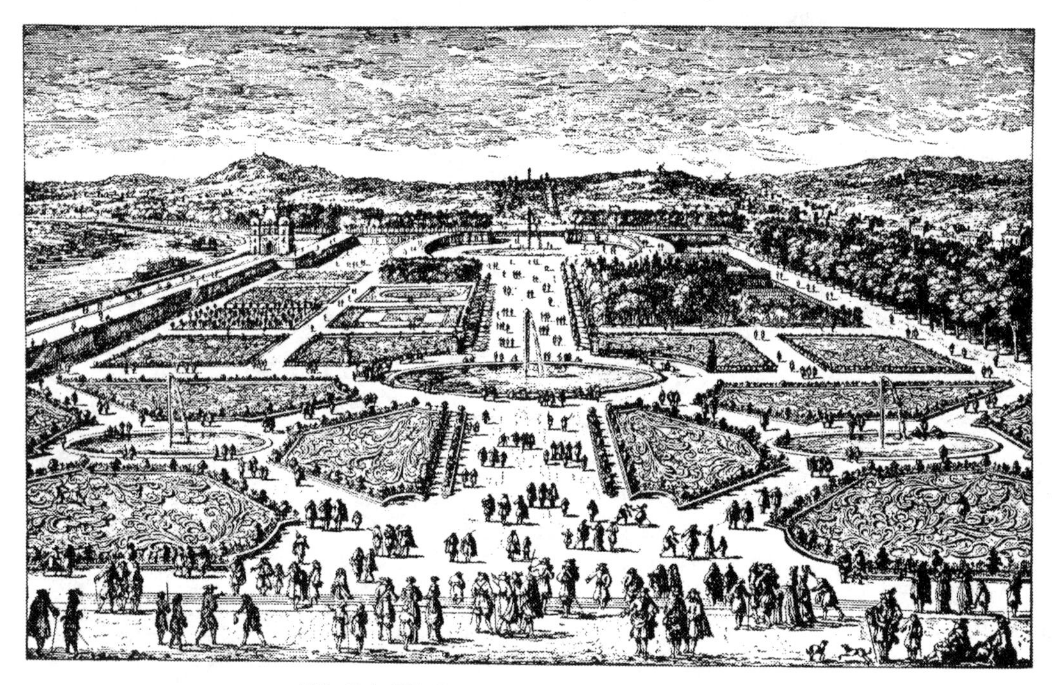

FIG. 335. THE TUILERIES GARDENS—THE PARTERRE

ing effect, while men were also attempting to produce something equally beautiful in the parterres. The two arts were so closely allied that the fashionable and profitable trade of the embroiderer was often carried on with the gardener's business by the same person. Some of the most important French collectors of plants, such as Jean Robin, laid out foreign gardens with the express purpose of providing new flower-patterns for embroidery; and on the other hand coloured drawings were a help in spreading the knowledge of plants. Thus it came about that the vanities of fashion as well as scientific botany gave an impulse to the art of gardening.

The chief artist of the parterre was Boyceau. His book, *Traité du Jardinage*, was written in his last years, and was published after his death in 1638. He stood on the threshold of the new age: the strict division made by Olivier de Serres into four or five separated gardens he discarded; and for him it was enough to have two gardens, one for the kitchen and the other for pleasure; in very small places these might be mixed together.

FIG. 336. SAINT-GERMAIN-EN-LAYE—PLAN OF THE GARDENS

In the pleasure-garden (the part for parterres, avenues, and shrubberies), he insists before all things on proportion; the height of the trees and hedges must be in well-considered plans proportional to the length and breadth of the paths; and the parterre required a surround of raised terraces, galleries and *berceaux*, which he classes all together under the one expression, *corps de relief*. Next to proportion he ranks variety, in spite of his desire for the most severe symmetry. "It tires me inexpressibly," he says, "when I find every garden laid out in straight lines, some in four squares, others in nine or in six, and nothing different anywhere." With ideas like this Boyceau planned the new parterres for the Tuileries Gardens, and his specially beautiful design greatly excited the admiration of the next generation (Fig. 335). Boyceau is regarded as the precursor of the great development of French gardening, but the earliest and most important time of his activities was really in the reign of Henry the Great.

The king, who impressed everyone about him as a brilliant and many-sided character, gave a fresh impulse to gardening in that long-desired period of peace which he conferred on his country. Olivier de Serres is full of admiration for the beauty in Henry's time of the remarkable gardens scattered about over the whole kingdom, but especially those he had had made at the castles of Fontainebleau, Saint-Germain-en-Laye, the Tuileries, Blois, and others. It is astonishing to behold the ordering of the plants, "which speak to you in letters of the alphabet, mottoes, sign-manuals, armorial bearings, and pictures; marvellously imitated are the gestures of men and animals, the shapes of buildings, ships, boats, and all sorts of things, in bushes and plants. There is no need to go to Italy or anywhere else to look at the best gardens, for our own France takes the prize, and serves as a school of learned men to give us the instruction we need."

As a fact, however, in Henry the Fourth's reign things were still at sixes and sevens, especially when Maria de' Medici brought her own country's art over the Alps, for then we can see very clearly the fresh inspiration from Italy. Henry's chief building was the new castle of Saint-Germain-en-Laye. Francis I. had made a great palace on old foundations on the high bank of the Seine, a little inland; and this was enlarged by Henry II. by the addition of a small central building from which one could enjoy the wonderful view from a broad terrace close to the river. The gardens at these old places seem to have been insignificant, for Du Cerceau does not mention them, and they are scarcely indicated on his plan. Henry IV. now put up a palace in place of the small theatre building, with fine pavilions at the corners, and raised high above the steep bank of the Seine (Fig. 336). The architect of the actual palace was a Frenchman; but for the garden, which had to slope down in steep terraces to the river, the king, acting on the advice of Maria de' Medici, called in an Italian, Francini by name.

The castle has disappeared except for one pavilion, and the gardens have vanished, leaving no trace. The place fell down after Henry's death, perhaps because it was built too quickly, or perhaps because the French architect could not sufficiently master the difficulty of laying foundations on such a steep slope; while as for the garden, a change in taste turned men's thought to very different tasks. From the drawings and the descriptions we get a complete Italian garden, such as France had seldom seen: by great steps, set in a perfect line, the descent to the river is made in six terraces. The picture seen from below ends with an enormous palace façade, in the same style as the ascent to the Farnese Gardens. Below by the river is a parterre showing on our picture a geometrical plan,

FIG. 337. FONTAINEBLEAU—THE ORANGERY

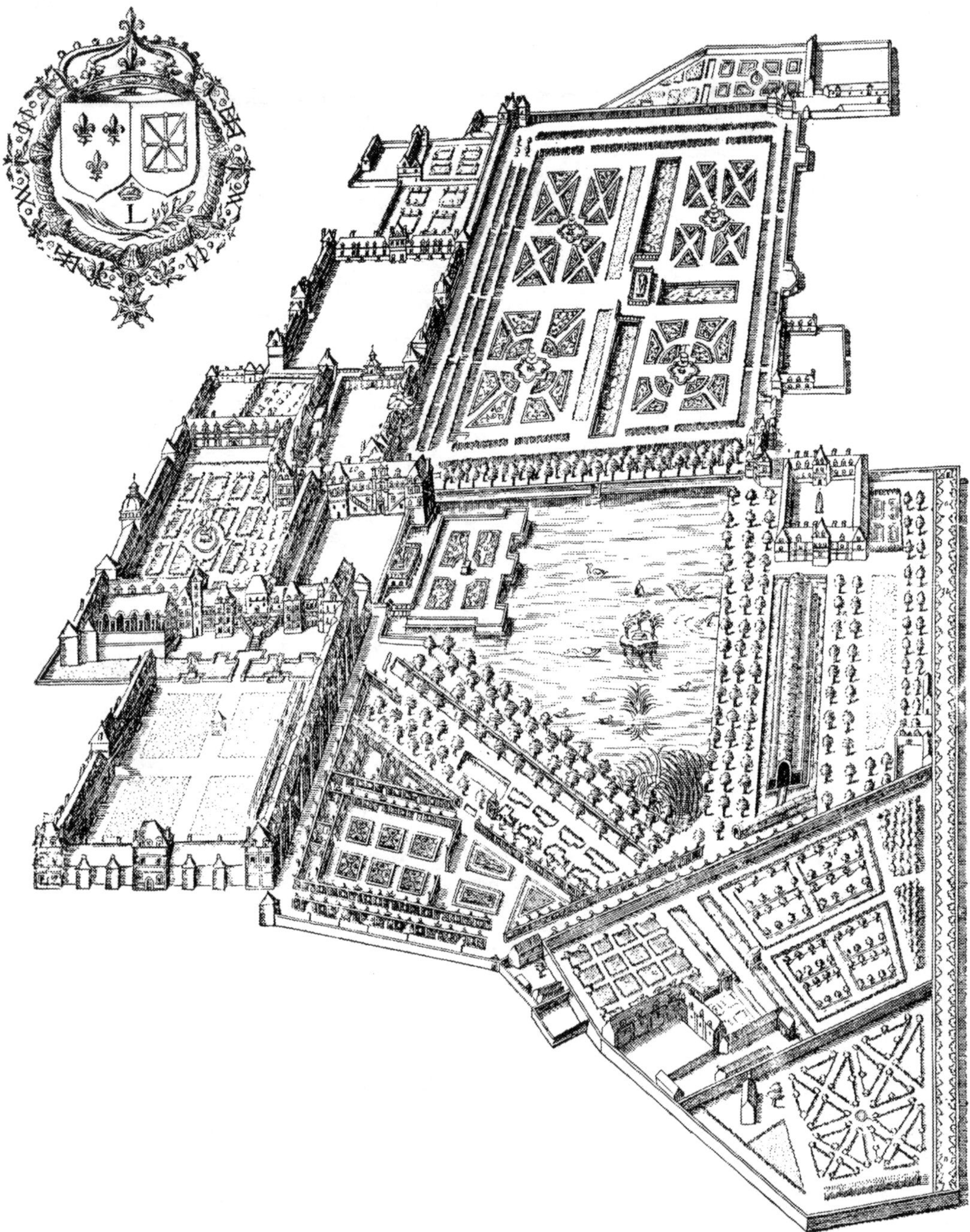

FIG. 338. FONTAINEBLEAU AT THE TIME OF HENRY IV.

and armorial bearings, with hedges round it and ornaments of basins and fountains. The terraces are partly filled with figures cut out of shrubs, and the dividing walls have openings in them that lead to long rows of grotto rooms, repeating the Italian ideas of water-games. Orpheus is there, with trees and beasts moving about to the sound of his lyre; also dragon-fountains, singing birds, and many other things. Beside the castle there are *giardini secreti* arranged two and two. The best thing of all is the lovely view from the upper terrace. The place where the cascades were, which Evelyn mentions, is not recognisable from the engravings. Later on Boyceau designed a series of parterres for this garden, which were made in those parts that lie at the foot of the hill.

Next to Saint-Germain-en-Laye, Henry especially loved Fontainebleau; and here we must look on him as the real creator of the gardens, though of course he found plan and site ready to his hand. The little parterre of Francis I. was altered and enlarged by Henry to make a specially charming enclosed garden (Fig. 337). On three sides of it there were galleries, and the fourth ended with a large aviary, imitating the Italian style, with tall trees and green shrubs below the copper wire which roofed it. The Diana of Francis

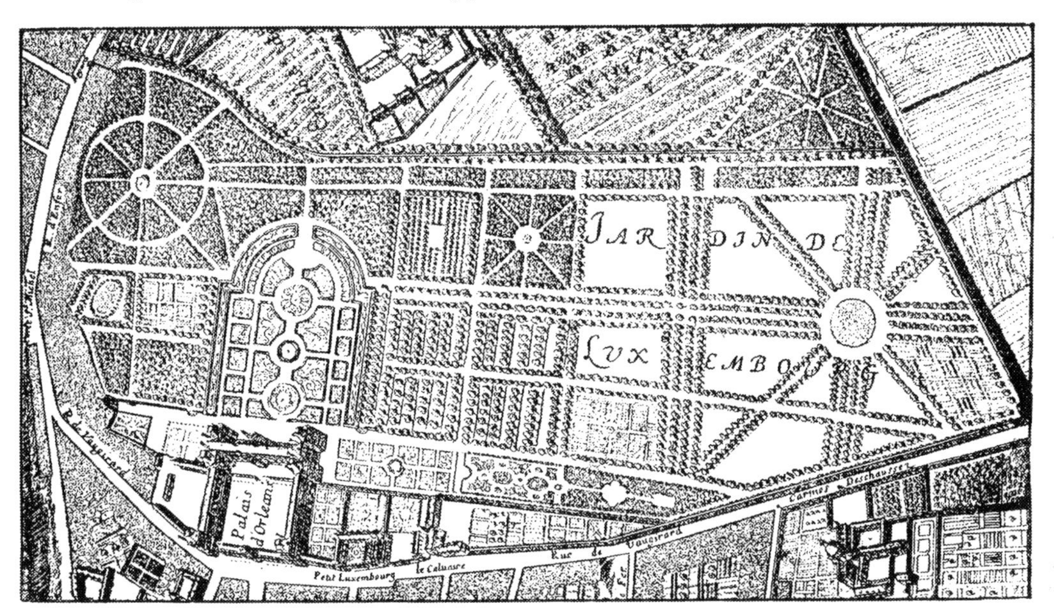

FIG. 339. LUXEMBOURG—THE GARDENS IN 1652

was still the pride of the central fountain, around which lay the beds of the parterre, with the terraces running round, high above. The chief adornment of this pretty little garden was a great number of choice statues, showing more than anything an Italian influence. Later on, when Louis XIV. substituted an orangery for the aviary, oranges were cultivated (as the drawing shows) in this garden, which otherwise remained unaltered.

On the other side of the great pond, in the middle of which Henry had set a pavilion as a sort of retreat, he made a large parterre (Fig. 338), which in essentials is the same now, except that then there was a broad canal cutting through the square, and in the middle of it a colossal figure of Father Tiber reposing on a rock. At the corners of this parterre there were other fountains arranged in pairs. Henry IV. made another innovation, of paramount importance, in the situation of this canal, which starts from the parterre on a higher terrace above it, and unites this chief garden with the real park. We shall see later how exactly the plan fits in with the novel arrangement of gardens in the time of Louis XIV. Whether it was, as some have thought, also designed by Francini, is more a guess than a certainty.

It was certainly a French artist who planned the castle and garden of the Luxembourg (Fig. 339). Obedient to the wish of Maria de' Medici, a piece of Italy rose on French soil in the place where she made her home after her husband's death in 1615. She suggested models from her own country, the Pitti Palace and the Boboli Gardens, to the architect, Salomon de Brosse; but in the castle itself he could not avoid French taste, however hard he may have tried to please the lady. In spite of the *rustica* façade which is reminiscent of the court of Ammanati, the French pavilion-building dominates the scene. In the garden the somewhat level grounds stood in the way of a complete likeness to the Boboli Gardens, but in the original lay-out (which is not much altered) there were many features to remind Maria of her gardens at home. The actual boundaries of the place at

FIG. 340. LUXEMBOURG GARDEN, PARIS—THE GREAT PARTERRE

the beginning were different from what they now are; they extended somewhat farther towards the south in length; and on a piece of ground, which at that time enclosed a Carthusian cloister, were shrubberies stretching to the west from the main axis. Now they are only half the size, while on the east also the garden has suffered somewhat.

Round the parterre (Fig. 340) there was a depression which was like the Boboli amphitheatre, and there were terraces thrown up, also two steps, with vases and statues on the upper one. These ran straight, and not as they are now, on both sides, ending in a semicircular shape at the back, and leaving an open way through in the middle. Above there were hedges made of box and yew, and this gave a pleasing colour as background to the rest of the plantation, which was mostly of thick foliage. Any further likeness to the Boboli Gardens was disturbed by the deviating middle axis, and still more by the *parterre de broderie*, which a French garden could no longer do without. The ornamental garden was at first adorned with a triton squeezing a fish so that a powerful stream of water gushed out from its mouth. In the complicated patterns of the parterre (Fig. 334) flowers were found less and less. (Boyceau made a set of his very finest designs for the Luxembourg.) Evelyn, who saw the garden in 1644, and cannot sufficiently express the pleasure this "sweet retirement" gave him, notices especially that the parterre was entirely laid out with box, and that this had a wonderful effect as seen from the castle. The flowers, banished from here, were reared in a special walled part, a small garden "on which the Duke spends many thousand pistoles."

Maria made use of another resemblance to the Boboli Gardens (caused by an accidental enlargement of the boundaries through the extension of the shrubbery part) to construct a great crossway avenue at right angles with the main axis. The end of this may have suggested the Isolotto to her mind. Evelyn says that a large fish-pond was there, but it was not finished at that date, 1644. Close by was a medicinal garden, and a

FIG. 341. LUXEMBOURG GARDENS, PARIS—THE FOUNTAIN OF THE MEDICIS

conservatory; and the other shrubberies, with their hedges, enclosed many a meadow or field. Evelyn also admired in the small park on the east a star-shaped shrubbery with a fountain in the middle of it. The chief ornament at the present time, the Medici fountain, was at first in a different place at the end of the crossway that leads to the façade; and instead of the group that we now see, which is Polyphemus watching Acis and Galatea, there was a nymph: the place at the top end of the canal has only been the home of this fountain since the nineteenth century (Fig. 341), when the garden, after many chances and changes, received its final form under the Second Empire.

The life to be seen nowadays, in such different circumstances, seems to be really much the same as what Evelyn describes with appreciation. Anyone who has gone to the park on a Sunday has seen his picture repeated in every detail, but in Evelyn's day the garden for merry Parisians was not so sure to be open as it is now, for Sauval writes in his work on the antiquities of Paris in 1650 that it was often open but often not. We need not speak of its shocking appearance in the time of the Revolution, when the Luxembourg was turned into a prison.

The jealous unending rivalry of Maria and her great opponent Cardinal Richelieu spurred them both on to a bloodless strife in the making of their gardens. There existed only one garden that could compete with the Luxembourg, and this was the one that Richelieu made beside his country house at Ruel. The house was not large, and therefore so much the larger could the fenced-in garden be, and as a fact it included, within the hedged squares of the park, cornfields, meadows, and even vineyards; also there were shrubberies with evergreen trees and avenues that Evelyn greatly admired for the way they were cut. Beautiful fountains were set up as *points de vue* of the avenues, one of them shaped like a basilisk that shot up its water to the height of sixty feet and twisted round so quickly that you could not escape being wet if you were anywhere near. From this place one passed on to the Citron Garden, on which the Arch of Constantine was so artfully painted that swallows tried to fly through it. With this pleasing illusion Evelyn was delighted. In front of the house there was a parterre with beautiful fountains and statues in bronze. The grotto enjoyed a great reputation because of its fine shell-work, the fall of water like rain, and the two musketeers in front, who scared visitors, as they hurried off, with a water-salvo—an idea that reminds us of Villa Lante.

FIG. 342. RUEL—THE GROTTO

But the real pride of the garden was a great cascade, which plunged down from a steep acclivity, checked by several basins and pouring over marble steps, "with amazing fury and noise." At each basin there was a fountain, the last one a gigantic shell in lead. Thence the water flowed in a quiet narrow canal to a grotto (Fig. 342). A great number of ideas were exhibited in a garden like this, compared with the simple and imposing but monotonous plans of Du Cerceau's time. But all the same there is no complete mastery in this kingdom of art, for they have not yet learned to utilise important regions with a view to the whole scheme; for instance, the cascade, which should be the chief feature, is not in the main axis of the house, but at the side, at the end of a wide avenue which leads to the Arch of Constantine.

Cascades were a new contribution from Italy, and were introduced into French gardens in the first half of the seventeenth century. Where it was not possible to find enough fall for them, terraces of turf were built, as for instance on the flat ground at Liancourt, a pleasant garden which the wife of Marshal Schomberg had enjoyed making. The garden of St. Cloud boasted of the finest cascade then known. Among the many Italians who came to France in the train of Maria de' Medici, and who with their great love of adventure and picturesque appearance were characteristic of this court, the banking family of the Gondi is conspicuous. They quickly acquired property and built a villa on the high bank of the Seine near the boundaries of the town; and by the side of the house they made the famous cascade, in Italian fashion, which Evelyn describes in its early condition. But several decades later the place was destined to become far more magnificent, and the great laying-out of the gardens took place in the reign of Louis XIV.

Evelyn's straightforward descriptions lead us easily through the first half of the seventeenth century, which may conveniently bring to a close our survey of the art of the Renaissance, little though France has to show of any kind of check or sudden upheaval in the course of development. Evelyn visited not only the large towns, but also the little town-gardens, whose limited space the owners tried to make look larger with painted perspective. To Evelyn the small garden of the Comte de Liancourt seemed to look a great deal larger by using this method, and by the help of paintings they made their little gardens "flow for some miles." And a small doll-theatre at the end increased this childish pleasure. Another garden which Evelyn liked very much belonged to a man called Morine, who began as a simple gardener. He laid out his garden as a perfect oval surrounded by thuyas, clipped smooth to look like walls, and within it he introduced very beautiful and at that time unusual plants, such as tulips, ranunculuses, crocuses and anemones. He himself, a very old collector, lived in a small garden-house at the end. It is evident that in this type of small town-garden the love of surprises was particularly lively, for at "an house called Maison Rouge" Evelyn hears a noise that "resembles the noise of a tempest, batailles of guns, etc., at its issue," and much beside.

On the threshold of this period the French garden found its earliest poets. René Rapin, the learned Jesuit who wrote so many books, published in 1665 a poem called *Hortorum libri quattuor*. This was in undisguised imitation of Virgil's Georgics, but aimed at giving a supplement on the province of the garden, which the Roman poet had neglected. Rapin had already seen something of the development of gardens in the *grand siècle*, and this appears in much of what he wrote; but he chiefly concerns himself with the gardens of the time of Maria de' Medici and Henry IV. All that he writes in his first book, *The Flower Garden*, is what is demanded by the usual theorists; and about flowers he only gives pretty myths in the style of Ovid. But in the second book he is inspired with new ideas, the exhaustive treatment of the park (*nemus*): he says that when you step out of the garden the park must at once appear as a stage formally arranged. The trees must be in the form of the quincunx, with straight lines and right angles, although he also likes slanting lines and circular arches in some places. All the paths have to be provided with fine sand or closely mown grass, and at the sides the beeches or cypresses have to be clipped so as to make straight walls, but the lighter branches may be worked into a thousand shapes and mazes of any kind. He is full of praise for the oak as a good forest tree, and he is emphatically opposed to the clipping of oaks, using all his rhetoric and calling down all mythological punishments upon the heads of the desecrators of trees.

This care for the park and its cultivation is by no means unfamiliar in Renaissance times, but the close connection of its main design with that of the garden points to a period which is still to come.

In the same way the third book, which treats of water, adhering to all the fantastic ideas of the Renaissance, rejoicing in every one of the innumerable water-devices, and demanding a great extension and development of cascades, does try most energetically to enforce strict regularity even in the domain of water. It is admitted that tricks of teasing waters in grottoes, and pumice and shells for decoration, are all wonderful attractions for the people on festival days; but anyone who is sensible, large-minded and thoughtful is fond of great sheets of water, and canals like rivers: it is to sing of these that his lyre shall be strung. To this poem of Rapin's, which is by far the best of his works, and long enjoyed a high renown, he later added a treatment of the same subject in prose. Here he shows learning both fundamental and far-reaching, and in the form of a representation of the *Quarrel between Ancients and Moderns* he puts forward the modern view; men of the old style, he admits, were full of enthusiasm for garden architecture, but remained stationary at one stage of their art; whereas all progress, all individual art, belonged to the moderns. The garden of Alcinous was nothing more than a peasant's garden, and even there we found very few kinds of fruits.

Now, however, the garden is the glory of the age, and its noblest art, and there is no house worth looking at that cannot boast of its show garden. "What once was a servant's work is now the achievement and delight of the master."

These words may serve as a motto for the period of garden art upon whose threshold Rapin stood, the age of Louis XIV.

CHAPTER X

ENGLAND IN THE TIME OF THE RENAISSANCE

CHAPTER X

ENGLAND IN THE TIME OF THE RENAISSANCE

ALTHOUGH France still shows traces, all too slight, of what her gardens were in Renaissance days, we find that England gives us an opportunity only now and then, and only in individual cases, of seeing on the spot what the earlier developments were really like, and the reason is partly the great affection for horticulture that has been felt for hundreds of years by the whole nation. The English have been much preoccupied with current events; a landowner has generally been a very rich man, holding his house and garden for centuries in an unbroken family line: from one generation to another these have assumed totally different aspects.

There has been, moreover, a conscious reaching back to a former state of things, and many a supposed novelty—which is often a mere caprice—has been introduced only to prove similar to the old style. This is the present condition of affairs almost everywhere in the rather overwhelming abundance of English gardens, where flowers that suit a late kind of horticulture are combined with the modern fancy for what is old-fashioned: we find an inherited art, which is very attractive, but not historically sound.

The want of pictures makes it particularly difficult to decide about English gardens of the Renaissance, for in the sixteenth and seventeenth centuries there were very few pictures or engravings. It was not till the last quarter of the seventeenth century and the beginning of the eighteenth that the copperplate engravers (who were making copies from nature and from every human art with amazing skill) streamed over from abroad and a few Englishmen joined them. There were, however, many literary works at the time of the Renaissance, and they make up to us more or less for the want of pictures. In their happy descriptions the English garden lives again, and takes a place in the story of the great revival.

During the time when Italy was rapidly developing, and France, a joyous competitor, was emulating her learning in early work, England was so exhausted by devastating civil wars that no recognisable trace of important horticulture appears until the second decade of the sixteenth century. Henry VII. took the home government into his own hands so entirely that he had neither time nor desire to cast his eyes abroad and see what could be done. And yet it was from there that the revival was to come. There is certainly a report of a series of royal gardens which his son Henry VIII. inherited in 1509. At the Tower of London, at Westminster, at Woodstock, and at numerous other castles, there were gardens with their own gardeners. Of these, no doubt, the greater part were still used merely for vegetables, and the small ornamental plots were just the castle-gardens of the Middle Ages, very little altered after the unrest of a hundred years.

The garden described by James I. of Scotland, which he wrote about when he was a prisoner in Windsor Castle from 1413 to 1424, can undoubtedly serve as a portrait of the same place a hundred years later. The prince in his prison chooses his words minstrel-fashion: beside the towers, he says, lies the lovely garden, in every corner are bowers of lattice-work, overshaded by junipers. Hedges of whitethorn shield the path from the beholder's glance, and from the branches the nightingale's song fills the whole garden, so that the very walls ring again.

> Now was there mayde fast by the touris wall
> A garden faire, and in the corneris set
> Ane herbere grene, with wandis long and small
> Raillit about, and so with treeis set
> Was all the place and hawthorn hedges knet,
> That lyfe was non, walkyn there for bye
> That myght within scarce any wight espye.

The first person who with full comprehension turned to the new art which had arisen on the other side of the Alps and was now pressing over the Channel, was Cardinal Wolsey, the all-powerful and much-beloved minister of Henry VIII. Wolsey was at the height of his reputation; and the king, young, socially inclined, and addicted to the lighter side of life, trusted him completely, and in a gay impulsive moment proposed to have him show another sort of ability, by devising wonderful buildings. As a fact, Wolsey had already had extensive alterations carried out in the gardens of York Place (later Whitehall), his London episcopal residence, but now he wanted to build himself a country house.

We can see, by the circumstantial way in which the question of site was handled—it was said that he summoned the best physicians in London to advise him, and also sent to learned men in Padua—how little one would have dreamed of building a house (about 1514) that was not under the protection of the state, or in a strong situation of its own. Wolsey finally selected Hampton Court, the most healthy place within twenty miles of London, on the highest bank of the Thames south of the city. It was taken for granted that the outside of this country house must be crowned by towers and battlements, and also that it should be surrounded by a moat. English architecture, like French, cut loose from the Gothic tradition quite late, and then only for a short time. But the ways of the Middle Ages were only given up very gradually, and even half-way through the century the nobility, especially in the North, entrenched themselves in their fortresses behind walls, towers, moats, and drawbridges. But although Germany adhered firmly to the tradition of the moat in the Middle Ages—even to-day we see a few examples here and there—in England and in France they ceased to be more than part of a scheme of decoration for a house, and later on they were filled up, since after they were no longer a protection, the science of the day said they were unhealthy in such a damp climate.

Wolsey worked feverishly at his castle, hundreds of labourers being set to work on it, so that by 1516 the place was so far finished that Wolsey could receive his master for the first time. For years Hampton Court was the centre of gay festivities, especially masquerades, which Henry liked better than anything. More than once the king paid his favourite a surprise visit with a company disguised under masks, among whom the clever host no doubt recognised his royal guest. Shakespeare, in his *Henry VIII.*, gives a scene of the kind at Wolsey's palace at York Place, but at Hampton Court they were more free, and had the advantage of larger gardens for music and revelry. But when the statesman was alone, he loved to take long walks in these gardens, to make plain to himself, and so to strengthen, all those plans for the guidance of the state that were entrusted to him. Thus does he appear before us in Cavendish's biography:

> My galleries ware fayer both large and long,
> To walk in them whan that it lyked me best;
> My gardens sweet, enclosed with walles strong,
> Embanked with benches to sytt and take my rest;
> The knotts [1] so enknotted, it cannot be exprest,
> With arbors and alyes so pleasant and so dulce,
> The pestylent ayers with flavors to repulse.

These gardens lay to the south-east, between the house and the river and on the other side there were orchards and vegetable-gardens. To the north, on either side of the wide road, there was a park, one enclosed with a wall, the other with a wooden fence, the whole embracing two thousand acres.

But Wolsey in the long run had not the strength to keep in check the overweening pride of the king. Henry soon outgrew his instructions, and Wolsey had always made himself hated by his contemporaries because of his ambition and his extravagant love of display. He had particularly shown this whenever he was travelling to his country place, and every man who had like ambitions now became his enemy. He was bound to fall as soon as these men could induce the king (whose favour was Wolsey's sole support) once to behold his favourite with their eyes—that is, with jealousy and envy; and they knew that Hampton Court was a very special attraction to their royal master. Wolsey did what he could to stay the impending storm by making a present of his estate to Henry. There is a story (misplaced before 1526) which says that when the king asked angrily, "Why should a subject build such a gorgeous palace?" the cardinal, who was prepared for the blow, replied, "To give it to his master." But Wolsey continued to live at Hampton Court till his fall in 1529.

After the death of Wolsey, Henry made the greatest possible haste to take over the palace which he so eagerly desired, and from that moment it is closely connected with his life. It was the home of nearly all his wives, and rumour says that the restless spirit of Catherine Howard still wanders in torment through the halls. The king's first care was to have the cardinal's coat of arms removed, and the Tudor arms carved instead; and, as he was passionately fond of every kind of sport, he had two closed courts on the north side made for tennis. The games, especially tennis, were importations from France, but they became extraordinarily popular in England. The government grew to look on them with an unfriendly eye, because they feared, not unreasonably, that men would be led away from archery, which was really useful to the state. But no restrictions made the slightest difference; and in 1541 games were strictly prohibited in public places, and only allowed at private houses if the owners got a licence, which cost £100 a year. Of course rich people, like the king, laid down parts of their garden for these games. The king had also confiscated Wolsey's town palace, unmindful of the fact that it belonged to the Archbishopric of York, and had given it a new name, White Hall (Whitehall), which it bears to this day; and here first of all a bowling-green was laid out. These places for games were sometimes covered rooms, sometimes open, turfed, and bordered with hedges. They now play an important part in the English garden.

[1] Beds of interwoven pattern; see remarks on a following page.

After the king became the owner of Hampton Court, its appearance was greatly improved. The private pleasure-gardens were kept in the old place on the south-west front, with their lovely view of the Thames. This view one could get not only from the windows, but from a hill called the Mount, standing at the end of the chief pleasure-garden, at that time called the King's Garden. In our story we have often met with mounts like this, and in the Northern countries they have remained true to their original purpose, and have therefore been kept up longer. In England the name "Mount Pleasant" has remained in cases where the actual hill has spread out wider, and ultimately disappeared. Leland has a great deal to say on this subject in the book of travels which he wrote about England in the middle of the sixteenth century. In the tree-garden at Wressell (or Wrassal) Castle in Yorkshire there were apparently several of these spiral hills with topiary work, winding round, with steps leading to the top, like the windings of a shell, to make an easy walk. It is not known whether the wish of Olivier de Serres to have these paths laid out as botanical gardens was ever carried out.

FIG. 343. HAMPTON COURT—OLD GARDEN SCENES

The Mount at Hampton Court adjoined what now goes by the name of the Privy Garden. Hedged paths led up in a spiral, and at the top there was an arbour, or at any rate a seat of some kind, with a lion bearing arms as its chief ornament. In the garden itself there were beds of interwoven patterns—"knotted beds" is the technical name; these were filled with garden flowers of every season—violets, primroses, pinks, mint, and other sweet-smelling kinds. Roses were bought for it "at fourpence per hundred, sweet williams at threepence per hundred." But beside the geometrical patterns they placed figures of animals in quite the old-fashioned way. Stephen Hawes, in his poem, *Pastime of Pleasure*, describes a garden of the beginning of the seventeenth century with interwoven beds of immense size, and tells how:

> Rampande lyons stode by wonderfly
> Made all of herbes, with dulcet swetenes
> With many dragons, of marveylous likenes
> Of divers floures, made full craftely
> By Flora coloured with colours sundrye.

The separate beds were bordered in different ways—in the King's Garden with horizontal stakes, striped green and white, as we can see (in a picture of the royal family) on the side wing at Hampton Court (Fig. 343).

Here we see yet another of the king's darling devices—one he carried out in a most exaggerated way, viz. heraldic animals, stuck up on green and white striped poles: his bills have plenty to say about these costly ornaments, which are scattered everywhere in the orchard. Perhaps the animals in carved wood in the garden at Gaillon were of a similar kind. From the bills we also learn that between the beds there were little cobbled paths, and every here and there slight elevations for bronze sundials. The heraldic animals themselves were gilt, and sat upright, holding little banners.

The King's Garden was at the back and apparently was closed in behind the hill by two summer-houses connected by a gallery. The so-called Pond Garden (of the present day) is very likely much the same as it was originally (Fig. 344), for it shows its old characteristics in a marked manner. It was a sunk parterre, with terraces at the sides, answering to the turfed seats of the gardens of the Middle Ages. In the centre was a large tank for fish, and round this again were the heraldic animals on their sticks. The actual arrangement of the plants that are there to-day, though old-fashioned and attractive, belongs to a later date.

Much as these, the first Renaissance gardens in England, are still admired, they were no doubt modest and very small. In many things people had to go short, and especially in water, which had to be fetched out of the Thames by night, if only to get the tanks filled.

The tree-gardens on the other side of the house were larger, and planted more like orchards; whereas trees found no room, as a rule, in small flower-gardens. There were summer-houses among the trees, and perhaps they were set against the walls, as they were called towers. The gardens were intersected by canals, just as in France, and across them charming bridges were thrown, decorated with the inevitable heraldic beasts. The high brick walls, dividing the different parts of these tree-gardens, were covered with creepers, so prettily fastened to the walls that they completely concealed them, to the wonder and admiration of foreign visitors. These "nut-gardens" with their high walls, afterwards used as espaliers, are a pleasant feature which exists in the English "kitchen-gardens" of the present day, and is almost the only one that has been able to survive the vanished past.

Flower- and fruit-gardens were not so completely separated that all ornament was reserved for the former. The orchard, or rather the tree-garden, was the shady place, and therefore the place chosen for walks. Leland in his description of England during the middle of the century has more to say about orchards, where the paths are decorated with topiary work, than about other gardens. Also in the official survey (1526) of Thornbury much more weight is attached to the tree-garden. Thornbury belonged to the unfortunate Duke of Buckingham, who was one of the first to excite the displeasure of the king, and paid for the impetuosity of his temperament on the scaffold. When his property was confiscated, the official valuer found on the south side, beside the inner fortifications, a garden surrounded by fine galleries, along which one went through to the church from the rooms, either above or below.

This Privy Garden, answering to the Italian *giardino secreto*, betrayed by its situation

FIG. 344. HAMPTON COURT—THE OLD POND GARDEN

a relationship with the castle-garden of the Middle Ages. But it cannot have contained many things of value, for it is not mentioned again. On the east of it, however, was a lovely piece of ground, which one could reach either by way of the galleries or by a "private path," and which was at any rate farther away from the castle. It was an orchard full of young saplings, with choice fruits, and many roses and other pretty things. There were avenues, too, for walks in the open. Round about, at a suitable height, there were other paths with seats, and thorn and maple trees, also terraces from which there were two views, one of the garden and one above and beyond it. From the outer side these terraces were enclosed, first by a trellis hedge, and then by quickset and sunk hedges. From the tree-garden several doors led into the different parts of a new park.

This picture is very like what we saw in the early French gardens: different sections, flower-gardens with galleries round them (and thus joined on to the castle), on the far side a plot of land larger than orchard or tree-garden, enclosed in raised ground, with a small fence beyond. But we must think of these gardens as small; and at the beginning of the seventeenth century we find the critic Gervase Markham recommending that they should purposely be made small, because "large cages make it no better for the birds." It was long before Englishmen went so far as to consider that, when a new castle was being built, the garden was of primary importance. Henry VIII., who never ceased to regard with envy and concealed admiration his splendid rival Francis I., decided to build a place to compete with that marvel, Chambord on the Loire, and in its very name, Nonsuch (bestowed as a caliph might have bestowed it), he indicated that it was to excel all others. The building of this wonderful place began in 1538 under the direction of an Italian.

But in England the influence of Italy was limited to architecture, as in France to ornamentation. In non-clerical building, England clung to Gothic ideas as fundamental, and carried them out more strictly than any other country and with better results. When Henry died in 1547 the castle was not finished, and so far there was not a word about gardens, though the king had had two parks enclosed. In Queen Mary's reign the edifice was nearly destroyed, and during that joyless time the court seemed to have no inclination for cheerful society. But Mary herself loved flowers, and whenever she as a princess visited her little brother Edward at Hampton Court (his usual residence in his childhood) the gardener would hand her a bunch of flowers, and the accounts of her private purse invariably note that she gave a five-shilling piece to the man. This is a little genre picture that lends an attraction, slight, yet touching, to the sad figure of this pathetic woman. Nonsuch escaped destruction because it was bought by Henry FitzAlan, Earl of Arundel, who was one of the most conspicuous lovers of art and science in his own generation. His library and collections were famous far and wide. By him Nonsuch was raised to the glorious condition the king had aimed at; and he also laid out the gardens and made the place a "pearl of the kingdom" (Fig. 345).

Elizabeth, with all her love of display, did not share her father's greed, and did not desire to be the first in the kingdom in this matter of gardens and buildings, never to be surpassed by any of her nobles, whose relations with their queen were now of an utterly different kind. It would have been dangerous to rouse the king's jealousy, for his pride could not brook that a subject should live in more splendid surroundings than his own. Elizabeth felt differently, and she encouraged the nobles to go on building—indeed she tempted them by proposing visits. Now although it was honour and glory to receive a visit from the queen, it caused many a sigh to some of her hosts, who were not over-burdened with wealth. For it was like a heavy tax; the queen travelled with her retinue, who ate up all the stores like a swarm of locusts; and after a royal visit there must often have been a long period of privation.

The queen loved Nonsuch particularly, though there may have been some slight regret when she remembered that it had once been a royal seat; but still in her usual way she allowed the earl to complete the building, and only bought it back after he was dead. Then it was her favourite till the last years of her reign. Hunting was at its very best in the park at Nonsuch, which was well stocked with wild animals. She loved to have hunting-parties on festive occasions, and even in her sixty-seventh year she was to be seen every day at the chase. It was at Nonsuch that she died. She would not check all the new growth in "the splendid gardens, woods with their trees pruned in all manner

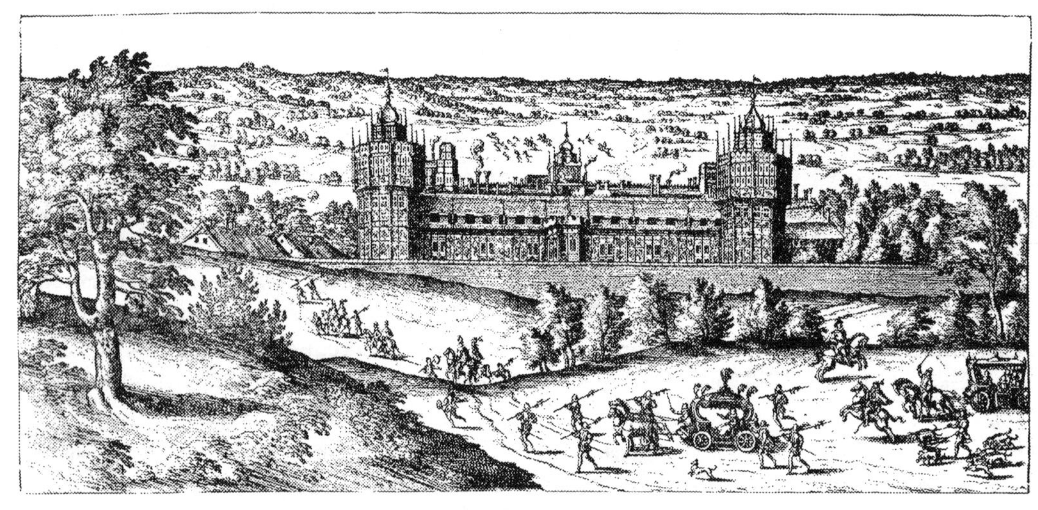

FIG. 345. NONSUCH CASTLE

of ways, meadows and paths" (of which Robert Laneham writes), "so shaded from above that you fancy Pleasure has chosen it for her seat, there to dwell with Health." Paul Hentzner saw the place on his travels in September 1558, and especially praises the many pillars and pyramids made of marble, and the two fountains, one round, the other pyramidal, with birds spurting out water.

The pleasure-garden was bounded on three sides by the main building, on the fourth by a high brick wall. It was divided in the usual way into square beds bordered with hedges, which in the second half of the century were beginning to take the place of the old wooden trellis. The avenues round were still for the most part fruit-trees, and there was no lack of water-tricks, as for example "a pyramid, which splashes people as they go by." In one little wood Hentzner admires an Actæon fountain.

The gossiping pen of Robert Laneham, a junior court official, describes the extravagant pomp and magnificence shown by a great man when he received the queen on a visit. In a letter to a London friend he tells of the festivities at Kenilworth arranged by Lord Leicester, the queen's favourite, in honour of her visit in 1575; and Sir Walter Scott has given the account, in a poetical way, in his novel. The queen had presented

Kenilworth to her favourite in the fifth year of her reign, and Leicester had added to the old castle, which was surrounded by a very wide moat, a new wing, and furnished it with wonderful things.

Laneham says that beside the new wing Leicester's plan was carried out for a garden, which embraced an acre or more, and lay to the north.

> Close to the wall is a beautiful terrace, ten feet high and twelve feet broad, quite level, and freshly covered with thick grass, which also grows on the slope. There are obelisks on the terrace at even distances, great balls, and white heraldic beasts, all made of stone and perched on artistic posts, good to look at. At each end is a bower, smelling of sweet flowers and trees. The garden ground below is crossed by grassy avenues, in straight lines on both sides, some of the walks, for a change, made of gravel, not too light and dusty, but soft and firm and pleasant to walk on, like the sands by the sea when the tide has gone out. There are also four equal parterres, cut in regular proportions; in the middle of each is a post shaped like a cube, two feet high; on that a pyramid, accurately made, symmetrically carved, fifteen feet high; on the summit a ball ten inches in diameter, and the whole thing from top to bottom, pedestal and all, hewn out of one solid block of porphyry, and then with much art and skill brought here and set up. Flowering plants, procured at great expense, yield sweet scent and beauty, with fresh herbs and flowers, their colours and their many kinds betraying a vast outlay; then fruit-trees full of apples, pears, and ripe cherries—a garden, indeed, so laid out that, either on or above the lovely terrace paths, one feels a refreshing breeze in the heat of summer, or the pleasant cool of the fountain. One can pluck from their stalks, and eat, fine strawberries and cherries.

Thus Laneham, and he cannot sufficiently praise the song of the birds and the view of field and river beyond the flowers and trees. "It is Paradise, in which the four rivers are wanting, but so is the fatal tree. Certainly there is herein a witness to a noble mind that can in such wise order all."

Sir Walter Scott makes Elizabeth and Leicester stroll through the pleasaunce in confidential talk; but his poetic fancy carries him too far when he leads us from terrace to terrace, from parterre to parterre. We can to-day identify the site of this castle-garden with its grass terrace and the four main beds north of the castle. The fruit-trees, at that time kept strictly to the orchard in Italy, are here an ornament of the pleasure-garden.

Lord Burleigh, clever and prudent, for many years Elizabeth's Prime Minister, made for himself one of the most striking gardens in England at Theobalds, to the north of London, in Hertfordshire. Lord Burleigh, as he said himself, had only wanted to build a small house, but the constant visits of the queen forced him to enlarge it more and more, and in the end it became one of the stateliest castles of his time. He was sober-minded and sensible, and he succeeded in leaving behind him a great property for his successors, which no other minister of Elizabeth did—a property not won by robbery and oppression, but from regular income and economy. But, economical as he was, he passionately loved laying out his gardens, his walks, his fountains, and this was done at Theobalds most delightfully and at great expense; the avenues were so long that one could walk for as much as two miles without coming to the end. Fortunately many descriptions of the place are extant, though they are of various dates. Round the castle there were several gardens, all separate, and in no special relation to one another. The south front looked on the main garden, of very different dimensions from the one at Kenilworth, for it covered seven acres. The brick walls on three sides of it had a pleasing appearance, because of the light violet colour that some English bricks take on.

At first, as Hentzner notices in 1598, this garden was almost entirely surrounded by a moat, which was so wide that one could drive along it, but later this was closed in, or at any rate one never hears any more about it. A great porch looks out on this garden, with portraits of all the kings of England; and another porch is made like a grotto. Here "there is water streaming out of a rock into a basin supported by the figures of two slaves; on the ceiling is painted the Zodiac with sun and moon in their courses, and on each side six trees, with bark, leaves and birds'-nests, all complete and natural." The parterre is laid out in nine beds with hedges round them, and by the side are the trees of a lovely avenue: the leading feature is a fountain of white marble, with pillars and pyramid in wood. There is a labyrinth with a slight mound in the middle, called the Hill of Venus, "one of the fairest places in the world." At the end of the garden is a summer-house, in front semicircular, and inside it are marble figures of the twelve emperors of Rome; on the

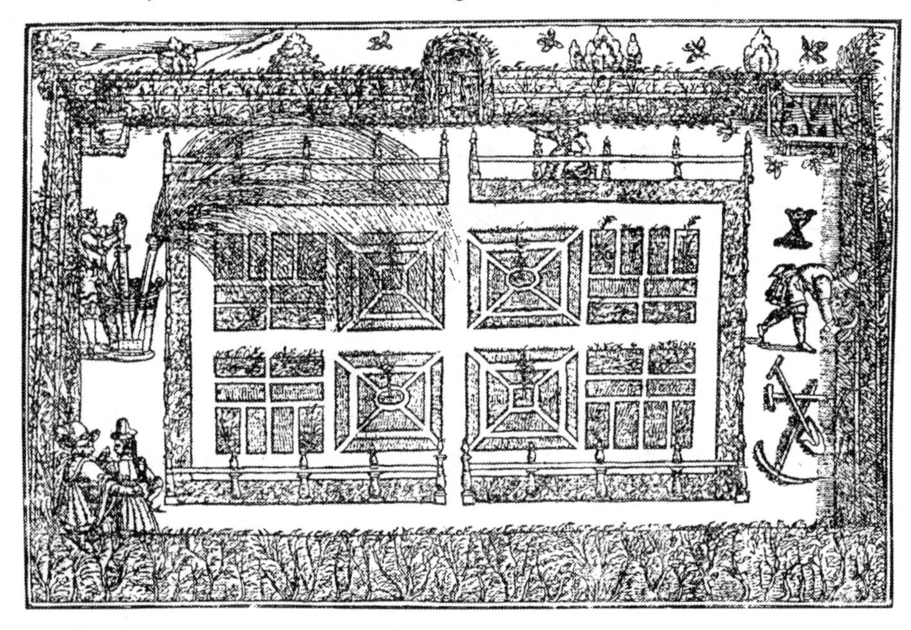

FIG. 346. A HEDGED GARDEN

other side is a basin that partly serves as a fish-pond, and is also used as a cold bath in the hot season of the year. A little bridge leads from this small summer-house to another.

Hentzner was not able to see the house itself, because he arrived on 8 September, 1598, the very day of the funeral of its owner. Many a splendid fête had this garden seen, many a time had the queen been received here with the utmost magnificence, even on that occasion when she came to condole with her sorely-stricken host in his domestic grief. With a view to this visit George Peele, a contemporary of Shakespeare's, and himself a poet, wrote a sort of masque, and in the "Gardener's Speech" the queen is thus addressed: "The hillocks removed and the plot levelled, I cast it into four quarters. In the first I framed a maze, not of hyssop and thyme, but that which maketh time itself wither with wondering; all the Virtues, all the Graces, all the Muses winding and wreathing about Your Majesty."

In the introduction to this work Peele has given the customary description of a formal garden of the period, if possible perfectly level, a square, and this square further divided into four parts by cross-paths. Garden books of the time often describe places of this sort, but the literature is meagre enough in its content; for most authors are satisfied with quotations from ancient and perhaps a few modern works, often copied out quite without thought. When Andrew Borde and Thomas Hill, the latter a very prolific writer of that time, recommend the east and north sides of a house as most suitable for the garden, it is pointed out by Gervase Markham, writing somewhat later, that this advice is only taken from Italian writers without the least consideration of local conditions. The accompanying woodcuts make it abundantly clear what special stress is laid upon fences, of which there are two or even three (Fig. 346). These drawings also show the hedges and arbours (Fig. 347), and how the "knotted beds," like ribbons interwoven, have earned their name, and again how a tree is almost always the centre piece of a labyrinth, and various other trifling characteristics. Such is the garden of which we get a fleeting glimpse every now and then in Shakespeare—the pretty knotted beds, the brick walls, the summer-house where the shoots that grow too fast are cut away, and the arbour of box where an eavesdropper can find good cover.

FIG. 347. A HOUSE GARDEN

In Elizabeth's time, that age of navigators and explorers, the simple scheme was much enriched by the introduction of new plants, and the culture of flowers was carried on zealously. When Harrison in 1587 re-edited Holinshed's *Chronicle*, he added a spirited passage about the introduction of herbs, plants, and fruits which were being brought in every day from India, America, the Canary Islands, and all parts of the world. He says that there is scarcely one nobleman and hardly one merchant who does not possess a stock of these flowers, which are being one after another entrusted to our soil, so that we begin to think of them as our usual plants. He goes on to say that he has seen in many a garden three or four hundred or even more novelties, and nobody would have heard their names forty years ago. So now if people look at the gardens beside their houses, they find wonderful beauty there; and it is not only flowers, called by Columella *sidera terrena*, but there are strange medicinal herbs, that have been collected for the past forty years. People who have these, he says, can only think of their old gardens as dung-heaps or morasses. It was almost incredible, he thought, how much art could do to help nature, by enlarging flowers, doubling them, changing their colour. Our gardeners were so thoughtful and so skilled that they seemed like going round with Nature herself, and setting her in the right course as though they were her superiors. He says finally, and with evident pride, that although the gardens of the Hesperides have been renowned for their beauty, he cannot but believe that if the two could be compared by an impartial judge, the gardens of his day would bear off the prize.

This botanical interest, common to all countries, created a strong bond in commerce, partly at home and partly abroad: in Germany this was especially felt. Princes, nobles, and men of learning stood on the same ground. In England the learned men took the lead, cultivating special botanic gardens for rearing foreign medicinal herbs; and John Gerard, the author of several botanical books, had a famous scientific garden in Holborn at the end of the sixteenth century. The Tradescants were a family of educated gardeners, who in Elizabeth's reign had come over from Holland, and had won great esteem for the acclimatisation of foreign plants in England. Both father and son were commissioned by Lord Salisbury, Lord Burleigh's son, to travel for him in countries over the sea; but Tradescant's own wish to explore led him farther and farther, and his garden became a great sight: it was visited by king and queen, and was in existence till 1749. On the Tradescant tombstone we read that they

> Lived till they had travell'd Art and Nature through;
> As by their choice collections may appear
> Of what is rare in land, in sea, in air.

This collection formed the basis of the Ashmolean Museum, founded at the same time as the Jardin des Plantes in Paris, but in England the public Botanic Gardens started much later.

With the seventeenth century the interest in gardens begins to make an appearance in *belles lettres*, quite independently of real practical work on the one hand and theoretical professional advice on the other. It is well worth noting, that the most important and the most far-seeing spirit of his age, Francis Bacon, was the first to direct attention to the matter in this way, though he was neither architect nor gardener. On the path now smoothed by literary dilettanti the development of horticulture in England began to make progress, which a hundred years later led to the real revolution in style.

Bacon's essay "Of Gardens" was written in his lively conversational style, full of his own personal ideas and fancies. His notion was to put forward a scheme in better taste for the gardens he saw about him; he was always practical, and bore in mind what it was possible to do; moreover his aim was avowedly educational. The demand of Homer, which he puts at the head of the essay, that the garden should always show something in flower, is at once followed by a list of plants, soberly and methodically allotted to each particular month.

The place he plans is expressly intended for a princely owner, and the thirty acres he demands for the whole is for those days a remarkably large area. We must here recall that Kenilworth had only one acre, and the large garden at Theobalds seven acres. But Bacon has only reserved twelve for his flower-garden proper, as he breaks up the whole estate into three parts: the house, as he insists in the essay, "Of Building," must have a way to the garden by open galleries—"The Row of Return, on the Banquet Side, let it be all Stately Galleries"—but this demand was not carried out at Bacon's own place, and the open veranda has never played a great part in English country houses, for an Englishman likes to go straight out of a room into the open air and sit there.

Next to the house there is to be a lawn, with an avenue of trees in the middle, and covered shady walks on either side. "Nothing is more pleasant to the Eye, than green Grass kept finely shorn." Bacon prefers this to a parterre proper, for the knotted beds cut separately "with divers Coloured Earths," seem to him childish: "they be but Toys; you may see as good Sights many Times in Tarts."

The main garden is in the middle of the estate. It is exactly square, "encompassed on all the four sides with a Stately Arched Hedge . . . over the Arches let there be an

Entire Hedge . . . and upon the Upper Hedge, over every Arch, a little Turret with a Belly, enough to receive a Cage of Birds; and over every Space, between the Arches, some other little Figure, with Broad Plates of Round Coloured Glass, gilt, for the Sun to play upon. But this Hedge I intend to be raised upon a Bank, not steep, but gently slope, . . . set all with Flowers . . . on either side Ground enough for diversity of Side Alleys."

The trees are to be of different kinds, and some of them fruit-trees. The walks are enclosed by hedges, and there is a mound at each end for the sake of getting a view over the wall. In these avenues one can walk if shade is wanted, for the main garden itself must not be too full of bushes, but should be open and airy, while on both sides there ought to be rather sunny walks with fruit-trees and pretty arbours. This part should be intersected by wide dignified paths, and round the beds there may be a very low hedge and little pyramids. "I, for my part, do not like Images cut out in Juniper, or other Garden Stuff; they be for Children."

In the centre there is to be a large mound, with spiral paths, wide and easy of ascent, and on the top "some fine Banqueting House with some Chimneys neatly cast." Water is a great ornament whether as an artistic fountain or as a bath, and the chief requisite is that the water shall always be kept clean, "for Pools mar all." Marble or gilt statues may be good, but are of secondary importance.

The third division of Bacon's garden is what he calls "the Heath." This is to be half as big as the main garden, and as far as possible is to be of a natural wildness.

> Trees I would have none in it, but some Thickets made only of Sweetbriar, and Honeysuckle, and some Wild Vine amongst; and the Ground set with Violets, Strawberries, and Primroses. For these are sweet, and prosper in the Shade. And these to be in the Heath, here and there, not in any Order. I like also little Heaps, in the Nature of Mole-hills (such as are in Wild Heaths) to be set, some with Wild Thyme; some with Pinks; some with Germander, that gives a good Flower to the Eye; some with Periwinkle; some with Violets; some with Strawberries; some with Cowslips; some with Daisies; some with Red Roses; some with Lilium Convallium; some with Sweet Williams red; some with Bears-Foot; and the like Low Flowers, being withal Sweet and Sightly. Part of which Heaps, to be with Standards, of little Bushes, pricked upon their Top, and Part without. The Standards to be Roses; Juniper; Hollies; Bear-Berries (but here and there, because of the Smell of their Blossom); Red Currants; Gooseberries; Rosemary; Bays; Sweetbriar; and such like. But these Standards, to be kept with Cutting, that they grow not out of Course.

This description must be compared with that of an important Italian garden of about the same time, if one is to see how entirely the two chief factors in the south, viz. stone building and (all-important) water are kept in the background by Bacon. In a country like Italy it is stone building that keeps house and garden united, and the arrangements for water are the connecting link between architecture and the world of plants. But all that Bacon asks for as extra trimmings—mostly frames or glass—is undeniably barbaric in its character; and this is all the more evident if compared with his refined taste in the arrangement of the plants themselves, where there is always delicacy in details. Refinement, rest and peace are the secret of the ideal, but in a real garden it would have been hard to find the proportions quite satisfactory. In any case, the very high hill in the middle must have made an uncomfortable break in the great parterre, and a fountain with sculpture would have been more attractive; but the raised side-paths round the middle garden are a happy thought.

The entrance with the closely-mown lawn points the way to the English style, for in the treatment of superlatively beautiful lawns there has been, and still is, a very lovely and effective feature of English horticulture. Bacon's idea of a heath is quite novel and very surprising; there seems to have been no living example of it, and it remained unique. For all its charm as an idea, it seems a great waste of space. Only in very recent days was this notion taken up and carried out, and then the interest in wild gardens has always been in the smaller places only.

An unusual sidelight is thrown on this design by another literary record of the period. In 1613, on the eve of the Feast of the three Kings, the lawyers of Gray's Inn had a masque in honour of Lord Somerset's wedding. The stage directions describe a garden "of wonderful beauty" which has more than one point in common with Bacon's. Here are the four quarters into which the whole garden is divided, with walks all the way round, but in the centre of the crossways there is a fine Neptune fountain. The god with his trident shakes water into a shell held up by three figures which are standing on a pillar. This garden has not only a brick wall covered with fruit-trees, but also a pretty hedge inside the wall, with balustrades and lions and unicorns, serving as torch-holders. The great squares are enclosed with cypresses and junipers, and adorned with flowers and pyramids. At all the four corners there are pots full of pinks. At the end of the garden rises the mound, so steep that the steps are like seats covered with grass. Above there is a triple-arched arbour, covered with roses and honeysuckle and ornamented with turrets. Above it protrude the tops of fruit-trees.

The fact that the flowers are artificial, and lighted by concealed lamps, and the trees and walls merely painted, is only the necessary presentation of a stage garden, which is very excellently imitated from the real thing. The mound also, made with steps, at the end of the garden, was common at that time, and has been kept here and there till the present day, as for example in the Rockingham garden, which in other respects is greatly changed.

When in 1603 James I. ascended the English throne, there were already certain traces of Italian Renaissance influence to be seen in Scotland, his native land. Indeed in some ways the castles of Scotland, for the most part on high ground rising steeply from the neighbouring country, were more inclined to learn a lesson from the Italian terraces than the English were. It is almost impossible to get a really clear picture of the Scotch Renaissance gardens, because the descriptions are so meagre, but all the same the effect of these terrace structures, with their firm base-lines so hard to remove, was no doubt to protect many of these gardens in the North during the destructive wars of the eighteenth century. And if it is true that the gardens of to-day, which are proud of their marks of antiquity, have nothing to show but the framework, it is also the fact that a place like Drummond Castle in Perthshire cannot deny the influence of Italy in the terraces, whose lofty protecting walls are intersected by graceful sloping steps, or in the doors that lead out from the terraces into the open. Barncluith in Lanarkshire calls to mind Italian villas, partly by the row of five narrow terraces overlooking the high bank of the Avon, partly by the stairway at the side, and partly by the delightful summer-house at the end with the semicircular steps that lead up to it.

The gardens in Scotland later on became more like those of Italy, because of a growing taste for the evergreen yew hedges, which were introduced in the seventeenth century, and were excellently acclimatised, and so remained in favour in the eighteenth. One must not, of course, be led away by present-day appearances to fancy one has a picture of what things looked like at the beginning of the seventeenth century.

King James himself was an ardent garden lover, and when he was King of England he took an active part in introducing foreign plants. He made manifest his interest in the castles and gardens round about him in the lordly fashion (we must admit) of his forefathers. It is at all times a dangerous undertaking for a subject to play host to his king. Burleigh's son, the first Lord Salisbury, in the summer of 1606 entertained James and his guest, the King of Denmark, with extraordinary festivities at his family seat, Theobalds. Ben Jonson enhanced the glory of the foreigner's visit by producing one of his masques, in which three "Hours," Law, Justice, and Peace, present an address in Latin verse.

The eyes of James were dazzled by the splendour of it all, and his desire to call this beautiful country place his own became irresistible. Salisbury made the best of a bad job, and accepted the proposed exchange of Hatfield, at that time the king's property. Less than a year afterwards, on 22 May, 1607, the new master entered Theobalds with great pomp and ceremony, and once more Ben Jonson's muse honoured the feast: the grieving Spirit of the House is brought into the presence of the new mistress, Queen Anne, and the beauty and glamour of the lady change lament to joyfulness. But Ben Jonson is a bad prophet when he makes Mercury reply to the Genius, who has asked what induced the former master to leave his home:

> Nor gain, nor need; much less a vain desire
> To frame new roofs, or build his dwelling higher;
> He hath with mortar busied been too much,
> That his affections should continue such.

For Lord Salisbury had hardly turned his back on the home of his fathers when he started with undiminished energy at Hatfield "to frame new roofs, and build his dwelling higher" (Fig. 348).

In his time the fine house stood almost as it is still inhabited by the successors of the man who founded it. True, the greater part of the garden itself is made on a site that was acquired later, but it follows the old style, only on a larger scale. All the gardens on the east side are fairly new: one of them leads from the terrace next the house to a parterre, from which you first come to the bowling-green, then to a labyrinth somewhat sunk, and on the other side to a charming water-garden, which has lately been remade. It is only the western sections that have kept their old appearance, at any rate in the main features (Fig. 349). One square of about 250 feet is shut in by a kind of arcade of clipped limes. The parterre is confined by a rather low hedge, and among other ornaments has a simple, beautiful fountain. A small rose-garden, to-day in front of the stables, very likely dates from the king's time, for we find part of the Elizabethan house built up into the stable.

It is obvious that these were not the only gardens. In the records of accounts we find mention of a handsome Neptune fountain in marble, which Salomon de Caus erected for £113; and another Frenchman of the name of Simon Sturtevant was to have added an important water-piece, but this was put a stop to by the death of Lord Salisbury in 1612. The accounts also speak of a garden at the foot of the declivity, and its flowers and avenues: it was called the Valley Garden, and had pretty bridges across its stream.

On the other side was a fine vineyard, which Lord Salisbury, who was full of enter-

FIG. 348. HATFIELD HOUSE, HERTFORDSHIRE—GARDEN AND LABYRINTH

prise, laid out himself. He also zealously supported the cult of mulberry-trees, which James had successfully imported. The mulberry was well known in England in Queen Elizabeth's reign, and tradition even says that the four noble specimens at Hatfield were planted by the queen herself. Shakespeare was familiar with the mulberry, and makes Volumnia say (*Coriolanus*, Act III. Scene ii.):

> Correcting thy stout heart,
> Now humble as the ripest mulberry
> That will not hold the handling.

[EDITOR'S NOTE: There were other references:

> When he was by, the birds such pleasure took,
> That some could sing, some others in their bills
> Would bring him mulberries and ripe-red cherries;
> He fed them with his sight, they him with berries.
>
> *Venus and Adonis.*

> Feed him with apricocks and dewberries,
> With purple grapes, green figs and mulberries.
>
> *A Midsummer Night's Dream,* Act III. Scene i.

> And Thisby, tarrying in mulberry shade.
>
> *A Midsummer Night's Dream,* Act V. Scene i.

Shakespeare not only knew the mulberry but grew it. He planted it in his garden at New Place, Stratford-on-Avon. Those who doubt this may note Malone's assurance:

> That Shakespeare planted this tree is as well authenticated as anything of that nature can be . . . and till this was planted there was no Mulberry tree in the neighbourhood. The tree was celebrated in many a poem, one especially by Dibdin, but about 1752, the then owner of New Place, the Rev. Mr. Gastrell, bought and pulled down the house, and wishing, as it should seem, to be damned to everlasting fame, he had some time before cut down Shakespeare's celebrated Mulberry tree, to save himself the trouble of showing it to those whose admiration of our great poet led them to visit the poetick ground on which it stood.

It should be noted that Malone was a highly diligent and competent student of Shakespeare; see his supplement to Steevens's edition, 1778, and his own admirable edition of 1790. The famous mulberry is mentioned in "Boswell." Thus: "This joint expedition of these two eminent men [Samuel Johnson and David Garrick] to the metropolis was many years afterwards noticed in an allegorical poem on Shakespeare's mulberry-tree, by Mr. Lovibond, the ingenious author of *The Tears of Old-May-Day*." The reference under date March 25, 1776, should also be read. See *The Life of Samuel Johnson, LL.D.*, by James Boswell, "Temple Classics," J. M. Dent and Sons Ltd.]

It was Salisbury who sent Tradescant abroad as his gardener, to collect fruits and flowers for Hatfield from Holland, Belgium, France and Italy. The whole estate is a

FIG. 349. HATFIELD HOUSE—THE OLD GARDEN

splendid example of an English country house of the period: in front of the mansion with its three wings is a court closed in with balustrades and fine trellis-work, through which originally ran the main approach to the house, with a broad paved path and plots of grass on either side. Behind this at Hatfield are the domestic offices and kitchens, but in most houses the garden is close adjoining, and often has wide terraces overhanging on both sides.

There is a very similar place at Montacute (Fig. 350) in Somersetshire, built at the end of the sixteenth century, and until recently the home of the family of its original founder, Sir Edward Philips, but afterwards of Lord Curzon of Kedleston. Two inscriptions are left that point to a hospitable spirit: at the entrance porch the words, "For you, my friends," and at the garden gate, "Through this wide open door none treads too early

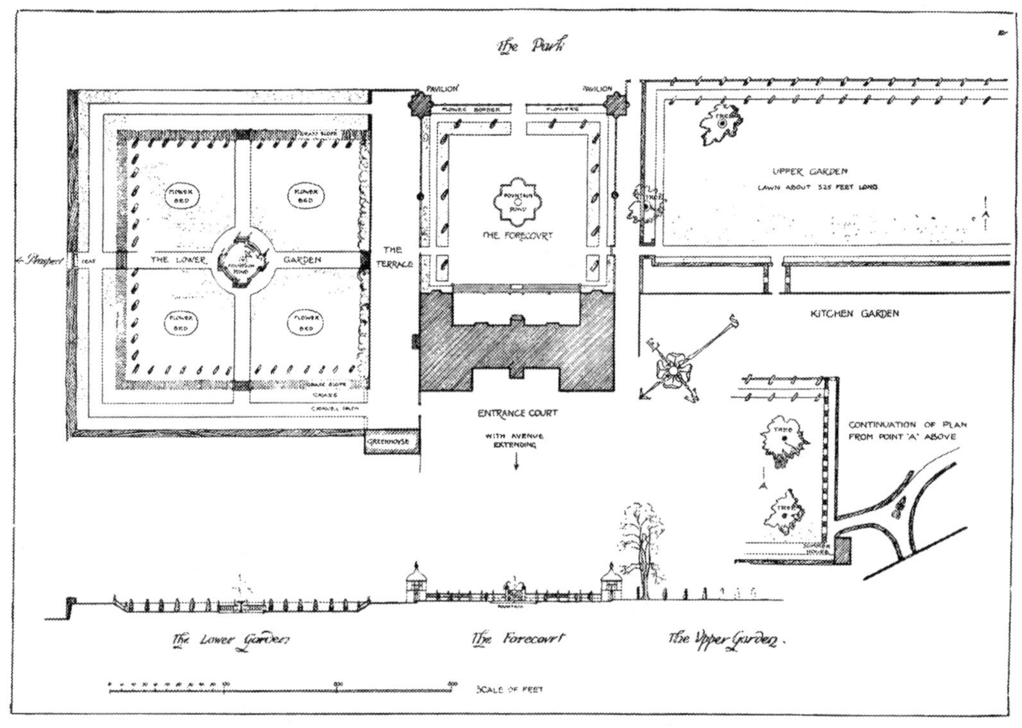

FIG. 350. MONTACUTE, SOMERSETSHIRE—PLAN OF THE SITE

and none leaves too late." At this house, as at Hatfield, the main entrance has been changed to the opposite side. Originally one approached the house through the fine front court, which had a balustrade-top to the walls and also a fountain. On this balustrade there are obelisks, two pretty pavilions in the middle, and also, in the corners flanking the entrance, unusually fine summer-houses. On the side of the court and house runs a wide terrace, from which steps lead to the flower plot—a sunk parterre, overlooked by high walks, partly gravel, partly grass. At the cross-paths stands the chief ornament, a basin with a balustrade in the best Renaissance style, the simple shell fountain of spring water in the middle. The parterre is nowadays laid out in lawns with pyramids of box; but this plantation belongs to a later date, taking the place of geometrical patterns worked out in box and flowers. On the other side of the house there is a huge lawn, where we

FIG. 351. HADDON HALL—CEDARS ON THE UPPER TERRACE

might expect to find kitchen- and tree-gardens, 350 feet by 150 feet, but as a fact planted with only a few big trees; perhaps it began as a bowling-green between two avenues.

There is another garden, which has kept only its outlines, but it is enough to show how lively was Italian influence at the end of the sixteenth century. This is Haddon Hall (Fig. 351), in Derbyshire, the seat of the Manners family, dukes of Rutland. The house belongs to very different architectural periods, and forms a large irregular mass, with four terrace gardens at the side, accommodated to the height of the hill—the higher one a broad strip of grass with a row of trees on either side. This kind of plantation is one of the English Renaissance ideas, with steps at the side leading to the first terrace by which we pass straight on to the house. The corners of the lawn are emphasised by lofty yews, and a balustrade at the front end has a stairway through it leading to the chief parterre, which is sunk near the house; very high walls stand between the two lower terraces and the actual dwelling, the terraces in no regular arrangement, lying at the foot of the hill and connected with kitchen-garden and orchard.

Under the earlier Stuarts the garden pursued the even tenor of its way. People began once more to look eagerly towards France, the court and nobles even to call in French experts. One of the most interesting figures among artists was Salomon de Caus. He was an engineer and architect, a native of Normandy, who had travelled far and wide, and accumulated, chiefly in Italy, a great deal of knowledge, more especially of water schemes and devices. James I. brought this well-informed man to the court as drawing-master to his children, Elizabeth and Henry Frederick. The young Prince of Wales lived during the last years of his life, at Richmond, where he died young and unexpectedly in 1612. It was to please the restless merry boy, who always loved to live and play out of doors, that Salomon de Caus had several different projects carried out at Richmond, and more particularly a series of ingenious inventions for water novelties, so as to "satisfy that fine appetite for knowledge, always striving after something new." It was now that he discovered how water could be shot upward by the expansion of steam, and this conferred on him the great honour of being reckoned among the discoverers of the steam engine. He had commissions to construct fountains in country places, as for example at Hatfield House for Lord Salisbury. Unfortunately we know very little about the Richmond garden, for when the young heir died there was an end to the activities of de Caus in England. His pupil Elizabeth, after her marriage with the Count Palatine Frederick, summoned Salomon to Heidelberg, where there was great scope for his work in the construction of the garden at the castle. But he left his son Isaac in England, and in 1615 the Earl of Pembroke employed him at Wilton House in Wiltshire, where he created one of the most important gardens in all England. The father and the son both had literary leanings, and they made their works known with pen and pencil. The garden at Wilton House (Fig. 352) and also the Heidelberg garden were pictured in full detail in twenty-four copperplates, and a book was published (in 1615) under the title of *Hortus Pembrochianus*.

The wide stretch of ground, four thousand feet long and four hundred feet broad, was divided into three sections, one behind the other, cut by generous paths. The first consists of parterres, with beds cleverly hemmed in by low hedges, every four with one fountain and its waterspout, and connected in fours with one marble statue set up in

FIG. 352. THE GARDEN AT WILTON HOUSE, NEAR SALISBURY ("HORTUS PEMBROCHIANUS")

the middle of the group. The designation, "embroidered parterres," proves that this particular kind had already come over from France. Possibly Isaac de Caus had himself seen them there, but it is more likely that André Mollet, Claude's son, who was also in King James's service, had brought them over.

This *parterre de broderie* was apparently laid out in box only, without any flowers, for Isaac still prefers to keep his flowers on the side separately. At the end of this section is a narrow lower terrace, put there "to give a better view over the parterre." Next come two large thickets, through which the river Nadder—at this point forty-four feet wide

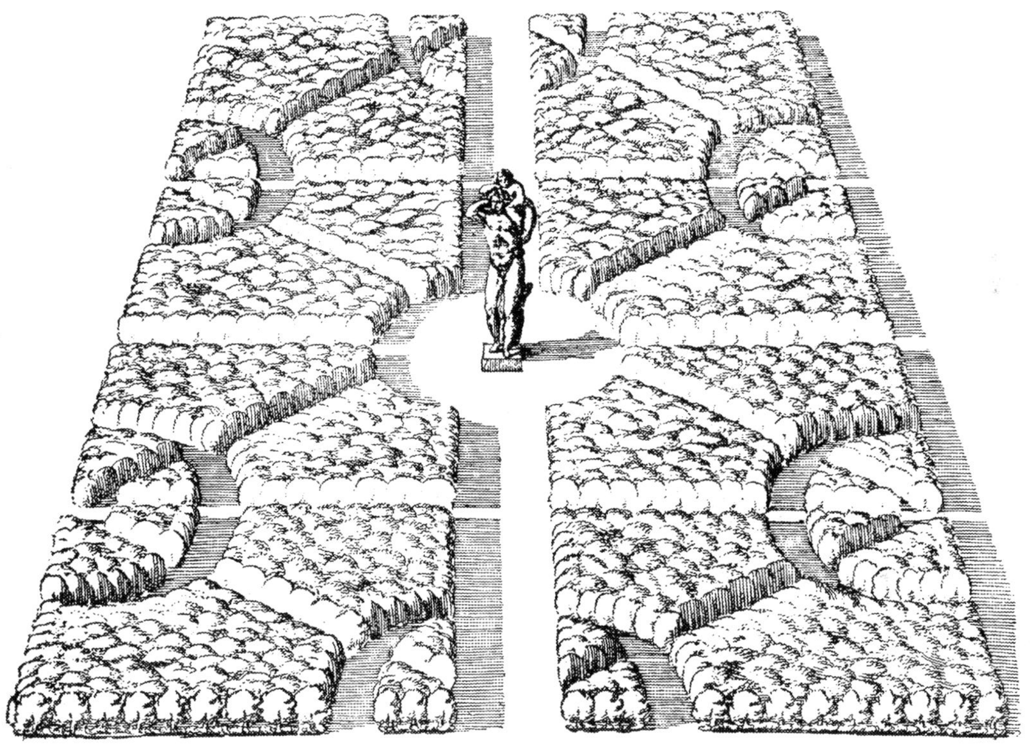

FIG. 353. THE THICKET AT WILTON HOUSE ("HORTUS PEMBROCHIANUS")

—flows in such a remarkable way that its course is not altered at all. The wood is here strictly measured out, and in its centre are the usual statues of Flora and Bacchus (Fig. 353), but no attention is paid to the stream. On both sides of the thicket there are covered walks three hundred feet long. The wide road in the middle crosses the river with a fine bridge which was at a later time ornamented with figures of lions. At the extremity of the thicket there are on each side of the way two great water-tanks, with a column in the middle, and bright glittering *jets d'eau*. The third great section has walks made in concentric ovals, with cherry-trees planted at the edge of strips of grass; and in the centre of the space left stands the Borghese Gladiator, which was considered by de Caus to be "the most famous statue of antiquity." On both sides there are again covered ways with pavilions where they cross. At the very end is a terrace extending the whole width of the garden, with a balustraded wall as a boundary. In the middle of this terrace you pass under the arches into a grotto, where marble statues and niches and pillars are ranked along the inside wall. Here there is the well-known type of grotto architecture, no doubt a design of Inigo Jones. On both sides stairs lead up, and on their balustrade sit sea-monsters, spurting water upward to the terrace. Above the grotto is a large basin with a well, and these various singularities remind one of the garden at Heidelberg. There is nothing left now of what is here described, for the Pembroke family adopted every new style as it came up. Grotto, terrace, parterre, all have vanished, and the river course is now spanned by a bridge of the classical type. The picturesque

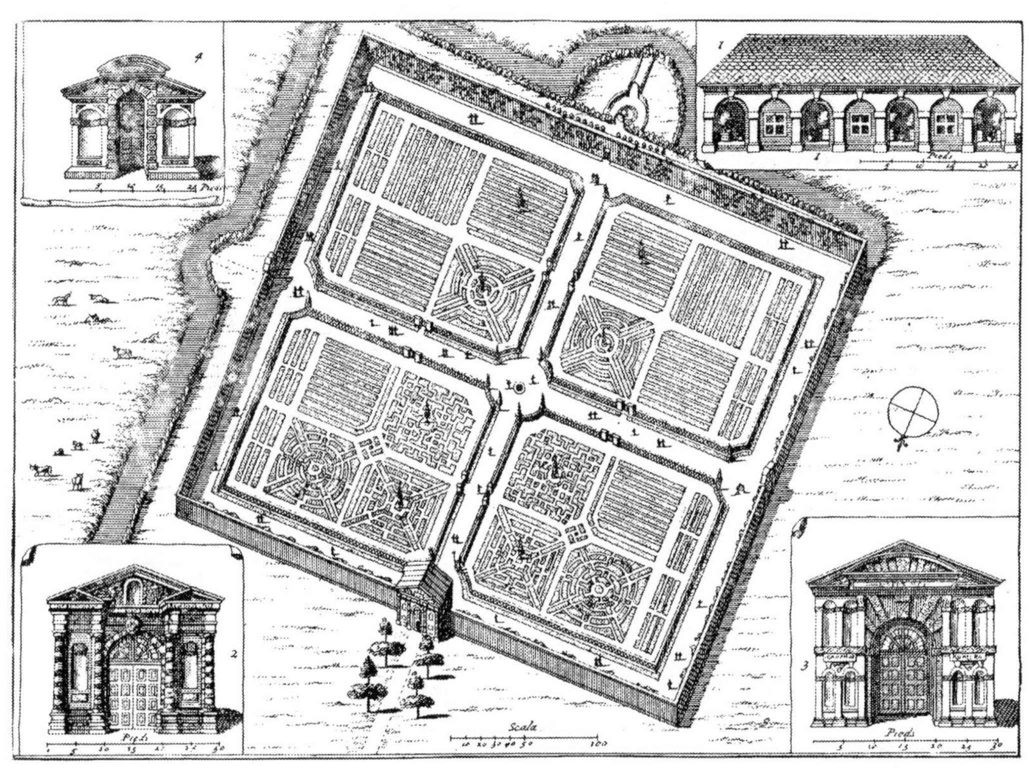

FIG. 354. OXFORD BOTANIC GARDEN

style had obliterated all, when in the nineteenth century there came a fresh wave of fashion, and the place was once more made architectural and formal in design.

Charles I. as well as his queen patronised gardening. It was during his reign that the Botanic Garden at Oxford (Fig. 354) was founded by Henry Danvers, Earl of Danby. The year (incorrectly given by most authorities) was 1621, the first stone having been laid on 25 July of that year. The whole ground of five acres is bounded on both sides by a canal. Inside this there is a wall with three monumental entrance gates, attributed to Inigo Jones, and on the inner face of this wall espalier fruit is grown. The whole square is laid out on a very simple plan: every four beds are united by an encircling trellis so as to form one "quarter" of the garden. For this quarter in the year 1648 no fewer than 1600 different plants were brought. In winter the pot plants were kept in a simple orangery. At that date the practice of having conservatories had hardly begun, but they had become especially necessary since the introduction of the orange. In the first half of the century hothouses were very rare, and only to be found in a few princely gardens.

Sir William Temple expressly says that in the garden of Moor Park in Hertfordshire, which he saw for the last time about the year 1655, the galleries round the principal parterre would have been very suitable for oranges, myrtles, etc., "if this part had been as well kept up as it is to-day." This he wrote in 1685, but the "lovely perfect garden," which he had seen for the last time thirty years previously, still seems to him the best possible model; and indeed his description is of a thoroughly characteristic middle-sized garden of the first half of the century.

It lies on the slope of a not very high hill, on which stands the house: the wide front where the best rooms are, and those that are most in use, looks out on the garden, and the large living-room opens immediately on a gravelled terrace, about a hundred feet long and correspondingly broad. On the edge of the terrace stand laurel-trees, at wide distances from each other; these he thinks have "the beauty of oranges, without their fruit or flower." Three stone steps lead down to a large parterre, divided into squares by gravel walks, and ornamented with two fountains and eight statues. At the end of the terrace by the house there are two summer-houses, and at the sides of the parterre broad galleries with stone arches, which also end in summer-houses, and serve as shady paths for the parterre. Over these arched corridors there are two terraces with balustrades and lead roofs. The approach to these airy walks is by way of the two summer-houses at the end of the first terrace.

From the middle of the parterre a high stair on both sides of a grotto leads into the lower garden, which is well planted with fruit-trees, in squares round a "wilderness" (this was the name for interwoven paths in clipped shrubberies). Sir William added that if the hill had not come to an end at this point, and the wall abutted on a country road, there might have been a third garden added with all sorts of things in it; but this want was compensated for by a garden on the other side of the house, wild and shady, and adorned with water-work and rough rock.

In the middle of the century the development of the garden was threatened by a sharp crisis. The revolution which robbed Charles of throne and life, ruined his family, and all the Royalist nobility, also threatened their possessions with utter destruction, and before all else their pleasure-gardens. Not that Cromwell was hostile to gardening as such, for kitchen- and fruit-gardens were actively encouraged in these years. A certain Hartlib, a Pole by birth, earned a pension of £100 from Cromwell in recognition of the work he had done for the furtherance of agriculture, and especially because he had encouraged gardeners in trade enterprise on a large scale, which was extremely rare, except quite near London. The earth was to be made productive, the labour was to be sober and intelligent, here as in every other department where the masters were earnest men who abhorred all joy and gaiety. Gardens for beauty or pleasure were quite unnecessary, and was it not better that they should vanish off the face of the earth, more especially those that had belonged to the hated race of kings? Theobalds, Nonsuch, Hampton Court, and Wimbledon were at that time the finest of the royal seats. So a commission was sent down, to make an inventory and value the whole property; and it is to its business-like, official, summary report that we owe our clear picture, generally hard to get, of the royal gardens of the period.

Hampton Court escaped the fate of the others, for the Protector selected it for himself and his family to live in. Therefore the order to have it parcelled out and then sold—the fate of the other royal houses—was "left open for a time, till Parliament should deal with it again." Nonsuch was confiscated and brought under the hammer, but escaped destruction by a sort of accident. Only the trees were, for the most part, cut down by "those destructive and greedy rebels" (as Evelyn calls them in 1665) who, he says,

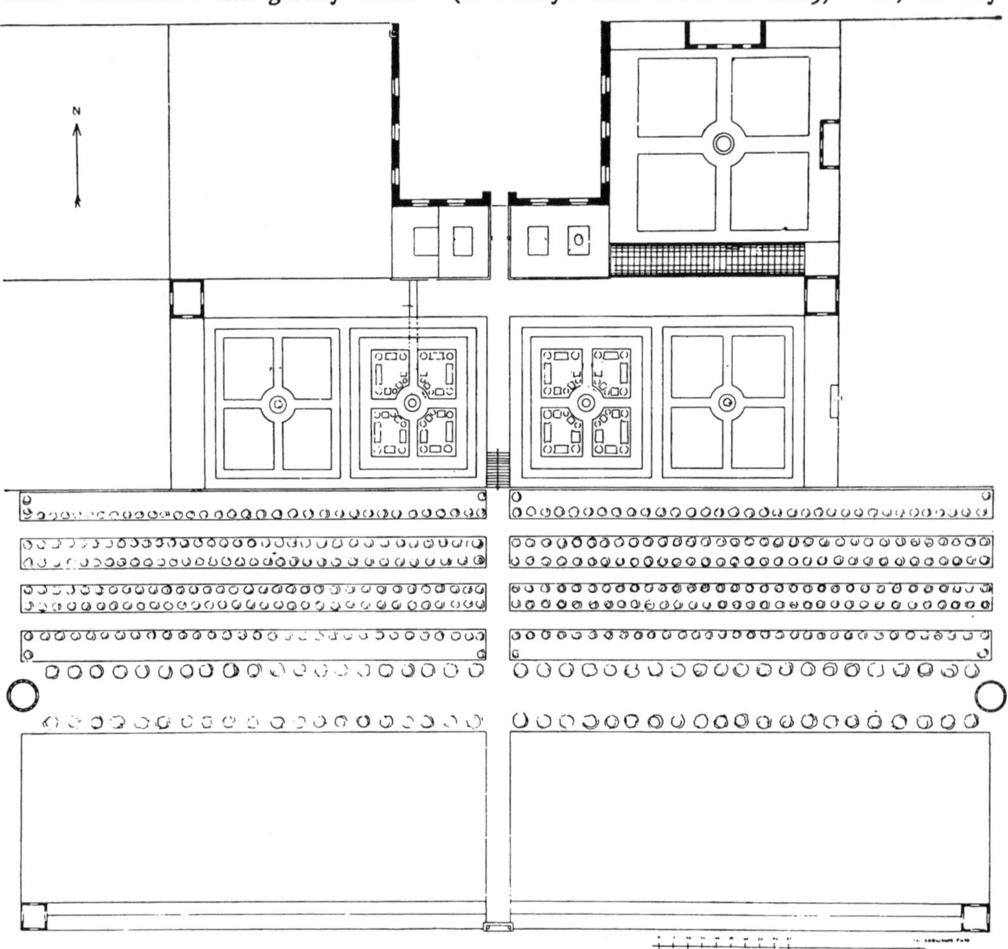

FIG. 355. THE OLD GARDEN AT WIMBLEDON—GROUND-PLAN

desecrated one of his Majesty's loveliest country seats. But the estate that escaped the Commonwealth fell a victim to baser greed, for Charles II. presented it after the Restoration to his mistress, the Countess of Castlemaine, and she, having learned a lesson from Parliament, parcelled out the royal estate and sold it. Theobalds, however, which before the revolution was standing in all its glory, then vanished from the face of the earth.

The same fate befell Wimbledon, the darling home of Queen Henrietta. It is really an instance of the irony of fate that the only description available of this garden (Fig. 355), which exhibited the finest stage of art in the middle of the century in England, is that set out with meticulous nicety in the official valuation that was made with a view to its complete destruction.

When we read the description, we cannot but feel that a real love and sympathy for the garden guided the pens of the parliamentary officials, and yet the object of the nine signatories was only a money valuation, and the end was the inevitable hammer. All that had been collected with endless love and pains was now divided among small owners, who were taught, when all the pride and beauty was in ruins, how they could utilise the ground by raising vegetables.

END OF VOL. I

MADE AT THE TEMPLE PRESS LETCHWORTH IN GREAT BRITAIN

Printed in the United States
By Bookmasters